MATHEMATICAL THEORY OF COMPRESSIBLE FLUID FLOW

RICHARD VON MISES
late Gordon McKay Professor of Aerodynamics
and Applied Mathematics, Harvard University

Completed by
HILDA GEIRINGER and G. S. S. LUDFORD

DOVER PUBLICATIONS, INC.
Mineola, New York

Editing supported by the Bureau of Ordnance, U.S. Navy, under Contract NOrd 7386.

Bibliographical Note:

This Dover edition, first published in 2004, is an unabridged republication of the work first published by Academic Press Inc., New York, in 1958.

Library of Congress Cataloging-in-Publication Data

Von Mises, Richard, 1883–1953.
Mathematical theory of compressible fluid flow / Richard von Mises ; completed by Hilda Geiringer and G.S.S. Ludford.
p. cm.
Reprint. Originally published: New York : Academic Press, 1958.
Includes bibliographical references and index.
ISBN 0-486-43941-0 (pbk.)
1. Aerodynamics. 2. Compressibility. 3. Fluid dynamics. I. Geiringer, Hilda, 1893–1973. II. Ludford, G. S. S. III. Title.

QA930.V63 2004
532'.58—dc22

2004056020

Manufactured in the United States of America
Dover Publications, Inc., 31 East 2nd Street, Mineola, N.Y. 11501

PREFACE

When Richard von Mises died suddenly in July, 1953, he left the first three chapters (Arts. 1–15) of what was intended to be a comprehensive work on compressible flow. By themselves these did not form a complete book, and it was decided to augment them with the theory of steady plane flow, which according to von Mises' plan was the next and last topic of the first part.

This last work of Richard von Mises embodies his ideas on a central branch of fluid mechanics. Characteristically, he devotes special care to fundamentals, both conceptual and mathematical. The novel concept of a specifying equation clarifies the role of thermodynamics in the mechanics of compressible fluids. The general theory of characteristics is treated in an unusually complete and simple manner, with detailed applications. The theory of shocks as asymptotic phenomena is set within the context of rational mechanics.

Chapters IV and V (Arts. 16–25) were written with the author's papers and lecture notes as guide. A thorough presentation of the hodograph method includes a discussion and comparison of the modern integration theories. Shock theory once more receives special attention. The text ends with a study of transonic flow, the last subject to engage von Mises' interest.

In revising the existing three chapters great restraint was exercised, so as not to impair the author's distinctive presentation; a few sections were added (in Arts. 7, 9, 15). More than forty pages of Notes and Addenda, partly bibliographical and historical, and partly in the nature of appendices, follow the text. This is in line with von Mises' practice of keeping text free from distraction, while at the same time providing a fuller background. The text is, however, completely independent of the Notes.

Every facet of the work was studied jointly by us, in an attempt to continue in the author's spirit. Final responsibility for the text and the Notes to Arts. 16–21, 25 and the Notes to Chapters I, II lies, however, with Hilda Geiringer (Mrs. R. v. Mises) and for the text and Notes to Arts. 22–24 and the Notes to Chapter III with G. S. S. Ludford.

The present book contains no extensive discussion of the approximation theories, which have proved to be so fruitful. It was the author's intention to discuss these in the second part of his work, along with various other topics. The book has been written as an advanced text-book in the hope that both graduate students and research workers will find it useful.

We are greatly indebted to many people for help given in various phases of our task. Helen K. Nickerson, who was the much appreciated assistant of von Mises in the writing of the first three chapters, helped in their later revision and read Chapters IV, V with constructive criticism. The whole manuscript was read by G. Kuerti, who suggested important improvements. S. Goldstein at times gave us the benefit of his unique insight into the whole field of mechanics. M. Schiffer was always ready with discussion and advice on delicate questions of a more mathematical nature. The influence of C. Truesdell's important contributions to the history of mechanics is obvious in many Notes; he also readily provided more specific information.

Very able assistance was rendered by W. Gibson and S. Schot, whom we thank cordially for their dedicated interest and valuable help. Thanks are also due to M. Murgai who prepared the subject index and to D. Rubenfeld who made the final figures.

We are particularly grateful to F. N. Frenkiel as an understanding and patient advisor. The work of Hilda Geiringer at the Division of Engineering and Applied Physics, Harvard University, was generously supported by the Office of Naval Research; that of G. S. S. Ludford, was carried out under the sympathetic sponsorship of the Institute for Fluid Dynamics and Applied Mathematics, University of Maryland, and its director M. H. Martin.

Finally, much more than a formal acknowledgment is due to Garrett Birkhoff, who enabled H. Geiringer to carry out her task under ideal working conditions. It is mainly due to his vision and understanding that this last work of Richard von Mises has been preserved.

HILDA GEIRINGER
G. S. S. LUDFORD

Cambridge, Massachusetts
Fall 1957

The leitmotif, the ever recurring melody, is that two things are indispensable in any reasoning, in any description we shape of a segment of reality: to submit to experience and to face the language that is used, with unceasing logical criticism.

from an unpublished paper of R. v. Mises

CONTENTS

CHAPTER V

INTEGRATION THEORY AND SHOCKS

CHAPTER I

INTRODUCTION

Article 1

The Three Basic Equations

1. Newton's Principle

The theory of fluid flow (for an incompressible or compressible fluid, whether liquid or gas) is based on the Newtonian mechanics of a small solid body. The essential part of Newton's Principle can be formulated into the following statements:

(a) To each small solid body can be assigned a positive number m, invariant in time and called its mass; and

(b) The body moves in such a way that at each moment the product of its acceleration vector by m is equal to the sum of certain other vectors, called forces, which are determined by the circumstances under which the motion takes place (Newton's Second Law).[1]

For example, if a bullet moves through the atmosphere, one force is gravity $m\mathbf{g}$,* directed vertically downward ($g = 32.17$ ft/sec^2 at latitude 45°N); another is the air resistance, or drag, opposite in direction to the velocity vector, with magnitude depending upon that of the velocity, etc.

By means of a limiting process, this principle can be adapted to the case of a continuum in which a velocity vector $\mathbf{q}$ and an acceleration vector $d\mathbf{q}/dt$ exist at each point. Let P be a point with coordinates (x, y, z), or position vector $\mathbf{r}$, and dV a volume element in the neighborhood of P; to this volume element will be assigned a mass $\rho\, dV$, where ρ is the density, or mass per unit volume. Density will be measured in slugs per cubic foot. For air under standard conditions (temperature 59°F, pressure 29.92 in. Hg, or 2116 lb/ft^2), $\rho = 0.002378$ slug/ft^3, as compared with $\rho = 1.94$ slug/ft^3 for water. The forces acting upon this element are, the external force of

* Vectors will be identified by means of boldface type **a**, **v**, etc.; the same letter in lightface italic denotes the absolute value of the vector: $a = |\mathbf{a}|$; components will be indicated by subscripts, as a_x, a_y. The scalar and vector products of **a** and **b** will be represented by $\mathbf{a}\cdot\mathbf{b}$ and $\mathbf{a} \times \mathbf{b}$ respectively. In the figures the bar notation for vectors will be used namely $\bar{a}$, $\bar{g}$, etc.

gravity $\rho\mathbf{g}\,dV$ and the internal forces resulting from interaction with adjacent volume elements. Thus, after dividing by dV, the relation

$$\rho\frac{d\mathbf{q}}{dt} = \rho\mathbf{g} + \text{internal force per unit volume} \tag{1}$$

is a first expression of statement (b).

To formulate part (a) of Newton's Principle, note that the mass to be assigned to any finite portion of the continuum is given by $\int\rho dV$ and therefore, since this mass is invariant with respect to time,

$$\frac{d}{dt}\int\rho\,dV = 0. \tag{2}$$

These two relations will be developed further in succeeding sections, but first the meaning of the differentiation symbol d/dt occurring in Eqs. (1) and (2) must be clarified.

The density ρ and the velocity vector $\mathbf{q}$ are each considered as functions of the four variables x, y, z, and t, so that partial derivatives with respect to time and with respect to the space coordinates may be taken, as well as the directional derivative corresponding to any direction l, given by

$$\frac{\partial}{\partial l} = \cos(l, x)\frac{\partial}{\partial x} + \cos(l, y)\frac{\partial}{\partial y} + \cos(l, z)\frac{\partial}{\partial z},$$

where $\cos(l, x)$, $\cos(l, y)$, and $\cos(l, z)$ are the direction cosines defining the direction l. In particular, if l be taken as the direction of $\mathbf{q}$, the direction cosines may be expressed in terms of $\mathbf{q}$, to give

$$q\frac{\partial}{\partial s} = q_x\frac{\partial}{\partial x} + q_y\frac{\partial}{\partial y} + q_z\frac{\partial}{\partial z}, \tag{3}$$

where s is used in place of l to designate the direction of the line of flow; for this direction $ds = q\,dt$. By d/dt in Eqs. (1) and (2) is meant, not partial differentiation with respect to t at constant x, y, z, but rather differentiation for a given particle, whose position changes according to Eq. (3):

$$\begin{aligned}\frac{d}{dt} &= \frac{\partial}{\partial t} + \frac{dx}{dt}\frac{\partial}{\partial x} + \frac{dy}{dt}\frac{\partial}{\partial y} + \frac{dz}{dt}\frac{\partial}{\partial z}\\ &= \frac{\partial}{\partial t} + q_x\frac{\partial}{\partial x} + q_y\frac{\partial}{\partial y} + q_z\frac{\partial}{\partial z} = \frac{\partial}{\partial t} + q\frac{\partial}{\partial s}.\end{aligned} \tag{4}$$

The acceleration vector $d\mathbf{q}/dt$ is the time rate of change of the velocity vector $\mathbf{q}$ for a definite material particle which moves in the direction of $\mathbf{q}$ at the rate $q = ds/dt$. The operation d/dt may be termed particle differentiation, or material differentiation, as distinguished from partial differentiation with respect to time at a fixed position. An alternative form of Eq. (4) is

$$\frac{d}{dt} = \frac{\partial}{\partial t} + (\mathbf{q}\cdot\text{grad}). \tag{4'}$$

Here, grad is considered as a symbolic vector with the components $\partial/\partial x$, $\partial/\partial y$, and $\partial/\partial z$ in accordance with the well-known notation of grad f for the vector with components $\partial f/\partial x$, $\partial f/\partial y$, and $\partial f/\partial z$, which is called *the gradient of* f, and the scalar product $\mathbf{q}\cdot$grad means the product q times the component of grad in the $\mathbf{q}$-direction, i.e., $q\,\partial/\partial s = q_x\,\partial/\partial x + q_y\,\partial/\partial y + q_z\,\partial/\partial z$. Equations (4) or (4′) will be referred to as the *Euler rule of differentiation.*[2]

When conditions are such that the partial derivative with respect to time, $\partial/\partial t$, vanishes for each variable, the phenomenon is called *steady*, as in steady flow, steady motion. Particle differentiation then reduces to $q\,\partial/\partial s$.

2. Newton's equation for an inviscid fluid

The above equation (1) holds for any continuously distributed mass in which the density ρ is defined at each point and at each moment of time. Different types of continua may be characterized by the form of the term which represents the forces arising from interaction of neighboring elements. An *inviscid* fluid is one in which it is assumed that the force acting on any surface element $d\mathcal{S}$, at which two elements of the fluid are in contact, acts in a direction *normal to the surface element.* It can be shown (see, for example, [16],* p. 2) that at each point P the stress, or force per unit area, is independent of the orientation (direction of the normal) of $d\mathcal{S}$. The value of this stress is called the *hydraulic pressure*, or briefly *pressure*, p, at the point P. The word "pressure" indicates that the force is a thrust, directed toward the fluid element on which it acts: negative values of p are not admitted.

For a small rectangular cell of fluid of volume $dV = dx\,dy\,dz$ (see Fig.1), two pressure forces act in the x-direction, on surface elements of area $dy\,dz$. Taking the x-axis toward the right, the left-hand face experiences a force $p\,dy\,dz$ directed toward the right, while the right-hand face experiences a force $(p + dp)\,dy\,dz$ directed toward the left; here $dp = (\partial p/\partial x)\,dx$. The resultant force in the x-direction is thus

$$-dp\,dy\,dz = -\frac{\partial p}{\partial x}\,dx\,dy\,dz = -\frac{\partial p}{\partial x}\,dV.$$

Therefore the internal force per unit volume, appearing in Eq. (1), has

* Numbers in brackets refer to the bibliography of standard works, p. 502 ff.

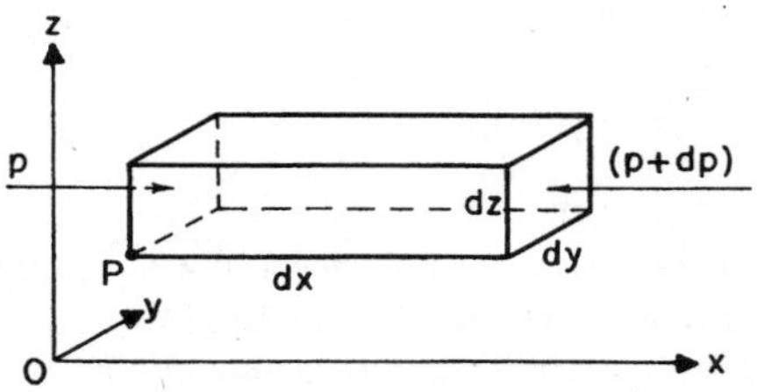

FIG. 1. Rectangular volume element with pressure forces in x-direction.

x-component $-\partial p/\partial x$; similarly, the remaining components are found to be $-\partial p/\partial y$ and $-\partial p/\partial z$. Hence

$$\rho \frac{d\mathbf{q}}{dt} = \rho \mathbf{g} - \operatorname{grad} p \tag{I}$$

expresses part (b) of Newton's Principle. This equation was first given by Leonhard Euler (1755), but is usually known as *Newton's equation.*[3]

The vector Eq. (I) is equivalent to the three scalar equations

$$\rho \frac{dq_x}{dt} = \rho g_x - \frac{\partial p}{\partial x}$$

$$\rho \frac{dq_y}{dt} = \rho g_y - \frac{\partial p}{\partial y} \tag{I$'$}$$

$$\rho \frac{dq_z}{dt} = \rho g_z - \frac{\partial p}{\partial z}.$$

Equations (I) and (I′) are valid only for inviscid fluids. If viscosity is present, additional terms must be included in the expression for the internal force per unit volume, thus generalizing Eq. (I). This will be discussed later (Sec. 3.2).

The motion of the fluid may also be described by following the position of each single particle of the fluid for all values of t. From this point of view the so-called *Lagrangian equations* of motion are obtained, which determine the position (x, y, z) of each particle as a function of t and of three parameters identifying the particle, e.g., the position of the particle at $t = t_0$.[4] Except in certain cases of one-dimensional flow, the Eulerian equations are more manageable.

In either case the following basic concepts apply. The space curve described by a moving particle is called its *path* or *trajectory*. The family of such trajectories is defined by the differential equation $d\mathbf{r}/dt = \mathbf{q}(\mathbf{r}, t)$. On the other hand, at each fixed time $t = t_0$, there is a (two-parameter) family of *streamlines*: for $t = t_0$ each streamline is tangent at each point to the velocity vector at this point, its differential equation being $d\mathbf{r} \times \mathbf{q}(\mathbf{r}, t_0) = 0$, or $dx:dy:dz = q_x:q_y:q_z$, where $q_x = q_x(x, y, z, t_0)$, etc. *Thus in general the family of streamlines varies in time.* Clearly the streamline through **r** at $t = t_0$, is tangent to the trajectory of the particle passing through **r** at that instant.

If, in particular, the motion is *steady*, i.e., at each point P the same thing happens at all times, then there is only one family of streamlines, which then necessarily coincides with the trajectories.

In the general case it is convenient to consider, in addition, the *particle lines* (or "world lines"). They are defined in four-dimensional x, y, z, t-space, each line corresponding to one particle. They appear in Chapter III as the x, t-lines. A particle path is the projection onto x, y, z-space of the respective particle line.

3. Equation of continuity

In order to express part (a) of Newton's Principle—conservation of mass—in the form of a differential equation, the differentiation indicated in Eq. (2) could be carried out by transforming the integral suitably (see below, Sec. 2.6). It is simpler, however, to consider again the rectangular cell of Fig. 1. Fluid mass flows into the cell through the left-hand face at the rate of $\rho q_x\, dy\, dz$ units of mass per second, and out of the right-hand face at rate $[\rho q_x + d(\rho q_x)]\, dy\, dz$, where $d(\rho q_x)$ is the product of dx and the rate of change of ρq_x in the x-direction, or $[\partial(\rho q_x)/\partial x]\, dx$. Thus the net increase in the amount of mass present in this volume element caused by flow across these two faces is given by $-[\partial(\rho q_x)/\partial x]\, dx\, dy\, dz$, with analogous expressions for the other pairs of faces. If we use the expression div **v**, *divergence of the vector* **v**, for the sum $(\partial v_x/\partial x) + (\partial v_y/\partial y) + (\partial v_z/\partial z)$, the total change in mass per unit time is $-\text{div}\,(\rho \mathbf{q})\, dV$. Now if the mass of each moving particle is invariant in time, the difference between the mass entering the cell and that leaving the cell must be balanced by a change in the density of the mass present in the cell. At first, the mass in the cell is given by $\rho\, dV$, and after time dt by $(\rho + d\rho)\, dV$, where $d\rho = (\partial\rho/\partial t)dt$. Thus the rate of change of mass in the cell, per unit time, is given by $(\partial\rho/\partial t)\, dV$, so that

$$\text{(II)} \qquad \text{div}\,(\rho\mathbf{q}) = -\frac{\partial\rho}{\partial t}.$$

This relation, valid for any type of continuously distributed mass, is known as the *equation of continuity.*[5]

A slightly different form of (II) is obtained by carrying out the differentiation of $\rho\mathbf{q}$, giving

$$q_x \frac{\partial\rho}{\partial x} + q_y \frac{\partial\rho}{\partial y} + q_z \frac{\partial\rho}{\partial z} + \rho\,\text{div}\,\mathbf{q} = -\frac{\partial\rho}{\partial t}$$

or, using the Euler rule (4),

$$\text{(II}')\qquad \rho \operatorname{div} \mathbf{q} = -\frac{d\rho}{dt}.$$

Either form, (II) or (II′), will be used in the sequel. In the case of a steady flow, Eq. (II) reduces to

$$\operatorname{div} (\rho\mathbf{q}) = 0,$$

while in the case ρ = constant, that of an *incompressible fluid*, Eq. (II′) gives

$$\operatorname{div} \mathbf{q} = 0,$$

whether or not the flow is steady. The equation of continuity remains unaltered when viscosity is admitted.

4. Specifying equation

The equations (I) and (II), which express Newton's Principle for the motion of an inviscid fluid and are usually referred to as *Euler's equations*, include one vector equation and one scalar equation, or four scalar equations. There are, however, five unknowns: q_x, q_y, q_z, ρ, and p, in these four equations. It follows that one more equation is needed in order that a solution of the system of equations be uniquely determined for given "boundary conditions". Boundary conditions, in a general sense, are equations involving the same variables, holding, however, not in the four-dimensional x, y, z, t-space, but only in certain subspaces, as at some surface $\Phi(x, y, z) = 0$, for all t (boundary conditions in the narrower sense), or at some time $t = t_0$, for all x, y, z (initial conditions).

There exists no general physical principle which would supply a fifth equation to hold in all cases of motion of an inviscid fluid, as do Eqs. (I) and (II). What can and must be added to (I) and (II) is some assumption that specifies the particular type of motion under consideration. This fifth equation will be called the *specifying equation*. Its general form is

$$\text{(III)}\qquad F(p, \rho, \mathbf{q}, x, y, z, t) = 0,$$

where it is understood that derivatives of p, ρ, and $\mathbf{q}$ may also enter into F.[6]

The simplest form of specifying condition results from the assumption that the density ρ has a constant value, independent of x, y, z, and t. It is evident that if ρ is a constant, the number of unknowns reduces to four, and (I) and (II) are sufficient. This is the case of an incompressible fluid.

The most common form of a specifying equation consists in the assumption that p and ρ are variable but connected at all times by a one-to-one relation of the form

(IIIa) $$F(p, \rho) = 0.$$

This means that if the pressure is the same at any two points, the density is also the same at these two points, whether at the same or different moments in time. Examples of such (p, ρ)-relations are

(5a) $$\frac{p}{\rho} = \text{constant},$$

(5b) $$\frac{p}{\rho^{\kappa}} = \text{constant},$$

(5c) $$p = A - \frac{B}{\rho}, \qquad B > 0,$$[7]

where κ, A, and B are constants. In the first of the examples (5) pressure and density are proportional; in general, it will be assumed that ρ increases as p increases, and vice versa, so that $dp/d\rho > 0$. If the specifying equation is of the form (IIIa), the fluid is called an *elastic fluid*, because of the analogy to the case of an elastic solid where the state of stress and the state of strain determine each other. A large part of the results so far obtained in the theory of compressible fluids holds for elastic fluids only.

The special assumption that the specifying equation is of the form (IIIa) is, however, too narrow to cover, for example, the conditions of the atmosphere in the large. It is well known from thermodynamics that for each type of matter a certain relation exists among the three variables, pressure, density, and temperature, the so-called *equation of state*. Thus the temperature can be computed when p and ρ are known. If the atmosphere were assumed to be elastic, so that a specifying equation of the form (IIIa) held, it would then be sufficient to measure p in order to know the temperature as well. This is obviously not the case, so a specifying equation of the form (IIIa) cannot hold for the atmosphere in general. [The equation of state is not a specifying equation, even of the more general type (III), since it implies temperature as a new variable.] In many aerodynamic problems, only comparatively small portions of the atmosphere need be considered, such as the vicinity of the airplane. In such cases, there is no objection *in principle* to the use of a specifying equation of the form (IIIa), if this brings the solution of the problem within reach.

The particular cases corresponding to the examples of (p, ρ)-relations given in (5) can be interpreted in terms of certain concepts from thermo-

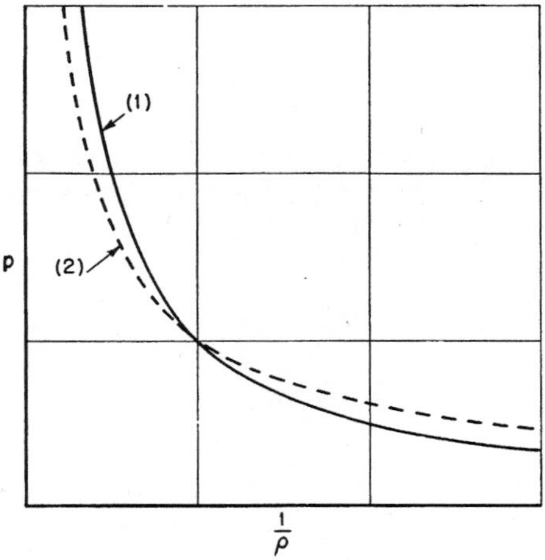

FIG. 2. p versus $\frac{1}{\rho}$ for (1) isothermal and (2) isentropic flow.

dynamics. For a so-called *perfect gas*,* the equation of state is

$$\frac{p}{\rho} = gRT, \tag{6}$$

where T is the absolute temperature (°F + 459.7) and R is a constant depending upon the particular gas.[8] For dry air, if considered as a perfect gas, the value of R can be taken as 53.33 ft/°F. From Eq. (6) it follows that for a perfect gas the condition (5a) implies a flow at constant temperature, or *isothermal* flow.

The entropy S of a perfect gas is defined by

$$S = \frac{gR}{\gamma - 1} \log \frac{p}{\rho^\gamma} + \text{constant}, \tag{7}$$

where γ is a constant, having the value 1.40 for dry air. Thus the motion of a perfect gas with the condition (5b) as specifying equation and $\kappa = \gamma$ is *isentropic*. The motion of any fluid having the specifying equation (5b), with $\kappa > 1$, will be termed *polytropic*.

If p is plotted against $1/\rho$, the specifying equation for an isothermal flow is represented by (1), an equilateral hyperbola (see Fig. 2); the curve for an isentropic flow is shown as the dotted line (2). Whenever the variation of p and of ρ is confined to a small range of values, the relevant part of these (or other) curves can be approximated by a straight line, giving the third type

* A perfect gas is not necessarily inviscid (see [16], p. 83).

of specifying condition (5c). This linearized form of the (p, ρ)-relation (see Sec. 17.5) often facilitates the solution of a problem. Specifying equations which are not of the form (IIIa) will be discussed later (Arts. 2 and 3).

In the case of steady flow, which was defined by the added condition $\partial/\partial t = 0$ for all five dependent variables, it would appear that there are more differential equations than unknowns. Actually, if t does not occur in the specifying equation, which is true *a fortiori* when this equation is of the form (IIIa), the system consisting of Eqs. (I), (II), and (III) does not include t explicitly. For such a system the assumption $\partial/\partial t = 0$ at $t = 0$ leads to a solution totally independent of t. Thus, the assumption of steady motion is a boundary condition only, according to the definition at the beginning of this section. The same is true in the case of a *plane motion*, which is defined by the conditions $q_z = 0$ and $\partial/\partial z = 0$ for all other variables, provided z does not occur explicitly in the specifying equation.

5. Adiabatic flow

In many cases the specification of the type of flow is given in thermodynamic terms. It is then necessary, in order to set up the specifying equation (III), to express these thermodynamic variables in terms of the mechanical variables. Two examples have already been mentioned above. If it is known (or assumed) that the temperature is equal at all points and for all values of t, then the equation of state, which is a relation between T, p, and ρ, supplies a relation of the form (IIIa), $F(p, \rho) = 0$; or, if the condition reads that the entropy has the same value everywhere and for all times, the definition (7) supplies another relation of the type (IIIa).

The most common assumption in the study of compressible fluids is that no heat output or input occurs for any particle. If this refers to heat transfer by radiation and chemical processes only, the flow is called *simply adiabatic*. If heat conduction between neighboring particles is also excluded we speak of *strictly adiabatic* motion. In order to translate either assumption into a specifying equation, the First Law of Thermodynamics must be used, which gives the relation between heat input and the mechanical variables. If Q' denotes the total heat input from all sources, per unit of time and mass, the First Law for an inviscid fluid can be written

$$Q' = c_v \frac{dT}{dt} + p \frac{d}{dt}\left(\frac{1}{\rho}\right), \tag{8}$$

where c_v is the specific heat of the fluid at constant volume, and quantity of heat is measured in mechanical units. The first term on the right represents the part of the heat input expended for the increase in temperature; the second term corresponds to the work done by expansion. Equation (8) is equivalent to the more familiar equation

$$dQ = c_v\,dT + p\,dv,$$

which is derived from (8) by multiplying by dt and writing v (specific volume) in place of $1/\rho$.

If it is known that a flow is strictly adiabatic, i.e., that the total heat input from any source (conduction, radiation, etc.) is zero, then Q' has to be set equal to zero in (8) while T may be expressed in terms of p and ρ by means of the equation of state. Finally, an expression for c_v in terms of the variables T, p, and ρ is needed. For a perfect gas, where the equation of state is (6), it is generally assumed that c_v is a constant given by

$$c_v = \frac{gR}{\gamma - 1}, \tag{9}$$

where, for dry air, $\gamma = 1.4$. We note for further reference that from Eqs. (6) and (9)

$$c_v T = \frac{1}{\gamma - 1}\frac{p}{\rho}. \tag{9'}$$

Thus, from (6), (8), and (9)

$$\begin{aligned} Q' &= \frac{gR}{\gamma - 1}\frac{1}{gR}\frac{d}{dt}\left(\frac{p}{\rho}\right) - \frac{p}{\rho^2}\frac{d\rho}{dt} \\ &= \frac{1}{\gamma - 1}\left(\frac{1}{\rho}\frac{dp}{dt} - \frac{p}{\rho^2}\frac{d\rho}{dt}\right) - \frac{p}{\rho^2}\frac{d\rho}{dt} \\ &= \frac{1}{\gamma - 1}\frac{p}{\rho}\left(\frac{1}{p}\frac{dp}{dt} - \frac{\gamma}{\rho}\frac{d\rho}{dt}\right), \end{aligned}$$

or

$$Q' = \frac{1}{\gamma - 1}\frac{p}{\rho}\frac{d}{dt}\left(\log\frac{p}{\rho^\gamma}\right) \tag{10}$$

holds for a perfect inviscid gas. With the assumption that $Q' = 0$, Eq. (10) reduces to the specifying equation

$$\frac{d}{dt}\left(\log\frac{p}{\rho^\gamma}\right) = 0, \tag{11}$$

holding for strictly adiabatic flow of a perfect inviscid gas.

By means of (6) and (7), Eqs. (10) and (11) may be expressed in terms of the entropy S, giving

$$Q' = T\frac{dS}{dt}, \qquad \frac{dS}{dt} = 0 \text{ when } Q' = 0. \tag{12}$$

Thus, S (or p/ρ^γ) keeps the same value for each particle at all times when $Q' = 0$. Nevertheless, this constant value of S may be different for distinct particles, so that strictly adiabatic flow need not also be isentropic in the sense of Sec. 4.[9] Only under the additional assumption that all particles have a common value for S at some particular time t, does the condition (11) lead to a (p, ρ)-relation of the form (5b). Actual cases of fluid motion will be met later (Sec. 15.6) in which the specifying equation is exactly Eq. (11), while p/ρ^γ has different values for different particles. The specifying equation (11) is not of type (IIIa); a perfect inviscid gas in adiabatic flow does not necessarily behave like an elastic fluid.

Article 2

Energy Equation. Bernoulli Equation

1. Some transformations

Newton's equation (1.I), the equation of continuity (1.II), and an appropriate form of specifying equation (1.III) constitute a complete basis for the theory of inviscid compressible fluid flow. In the present article a scalar equation, known as the *energy equation*, will be derived from Eqs. (1.I) and (1.II), independently of Eq. (1.III). This equation holds in all cases, regardless of any hypothesis that the fluid is elastic or the flow adiabatic, etc., since these assumptions enter only with Eq. (1.III). Only the assumption that the fluid is inviscid must be retained since it was used in the derivation of (1.I). The effect of viscosity on the energy equation will be discussed in Art. 3. It is to be emphasized *that the energy equation is a mathematical consequence of* (1.I) *and* (1.II) and, therefore, does not impose any restriction upon the system of functions defined by (1.I) and (1.II); in particular, the energy equation, as such, cannot serve in place of the specifying equation (1.III). If, however, the energy balance is subject to some condition, which is equivalent to a specifying equation, one can use the energy equation to express this condition in the form of a specifying equation.

Taking the scalar product with $\mathbf{q}$ on the two sides of the vector equation (1.I) we obtain the scalar equation

$$(1) \qquad \rho\mathbf{q}\cdot\frac{d\mathbf{q}}{dt} = \rho\mathbf{q}\cdot\mathbf{g} - \mathbf{q}\cdot\text{grad } p.$$

Each term will now be suitably transformed.

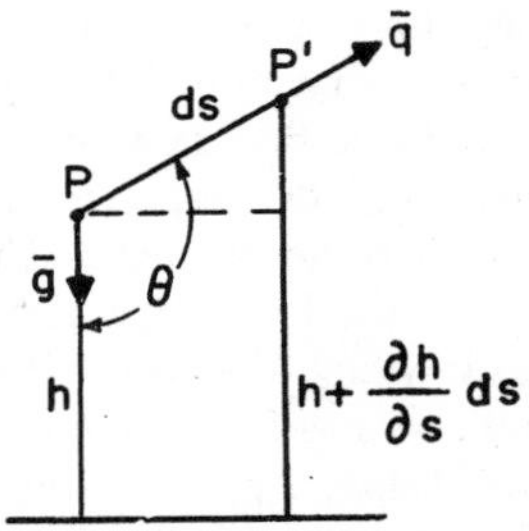

FIG. 3. Change of geometric height.

(a) From the fact that $2\mathbf{q}\cdot d\mathbf{q}/dt$ is the derivative of $\mathbf{q}\cdot\mathbf{q} = q^2$ follows

$$\rho\mathbf{q}\cdot\frac{d\mathbf{q}}{dt} = \rho\frac{d}{dt}\left(\frac{q^2}{2}\right). \tag{2}$$

(b) From the geometrical definition of a scalar product, the value of $\mathbf{q}\cdot\mathbf{g}$ is $qg \cos \theta$, where θ is the angle between the directions of $\mathbf{q}$ and $\mathbf{g}$ (see Fig. 3). If h denotes the function of x, y, z giving the height of each point P above some arbitrary horizontal plane of reference, then the height of P', reached after displacement from P by an amount ds in the direction of $\mathbf{q}$, is $h + dh = h + (\partial h/\partial s)\, ds$. Hence $\cos \theta = -\partial h/\partial s$ from Fig. 3. Since h is a function of position only, $\partial h/\partial t = 0$, and the Euler rule of differentiation, (1.4) gives $dh/dt = q\, \partial h/\partial s$. Thus

$$\rho\mathbf{q}\cdot\mathbf{g} = -\rho g q\frac{\partial h}{\partial s} = -\rho g\frac{dh}{dt} = -\rho\frac{d}{dt}(gh). \tag{3}$$

(c) From the analytic definition of a scalar product, we have

$$\mathbf{q}\cdot\text{grad } p = q_x\frac{\partial p}{\partial x} + q_y\frac{\partial p}{\partial y} + q_z\frac{\partial p}{\partial z}.$$

The first term on the right may be written as

$$q_x\frac{\partial p}{\partial x} = \frac{\partial}{\partial x}(pq_x) - p\frac{\partial q_x}{\partial x},$$

and similarly for the other two terms, so that

$$\mathbf{q}\cdot\text{grad } p = \text{div }(p\mathbf{q}) - p \text{ div } \mathbf{q} = \text{div }(p\mathbf{q}) + \frac{p}{\rho}\frac{d\rho}{dt}, \tag{4}$$

where the last term is obtained by substituting for div $\mathbf{q}$ from the equation of continuity (1.II′).

When (2), (3), and (4) are substituted in (1), the result is

$$\rho \frac{d}{dt}\left(\frac{q^2}{2}\right) = -\rho \frac{d}{dt}(gh) - \text{div}(p\mathbf{q}) - \frac{p}{\rho}\frac{d\rho}{dt}$$

or, setting

$$w = \text{div}(p\mathbf{q}) = \frac{\partial(pq_x)}{\partial x} + \frac{\partial(pq_y)}{\partial y} + \frac{\partial(pq_z)}{\partial z}, \tag{5}$$

rearranging terms and dividing by ρ,

$$\frac{d}{dt}\left(\frac{q^2}{2} + gh\right) + \frac{w}{\rho} = -\frac{p}{\rho^2}\frac{d\rho}{dt} = p\frac{d}{dt}\left(\frac{1}{\rho}\right). \tag{6}$$

This relation is valid for any functions p, ρ, and $\mathbf{q}$ of the variables x, y, z, t satisfying Eqs. (1.I) and (1.II).

2. The energy equation for an element of an inviscid perfect gas

Physical interpretation of the term w will show that (6) may rightly be considered as a preliminary form of the energy equation for a fluid element. Consider again an infinitesimal portion of the moving fluid included in a rectangular cell with sides of length dx, dy, dz (see Fig. 1). The force due to pressure on the left-hand face has the magnitude $p\,dy\,dz$ and the direction of the positive x-axis, so that the work done per unit time by this force is $q_x p\,dy\,dz$. On the opposite face the pressure has the direction of the negative x-axis, so that the work done against pressure in unit time is given by $[q_x p + d(q_x p)]\,dy\,dz$, where $d(q_x p) = [\partial(q_x p)/\partial x]\,dx$. Thus $[\partial(q_x p)/\partial x]\,dx\,dy\,dz$ is an expression for the work done against pressure on this pair of faces. Adding the analogous expressions for the other two pairs of faces, the total work done per unit time against pressure is given by $\text{div}(p\mathbf{q})\,dx\,dy\,dz$. Thus $w = \text{div}(p\mathbf{q})$ represents the *work per unit volume*, and w/ρ the *work per unit mass*, both referred to unit time, *against pressure* on the surface of an element of fluid.

Since $q^2/2$ is the kinetic energy and gh the potential energy (except for an additive constant) per unit mass, the left-hand member of Eq. (6) may be interpreted as the change per unit time of the sum of the kinetic and potential energies, plus the work done per unit time against the pressure on the surface of the fluid element, all per unit mass. The right-hand member of (6) occurs in Eq. (1.8), which expresses the First Law of Thermodynamics, from which $p\,d(1/\rho)/dt = Q' - c_v\,dT/dt$. Thus Eq. (6) may be written

$$\frac{d}{dt}\left(\frac{q^2}{2} + gh + c_v T\right) + \frac{w}{\rho} = Q', \tag{7}$$

where it has been assumed that c_v is a constant, which is true for a perfect gas; the case of a variable c_v will be discussed in Sec. 3. For a perfect gas, the product c_vT is called the (specific) *internal energy*. The content of Eq. (7) may then be expressed by this statement: *For an inviscid perfect gas, the heat input per unit time, Q', is equal to the time rate of change of the total energy (kinetic, potential, and internal) plus the work done per unit time at the surface of the element against pressure; all quantities are expressed per unit mass.*

Under the assumption that the flow is strictly adiabatic, $Q' = 0$, the left-hand member of (7) must vanish. In this case (7) serves to express the condition $Q' = 0$ in terms of mechanical variables [using (1.9′)]. The reader may verify that equating the left-hand side of (7) to zero leads back to (1.11), which was given as the specifying equation for strictly adiabatic flow.

Another expression for Q' has been obtained in (1.12), and (7) then becomes

$$\frac{d}{dt}\left(\frac{q^2}{2} + gh + c_vT\right) + \frac{w}{\rho} = T\frac{dS}{dt}. \tag{8}$$

Either of the equations (7) or (8) is the classical energy equation for a fluid element of a perfect gas. In Eq. (8) three thermodynamic quantities occur: T, S, c_v. If these are replaced by their values from the equation of state (1.6) and the definitions (1.7) and (1.9), respectively, (8) must reduce identically to (6). Various alternative forms of the energy equation can be given if thermodynamic variables other than c_v, T, and S are used. For example, if the specific heat at constant pressure, c_p, is introduced (for a perfect gas) by $c_p = c_v + gR$,* Eq. (1.8) with $T = p/\rho gR$ becomes

$$Q' = \frac{c_v}{gR}\frac{d}{dt}\left(\frac{p}{\rho}\right) + p\frac{d}{dt}\left(\frac{1}{\rho}\right) = \frac{1}{\rho gR}\left(c_v\frac{dp}{dt} - \frac{pc_p}{\rho}\frac{d\rho}{dt}\right).$$

This expression may replace the right-hand member of Eq. (8).

3. Nonperfect (inviscid) gas

Equations (7) and (8) have been derived under the hypotheses that the equation of state is (1.6) and that S and c_v are given by (1.7) and (1.9). Slightly generalized forms of Eqs. (7) and (8) hold also in the case of a nonperfect gas. It is assumed that some equation of state holds, in the form

$$T = T(p, \rho). \tag{9}$$

The nature of the gas is determined by the form of the function $T(p, \rho)$ and that of two other functions: the *entropy* $S(p, \rho)$ and the *internal energy*

* It then follows from (1.9) that $\gamma = c_p/c_v$, the ratio of the specific heats.

$U(p, \rho)$. These two functions, however, cannot be chosen independently of $T(p, \rho)$ and of each other. It is required that for a nonperfect gas, too, Eq. (1.12) is valid—which means that the heat input Q' per unit time and mass equals $T\,dS/dt$—and that the First Law of Thermodynamics holds for any change in p and ρ in the form

$$T\frac{dS}{dt} = Q' = \frac{dU}{dt} + p\frac{d}{dt}\left(\frac{1}{\rho}\right) \tag{10}$$

corresponding to (1.8). Since

$$\frac{dU}{dt} = \frac{\partial U}{\partial p}\frac{dp}{dt} + \frac{\partial U}{\partial \rho}\frac{d\rho}{dt} \quad \text{and} \quad \frac{dS}{dt} = \frac{\partial S}{\partial p}\frac{dp}{dt} + \frac{\partial S}{\partial \rho}\frac{d\rho}{dt},$$

the relation (10) can be correct for arbitrary dp/dt and $d\rho/dt$ only if

$$T\frac{\partial S}{\partial p} = \frac{\partial U}{\partial p} \quad \text{and} \quad T\frac{\partial S}{\partial \rho} = \frac{\partial U}{\partial \rho} - \frac{p}{\rho^2}. \tag{11}$$

Eliminating U, we obtain

$$\frac{\partial}{\partial \rho}\left(T\frac{\partial S}{\partial p}\right) = \frac{\partial}{\partial p}\left(T\frac{\partial S}{\partial \rho} + \frac{p}{\rho^2}\right).$$

This leads to the condition

$$\frac{\partial(S, T)}{\partial(p, \rho)} \equiv \frac{\partial S}{\partial p}\frac{\partial T}{\partial \rho} - \frac{\partial S}{\partial \rho}\frac{\partial T}{\partial p} = \frac{1}{\rho^2}, \tag{12}$$

or

$$\frac{\partial(S, T)}{\partial\left(\dfrac{1}{\rho}, p\right)} = 1. \tag{12'}$$

It can easily be verified that (12) is satisfied by the expressions for T and S given in Eqs. (1.6) and (1.7) for a perfect gas, while the corresponding value of U is found to be, up to an additive constant

$$U = \frac{1}{\gamma - 1}\frac{p}{\rho}, \tag{13}$$

and this equals $c_v T$ by (1.9′).

For a given equation of state (9), there exists an infinity of functions $S(p, \rho)$ satisfying Eq. (12). When one of these has been chosen, the internal energy $U(p, \rho)$ is determined, except for an additive constant, by (11). The right-hand side of (6) is then to be replaced from (10) by $Q' - dU/dt$

or $T(dS/dt) - dU/dt$, and the resulting equation differs from (7) or ([illegible]) only in having U in place of c_vT:

$$\frac{d}{dt}\left(\frac{q^2}{2} + gh + U\right) + \frac{w}{\rho} = Q' = T\frac{dS}{dt}. \tag{14}$$

The condition of adiabatic flow is again $dS/dt = 0$, and this becomes a (p, ρ)-relation is the value of S is supposed to be the same for all particles.

For example, if the equation of state is taken as

$$T = A_1p + \frac{B_1}{\rho},$$

where A_1 and B_1 are constants, then an admissible choice for S is

$$S = Ap + \frac{B}{\rho} \qquad \text{with} \qquad A_1B - AB_1 = 1,$$

and the corresponding expression for U is

$$U = \frac{AA_1}{2}p^2 + AB_1\frac{p}{\rho} + \frac{BB_1}{2\rho^2} + \text{constant},$$

as can easily be verified by differentiation.

4. Energy equation for an elastic fluid

If the specifying equation is a (p, ρ)-relation, the energy equation may be given another form. Since only a single fluid element is under consideration, it is not necessary to know that a (p, ρ)-relation holds for the entire fluid; it is sufficient to assume that such a relation holds for the particle in question, i.e., that for any two states of the particle with the same value for p the value of ρ is also the same.

Since p is a function of ρ, the right-hand member of Eq. (6) can be expressed entirely in terms of ρ. Let e be a function of ρ (or of p) whose derivative with respect to ρ is p/ρ^2:

$$e = \int^{\rho} \frac{p}{\rho^2}\, d\rho. \tag{15}$$

Then

$$\frac{de}{dt} = \frac{p}{\rho^2}\frac{d\rho}{dt} = -p\frac{d}{dt}\left(\frac{1}{\rho}\right).$$

so that Eq. (6) can be written in the form

$$\frac{d}{dt}\left(\frac{q^2}{2} + gh + e\right) + \frac{w}{\rho} = 0. \tag{16}$$

The quantity e may be called the *expansion energy* or *strain energy*; then Eq. (16) states: *The time rate of change of the sum of the kinetic, potential, and strain energies is equal to the work done by the pressure forces on the surface per unit time and mass.* It is to be noted that the strain energy exists only in the case of a relation between p and ρ for the particle in question. For the important examples of over-all (p, ρ)-relationships given in Sec. 1.4 the corresponding strain energies are:

(a) Incompressible flow, or $\rho = \rho_0$:

$$e = \text{constant};$$

(b) Isothermal flow of a perfect gas, or $\frac{p}{\rho} = \frac{p_0}{\rho_0}$:

$$e = \int^{\rho} \frac{p_0}{\rho_0} \frac{d\rho}{\rho} = \frac{p_0}{\rho_0} \log \rho + \text{constant};$$

(c) Polytropic flow, or $\frac{p}{\rho^\kappa} = \frac{p_0}{{\rho_0}^\kappa}$:

(17)
$$e = \int^{\rho} \frac{p_0}{{\rho_0}^\kappa} \rho^{\kappa-2}\, d\rho = \frac{p_0}{{\rho_0}^\kappa} \frac{\rho^{\kappa-1}}{\kappa - 1} + \text{constant}$$

$$= \frac{1}{\kappa - 1} \frac{p}{\rho} + \text{constant};$$

(d) Linearized condition, $p = p_0 - \frac{B}{\rho}$:

$$e = -\frac{p_0}{\rho} + \frac{B}{2\rho^2} + \text{constant}.$$

The lower limit in (15) has been omitted, which means that there remains an undetermined additive constant in e. This is of no importance since only the derivative of e appears in (16).

5. Bernoulli equation

If we use Eqs. (2) and (3), Eq. (1) can be written

$$\frac{d}{dt}\left(\frac{q^2}{2} + gh\right) = -\frac{1}{\rho}\, \mathbf{q} \cdot \text{grad}\, p.$$

With the further *assumption that the motion is steady*, so that $\partial p/\partial t = 0$, the Euler rule (1.4) gives

$$\mathbf{q} \cdot \text{grad}\, p = \frac{dp}{dt},$$

and the above relation becomes

$$\frac{d}{dt}\left(\frac{q^2}{2} + gh\right) = -\frac{1}{\rho}\frac{dp}{dt}. \tag{18}$$

Under the hypothesis that a (p, ρ)-relation exists for the particle in question, ρ can be regarded as a function of p. A new function can be introduced, say P, whose derivative with respect to p is $1/\rho$:

$$P = \int^{p} \frac{dp}{\rho}. \tag{19}$$

Then $dP/dt = (1/\rho)(dp/dt)$ and (18) becomes

$$\frac{d}{dt}\left(\frac{q^2}{2} + gh + P\right) = 0, \tag{20}$$

which is known as the *Bernoulli equation.*[10]

Usually P/g (which has the dimension of length) is called the *pressure head*, and $q^2/2g$ the *velocity head*. Equation (20) is equivalent to

$$\frac{d}{dt}\left(\frac{q^2}{2g} + h + \frac{P}{g}\right) = 0, \qquad \text{or} \qquad \frac{q^2}{2g} + h + \frac{P}{g} = H, \tag{20$'$}$$

where H is a constant (which may be different for each particle), or in words: *During the steady motion of a fluid that is either elastic or subject to a particle-wise (p, ρ)-relation the sum H of the velocity head, the geometrical elevation, and the pressure head, has a constant value for each particle.* For all particles on the same streamline this constant is the same. The quantity H is known as the *total head* or as the *Bernoulli constant* or the *Bernoulli function* (since it may vary from one streamline to the next).[11]

The Bernoulli equation can also be given in differential form, by multiplying through by dt in Eq. (18):

$$q\,dq + g\,dh + \frac{1}{\rho}\,dp = 0 \tag{21}$$

or, neglecting gravity,

$$\rho q\,dq + dp = 0, \tag{21$'$}$$

where the differentials refer, of course, to the changes in q, h, and p for a definite particle.

The Bernoulli equation in differential form can also be derived directly from Eq. (1.I) as the component of Newton's equation in the direction of $\mathbf{q}$, where d/dt is to be identified with $q(\partial/\partial s)$.

For the four cases listed previously, the function P is given by

(a) Incompressible flow, or $\rho = \rho_0$:

$$P = \frac{p}{\rho_0} + \text{constant};$$

(b) Isothermal flow of a perfect gas, or $\frac{p}{\rho} = \frac{p_0}{\rho_0}$:

$$P = \frac{p_0}{\rho_0} \log p + \text{constant};$$

(22)

(c) Polytropic flow, or $\frac{p}{\rho^\kappa} = \frac{p_0}{\rho_0^\kappa}$:

$$P = \frac{\kappa}{\kappa - 1} \frac{p_0^{1/\kappa}}{\rho_0} p^{(\kappa-1)/\kappa} + \text{constant} = \frac{\kappa}{\kappa - 1} \frac{p}{\rho} + \text{constant};$$

(d) Linearized condition, $\rho = \frac{B}{p_0 - p}$:

$$P = \frac{2p_0 p - p^2}{2B} + \text{constant}.$$

Again, the additive constant in P is arbitrary.

If the Bernoulli equation (20) is compared with the energy equation (16) for an elastic fluid, it is seen that P includes both the strain energy and the surface work w. In fact, the equality

$$\frac{dP}{dt} = \frac{de}{dt} + \frac{w}{\rho}$$

follows for steady motion directly from (4), using (15), (5), and (19). Thus the pressure head P/g is not an energy, but the sum of an energy term and a work term.

In thermodynamics the sum of internal energy U plus the quotient p/ρ is known as *enthalpy*:

$$I = U + \frac{p}{\rho}. \tag{23}$$

In adiabatic flow, where $Q' = 0$, it follows from Eq. (10) that

$$\frac{dI}{dt} = \frac{dU}{dt} + \frac{d}{dt}\left(\frac{p}{\rho}\right) = -p \frac{d}{dt}\left(\frac{1}{\rho}\right) + \frac{d}{dt}\left(\frac{p}{\rho}\right) = \frac{1}{\rho}\frac{dp}{dt} = \frac{dP}{dt},$$

since each particle possesses a (p, ρ)-relation expressed by the constancy of its entropy. Thus, the quantity P in the case of adiabatic flow differs from the enthalpy I only by an additive constant (which still may be different for each particle). If the motion is isentropic, P as well as I is an over-all function of the state variables, and in this case they differ by an over-all constant.

6. Two integral theorems

In order to extend the energy equation (7) to a volume of finite dimensions, some integral transformation formulas are needed. These will now be derived.

(a) Let V be a volume bounded by a closed surface $\mathcal{S}$. Let F be a function of x, y, z having first partial derivatives which, together with F, are continuous in V and on $\mathcal{S}$.

The formula to be proved is

$$\int_V \frac{\partial F}{\partial z}\, dV = \int_{\mathcal{S}} F \cos(n, z)\, d\mathcal{S}, \tag{24}$$

where (n, z) denotes the angle between the z-axis and the outer normal to $\mathcal{S}$. This formula transforms the volume integral over V into a surface integral over $\mathcal{S}$. The surface $\mathcal{S}$ may consist of a finite number of sections, on each of which $\cos(n, z)$ varies continuously.

Suppose first that V is such that a parallel to the z-axis intersects $\mathcal{S}$ in at most two points 1 and 2. The volume integral may then be evaluated as an iterated integral:

$$\int_V \frac{\partial F}{\partial z}\, dV = \int_A dA \int \frac{\partial F}{\partial z}\, dz = \int_A (F_2 - F_1)\, dA, \tag{25}$$

where the region of integration, A, for the double integral is the projection of V onto the x, y-plane, while the integral with respect to z is taken from the point 1 where the horizontal cylinder with cross section dA first cuts V, to the point 2 where it leaves V (see Fig. 4); F_1 and F_2 denote the values of F at these points. Let $d\mathcal{S}_1$ and $d\mathcal{S}_2$ denote the areas of the parts of $\mathcal{S}$ cut off by this cylinder, and n_1 and n_2 the directions normal to these surface elements. Then since dA, $d\mathcal{S}_1$, and $d\mathcal{S}_2$ are positive quantities,

$$dA = -\cos(n_1, z)\, d\mathcal{S}_1 = \cos(n_2, z)\, d\mathcal{S}_2 .$$

Thus, the last integral in (25) may be rewritten as

$$\begin{aligned} \int_A (F_2 - F_1)\, dA &= \int F_2 \cos(n_2, z)\, d\mathcal{S}_2 + \int F_1 \cos(n_1, z)\, d\mathcal{S}_1 \\ &= \int_{\mathcal{S}} F \cos(n, z)\, d\mathcal{S}. \end{aligned} \tag{26}$$

Equations (25) and (26) give exactly (24). One can easily see that the restriction to a surface with only two intersections with a parallel to the z-axis is unessential.

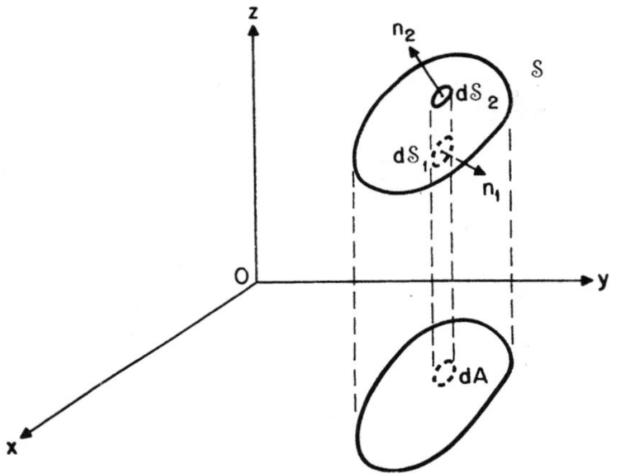

FIG. 4. Integral transformation from volume to surface integral.

Analogous formulas hold for the x- and y-directions. In particular, these formulas may be applied to the cases $F = v_x$, v_y, and v_z, where v_x, v_y, and v_z are the components of a vector $\mathbf{v}$, each formula being taken with differentiation in the direction of the component. When these results are added, the integrand on the left in Eq. (24) is div $\mathbf{v}$, while the integrand on the right is $v_x \cos(n, x) + v_y \cos(n, y) + v_z \cos(n, z)$, which is exactly the component v_n of $\mathbf{v}$ in the direction n. Thus

$$\int_V \operatorname{div} \mathbf{v}\, dV = \int_{\mathcal{S}} v_n\, d\mathcal{S}. \tag{27}$$

The right-hand member of Eq. (27) may be called the *flux of* $\mathbf{v}$ *through the surface* $\mathcal{S}$. The result (27) is known as *Gauss' Theorem*, or as the *Divergence Theorem.*

(b) Let f be a quantity associated with a fluid particle, i.e., a function of some or all of the variables q_x, q_y, q_z, p, and ρ. Then the following formula holds:

$$\frac{d}{dt}\int_V \rho f\, dV = \int_V \rho \frac{df}{dt}\, dV, \tag{28}$$

where ρ is the density and $d(\int_V \rho f\, dV)/dt$ means the rate of change of $\int_V \rho f\, dV$ as a certain portion of the fluid moves along.

Py the Euler rule (1.4′) and a computation similar to that used in deriving Eq. (4) we have

$$\rho \frac{df}{dt} = \rho \frac{\partial f}{\partial t} + \rho \mathbf{q} \cdot \operatorname{grad} f = \rho \frac{\partial f}{\partial t} + \operatorname{div}(\rho f \mathbf{q}) - f \operatorname{div}(\rho \mathbf{q}). \tag{29}$$

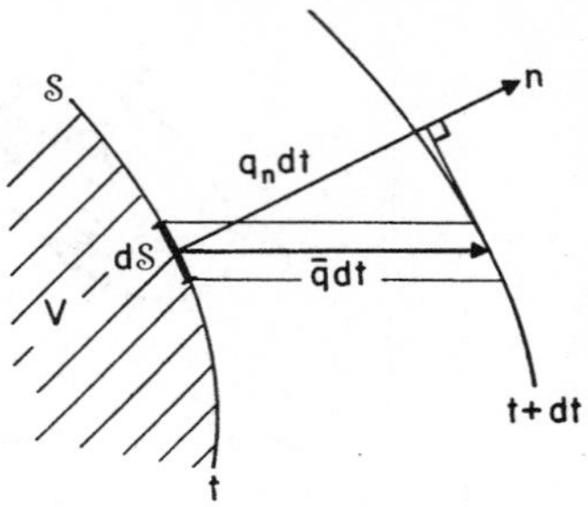

FIG. 5. Change of volume integral due to change and displacement of boundary surface.

From the equation of continuity (1.II), div $(\rho\mathbf{q}) = -\partial\rho/\partial t$, so that

$$(30) \quad \int_V \rho \frac{df}{dt}\, dV = \int_V \rho \frac{\partial f}{\partial t}\, dV + \int_V \text{div}\,(\rho f \mathbf{q})\, dV + \int_V f\left(\frac{\partial \rho}{\partial t}\right) dV.$$

Combining the two integrands involving $\partial/\partial t$ and applying (27) to the third integral, we obtain

$$(31) \qquad \int_V \rho \frac{df}{dt}\, dV = \int_V \frac{\partial(\rho f)}{\partial t}\, dV + \int_{\mathcal{S}} \rho f q_n\, d\mathcal{S}.$$

The left-hand side of Eq. (28) expresses the total rate of change of $\int \rho f\, dV$, due in part to the change within V of the integrand with respect to time (which produces the rate of change $\int_V \partial(\rho f)/\partial t\, dV$), and in part to the motion of the boundary $\mathcal{S}$ which changes the volume over which the integral is extended. This flow takes place across each surface element $d\mathcal{S}$; in time dt the surface element $d\mathcal{S}$ adds to the volume V a cylindrical element of volume $d\mathcal{S} \cdot q_n\, dt$ (see Fig. 5). Then $\int \rho f\, dV$ evaluated only for the additional volume equals $dt \int_{\mathcal{S}} \rho f q_n\, d\mathcal{S}$, and the rate of change is therefore $\int_{\mathcal{S}} \rho f q_n\, d\mathcal{S}$. It is thus seen that the left-hand side of Eq. (28) is precisely equal to the right-hand side of Eq. (31), and the proof is complete. Of course, with $f \equiv 1$, so that $df/dt = 0$, equation (28) reduces to the equation of continuity in its preliminary form (1.2).

7. Energy equation for a finite mass

With the aid of the foregoing integral theorems it is easy to derive from Eq. (7), which holds for a perfect gas, a relation between the energy, the work, and the heat input for a finite portion of the gas.

Both sides of Eq. (7) may be multiplied by $\rho\, dV$ and the result integrated over V. With the abbreviation $f = q^2/2 + gh + c_v T$, and the value of w from (5), the result is

$$(32) \qquad \int_V \rho \frac{df}{dt}\, dV + \int_V \text{div}\,(p\mathbf{q})\, dV = \int_V \rho Q'\, dV.$$

If we apply (28) to the first integral and (27) to the second, this becomes

$$\frac{d}{dt}\int_V \rho f\,dV + \int_{\mathcal{S}} pq_n\,d\mathcal{S} = \int_V \rho Q'\,dV. \tag{33}$$

Now the *total energy* of the fluid mass enclosed by the boundary surface $\mathcal{S}$ is given by

$$E = \int_V \rho\left(\frac{q^2}{2} + gh + c_v T\right) dV = \int_V \rho f\,dV, \tag{34}$$

while the work done, per unit time, against pressure on the surface $\mathcal{S}$ is

$$W = \int_{\mathcal{S}} pq_n\,d\mathcal{S}, \tag{35}$$

since the force has magnitude $p\,d\mathcal{S}$, and the rate of motion in the direction opposite to the force is q_n. Thus (33) gives

$$\frac{dE}{dt} + W = \int_V \rho Q'\,dV, \tag{36}$$

and the statement is the same for a finite portion of a perfect gas as for the infinitesimal portion considered in Sec. 2: *The total heat input per unit time is equal to the time rate of change of the total energy (kinetic, potential, and internal) plus the work done per unit time at the surface of the gas against pressure.*

The statement (36) holds independently of the type of specifying equation. For adiabatic flow, $Q' = 0$, and (36) gives

$$\frac{dE}{dt} + W = 0. \tag{37}$$

For an elastic fluid, the integral theorems may be applied as above, starting from Eq. (16), to give

$$\frac{dE'}{dt} + W = 0, \tag{38}$$

where

$$E' = \int_V \rho\left(\frac{q^2}{2} + gh + e\right) dV. \tag{39}$$

Article 3

Influence of Viscosity. Heat Conduction

1. Viscous stresses and hydraulic pressure

For an inviscid fluid the forces exerted on any fluid element by surrounding masses are normal to the surface element on which they act and have the same intensity p whatever the orientation of the surface element. This intensity p (force per unit area) was called the hydraulic pressure at the point under consideration. If viscosity is admitted, however, the stress vector on a surface element $d\mathbb{S}$ is no longer normal to $d\mathbb{S}$. The stress can be resolved into a normal component σ and tangential or shearing components τ. Considering surface elements normal to the coordinate axes (see Fig. 6), there are three *normal stresses*: σ_x, σ_y, and σ_z, and six *shearing stresses*, equal in pairs: $\tau_{xy} = \tau_{yx}$, $\tau_{yz} = \tau_{zy}$, and $\tau_{zx} = \tau_{xz}$.[12] Here, as is usual in the theory of elasticity, τ_{xy}, for example, denotes the component in the y-direction of the stress acting on a surface element with outward normal in the x-direction. More explicitly σ_x, τ_{xy}, and τ_{xz} denote the components of the stress vector $\mathbf{t}_x$ acting on the element with outward normal in the positive x-direction. Thus, a positive value of σ_x denotes a tensile stress, while a negative value corresponds to compression. If $\mathbf{t}_x$, $\mathbf{t}_y$, and $\mathbf{t}_z$ denote the stresses acting on surface elements with normals in the x-, y-, and z-directions, then equilibrium considerations show that the stress vector $\mathbf{t}_n$ acting on a surface element with outward normal in the direction n must be given by

$$(1) \qquad \mathbf{t}_n = \mathbf{t}_x \cos(n, x) + \mathbf{t}_y \cos(n, y) + \mathbf{t}_z \cos(n, z).$$

Also σ, which is the component of $\mathbf{t}_n$ in the direction n, is given by

$$(2) \qquad \sigma = \mathbf{t}_n . \hat{\mathbf{n}} = t_{nx} \cos(n, x) + t_{ny} \cos(n, y) + t_{nz} \cos(n, z),$$

where t_{nx}, t_{ny}, and t_{nz} are the x-, y-, and z-components of $\mathbf{t}_n$ and $\hat{\mathbf{n}}$ denotes the unit vector in the direction n, while the resultant shearing stress is

$$(3) \qquad \tau = \sqrt{t_n^2 - \sigma^2}.$$

In the particular case of an inviscid fluid, $\mathbf{t}_n$ is parallel to $\hat{\mathbf{n}}$ by hypothesis; then in (2), $\sigma = t_n$, (3) reduces to $\tau = 0$, and the component of (1) in the x-direction is $\sigma \cos(n, x) = \sigma_x \cos(n, x)$, so that $\sigma = \sigma_x$. Similarly, $\sigma = \sigma_y = \sigma_z$, as mentioned in Sec. 1.2, where the common value of the normal stresses was denoted by $-p$.

In general, however, the three normal stresses σ_x, σ_y, and σ_z are not necessarily equal, the differences being greater the larger the shearing

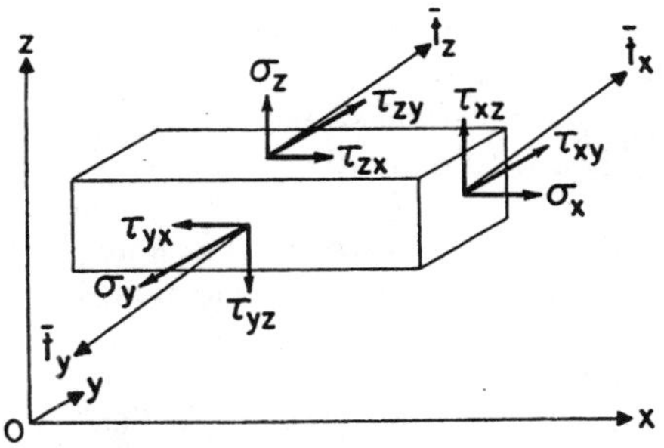

FIG. 6. Stress vectors and their components acting on surface of rectangular cell in viscous fluid.

stresses. In the theory of elasticity this lack of equality causes no difficulty, but in fluid mechanics the equation of state requires the existence of a hydraulic pressure p, equal for all directions. The question arises: What can be used in place of the variable p in the equation of state (and in other thermodynamic equations) when dealing with a viscous fluid?

An answer is suggested by this fact (following from Eqs. (2) and (1) by a simple computation which uses the elementary properties of the direction cosines of orthogonal directions): The sum of the normal stresses σ_x, σ_y, and σ_z is invariant under rotation of the coordinate axes, i.e., if x', y', and z' define a triple of orthogonal directions, and if $\sigma_{x'}$, $\sigma_{y'}$, and $\sigma_{z'}$ denote the normal stresses corresponding to these directions, then

$$\sigma_{x'} + \sigma_{y'} + \sigma_{z'} = \sigma_x + \sigma_y + \sigma_z . \tag{4}$$

For an inviscid fluid, this invariant quantity has the value $-3p$. Thus, it is natural to consider one-third of the sum (4) as the mean or *average tensile stress*, and its negative as the *mean pressure* at a given point. It is usual to take

$$p = -\tfrac{1}{3}(\sigma_x + \sigma_y + \sigma_z) \tag{5}$$

as the value of p in the thermodynamic equations. Whether or not this assumption is justified is an essentially experimental question, which cannot be settled definitely on the basis of the experimental evidence now available. Equation (5) will be used in this book as a working hypothesis.

Once p has been defined, the state of stress at any point may be thought of as being composed of a hydraulic pressure p, uniform in all directions, and a system of viscous stresses σ_x', σ_y', σ_z', τ_{xy}, τ_{yz}, and τ_{zx}, where

$$\sigma_x = -p + \sigma_x', \qquad \sigma_y = -p + \sigma_y', \qquad \sigma_z = -p + \sigma_z', \tag{6}$$

and consequently, using Eq. (5),

$$\sigma_x' + \sigma_y' + \sigma_z' = 0. \tag{7}$$

It is to be noted that the viscous stresses are not additional unknowns, over and above the five unknowns q_x, q_y, q_z, p, and ρ considered thus far. In the theory of viscous fluids the stresses σ' and τ are taken to be given functions of p, ρ, $\mathbf{q}$, and their derivatives, e.g., σ_x' may be given as proportional to $\partial q_x/\partial x$, etc. The exact form of this dependence will not be considered here, although a certain necessary restriction, due to the fact that viscous forces are "friction forces", will be mentioned in Sec. 3.

2. Newton's equation for a viscous fluid[13]

The general form of Newton's equation, holding for any type of continuum, was given in Eq. (1.1). For the "internal force per unit volume" there is first the contribution of the pressure p, which is $-\text{grad}\, p$, exactly as in the case of an inviscid fluid. In the rectangular cell of Fig. 6, the viscous forces acting in the x-direction on one set of faces, have the magnitudes

$$\sigma_x'\, dy\, dz, \qquad \tau_{yx}\, dx\, dz, \qquad \tau_{zx}\, dx\, dy,$$

and on the three opposite faces

$$\left(\sigma_x' + \frac{\partial \sigma_x'}{\partial x}\, dx\right) dy\, dz, \qquad \left(\tau_{yx} + \frac{\partial \tau_{yx}}{\partial y}\, dy\right) dx\, dz, \qquad \left(\tau_{zx} + \frac{\partial \tau_{zx}}{\partial z}\, dz\right) dx\, dy.$$

Upon subtracting and dividing through by the volume $dx\, dy\, dz$, we see that the x-component of the resultant viscous force, per unit volume, is

$$v_x = \frac{\partial \sigma_x'}{\partial x} + \frac{\partial \tau_{yx}}{\partial y} + \frac{\partial \tau_{zx}}{\partial z}.$$

Thus, the x-component of Newton's equation is

$$\rho\, \frac{dq_x}{dt} = \rho g_x - \frac{\partial p}{\partial x} + \frac{\partial \sigma_x'}{\partial x} + \frac{\partial \tau_{yx}}{\partial y} + \frac{\partial \tau_{zx}}{\partial z}. \tag{8}$$

Similarly, or by cyclic substitution on x, y, z in Eq. (8), the other components of Newton's equation are

$$\begin{aligned} \rho\, \frac{dq_y}{dt} &= \rho g_y - \frac{\partial p}{\partial y} + \frac{\partial \tau_{xy}}{\partial x} + \frac{\partial \sigma_y'}{\partial y} + \frac{\partial \tau_{zy}}{\partial z} \\ \rho\, \frac{dq_z}{dt} &= \rho g_z - \frac{\partial p}{\partial z} + \frac{\partial \tau_{xz}}{\partial x} + \frac{\partial \tau_{yz}}{\partial y} + \frac{\partial \sigma_z'}{\partial z}. \end{aligned} \tag{8}$$

The equation can also be given in vector form as

$$\rho\, \frac{d\mathbf{q}}{dt} = \rho \mathbf{g} - \text{grad}\, p + \mathbf{v}, \tag{Ia}$$

where $\mathbf{v}$ is the resultant viscous force per unit volume. If the stress is not split up into the uniform pressure p and the viscous stresses, the term

$-(\partial p/\partial x) + (\partial \sigma_x'/\partial x)$ in (8) is to be replaced by $\partial \sigma_x/\partial x$, etc. The reader familiar with the notation of tensor calculus will notice that Eq. (Ia) may also be written as

$$\rho \frac{d\mathbf{q}}{dt} = \rho \mathbf{g} - \operatorname{grad} p + \operatorname{grad} \Sigma', \tag{8'}$$

or

$$\rho \frac{d\mathbf{q}}{dt} = \rho \mathbf{g} + \operatorname{grad} \Sigma, \tag{8''}$$

where Σ denotes the stress tensor and Σ' the tensor of the viscous stresses.[14]

Equation (Ia) [or (8)] takes the place of (1.I). The addition of viscosity has no effect on the form of the equation of continuity (1.II) or (1.II′), and some specifying equation of the form (1.III) must be added to make the system of equations complete.

3. Work done by viscous forces. Dissipation

In Art. 2 the energy equation for an inviscid fluid was obtained from the vector equation (1.I) by scalar multiplication by $\mathbf{q}$. If we start from (Ia) instead of (1.I), the result of multiplying by $\mathbf{q}$ is the same as before, Eq. (2.6), except for additional terms on the right resulting from $\mathbf{q}\cdot\mathbf{v}$, namely,

$$\begin{aligned} &q_x\left(\frac{\partial \sigma_x'}{\partial x} + \frac{\partial \tau_{yx}}{\partial y} + \frac{\partial \tau_{zx}}{\partial z}\right) + q_y\left(\frac{\partial \tau_{xy}}{\partial x} + \frac{\partial \sigma_y'}{\partial y} + \frac{\partial \tau_{zy}}{\partial z}\right) \\ &\qquad + q_z\left(\frac{\partial \tau_{xz}}{\partial x} + \frac{\partial \tau_{yz}}{\partial y} + \frac{\partial \sigma_z'}{\partial z}\right). \end{aligned} \tag{9}$$

Each term in Eq. (9) can be split into two terms, e.g.,

$$q_x \frac{\partial \sigma_x'}{\partial x} = \frac{\partial}{\partial x}(q_x\sigma_x') - \sigma_x' \frac{\partial q_x}{\partial x}, \qquad q_x \frac{\partial \tau_{yx}}{\partial y} = \frac{\partial}{\partial y}(q_x\tau_{yx}) - \tau_{yx}\frac{\partial q_x}{\partial y}.$$

Next, let

$$\begin{aligned} -w' = &\frac{\partial}{\partial x}(q_x\sigma_x' + q_y\tau_{xy} + q_z\tau_{xz}) + \frac{\partial}{\partial y}(q_x\tau_{yx} + q_y\sigma_y' + q_z\tau_{yz}) \\ &+ \frac{\partial}{\partial z}(q_x\tau_{zx} + q_y\tau_{zy} + q_z\sigma_z') \end{aligned} \tag{10}$$

and, keeping in mind that $\tau_{xy} = \tau_{yx}$, etc.,

$$\begin{aligned} \theta = &\sigma_x' \frac{\partial q_x}{\partial x} + \sigma_y' \frac{\partial q_y}{\partial y} + \sigma_z' \frac{\partial q_z}{\partial z} + \tau_{xy}\left(\frac{\partial q_x}{\partial y} + \frac{\partial q_y}{\partial x}\right) \\ &+ \tau_{yz}\left(\frac{\partial q_y}{\partial z} + \frac{\partial q_z}{\partial y}\right) + \tau_{zx}\left(\frac{\partial q_z}{\partial x} + \frac{\partial q_x}{\partial z}\right). \end{aligned} \tag{11}$$

Then the equation that results in place of (2.6) can be written as

$$\frac{d}{dt}\left(\frac{q^2}{2} + gh\right) + \frac{w + w'}{\rho} = p\,\frac{d}{dt}\left(\frac{1}{\rho}\right) - \frac{\theta}{\rho}. \tag{12}$$

The definitions given in (10) and (11) have been chosen with a view to physical interpretation. It will be shown (a) that w' is the *work done* per unit of time and volume *against the viscous forces*, and (b) that θ is an essentially positive quantity having the dimensions of work per unit of volume and time, and may be called the *dissipation function*, or briefly *dissipation*, representing the rate of conversion of mechanical energy into heat, per unit of time and volume.

(a) With the above definitions of the components of stress, the viscous force acting on the left-hand face (with outward normal in the negative x-direction) of the rectangular cell of Fig. 6 has components

$$-\sigma_x'\,dy\,dz, \qquad -\tau_{xy}\,dy\,dz, \qquad \text{and} \qquad -\tau_{xz}\,dy\,dz$$

in the x-, y-, and z-directions, respectively. Thus, the work done by this force per unit time is

$$-q_x\sigma_x'\,dy\,dz - q_y\tau_{xy}\,dy\,dz - q_z\tau_{xz}\,dy\,dz.$$

At the opposite face (with outward normal in the x-direction), the work done is

$$\left[q_x\sigma_x' + \frac{\partial}{\partial x}(q_x\sigma_x')\,dx\right]dy\,dz + \left[q_y\tau_{xy} + \frac{\partial}{\partial x}(q_y\tau_{xy})\,dx\right]dy\,dz + \left[q_z\tau_{xz} + \frac{\partial}{\partial x}(q_z\tau_{xz})\,dx\right]dy\,dz.$$

Thus, the net work done at this pair of faces by the viscous forces, per unit of time and volume, is

$$\frac{\partial}{\partial x}(q_x\sigma_x') + \frac{\partial}{\partial y}(q_y\tau_{xy}) + \frac{\partial}{\partial z}(q_z\tau_{xz});$$

analogous expressions hold for the remaining pairs of faces, so that the total work for all faces is exactly the right-hand side of Eq. (10), and, therefore, $-w'$. Then w' is the work done per unit time and volume *against* the viscous forces. Since w is the work done against hydraulic pressure, the term $w + w'$ in (12) includes all work against surface stresses.[15]

(b) The velocity terms in (11) can be easily interpreted. For example, $d_{xx} = \partial q_x/\partial x$ is the time *rate of deformation* in the x-direction, since in time dt the left face of the rectangular cell undergoes the displacement $q_x\,dt$, while for the right face the displacement is $[q_x + (\partial q_x/\partial x)\,dx]\,dt$.

Also, since σ_x' is a normal stress due to the viscosity of the fluid, it is necessary to assume that σ_x' is positive (tensile stress) when a positive expansion occurs ($\partial q_x/\partial x$ positive) and negative in the case of compression ($\partial q_x/\partial x$ negative); but the first product in (11), $\sigma_x' \, \partial q_x/\partial x$, is nonnegative, as are the two following terms. It is known from the elements of the theory of elasticity (or the kinematics of a continuum) that the term $d_{xy} = \frac{1}{2}(\partial q_x/\partial y + \partial q_y/\partial x)$ represents the time *rate of shear* about the z-axis. The symmetric tensor D with components d_{xx}, d_{xy}, etc. is called *the rate of deformation tensor* or *rate of strain tensor*. Again, the nature of viscosity requires that a rate of shear strain be accompanied by a shearing stress of the same, not opposite, sign. Thus, the fourth term in Eq. (11), $\tau_{xy}(\partial q_x/\partial y + \partial q_y/\partial x)$, is nonnegative, as are the following terms, and consequently the value of θ itself is nonnegative.[16] The above assumptions as to the signs of the components of viscous stress may be summarized in the statement that the viscous stress tensor Σ', has the nature of a resistance force or friction, i.e., of a force that *consumes rather than produces energy*.[17]

The most usual assumption made in the theory of viscous flow is that each of the six viscous stresses is a linear function of the rates of deformation with coefficients constant or known functions of p and ρ. So specific a hypothesis is not adopted here, but the assumption concerning signs is essential, since the contrary case would correspond to a "friction force" between material elements that would tend to increase their relative velocity. A world in which friction forces accelerated moving bodies would contradict all our experience.

4. The energy equation for a viscous fluid

The energy equation in mechanical terms has already been given in Eq. (12). As in Art. 2, the heat input Q' may be introduced by using the First Law of Thermodynamics, which gives the relation between Q' and the mechanical variables. This time, however, the statement (1.8) of the First Law must be modified to allow for the rate of dissipation of energy, which is $\theta \, dV$ for a volume element dV. The modified equation, for a unit mass, reads

$$(13) \qquad Q' + \frac{\theta}{\rho} = c_v \frac{dT}{dt} + p \frac{d}{dt}\left(\frac{1}{\rho}\right),$$

where Q', as before, denotes the heat delivered to the particle per unit of time and mass by radiation or conduction from surrounding fluid elements, etc. When Eqs. (12) and (13) are combined, the energy equation reads

$$(14) \qquad \frac{d}{dt}\left(\frac{q^2}{2} + gh + c_v T\right) + \frac{w + w'}{\rho} = Q'$$

if c_v is considered to be a constant, as for a perfect gas. The difference between Eqs. (14) and (2.7) lies only in the additional term w'/ρ representing the work done per unit mass and time against the viscous stresses at the surface of the element.

If the flow is considered as being strictly adiabatic, the condition $Q' = 0$ supplies the specifying equation

$$\frac{d}{dt}\left(\frac{q^2}{2} + gh + c_vT\right) + \frac{w + w'}{\rho} = 0. \tag{15}$$

This equation does not lead to the simple condition $dS/dt = 0$ given in (1.12) as the specifying equation for strictly adiabatic flow in the case of an inviscid perfect gas; for, instead of (1.8), it follows from (13) that

$$Q' = c_v \frac{dT}{dt} + p \frac{d}{dt}\left(\frac{1}{\rho}\right) - \frac{\theta}{\rho}.$$

If T and c_v are replaced by the expressions given in (1.6) and (1.9), the result that replaces (1.12) is

$$Q' = T\frac{dS}{dt} - \frac{\theta}{\rho}, \tag{16}$$

where the definition (1.7) of the entropy S is used. Thus (15) is equivalent to

$$T\frac{dS}{dt} = \frac{\theta}{\rho}, \tag{17}$$

and either (15) or (17) may be taken as the specifying equation for the strictly adiabatic flow of a viscous perfect gas.

Incidentally, it follows from Eq. (17) that, since θ is nonnegative, $dS/dt \geqslant 0$ for strictly adiabatic flow, as compared with $dS/dt = 0$ in the case of an inviscid fluid, i.e., the entropy of a particle of a viscous perfect gas can never decrease if no heat input or output occurs. This is an expression of the Second Law of Thermodynamics.

The result (14) holds only if the fluid is such that c_v is constant. For a viscous fluid for which an equation of state of the form (2.9) holds, functions $U(p, \rho)$ and $S(p, \rho)$ can be found, as shown in Sec. 2.3, satisfying

$$T\frac{dS}{dt} = Q' + \frac{\theta}{\rho} = \frac{dU}{dt} + p\frac{d}{dt}\left(\frac{1}{\rho}\right) \tag{18}$$

and subject to the unchanged restriction (2.11). The energy equation is then

$$\frac{d}{dt}\left(\frac{q^2}{2} + gh + U\right) + \frac{w + w'}{\rho} = Q' = T\frac{dS}{dt} - \frac{\theta}{\rho} \tag{19}$$

rather than Eq. (2.14) as for an inviscid fluid. The specifying equation for strictly adiabatic flow is again (17), or (15) with c_vT replaced by U.

The integration of the energy relation over a finite volume can be carried out as in Sec. 2.7. One new term appears: W', the integral of $w'\,dV$. It can be shown by a formal transformation, or inferred from the physical meaning of w', that W' is the work done per unit time against the viscous stresses on the surface of the volume under consideration. Thus if $\mathbf{t}'_n$ is the viscous stress on a surface element $d\mathcal{S}$ with outward normal n, we have

$$(20) \qquad W' = \int_V w'\,dV = -\int_{\mathcal{S}} \mathbf{t}'_n \cdot \mathbf{q}\,d\mathcal{S}.$$ [18]

The integrated energy equation will then read

$$(21) \qquad \frac{dE}{dt} + W + W' = \int_V \rho Q'\,dV = \int_V \left(\rho T\frac{dS}{dt} - \theta\right) dV$$

in the case of a viscous perfect gas, as compared to Eq. (2.36); the function E is the same as that in (2.34).

5. Heat conduction

All relations so far discussed in this article are valid regardless of whether or not there is heat conduction within the fluid. Heat conduction can be a factor in the mechanics of a fluid only if it occurs explicitly or implicitly in the specifying condition. For example, the specifying condition may be that a particle experiences no heat input or output other than the exchange of heat with surrounding particles by means of conduction. In order to set up the specifying equation corresponding to this case, it is necessary first to discuss the mechanism of heat flow.

It is usual to assume that at each point of a continuously distributed mass the flow of heat in any direction is proportional to the derivative of the temperature function in that direction, and that the flow takes place from higher values of T toward lower values. Consider again the rectangular cell of Fig. 6. Supposing $\partial T/\partial x$ is positive, heat will flow through the left face of the cell in the negative x-direction at the rate of $k(\partial T/\partial x)\,dy\,dz$ per unit time, where k is the *coefficient of (internal) thermal conductivity* for the substance. The flux through the opposite face is

$$\left[k\frac{\partial T}{\partial x} + \frac{\partial}{\partial x}\left(k\frac{\partial T}{\partial x}\right)dx\right]dy\,dz,$$

so that the net gain of heat resulting from flow across these two faces is

$$\frac{\partial}{\partial x}\left(k\frac{\partial T}{\partial x}\right)dx\,dy\,dz,$$

and the net gain of heat per unit time and volume, resulting from flow across all faces, is

$$\frac{\partial}{\partial x}\left(k\frac{\partial T}{\partial x}\right) + \frac{\partial}{\partial y}\left(k\frac{\partial T}{\partial y}\right) + \frac{\partial}{\partial z}\left(k\frac{\partial T}{\partial z}\right) = \text{div }(k \text{ grad } T). \tag{22}$$

Here k may be a given constant, or a given function of T.

If a fluid motion occurs under *simply* adiabatic conditions (Sec. 1.5), i.e., there is no external heat exchange (by radiation, etc.), the total heat input per unit time and volume, $\rho Q'$, must be given by Eq. (22), or

$$Q' = \frac{1}{\rho}\text{ div }(k \text{ grad } T), \tag{23}$$

as compared with the relation $Q' = 0$ corresponding to *strictly* adiabatic flow. The specifying condition (23) may be expressed also by substituting for Q' from (23) in the energy equations (14) or (19):

$$\frac{d}{dt}\left(\frac{q^2}{2} + gh + U\right) + \frac{w + w'}{\rho} = \frac{1}{\rho}\text{ div }(k \text{ grad } T) \tag{24}$$

or

$$T\frac{dS}{dt} - \frac{\theta}{\rho} = \frac{1}{\rho}\text{ div }(k \text{ grad } T). \tag{24$'$}$$

These equations differ from those for strictly adiabatic flow in having $(1/\rho)$ div (k grad T) in place of zero on the right.

6. General form of specifying equation[19]

For some purposes it seems useful to have a general form of specifying equation that is not so restrictive as form (1. IIIa), yet more explicit than Eq. (1. III), and which includes both viscous and inviscid fluids, with or without heat conduction. Such a form of specifying equation is

$$A\frac{dp}{dt} + B\frac{d\rho}{dt} = C, \tag{IIIb}$$

and the various cases are obtained by specializing, in appropriate ways, the coefficients A, B, and C.

Take first $C = 0$ and A and B as the partial derivatives with respect to p and ρ of a function F of p and ρ. Then Eq. (IIIb) expresses the fact that $F(p, \rho)$ has a constant value for each particle. Combined with the boundary condition that at some time this value is the same for all particles, we obtain the specifying equation of an elastic fluid. In particular, if the function F is the entropy $S(p, \rho)$ then (IIIb) is the specifying equation for the adiabatic flow of an inviscid fluid.

A general class of conditions is expressed by (IIIb) if A, B, and C are assumed to be arbitrary functions of x, y, z, t, and of the flow variables p, ρ, $\mathbf{q}$. The corresponding wide class of fluid motions is characterized by common properties. This class does not include the case of viscous or heat-conducting fluids and may be designated as the class of *ideal fluid motions.* (We shall consider a particular case in Sec. 9.6.)

The specifying equation for the case of viscosity and heat conduction is included in (IIIb) if the term C on the right is allowed to depend also on derivatives of p, ρ, and $\mathbf{q}$. In fact, if $T(p, \rho)$ and $S(p, \rho)$ are given, and we take

$$A = T\frac{\partial S}{\partial p}, \qquad B = T\frac{\partial S}{\partial \rho}, \tag{25}$$

then the left-hand side of (IIIb) is $T(dS/dt)$ and according to (16) must equal $Q' + \theta/\rho$. Here θ is defined by (11) as a given function of the viscous stresses σ', τ and of the spatial derivatives of $\mathbf{q}$; and the σ', τ are in general assumed to be given functions of these spatial derivatives of $\mathbf{q}$ with coefficients depending on p and ρ. In the strictly adiabatic case $Q' = 0$, while in the simply adiabatic case when heat conduction is admitted with no other heat input or output, Q' is determined by (23) as a function of the spatial derivatives of $T(p, \rho)$. To cover more general cases, one would have to add to the right-hand side of Eq. (23) an appropriate function expressing the heat production per unit of time and mass.

If the total heat input [and, therefore, the quantity C in Eq. (IIIb)] is given as a function of x, y, z, t, p, ρ, $\mathbf{q}$, and the spatial derivatives of p, ρ, $\mathbf{q}$, then the system of equations that governs the motion consists of the following: Eq. (Ia), which generalizes Eq. (1.I), Eqs. (1.II′) and (IIIb). The system can then be written in the form:

$$\begin{aligned} \frac{d\mathbf{q}}{dt} &= \mathbf{g} - \frac{1}{\rho}\operatorname{grad} p + \frac{\mathbf{v}}{\rho} \\ \frac{d\rho}{dt} &= -\rho \operatorname{div} \mathbf{q} \\ \frac{dp}{dt} &= \frac{B}{A}\rho \operatorname{div} \mathbf{q} + \frac{C}{A}, \end{aligned} \tag{26}$$

where $\mathbf{v}$, as in (Ia), is the resultant viscosity force per unit volume. Since $\mathbf{v}$ depends on the spatial derivatives of $\mathbf{q}$ and possibly on the flow variables (through the viscosity coefficients) the Eqs. (26) can be interpreted as follows: *At each moment and for each particle the material derivatives of the five unknowns* $\mathbf{q}$, p, ρ *are determined as functions of the instantaneous values of* p, ρ, $\mathbf{q}$ *at the point* x, y, z *and in a neighborhood of the point.*

Article 4

Sound Velocity. Wave Equation

1. The problem

Suppose that the fluid mass extends indefinitely in all directions, so that no boundaries are encountered, and that the fluid is inviscid and elastic. The latter assumption means, as in Sec. 1.4, that a universal and one-to-one relation between p and ρ exists. This (p, ρ)-relation is assumed to be differentiable, and such that density increases with pressure, and vice versa. Then $dp/d\rho$ is nonnegative, and we may write

$$\frac{dp}{d\rho} = a^2. \tag{1}$$

The dimensions of the left-hand member are given by $(ML/T^2L^2)/(M/L^3) = L^2/T^2$. Thus a has the dimensions L/T, a velocity. With the notation of Eq. (1), any derivative of p can be expressed in terms of the corresponding derivative of ρ, e.g., $\partial p/\partial x = a^2 \, \partial\rho/\partial x$, and, in particular,

$$\operatorname{grad} p = a^2 \operatorname{grad} \rho. \tag{2}$$

A state of rest, $\mathbf{q} = 0$, with uniform pressure and density, $p = p_0$, $\rho = \rho_0$, is certainly compatible with the equation of motion (1.I), and the equation of continuity (1.II), if gravity is neglected, for all the other terms in these equations are derivatives.[20] The values $p = p_0$, $\rho = \rho_0$ are assumed to satisfy the given (p, ρ)-relation.

Now we consider a small *perturbation* of the state of rest, caused by an initial *disturbance*: to each (x, y, z, t) there will correspond small values of $\mathbf{q}$, $p - p_0$, and $\rho - \rho_0$. The qualification *small* means that any product of two or more of these quantities or their derivatives will be supposed negligible compared to first-order terms. The particle derivative, d/dt, of any of these quantities may then be replaced by the partial derivative $\partial/\partial t$, since the additional terms are all of second order. Moreover, on the left-hand side of Eq. (1.I), $\rho \, \partial\mathbf{q}/\partial t$ may be replaced by $\rho_0 \, \partial\mathbf{q}/\partial t$, the difference $(\rho - \rho_0) \, \partial\mathbf{q}/\partial t$ being of second order. On the right-hand side, grad p becomes, by Eq. (2), a^2 grad ρ, for which one has to substitute a_0^2 grad ρ, where a_0 is the value of a for $p = p_0$, $\rho = \rho_0$, the difference $(a^2 - a_0^2)$ grad ρ being small of second order. Thus, with gravity neglected Eq. (1.I) becomes

$$\rho_0 \frac{\partial \mathbf{q}}{\partial t} = -a_0^2 \operatorname{grad} \rho. \tag{3}$$

Similarly, the equation of continuity (1.II′) has to be replaced by

$$\rho_0 \operatorname{div} \mathbf{q} = -\frac{\partial \rho}{\partial t}. \tag{4}$$

These two equations, (3) and (4), are the determining equations for a small perturbation in an inviscid elastic fluid originally at rest. There is one scalar and one vector equation, with one scalar unknown, ρ, and one vector unknown, $\mathbf{q}$. In contrast to the exact equations (1.I) and (1.II), the present equations are *linear* in the unknowns ρ and $\mathbf{q}$, which makes them more accessible to solution. On the other hand, the full implications of this linearization are not easily determined. See, however, a few comments in Sec. 13.3.

2. One-dimensional case. D'Alembert's solution

Some information on the behavior of a small perturbation may be obtained quickly by a consideration of the one-dimensional problem: it is assumed that all particles move in the same fixed direction, say that of x, while all unknowns are independent of y and z; i.e.,

$$q_y = q_z = 0 \qquad \text{and} \qquad \frac{\partial}{\partial y} = \frac{\partial}{\partial z} = 0.$$

Then only the x-component of (3) remains, and in (4) div is to be replaced by $\partial/\partial x$:

$$\rho_0 \frac{\partial q_x}{\partial t} = -a_0^2 \frac{\partial \rho}{\partial x}, \qquad \rho_0 \frac{\partial q_x}{\partial x} = -\frac{\partial \rho}{\partial t}. \tag{5}$$

On multiplying the second equation by a_0^2, differentiating with respect to t in the first and x in the second, and subtracting, we obtain

$$\frac{\partial^2 q_x}{\partial t^2} = a_0^2 \frac{\partial^2 q_x}{\partial x^2}, \tag{6a}$$

while if the multiplication by a_0^2 is omitted and t and x are interchanged in the differentiation step, the result is

$$\frac{\partial^2 \rho}{\partial t^2} = a_0^2 \frac{\partial^2 \rho}{\partial x^2}. \tag{6b}$$

It can also be seen that any twice differentiable function u of the variables q_x and ρ satisfies an equation of exactly the same form,

$$\frac{\partial^2 u}{\partial t^2} = a_0^2 \frac{\partial^2 u}{\partial x^2}, \tag{6}$$

if terms of higher order in the derivatives of q_x and ρ are neglected. In fact, if $u = u\ (q_x,\rho)$, then

$$\frac{\partial u}{\partial x} = \frac{\partial u}{\partial q_x}\frac{\partial q_x}{\partial x} + \frac{\partial u}{\partial \rho}\frac{\partial \rho}{\partial x}, \qquad \frac{\partial^2 u}{\partial x^2} = \frac{\partial u}{\partial q_x}\frac{\partial^2 q_x}{\partial x^2} + \frac{\partial u}{\partial \rho}\frac{\partial^2 \rho}{\partial x^2},$$

where in the second equation terms involving $(\partial q_x/\partial x)^2$, etc., have been omitted. Similarly

$$\frac{\partial^2 u}{\partial t^2} = \frac{\partial u}{\partial q_x}\frac{\partial^2 q_x}{\partial t^2} + \frac{\partial u}{\partial \rho}\frac{\partial^2 \rho}{\partial t^2}.$$

These expressions, together with (6a) and (6b), give (6).

Equation (6) is the (one-dimensional) *wave equation*. The general solution of this equation was given by d'Alembert:[21] if $f_1(z)$ and $f_2(z)$ denote arbitrary twice differentiable functions of the real variable z, then (6) is satisfied by

$$u = f_1(x + a_0t) + f_2(x - a_0t), \tag{7}$$

where z has been replaced by $x + a_0t$ in the first function and by $x - a_0t$ in the second; for, letting f_1' denote $df_1(z)/dz$, etc., we get

$$\frac{\partial^2 u}{\partial x^2} = f_1'' + f_2'', \qquad \frac{\partial^2 u}{\partial t^2} = a_0^2(f_1'' + f_2''),$$

from which (6) follows.

The small perturbation under consideration is assumed to have been caused by an initial disturbance imposed on a state of rest. The disturbance may be represented by giving the values of q_x and $\rho - \rho_0$ at $t = 0$, so that $q_x(x, 0)$ and $\rho(x, 0)$ are given functions of x. Knowing q_x at some time t means knowing also $\partial\rho/\partial t$ at the same time, according to the second of Eqs. (5). Thus, if u stands for $\rho - \rho_0$, the initial conditions for u can be written in this form:

$$\text{at } t = 0, \qquad u = f(x), \qquad \frac{\partial u}{\partial t} = g(x), \tag{8}$$

where f and g are given (sufficiently differentiable) functions of x. If u stands for q_x, we derive from the given $\rho(x, 0)$ the value of $\partial q_x/\partial t$ by means of the first Eq. (5), so that again the initial conditions for u have the form (8).

The problem before us is to find an integral of Eq. (6) satisfying the conditions (8). We are going to show that the solution reads

$$\begin{aligned} u(x, t) &= \frac{1}{2}[f(x + a_0t) + f(x - a_0t)] \\ &\qquad + \frac{1}{2a_0}[G(x + a_0t) - G(x - a_0t)], \end{aligned} \tag{9}$$

where $G(x)$ is a function whose derivative is $g(x)$:

$$G(x) = \int^x g(\xi)\, d\xi \; ; \tag{9'}$$

the second part of the solution (9) can also be expressed by

$$\frac{1}{2a_0} \int_{x-a_0 t}^{x+a_0 t} g(\xi)\, d\xi.$$

Indeed, the solution (9) has the form (7) with $f_1 = \frac{1}{2}f + G/2a_0$ and $f_2 = \frac{1}{2}f - G/2a_0$, so that (9) satisfies the differential equation (6). Further, it is seen immediately that $u(x, 0) = f(x)$, while

$$\left(\frac{\partial u}{\partial t}\right)_{t=0} = \left\{\frac{1}{2}[a_0 f'(x + a_0 t) - a_0 f'(x - a_0 t)] + \frac{1}{2a_0}[a_0 G'(x + a_0 t) + a_0 G'(x - a_0 t)]\right\}_{t=0}$$

$$= \frac{1}{2a_0} 2a_0 G'(x) = g(x),$$

as required by (8).

Equation (9) shows that the value of u for a given position x at a given time t depends only on the values of the initial disturbance u and $\partial u/\partial t$ in the interval $(x + a_0 t,\, x - a_0 t)$ and is independent of the values of the initial disturbance at all other points. On the other hand, the initial disturbance on an interval (x_1, x_2) determines the perturbation $u(x, t)$ only for values (x, t) that lie in the triangle in the x, t-plane having base (x_1, x_2) and sides with slopes $1/a_0$ and $-1/a_0$, respectively.

To discuss the solution (9), we assume first that g is identically zero and that $f(x)$ vanishes outside some interval AB: $(-c \leq x \leq c)$. In Fig. 7 the values of $u = \frac{1}{2}[f(x + a_0 t) + f(x - a_0 t)]$ corresponding to certain values of t ($t = 0$, $t = t_1$, $t = t_2$) are illustrated. For a fixed position x to the right of AB ($x > c$) the perturbation u will be zero except for t such that $x - a_0 t$

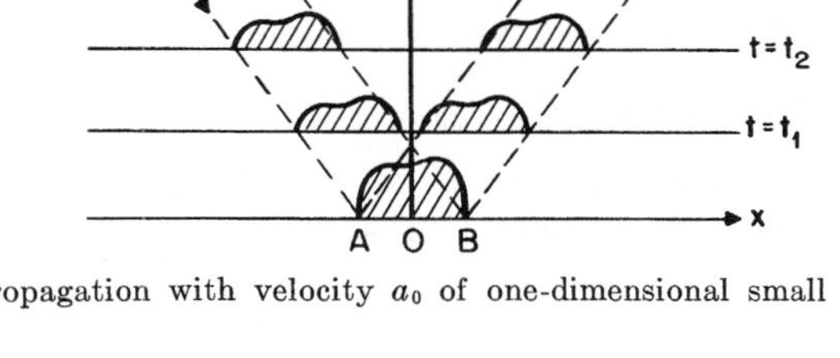

FIG. 7. Propagation with velocity a_0 of one-dimensional small disturbance.

falls between $-c$ and c; in other words, $u(x, t) \neq 0$ only for $(x - c)/a_0 < t < (x + c)/a_0$. Within this time interval u takes successively one-half of the values that $f(x)$ has in the interval $x = -c$ to $x = c$. For a fixed t, the perturbation is zero in the three intervals

$$(10) \quad x \leqslant -c - a_0t; \qquad c - a_0t \leqslant x \leqslant -c + a_0t; \qquad x \geqslant c + a_0t.$$

This whole situation can be described in the following terms: the initial disturbance $f(x)$ of width $2c$ splits into two equal parts, each again of width $2c$ but each half as high as the original $f(x)$, one moving to the right, the other to the left, and each with velocity a_0. Thus, a_0 may be called the *velocity of propagation of a small disturbance*. This velocity depends only on the nature of the fluid and the original state of rest; it is independent of the exact form of the disturbance. Typical disturbances of air at rest are noise or acoustic signals, so the shorter term *sound velocity* is used mostly. If the given (p, ρ)-relation is Eq. (1.5b): p/ρ^κ = constant, characterizing polytropic flow, then $a^2 = dp/d\rho = \kappa p/\rho$. For air, considered as a perfect gas and under standard conditions, i.e., $p_0 = 2{,}116\ \text{lb/ft}^2$, $\rho_0 = 0.002378\ \text{slug/ft}^3$, and $\kappa = \gamma = 1.4$, the value of a_0 is 1,116 ft/sec or 760.9 mph.[22]

The results of the above discussion are not essentially changed if the restriction $g(x) = 0$ is omitted. Suppose that q_x and $\rho - \rho_0$ at $t = 0$ both vanish outside a common interval AB. If u is identified with $\rho - \rho_0$, $f(x)$ again vanishes outside AB, and $g(x) = (\partial\rho/\partial t)_{t=0}$, which by (5) is proportional to $(\partial q_x/\partial x)_{t=0}$, must also vanish outside AB. Next, q_x must vanish at A and B, since it vanishes outside this interval and is continuous by hypothesis. Thus

$$\int_A^B \frac{\partial q_x}{\partial x}\, dx = q_x(B) - q_x(A) = 0.$$

But then also $\int_A^B g(x)\, dx = 0$; and from $G(x) = \int^x g(\xi)\, d\xi$ it follows that $G(A) = G(B)$. For a suitable choice of the constant of integration, this common value is 0. Then $G(x)$ has the further property $G(x) = 0$ for all x outside AB, which was the essential property of $f(x)$ used in the previous discussion. Again it follows that the perturbation is zero whenever Eq. (10) is satisfied, so that the disturbance is propagated to the right and left with velocity a_0 exactly as before. This time, however, the left and right half-waves are not equal to each other.

3. The wave equation in three dimensions

Before returning to the general case covered by Eqs. (3) and (4), two formal relations should be recalled. First, the divergence of a gradient equals the Laplacian operator, or, as applied to any scalar f,

$$\begin{aligned}\text{div (grad } f) &= \frac{\partial}{\partial x}\left(\frac{\partial f}{\partial x}\right) + \frac{\partial}{\partial y}\left(\frac{\partial f}{\partial y}\right) + \frac{\partial}{\partial z}\left(\frac{\partial f}{\partial z}\right) \\ &= \frac{\partial^2 f}{\partial x^2} + \frac{\partial^2 f}{\partial y^2} + \frac{\partial^2 f}{\partial z^2} = \Delta f.\end{aligned} \tag{11}$$

Secondly, as shown in Sec. 2.6, the flux of a vector through a closed surface $\mathcal{S}$ is the integral of the divergence of that vector over the volume enclosed by $\mathcal{S}$, or, as applied to some vector $\mathbf{v}$,

$$\int_V \text{div } \mathbf{v} \, dV = \int_{\mathcal{S}} v_n \, d\mathcal{S}. \tag{12}$$

With the use of (11), the equations (3) and (4) governing a small perturbation of an inviscid elastic fluid originally at rest can be suitably transformed. Taking the divergence of both sides of (3), applying (11), and differentiating (4) with respect to t, we obtain

$$\rho_0 \text{ div } \frac{\partial \mathbf{q}}{\partial t} = -a_0^2 \Delta \rho, \qquad \rho_0 \frac{\partial}{\partial t} \text{div } \mathbf{q} = -\frac{\partial^2 \rho}{\partial t^2}.$$

Since $\partial/\partial t$ and div are interchangeable, these give

$$\frac{\partial^2 \rho}{\partial t^2} = a_0^2 \Delta \rho. \tag{13}$$

On the other hand, if we differentiate (3) with respect to t and take the gradient of (4), there results

$$\rho_0 \frac{\partial^2 \mathbf{q}}{\partial t^2} = -a_0^2 \frac{\partial}{\partial t} \text{grad } \rho, \qquad \rho_0 \text{ grad (div } \mathbf{q}) = -\text{grad } \frac{\partial \rho}{\partial t}.$$

Here the right-hand members are equal, except for the factor a_0^2; therefore

$$\frac{\partial^2 \mathbf{q}}{\partial t^2} = a_0^2 \text{ grad (div } \mathbf{q}).$$

Taking the divergence of both sides of this equation, and applying (11) once more (with $f = \text{div } \mathbf{q}$), we obtain

$$\frac{\partial^2}{\partial t^2} (\text{div } \mathbf{q}) = a_0^2 \Delta(\text{div } \mathbf{q}). \tag{14}$$

The equation

$$\frac{\partial^2 u}{\partial t^2} = a_0^2 \Delta u = a_0^2 \left(\frac{\partial^2 u}{\partial x^2} + \frac{\partial^2 u}{\partial y^2} + \frac{\partial^2 u}{\partial z^2}\right) \tag{15}$$

is called the (three-dimensional) *wave equation*. From (13) and (14) it follows that this equation holds for $u = \rho$ and for $u = \text{div } \mathbf{q}$. As in the preceding section, it can be shown that any twice differentiable function of ρ and div $\mathbf{q}$ also satisfies Eq. (15), if terms of higher order in the perturbation variables and their derivatives are neglected. In particular, the total energy per unit volume, $\rho(q^2/2 + c_v T)$, is, at this level of approximation (neglecting q^2), a function of ρ only and thus also satisfies (15). (T depends only on ρ because, by the equation of state, T is a function of p and ρ, and here p is a function of ρ.) It can further be seen, by differentiation of (15), that if an arbitrary function u satisfies the wave equation, then so also do its derivatives $\partial u/\partial x$, $\partial u/\partial t$, e.g., any component of grad p, etc. The equation preceding (14) can also be written

$$\frac{\partial^2 \mathbf{q}}{\partial t^2} = a_0^2(\Delta \mathbf{q} + \text{curl curl } \mathbf{q}).$$

(We denote by curl $\mathbf{q}$ a vector with components $\partial q_z/\partial y - \partial q_y/\partial z$, $\partial q_x/\partial z - \partial q_z/\partial x$, and $\partial q_y/\partial x - \partial q_x/\partial y$, which will be considered in detail in Art. 6.) Moreover from Eq. (3)

$$\frac{\partial}{\partial t}(\text{curl } \mathbf{q}) = 0.$$

Hence in any region where curl $\mathbf{q} = 0$ initially, it is always zero, and consequently $\mathbf{q}$, as well as any component of $\mathbf{q}$, satisfies the wave equation.

4. Poisson's solution[23]

Functions $u(x, y, z, t)$ satisfying the wave equation may be found as follows. Let $f(x, y, z)$ be an arbitrary and sufficiently differentiable function of the space coordinates, P a point with coordinates x, y, z, and $\mathcal{S}$ the spherical surface with center at P and radius $R = a_0 t$. For any position (x, y, z) and time t, we define $\bar{f}(x, y, z, t)$ by

$$\bar{f}(x, y, z, t) = \frac{1}{4\pi R^2}\int_{\mathcal{S}} f(\xi, \eta, \zeta)\, d\mathcal{S} = \frac{1}{4\pi}\int f\, d\sigma, \tag{16}$$

where the first integral is to be taken over the total surface $\mathcal{S}$. Since the area of $\mathcal{S}$ is $4\pi R^2$, $\bar{f}$ represents the mean or average value of f on $\mathcal{S}$, the variable t entering only in the radius $R = a_0 t$ of the sphere. The third term in (16) is obtained from the second by writing $d\sigma = d\mathcal{S}/R^2$, where $d\sigma$ is the solid angle under which $d\mathcal{S}$ is viewed from P. Then the function of four variables

$$u(x, y, z, t) = t\bar{f}(x, y, z, t) \tag{17}$$

satisfies Eq. (15).

To prove this, we note that the operations of averaging over the sphere and of differentiating with respect to x, y, or z may be interchanged, so that

$$\Delta u = t\,\Delta \bar{f} = \frac{t}{4\pi R^2}\int \Delta f\, d\mathcal{S} = \frac{1}{4\pi a_0{}^2 t}\int \Delta f\, d\mathcal{S}. \tag{18}$$

As to $\partial u/\partial t$, it is seen that a change dt in t causes a change $dr = a_0\,dt$ in the radius of $\mathcal{S}$, and therefore a change $(\partial f/\partial r)\,dr$ in the value of f corresponding to a $d\sigma$ on the sphere, where $\partial f/\partial r$ is the directional derivative of f along the radius of the sphere, or *radial derivative*. Thus

$$\frac{\partial \bar{f}}{\partial t} = \frac{a_0}{4\pi}\int \frac{\partial f}{\partial r}\, d\sigma. \tag{19}$$

Therefore

$$\frac{\partial^2 u}{\partial t^2} = \frac{\partial}{\partial t}\left(t\,\frac{\partial \bar{f}}{\partial t} + \bar{f}\right) = t\,\frac{\partial^2 \bar{f}}{\partial t^2} + 2\,\frac{\partial \bar{f}}{\partial t} = \frac{1}{t}\,\frac{\partial}{\partial t}\left(t^2\,\frac{\partial \bar{f}}{\partial t}\right),$$

which may be written, using $R = a_0 t$ and (19), as

$$\begin{aligned} \frac{\partial^2 u}{\partial t^2} &= \frac{1}{t}\,\frac{\partial}{\partial t}\left(\frac{R^2}{a_0{}^2}\,\frac{\partial \bar{f}}{\partial t}\right) = \frac{1}{4\pi a_0 t}\,\frac{\partial}{\partial t}\left(R^2\int \frac{\partial f}{\partial r}\, d\sigma\right) \\ &= \frac{1}{4\pi a_0 t}\,\frac{\partial}{\partial t}\left(\int \frac{\partial f}{\partial r}\, d\mathcal{S}\right). \end{aligned} \tag{20}$$

Using Gauss' Theorem (12) with $\mathbf{v} = \operatorname{grad} f$, and observing that the normal and radial directions are the same when $\mathcal{S}$ is a sphere, we obtain

$$\int \frac{\partial f}{\partial r}\, d\mathcal{S} = \int \operatorname{div}\,(\operatorname{grad} f)\, dV = \int \Delta f\, dV,$$

where the volume integral is extended over the interior of $\mathcal{S}$. The time rate of change of this volume integral depends only on the change of the domain of integration, since the integrand is independent of t. Increasing t by an amount dt adds a spherical shell of thickness $a_0\,dt$ to the volume; for this shell $dV = a_0\,dt\,d\mathcal{S}$, so that

$$\frac{\partial}{\partial t}\left(\int \frac{\partial f}{\partial r}\, d\mathcal{S}\right) = \frac{a_0\,dt \displaystyle\int \Delta f\, d\mathcal{S}}{dt} = a_0 \int \Delta f\, d\mathcal{S}. \tag{21}$$

Then, substituting this in (20) and using (18), we get

$$\frac{\partial^2 u}{\partial t^2} = \frac{1}{4\pi t}\int \Delta f\, d\mathcal{S} = a_0{}^2\,\frac{1}{4\pi a_0{}^2 t}\int \Delta f\, d\mathcal{S} = a_0{}^2 \Delta u,$$

and the proposition is proved.

As in the one-dimensional case, the main problem is to adapt the solution (17) to given initial conditions. Suppose that

$$\text{(22)} \qquad \text{at } t = 0, \qquad u = f(x, y, z), \qquad \frac{\partial u}{\partial t} = g(x, y, z),$$

where f and g are differentiable. Consider

$$\text{(23)} \qquad u(x, y, z, t) = \frac{\partial}{\partial t}(t\bar{f}) + t\bar{g},$$

where $\bar{f}$ and $\bar{g}$ are functions of x, y, z, and t derived from f and g by the averaging process (16). This expression certainly satisfies the wave equation (15), since it is the sum of two solutions: the second term is exactly of the form (17), and the first is the derivative of (17), and we have already seen that derivatives of a solution of the wave equation are also solutions. It remains to be shown that the initial conditions are fulfilled.

Since $\partial(t\bar{f})/\partial t = \bar{f} + t(\partial\bar{f}/\partial t)$, the right-hand member of (23) reduces to $\bar{f}$ for $t = 0$; but $\bar{f}(x, y, z, 0)$ is precisely $f(x, y, z)$. Indeed, $\bar{f}(x, y, z, 0)$ represents the limit, as $R \to 0$, of the mean value of $f(x, y, z)$ on a sphere of radius R about (x, y, z), and this limit must be $f(x, y, z)$, since f is continuous. The first condition (22) is therefore satisfied. Computing $\partial u/\partial t$ from Eq. (23), one obtains

$$\frac{\partial u}{\partial t} = 2\frac{\partial \bar{f}}{\partial t} + \bar{g} + t\left(\frac{\partial^2 \bar{f}}{\partial t^2} + \frac{\partial \bar{g}}{\partial t}\right).$$

For $t = 0$, the last term is 0, and the second term gives $\bar{g}(x, y, z, 0) = g(x, y, z)$ as above for f. The term $\partial\bar{f}/\partial t$ may be evaluated from (19). As $t \to 0$, so also does the radius of the sphere on which the values of $\partial f/\partial r$ are to be taken. In the limit, the contributions from diametrically opposite points on the sphere cancel, since the values of $\partial f/\partial r$ are the directional derivatives of f at (x, y, z) for exactly opposite directions. Thus for $t = 0$, $\partial u/\partial t$ reduces to $g(x, y, z)$, as required.

The solution (23) of the initial value problem for small perturbations is known as *Poisson's formula*.

5. Discussion

Consider a perturbation that originates within a bounded region of three-dimensional space; i.e., f and g vanish identically outside a finite domain D (see Fig. 8). The functions $\bar{f}$ and $\bar{g}$ will certainly vanish whenever the surface of the sphere about P does not intersect D. For a fixed point $P(x, y, z)$ outside D, there is a minimum radius R_1 and a maximum radius R_2 for spheres about P which do intersect D; for P within D, $R_1 = 0$. Then for

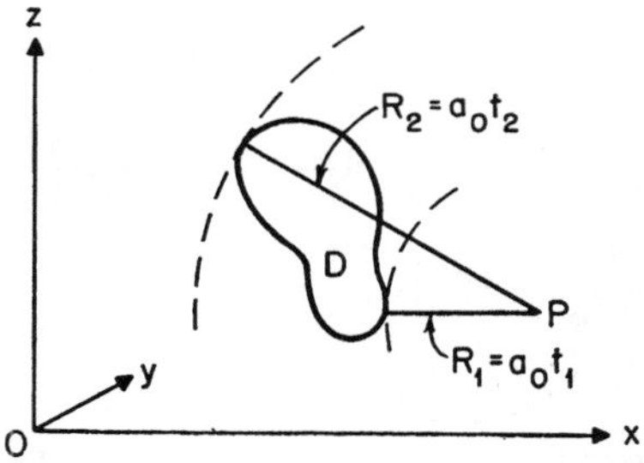

FIG. 8. Perturbation due to disturbance in D perceptible at point P only for $t_1(P) < t < t_2(P)$.

$R < R_1$ or $R > R_2$, the spheres do not intersect D; consequently, with $R_1/a_0 = t_1$, $R_2/a_0 = t_2$

$$u(x, y, z, t) = 0 \qquad \text{for } t < t_1 \text{ or } t > t_2, \tag{24}$$

where t_1 and t_2 vary with P. This means that the perturbation that originated in D is perceptible at P only within the time interval t_1 to t_2. On the other hand, for a fixed value of t, the perturbation is felt only at points P whose distance from some point of D is exactly a_0t: the perturbation can exist only in the domain D_t consisting of the surface points of all spheres of radius a_0t with center in D (see Fig. 9).

If D shrinks to an infinitesimal neighborhood of a single point O, the

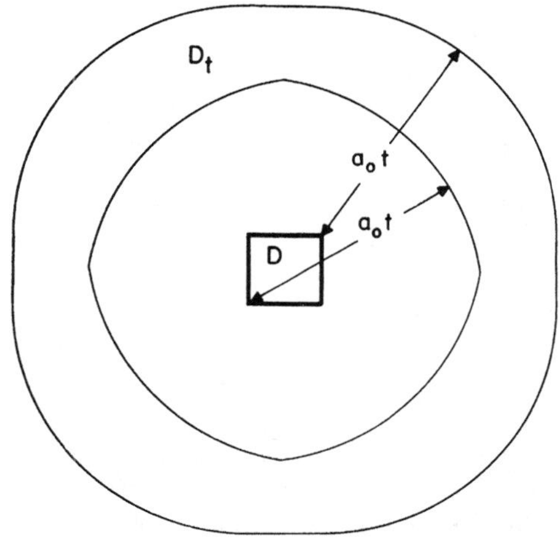

FIG. 9. Perturbation due to disturbance in D perceptible at time t in domain D_t only.

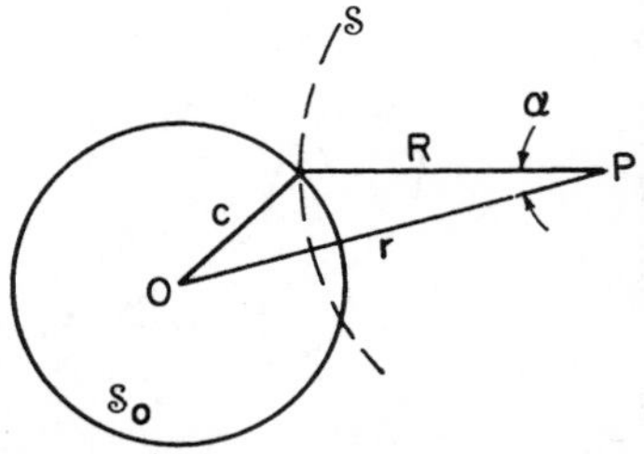

FIG. 10. Computation of a solid angle.

perturbation is perceptible at P only at the moment $t = OP/a_0$. On the other hand, at a given time t, only points at a distance $a_0 t$ from O are affected, that is, D_t becomes the surface of a sphere of radius $a_0 t$ and center O. Briefly, *a small disturbance in an inviscid elastic fluid originally at rest is propagated in all directions with constant velocity* $a_0 = \sqrt{dp/d\rho}$.

The manner in which the magnitude of the perturbation changes with time can be seen from a simple example. Let D be the interior of the sphere $\mathcal{S}_0 : x^2 + y^2 + z^2 = c^2$, and take $f(x, y, z) = U$, where U is a positive constant, for points in $\mathcal{S}_0$, while $f = 0$ elsewhere, and $g \equiv 0$. Since the problem is symmetrical, the perturbation $u(x, y, z, t)$ must be a function only of t and of $r = (x^2 + y^2 + z^2)^{\frac{1}{2}}$. Let P be a point outside $\mathcal{S}_0$, so that $r > c$ for P. From (16) it follows that the value of $\bar{f}$ at P for time t is given by $U/4\pi$ times the solid angle σ under which the intersection of $\mathcal{S}$ and $\mathcal{S}_0$ is viewed from P, where $\mathcal{S}$ is the sphere of radius $R = a_0 t$ and center P. The area on $\mathcal{S}$ cut out by a cone with vertex at P and vertex angle α is $2\pi R^2(1 - \cos\alpha) = \sigma R^2$; thus (see Fig. 10)

$$\sigma = \frac{\pi}{rR}[c^2 - (r - R)^2].$$

Then, substituting in the formula (23) for u, with $g = 0$, $R = a_0 t$, we have

$$(25) \quad u(x, y, z, t) = \frac{\partial}{\partial t}(t\bar{f}) = \frac{U}{4ra_0}\frac{\partial}{\partial t}[c^2 - (r - a_0 t)^2] = \frac{U}{2}\left(1 - a_0\frac{t}{r}\right).$$

This formula holds for $t_1 = (r - c)/a_0 \leqslant t \leqslant (r + c)/a_0 = t_2$ since R_1 and R_2 of (24) are obviously $r - c$ and $r + c$ in this case. For all other values of t we have $u = 0$ at P. At a given moment t, $u(x, y, z, t) = 0$ except at points for which $a_0 t - c \leqslant r \leqslant a_0 t + c$; these points fill a spherical shell of thickness $2c$. Within the shell the value of u increases steadily from the value $-Uc/2(a_0 t - c)$ at the inner surface to $Uc/2(a_0 t + c)$ at the outer surface, becoming zero at $r = a_0 t$. It can be seen that, although the values of u within the shell decrease as the shell expands with t, the integral of u over this volume is constant:

$$(26)\qquad \begin{aligned} I &= \int u\,dV = 4\pi \int_{a_0 t-c}^{a_0 t+c} ur^2\,dr \\ &= 4\pi \frac{U}{2} \int_{a_0 t-c}^{a_0 t+c} r(r - a_0 t)\,dr = \frac{4\pi U c^3}{3}. \end{aligned}$$

Thus the perturbation is dying out in time, while it spreads from the source. This behavior is different from that in the one-dimensional case.

For points within $\mathcal{S}_0$ one has $t_1 = 0$ as mentioned in Sec. 4, the perturbation lasting until $t_2 = (r + c)/a_0$. At the center O the perturbation dies out at $t = c/a_0$. By time $t = 2c/a_0$, it will have died out everywhere in $\mathcal{S}_0$. Thus, if a second disturbance is initiated in $\mathcal{S}_0$ after time $2c/a_0$, this perturbation will be propagated with the same velocity as the first wave and will reach each point with a constant time lag after the first wave. If c is very small, one can consider a succession of values of U, positive and negative, and this succession will be perceptible, on a reduced scale, at any distance r. This is the situation in the case of an acoustic signal. But this type of problem, usually studied in the theory of acoustics, will not be pursued here. In any case it follows from (25) that the intensity of a disturbance originating in a very small neighborhood of O is the same at all points of the spherical surface $\mathcal{S}$ around O reached at any time t.

6. Two-dimensional case[24]

If the motion is restricted to two dimensions, that is, if $q_z = 0$ and if all derivatives with respect to z vanish, then the wave equation (15) reduces to

$$(27)\qquad \frac{\partial^2 u}{\partial t^2} = a_0^2 \left(\frac{\partial^2 u}{\partial x^2} + \frac{\partial^2 u}{\partial y^2} \right).$$

In this case the behavior of a small perturbation can be inferred from the solution of the three-dimensional problem.

In order that the initial conditions be those of a two-dimensional flow, we assume that at the beginning the disturbance is confined to some infinite cylinder in the z-direction, with finite cross section C_0 in the x,y-plane, and that the given functions f and g in (22) are independent of z, that is, $f(x, y, z) = f(x, y)$, etc. Let $P = (x, y, 0)$ be a fixed point in the x,y-plane. The integral in the formula (16), for $\bar{f}(x, y, 0, t)$, extended over a sphere $\mathcal{S}$ of radius $R = a_0 t$ about P, may be expressed as an integral in the x,y-plane. If $d\mathcal{S}$ is an element of area at the point (ξ, η, ζ) on $\mathcal{S}$, and dA denotes the projection of $d\mathcal{S}$ onto the x,y-plane, then $d\mathcal{S}:dA =$

$R:\sqrt{R^2 - r^2}$ where $r^2 = (x - \xi)^2 + (y - \eta)^2$. The projection of $\mathcal{S}$ onto the x,y-plane is the interior of the circle C of radius R about P, covered twice, so that

$$\frac{1}{4\pi R^2}\int_{\mathcal{S}} f(\xi, \eta, \zeta)\, d\mathcal{S} = \frac{2}{4\pi R^2}\int_C f(\xi, \eta)\, \frac{R\, dA}{\sqrt{R^2 - r^2}}$$
$$= \frac{1}{2\pi R}\int_C f(\xi, \eta)\, \frac{dA}{\sqrt{R^2 - r^2}}.$$

For given f and g, the two-dimensional solution $u(x, y, t)$ may be given in the form (23), with the definition (16) of $\bar{f}$ replaced by

$$\bar{f}(x, y, t) = \frac{1}{2\pi R}\int_C f(\xi, \eta)\, \frac{dA}{\sqrt{R^2 - r^2}}. \tag{28}$$

Since the integrand in (28) is zero, except in the intersection of C and C_0, it is again true that at any time t there will be no perturbation at any point of the x,y-plane whose distance from the initial disturbance is greater than $a_0 t$. It cannot, however, be concluded that any region will become free of perturbation after a certain time has elapsed.

Article 5

Subsonic and Supersonic Motion. Mach Number, Mach Lines

1. Small perturbation of a state of uniform motion

The theory, developed in the preceding article, for a small perturbation of an elastic fluid initially in a state of rest can easily be extended to the case of an elastic fluid initially in a state of uniform motion. In the first place, constant values of velocity, pressure, and density, $\mathbf{q} = \mathbf{q}_0$, $p = p_0$, $\rho = \rho_0$, are compatible with the basic equations (1.I) to (1.III), provided merely that the values p_0 and ρ_0 satisfy (1.III) and that the effect of gravity is negligible. Further, the equations of motion are not different when referred to a coordinate system moving with constant velocity (inertia system). It follows, therefore, that for an observer moving with constant velocity $\mathbf{q}_0$, all phenomena will be exactly as described in Art. 4: a small disturbance at time $t = 0$ at a point O' of the inertia system is perceptible after time t on the surface $\mathcal{S}$ of a sphere of radius $R = a_0 t$ and center O'. An observer at rest, however, will at time t perceive the point O' to be displaced by an amount $q_0 t$ from its initial location O at $t = 0$. To him the spheres $\mathcal{S}$ successively reached by the perturbation will form a pencil of spheres with centers progressing along a line in the direction of $\mathbf{q}_0$ at the rate q_0 while their radii increase at the rate a_0.

The nature of this pencil depends upon the relative sizes of q_0 and a_0. In Figs. 11 to 13 are shown the three possible configurations of the pencil of spheres, with spheres corresponding to $t = 1, 2, 3$. The plane of the drawing is any plane containing O and $\mathbf{q}_0$; each circle represents the inter-

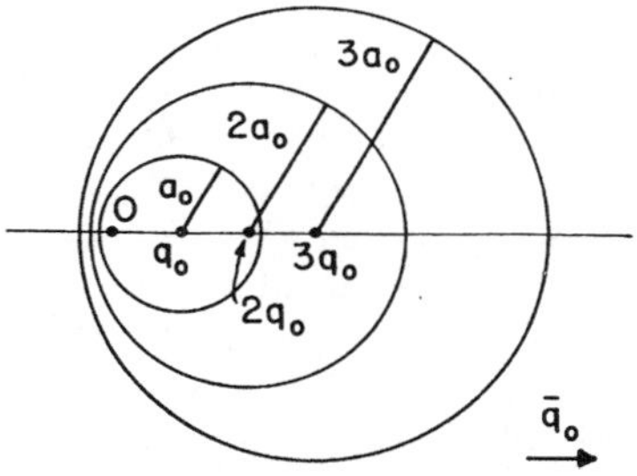

FIG. 11. Disturbance originated at O spreading in $t = 1, 2, 3, \cdots$ seconds with sound speed a_0 while gas moves with uniform horizontal speed $q_0 < a_0$, subsonic flow.

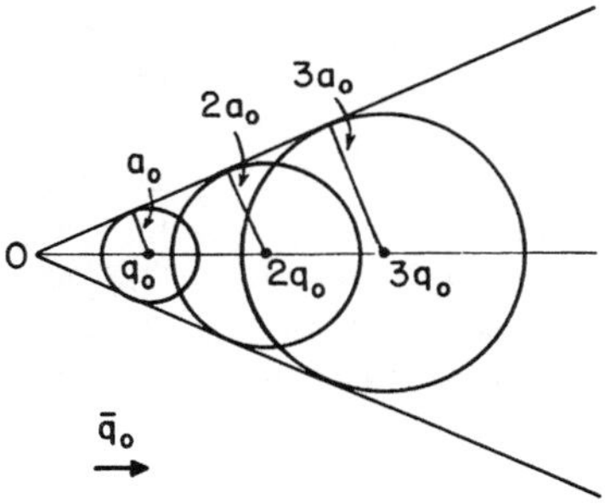

FIG. 12. See Legend to Fig. 11. $q_0 > a_0$, supersonic flow.

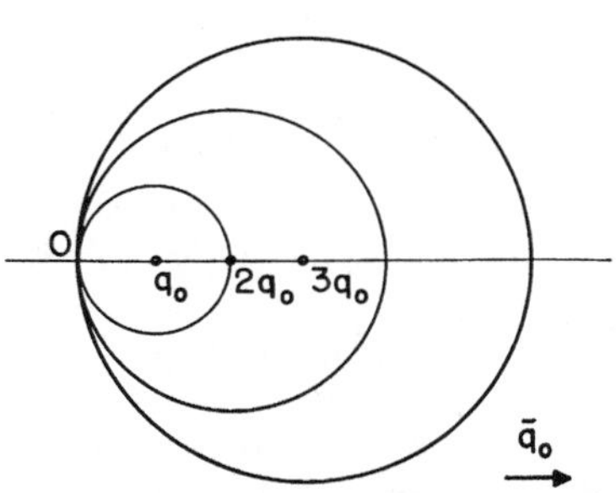

FIG. 13. See Legend to Fig. 11. $q_0 = a_0$, sonic flow.

section of a sphere S with this plane. The circle corresponding to time t has the radius a_0t and center at the abscissa q_0t. Thus the right-hand intersection of S with the horizontal axis has abscissa $q_0t + a_0t = (q_0 + a_0)t$ and is always to the right of O, moving to the right indefinitely as t increases. The left-hand intersection has abscissa $(q_0 - a_0)t$.

For $q_0 < a_0$ (Fig. 11), the left-hand intersection falls to the left of O, and farther to the left for larger t. All spheres include the source of the original disturbance, so that the perturbation is propagated in all directions from the source (although the speed of propagation is not the same on all rays emanating from this point) and eventually reaches all points of space. Thus, there is not too much difference between this phenomenon and the one discussed in the preceding article, where $q_0 = 0$.

If $q_0 > a_0$ (Fig. 12), the left-hand intersection with abscissa $(q_0 - a_0t)$ falls to the right of O and moves to the right as t increases; no sphere includes the source of the disturbance. All spheres are interior and tangent to a circular half-cone whose semivertex angle α_0 is given by

$$\sin \alpha_0 = \frac{a_0}{q_0}. \tag{1}$$

In this case, *the perturbation is not propagated in all directions from the source*, but only in such directions as lie interior to the half-cone determined by Eq. (1). This is a situation entirely different from that occurring in the cases $q_0 = 0$ and $q_0 < a_0$.

Finally, suppose $q_0 = a_0$ (Fig. 13). The half-cone given by Eq. (1) degenerates to the cone corresponding to $\alpha_0 = 90°$, i.e., the whole space to the right of the plane through O normal to the direction of $\mathbf{q}_0$. Here, the surface of any sphere S touches the boundary plane at the point O. The perturbation is propagated in all directions pointing to the right of this plane.

2. Terminology

The deeply rooted difference in the behavior of a small perturbation in the cases $q_0 \lesseqgtr a_0$ has led to a terminology generally adopted today in the theory of compressible fluids. The ratio q_0/a_0, flow velocity to sound velocity, is called the *Mach number*, M_0, of the flow (before perturbation), the half-cone determined by Eq. (1) a *Mach cone*, and the angle α_0 the *Mach angle*, all named after Ernst Mach, who was the first to observe and describe this type of phenomenon.[25] For uniform flow, the three cases

$$M_0 = \frac{q_0}{a_0} < 1, \qquad M_0 = 1, \qquad \text{and} \qquad M_0 > 1 \tag{2}$$

are referred to as *subsonic*, *sonic*, and *supersonic* flows, respectively. A Mach cone exists only in the case of supersonic flow; for subsonic flow, no real angle α_0 corresponds to Eq. (1). If $q_0 = 0$, the Mach number is zero, but this is also true for any q_0 if $a_0 = \infty$, i.e., in the case of an incompressible fluid ($d\rho = 0$, $dp/d\rho = \infty$). Thus the Mach number is, in a certain sense, also a measure of the comparative deviation of the actual behavior of a compressible fluid from that of an incompressible one.

For reasons that will become clear later, these definitions are also used in a wider sense. If q, p, and ρ are values of the velocity, pressure, and density at any point of a fluid in nonuniform motion in which a (p, ρ)-relation is defined, and if

$$(3) \qquad a = \sqrt{\frac{dp}{d\rho}}, \qquad M = \frac{q}{a} = q\sqrt{\frac{d\rho}{dp}}, \qquad \text{and} \qquad \sin\alpha = \frac{a}{q} = \frac{1}{M},$$

then a is called the *local sound velocity* (see Sec 4.2), M the *local Mach number*, and α (when it exists) the *local Mach angle*.

Any region of a fluid in motion in which $M < 1$ is described as subsonic, and in which $M > 1$ as supersonic. Points where $M = 1$ are known as sonic points. In the same sense we speak of subsonic, supersonic, or sonic speed, or flow. Sometimes the expression *transonic* (or transsonic) is used to describe a region in which $1 - M$ changes its sign, or a flow in which M is close to 1 everywhere. For the sake of completeness, it may be mentioned also that regions for which the value of M is exceptionally high are often referred to as *hypersonic.*

It is to be noted that all these definitions presuppose the definition of the derivative $dp/d\rho$. They certainly apply in the case of an elastic fluid, where by hypothesis a one-to-one (p, ρ)-relation holds throughout the medium. In all other cases, the symbols a, M, etc., may be used only in connection with an *ad hoc* definition of $dp/d\rho$. For example, it is possible to define

$$(4) \qquad \frac{dp}{d\rho} = \frac{dp}{dt} \bigg/ \frac{d\rho}{dt}$$

where d/dt has the same meaning as in (1.4); for the case of a steady motion, this definition reduces to

$$(4a) \qquad \frac{dp}{d\rho} = \frac{\partial p}{\partial s} \bigg/ \frac{\partial \rho}{\partial s}.$$

In particular, it may be that a (p, ρ)-relation holds for each particle, although not the same relation for all particles. An example of such behavior is a strictly adiabatic flow of a perfect inviscid gas, where the entropy $c_v \log (p/\rho^\gamma)$ changes from one particle to another (see Sec 1.5).

3. Propagation of the perturbation according to direction

For a fluid initially at rest, the perturbation is propagated *uniformly* in all directions, as seen in Sec. 4.5. Let us return to the special case where

$u = U$ initially within a sphere $\mathcal{S}_0$ of radius c; it was seen that u is different from zero at time t only in the concentric spherical shell of thickness $2c$ and average radius a_0t, and that the integral

$$I = \int u\, dV \tag{5}$$

over the spherical shell has a constant value, $I = 4\pi Uc^3/3$, for all values of t. In this section we need only the case where c is very small and U large so that the perturbation, at each time t, may be considered as concentrated on the surface of the sphere of radius a_0t about the small sphere $\mathcal{S}_0$, which may be considered to coincide with its center O. In any case, the perturbation u is the same at all points of any sphere about O, so that the *strength* I is uniformly distributed as to direction: to any bundle of rays through O filling a solid angle $d\sigma$ corresponds the perturbation strength $(I/4\pi)d\sigma$. This may be expressed by saying that the *intensity of propagation* in any direction is $I/4\pi$.

In the case of a moving fluid, however, the intensity of propagation is no longer the same for all directions from the initial point O, and, if the motion is supersonic, even vanishes for some directions, as seen in Sec. 1. Nevertheless, there is still symmetry about the direction $\mathbf{q}_0$: the directions lying on a cone with vertex O and axis parallel to $\mathbf{q}_0$ are indistinguishable with respect to intensity. If the semivertex angle of this cone is β, then all these directions make the angle β with the direction q_0. In computing how the perturbation strength varies with β, the appropriate element $d\sigma$ is the angular space between the two circular cones with semivertex angles β and $\beta + d\beta$, respectively. For a cone with semivertex angle β, the solid angle at the vertex is $\sigma = 2\pi(1 - \cos\beta)$; thus, for the angular space between the two cones,

$$d\sigma = 2\pi \sin\beta\, d\beta.$$

At any time t, the perturbation is uniformly distributed over the spherical surface $\mathcal{S}$ with center C (where $OC = q_0t$) and radius $R = a_0t$. If OP and OQ are rays on the two neighboring cones (see Fig. 14a), where P and Q lie on $\mathcal{S}$ and $OP = l$, say, then the area $d\mathcal{S}$ cut off between the two cones is given by

$$d\mathcal{S} = 2\pi(PN)\text{ arc } \widehat{PQ} = 2\pi l \sin\beta\, \frac{l\, d\beta}{\cos\delta},$$

where δ is the angle between $\widehat{PQ}$ and the normal to OP. The angle δ may be computed from the triangle OCP by the law of sines:

$$\sin\delta = \frac{q_0t}{a_0t}\sin\beta = M_0\sin\beta.$$

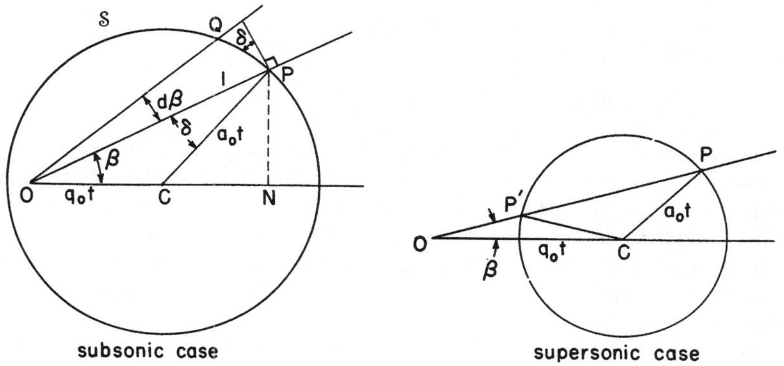

FIG. 14. Part of spherical surface (center C, radius a_0t) cut off by two neighboring cones (center O, side l).

This formula shows that the triangles OCP, for varying t, are similar triangles. For the same reason, the ratio l/a_0t is independent of t and depends only on β and on the given ratio $q_0/a_0 = M_0$. Then the strength of perturbation corresponding to $d\sigma$ is

$$\text{(6)}\qquad \begin{aligned}\frac{I}{4\pi}\frac{dS}{R^2} &= \frac{I}{4\pi}\frac{2\pi l^2 \sin\beta\, d\beta}{(a_0t)^2\sqrt{1 - M_0^2\sin^2\beta}} \\ &= \frac{I}{4\pi}\left(\frac{l}{a_0t}\right)^2 \frac{1}{\sqrt{1 - M_0^2\sin^2\beta}}\, d\sigma\end{aligned}$$

in the subsonic case, $M_0 < 1$. In the supersonic case, $M_0 > 1$, two such terms must be combined if $\beta < \alpha_0$, since the rays in $d\sigma$ meet the sphere S twice, corresponding to the two possible triangles OCP and OCP', for the given values of β and M_0 (Fig. 14b), and two possible values of the ratio l/a_0t. If $\alpha_0 < \beta$, the strength of perturbation corresponding to $d\sigma$ is zero.

In all cases the largest and smallest values of l, as β varies, are $q_0t + a_0t$ and $|\, q_0t - a_0t\,|$, respectively, so that the ratio $(l/a_0t)^2$ varies at most between $(M_0 - 1)^2$ and $(M_0 + 1)^2$. Thus, if $M_0 < 1$, the intensity of propagation [that is, the coefficient of $d\sigma$ in (6)] lies between finite positive limits, the difference between these limits tending to zero for $M_0 = 0$ (case of a fluid at rest). In the supersonic case, $M_0 > 1$, the factor $\sqrt{1 - M_0^2\sin^2\beta}$ in the denominator of (6) will vanish as β attains the value α_0, since $\sin\alpha_0 = 1/M_0$, and will be imaginary for all larger values of β. The intensity of propagation thus tends to infinity for rays adjacent to the Mach cone, and in this sense it is often said that small perturbations in a fluid moving at supersonic speed are *essentially propagated along the surface of the Mach cone.*

Formulas giving the total strength of perturbation for any cone coaxial with the Mach cone can be obtained by integrating Eq. (6).

4. Steady motion in two dimensions. Mach lines

As seen in Sec. 1.2, in any steady flow there exist fixed *streamlines*, i.e., curves fixed in space which are pathways of the fluid particles, and therefore have at each point the direction of the velocity vector. In a *two-dimensional steady* flow, the streamlines form a family of plane curves not intersecting each other, except at singular points. It is assumed also that the fluid is elastic, or, if it is not, that a suitable definition of $dp/d\rho$ at each point is given.

In a supersonic region of such a flow there is determined at each point not only a value of the sound velocity, a, and Mach number, M, but also a Mach angle, α, with $\sin \alpha = 1/M$, and therefore at each point P two directions forming the angles $+\alpha$ and $-\alpha$ with the streamline through P (see Fig. 15). At sonic points, where $M = 1$ and $\alpha = 90°$, the two directions are opposite to each other. As long as M is finite, α is different from zero, and the two directions cannot coincide. Thus, in any portion of the plane in which the motion is supersonic and M finite, two direction fields are defined which, under the usual continuity assumptions, determine two sets of curves. These curves are called *Mach lines*. If the angle between the velocity vector $\mathbf{q}$ and a fixed axis of reference be denoted by θ, then the angles for the Mach lines are $\theta + \alpha$ and $\theta - \alpha$, respectively. In the particular case of a uniform supersonic flow, with $\mathbf{q}_0$ in the x-direction, the Mach lines are the two sets of parallel straight lines forming the angles $+\alpha$ and $-\alpha$, respectively, with the x-axis.

In general, the Mach lines form a network of curvilinear quadrangles, since two lines pass through each point. In any suitably restricted region of supersonic flow, two sets of Mach lines can be distinguished, such that every line of the one set intersects each line of the other set and no line of its own set. The angle of intersection is 2α, and this angle is bisected by the streamline passing through the point of intersection. In general, the streamlines are not diagonal lines in the network of Mach lines.

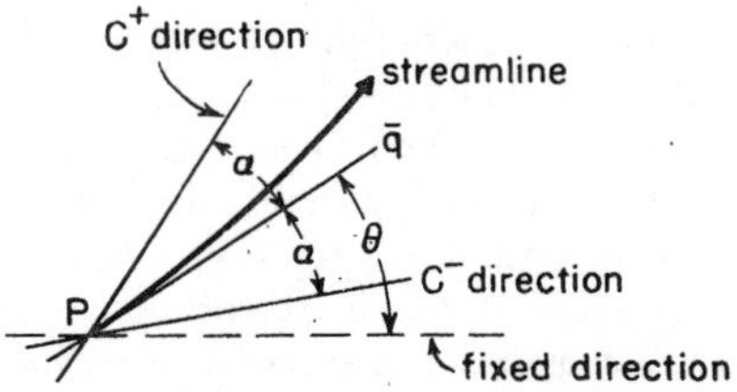

FIG. 15. Streamline and Mach lines in two-dimensional steady supersonic flow.

Analytically, the Mach lines can in some cases be represented by two equations, each containing a parameter. For example, in the case of uniform motion parallel to the x-axis, the Mach lines are given by

$$y - x \tan \alpha_0 = c_1 , \qquad y + x \tan \alpha_0 = c_2 ,$$

where α_0 is constant, and c_1 and c_2 are variable parameters. In other cases, a single equation with a single parameter holds for both sets. For example, all tangents to the unit circle,

$$x \cos \varphi + y \sin \varphi = 1,$$

where φ is the parameter, form a network of the required type in the region exterior to the circle.

Since α is acute (except at sonic points) the Mach lines can be given a definite orientation: the positive direction on any Mach line will be taken so that the line points toward the same side of the normal to the streamline as does **q**. Take counterclockwise angles as positive; then the line forming the angle $+\alpha$ with **q** will be called the C^+, or plus Mach line, while the other will be called the C^-, or minus Mach line. Thus an observer passing along a streamline will at each point have the C^+-line to his left and the C^--line to his right. At P (Fig. 15) the C^+-line forms the angle $\theta + \alpha$, and the C^--line the angle $\theta - \alpha$, with the x-axis.

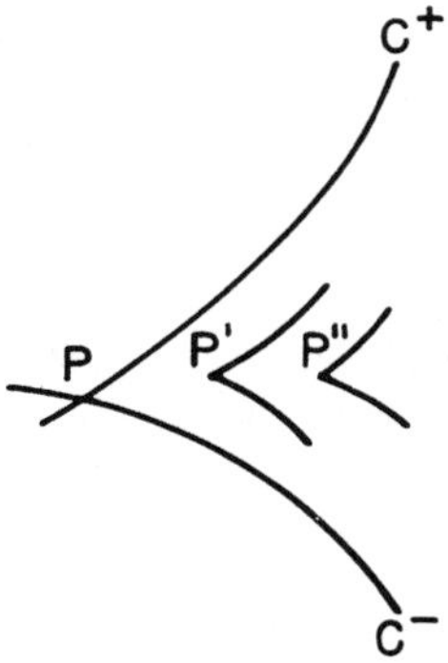

FIG. 16. Spreading of perturbations.

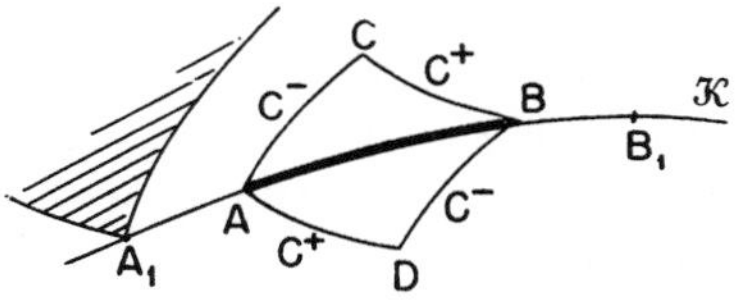

FIG. 17. Range of influence of disturbance at A_1, and interval of dependence. AB, of solution at C.

5. Significance of the Mach lines

The full significance of the Mach lines will become clear only as the analytical basis of the theory is developed in the succeeding chapters. Some interesting properties, however, can be discussed even at this stage.

Consider again a steady two-dimensional supersonic flow of an elastic fluid. In a sufficiently small neighborhood of a point P at which the velocity is $\mathbf{q}$, the flow can be considered as a uniform flow of velocity $\mathbf{q}$. Near P, therefore, a small perturbation at P can spread only in the angle between the tangents to the C^+-line and C^--line at P. Having reached a point P' in this neighborhood, the perturbation can spread further only between the two Mach directions at P' (see Fig. 16), and similarly for successive points P'', P''' $\cdots$. Since Mach lines of the same set do not intersect each other, this means that a small perturbation at P cannot spread to any point outside the region between the original C^+ and C^-, and can have no effect on the flow exterior to this region.

Now, let us consider a curve $\mathcal{K}$: A_1ABB_1, which is crossed in the same sense (at a nonzero angle) by all Mach lines, C^+ and C^-, through its points. The Mach lines are shown in Fig. 17. As we have just seen, a small disturbance at A_1 cannot have any effect outside the region between the two Mach lines through A_1. Also the C^--lines through A_1 and A cannot intersect, etc. Therefore, in particular, a disturbance at A_1 cannot influence the flow at any point interior to the quadrangle $ACBD$. The same is true for B_1 and for any point of $\mathcal{K}$ outside the arc AB. Thus the flow inside the quadrangle $ACBD$, formed by the four Mach lines through A and B, is independent of what happens on $\mathcal{K}$ outside the quadrangle. If we assume that the differential equations of the flow, together with a knowledge of conditions along $\mathcal{K}$, are sufficient to determine the flow in some area adjacent to $\mathcal{K}$, it follows from the above that the solution inside $ACBD$ is determined by values of the flow variables along AB, and is independent of values on $\mathcal{K}$ to the left of A and to the right of B. An analogous situation does not obtain in the case of an incompressible fluid or a compressible fluid in subsonic flow, where Mach lines do not exist. For these cases, a disturbance spreads in all directions; the flow in one region is never independent of what happens in another.

These very incomplete preliminary remarks on the role of the Mach lines are intended to give a rough idea of how different in their physical nature are supersonic flow and the more familiar subsonic flow. It is thus not surprising that specific mathematical methods have to be developed for solving flow problems in the two cases. On the other hand, the reader should not conclude that there is no common ground in the theory for the two cases of compressible fluid motion, or that some conspicuous phenomenon marks each transition of a particle from subsonic to supersonic flow.

CHAPTER II

GENERAL THEOREMS

Article 6

Vortex Theory of Helmholtz and Kelvin[1]

1. Circulation

A kinematic notion useful in many problems of hydrodynamics is that of circulation, which may be defined as follows. Consider a *circuit* $\mathcal{C}$, i.e., a simple closed curve in space together with a given sense of description (indicated in Fig. 18 by an arrow). On $\mathcal{C}$, each element of arc can then be considered as an infinitesimal vector $d\mathbf{l}$, having the direction of the tangent to $\mathcal{C}$. As usual, $\mathbf{q}$ denotes the instantaneous velocity at each point. If the scalar product of $\mathbf{q}$ and $d\mathbf{l}$ is integrated around the circuit, the line integral

$$(1) \qquad \Gamma = \oint_{\mathcal{C}} \mathbf{q} \cdot d\mathbf{l} = \oint_{\mathcal{C}} q \cos(\mathbf{q}, d\mathbf{l})\, dl = \oint_{\mathcal{C}} q_l\, dl$$

is called the *circulation around* $\mathcal{C}$. All values of $\mathbf{q}$ are to be taken for the same value of t.[2]

The circulation is *additive* in the following sense. Suppose we "bridge" the circuit $\mathcal{C}$ by some path AB (see Fig. 19) and give the two new circuits $ABDA$ and $BAEB$ the same sense of description as $\mathcal{C}$. Then the circulations Γ_1 and Γ_2, respectively, along the new circuits satisfy

$$(2) \qquad \Gamma = \Gamma_1 + \Gamma_2 .$$

In fact, the definition (1) shows that Γ_1 is the integral of $\mathbf{q} \cdot d\mathbf{l}$ along the path $ABDA$, which can be broken up into AB plus BDA. Similarly, Γ_2 is the integral along the path BA plus AEB. In the sum $\Gamma_1 + \Gamma_2$, the integrals along AB and BA cancel, since $\mathbf{q}$ is the same, while $d\mathbf{l}$ has opposite directions, along the two paths. Therefore the sum reduces to integrals along BDA and AEB, which is exactly the integral around $\mathcal{C}$, i.e., Γ.

The relation (2) can be generalized. Suppose $\mathcal{A}$ is any open two-sided surface spanning $\mathcal{C}$, i.e., having $\mathcal{C}$ as its rim. From one side of $\mathcal{A}$, the sense of description of $\mathcal{C}$ appears counterclockwise, and normals to $\mathcal{A}$ will always be drawn out from this side. On $\mathcal{A}$ draw two sets of curves forming a network, as in Fig. 20. Each mesh of the network is a closed circuit, the sense of description being taken counterclockwise as viewed from the normal to the surface, and has a value of the circulation corresponding to it: Γ_1, Γ_2, $\cdots$, etc. By repeated application of (2) we find

$$\Gamma = \Gamma_1 + \Gamma_2 + \cdots + \Gamma_m, \tag{3}$$

where m is the number of meshes in the network. We now increase the number of "bridges" in such a way that the network becomes more dense and all meshes become smaller, while the number of terms in Eq. (3) increases. Let a function γ be defined at each point P of $\mathfrak{A}$ as the limit of the quotient of the circulation along the contour of a mesh around P by the area of the mesh, for meshes about that point becoming steadily smaller in all directions. For the first mesh, the circulation Γ_1 is then approximately given by $\gamma_1 \, d\mathfrak{A}_1$, where $d\mathfrak{A}_1$ is the area of the mesh and γ_1 the value of γ

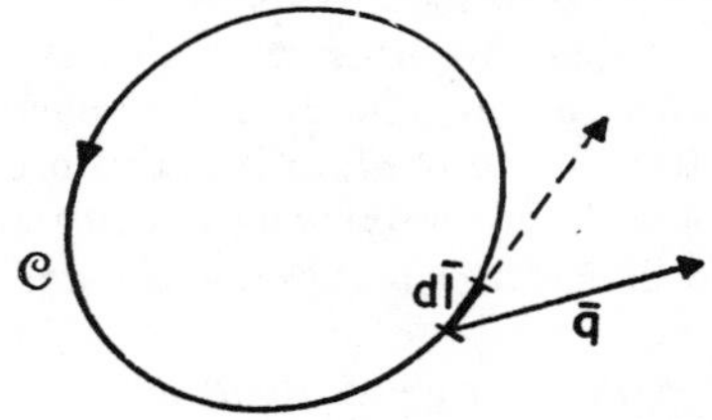

FIG. 18. Vector along circuit.

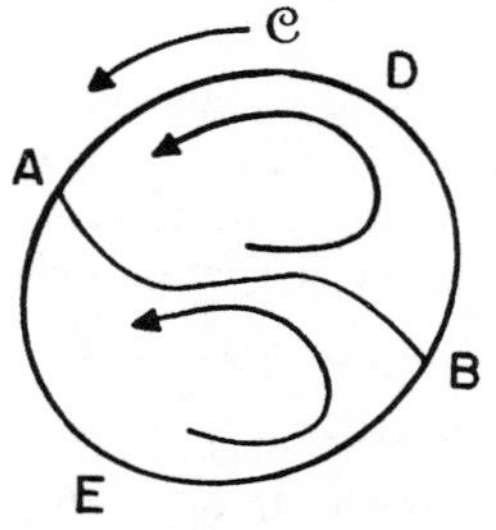

FIG. 19. Illustration of additivity of circulation.

FIG. 20. Circulation as surface integral.

at some point in the mesh, the approximation becoming better as $d\mathcal{A}_1$ gets smaller; and similarly for the other meshes. Thus, as the number of terms increases indefinitely, the right-hand member of (3) yields the surface integral of γ over $\mathcal{A}$, or

$$\Gamma = \int_{\mathcal{A}} \gamma \, d\mathcal{A}. \tag{4}$$

From the definition of γ, it is obvious that the value of this function at any point of $\mathcal{A}$ depends upon the distribution of the velocity $\mathbf{q}$ in the neighborhood of this point. In computing this relationship let us choose curves on $\mathcal{A}$ which always cross at right angles. Then at any point P we can set up a rectangular coordinate system, taking the z-axis in the direction of the normal to $\mathcal{A}$ at P and the x- and y-directions tangent to the two curves through P so as to form a right-handed coordinate system. Then an infinitesimal mesh starting at P is of the type illustrated in Fig. 21. In computing the line integral (1) for this mesh, the path may be broken up into four infinitesimal elements, and $q_l \, dl$ evaluated for each part. Along PP_1 the contribution is $q_x \, dx$, along P_1P_2 it is $[q_y + (\partial q_y/\partial x)dx]dy$, along P_2P_3 it is $-[q_x + (\partial q_x/\partial y)dy]dx$, and along P_3P it is $-q_y \, dy$. The sum of these terms gives the integral along the whole path, and the circulation along the contour of this infinitesimal mesh is therefore

$$d\Gamma = \left(\frac{\partial q_y}{\partial x} - \frac{\partial q_x}{\partial y}\right) dx \, dy.$$

Since $dx \, dy$ is the area of this mesh, the function γ must have the value $\partial q_y/\partial x - \partial q_x/\partial y$ at P. Now this quantity is exactly the z-component of the vector known as the *curl* of $\mathbf{q}$, defined by

$$\operatorname{curl} \mathbf{q} \equiv \left(\frac{\partial q_z}{\partial y} - \frac{\partial q_y}{\partial z}, \quad \frac{\partial q_x}{\partial z} - \frac{\partial q_z}{\partial x}, \quad \frac{\partial q_y}{\partial x} - \frac{\partial q_x}{\partial y}\right). \tag{5}$$

It is to be noted that this definition of curl $\mathbf{q}$ is valid in any rectangular right-handed coordinate system.[3] Since the z-direction is here that of the

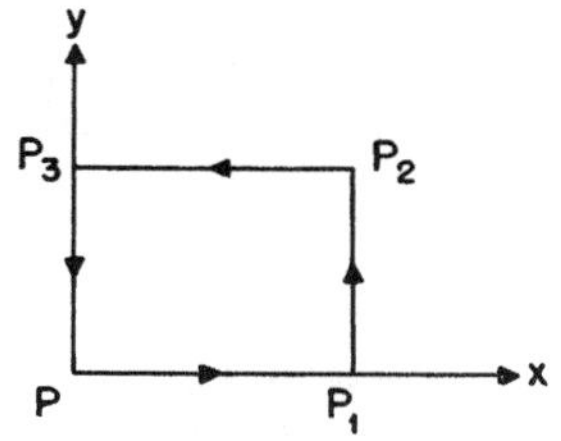

FIG. 21. Circulation around infinitesimal mesh.

normal to the surface $\mathcal{A}$, it appears that γ has the value of the component of curl $\mathbf{q}$ normal to the surface:

$$\gamma = (\text{curl } \mathbf{q})_n ;$$

Thus, (4) may be written as

$$\Gamma = \int_{\mathcal{A}} (\text{curl } \mathbf{q})_n \, d\mathcal{A}, \tag{6}$$

and this formula is independent of the coordinate system used. If a vector $d\bar{\mathcal{A}}$ be introduced, having magnitude $d\mathcal{A}$ and the direction of the normal to the surface, Eq. (6) may also be written as

$$\Gamma = \int (\text{curl } \mathbf{q}) \cdot d\bar{\mathcal{A}} \tag{6'}$$

When the two expressions (1) and (6′) are combined, the result is

$$\oint \mathbf{q} \cdot d\mathbf{l} = \int (\text{curl } \mathbf{q}) \cdot d\bar{\mathcal{A}}$$

This vector formula is known as *Stokes' Theorem.*[4] When $\mathbf{q}$ is the velocity of flow, it states that the circulation along any circuit is given by the surface integral of curl $\mathbf{q}$ over any surface spanning the circuit.

It is obvious that (6), or (6′), can be applied only if it is possible to find some surface that has the given circuit as rim and on which curl $\mathbf{q}$ is defined everywhere. For example, in the case of a flow around an infinite cylindrical obstacle, no such surface can be found for any circuit which surrounds the cylinder. Even here the theorem can be applied, to give a somewhat different result. As shown in Fig. 22, two such circuits, $\mathcal{C}_1$ and $\mathcal{C}_2$, can, by a bridge AB, be combined into a single circuit for which a suit-

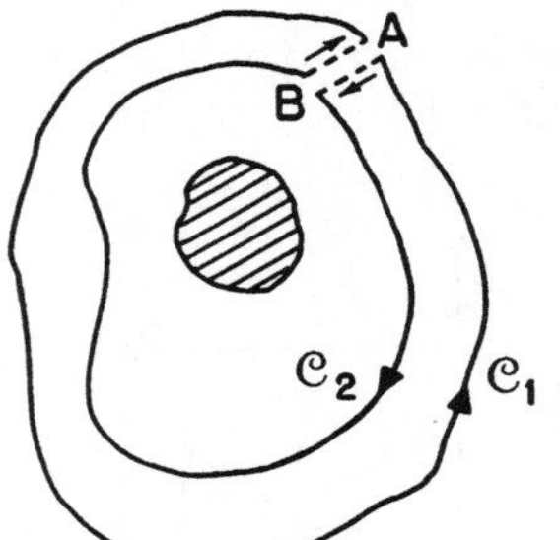

FIG. 22. Circuits surrounding obstacle.

able spanning surface exists. Then the integral (6) extended over this surface gives $\Gamma_1 - \Gamma_2$, since $\mathcal{C}_2$ is given a reverse orientation and the contributions from AB cancel. In particular, if curl $\mathbf{q} \equiv 0$ in the domain of flow, *irrotational flow* (see Art. 7), then $\Gamma_1 - \Gamma_2 = 0$, or $\Gamma_1 = \Gamma_2$: the circulation is equal for all circuits surrounding the obstacle.

2. Mean rotation

The vector curl $\mathbf{q}$ defined by (5) can be given a simple kinematic interpretation.

Let P be a point of the moving mass, and let Q be a neighboring point. In rigid-body rotation with angular velocity $\boldsymbol{\omega}$ about an axis through P, the curl of the velocity vector at Q is the same, namely $2\boldsymbol{\omega}$, no matter where Q is situated. This is not so in the case of a fluid or of any deformable mass.

We start by computing the angular velocity of PQ about an arbitrary given axis through P. Taking P as the origin, we choose a right-handed rectangular coordinate system such that the z-direction is that of the axis. Let $PQ = dr$ and let Q' be the projection of Q onto the x, y-plane (see Fig. 23), θ the angle between the z-axis and PQ (colatitude), and ϕ the angle between the x-axis and PQ' (longitude). Then the distance from the z-axis to Q is $PQ' = \sin\theta\, dr$, and the rectangular coordinates of Q, relative to P, are given by $dx = \cos\phi \sin\theta\, dr$, $dy = \sin\phi\sin\theta\, dr$, and $dz = \cos\theta\, dr$. Therefore, if the velocity at P is $\mathbf{q}$, the velocity vector at Q (relative to P) is

$$\left[\frac{\partial \mathbf{q}}{\partial x}\cos\phi\sin\theta + \frac{\partial \mathbf{q}}{\partial y}\sin\phi\sin\theta + \frac{\partial \mathbf{q}}{\partial z}\cos\theta\right] dr.$$

The angular velocity of the segment PQ about the z-axis is obtained by dividing the distance from the z-axis to Q, i.e., PQ', into the component of the velocity at Q in the direction which is perpendicular to PQ' and to the z-axis. The angles which this direction makes with the x-, y-, and z-axes

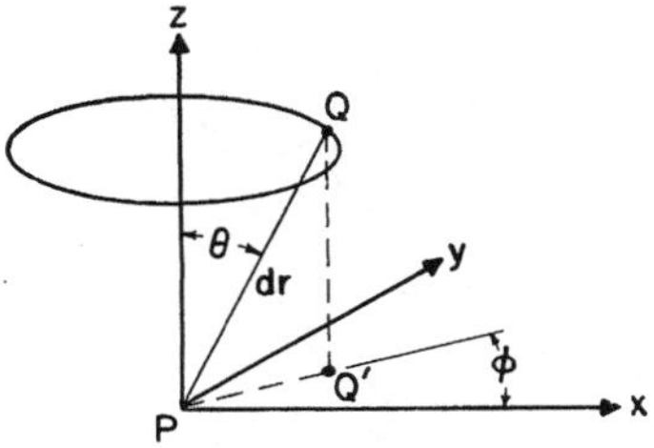

FIG. 23. Computation of mean rotation.

are $\phi + 90°$, ϕ, and $90°$, respectively, and the cosines of these angles are $-\sin\phi$, $\cos\phi$, and 0. Thus the required component is

$$\left[\frac{\partial q_y}{\partial x}\cos^2\phi - \frac{\partial q_x}{\partial y}\sin^2\phi - \left(\frac{\partial q_x}{\partial x} - \frac{\partial q_y}{\partial y}\right)\sin\phi\cos\phi\right]\sin\theta\, dr$$
$$+ \left[\frac{\partial q_y}{\partial z}\cos\phi - \frac{\partial q_x}{\partial z}\sin\phi\right]\cos\theta\, dr.$$

Division by $\sin\theta\, dr$ then gives the angular velocity of PQ about the z-axis, the value depending, in general, on the coördinates θ and ϕ of Q.

Now we compute the average angular velocity for all points Q on the same circle of latitude θ = constant, on the sphere dr = constant, by first integrating with respect to ϕ from 0 to 2π and then dividing by 2π. All integrals vanish except those of the first two terms, giving

$$\frac{1}{2\pi}\left[\frac{\partial q_y}{\partial x}\int_0^{2\pi}\cos^2\phi\, d\phi - \frac{\partial q_x}{\partial y}\int_0^{2\pi}\sin^2\phi\, d\phi\right] = \frac{1}{2}\left(\frac{\partial q_y}{\partial x} - \frac{\partial q_x}{\partial y}\right) = \frac{1}{2}\,(\text{curl } \mathbf{q})_z . \tag{7}$$

This result being independent of θ and dr, the same value is obtained for the average or mean angular velocity about the z-axis of the whole infinitesimal sphere at P. Also, the z-direction could be any direction, so the above result shows that *at any point P of the moving fluid, the vector $\frac{1}{2}$curl $\mathbf{q}$ represents the (instantaneous) mean angular velocity* or *mean rotation* for all segments PQ within an infinitesimal sphere of center P. We call it briefly the *mean rotation* or *mean angular velocity* of the fluid element around P.[5]

Excluding the case curl $\mathbf{q} \equiv 0$, Eq. (5) defines at each moment t at each point of the fluid a vector curl $\mathbf{q}$ which is twice the mean angular velocity of the fluid element around P. This vector is usually called the *vortex vector*. Any line within the fluid which at each of its points has the same direction as curl $\mathbf{q}$ is called a *vortex line*. All vortex lines passing through the points

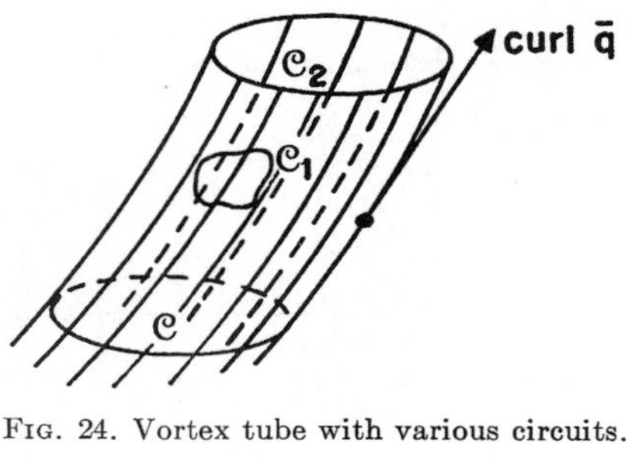

FIG. 24. Vortex tube with various circuits.

of a closed curve $\mathcal{C}$, not itself a vortex line, form a *vortex tube*. A vortex tube of infinitesimal cross section is called a *vortex filament*. The lateral surface of the tube will be called its *mantle*.[6]

For any circuit, such as $\mathcal{C}_1$ in Fig. 24, *which lies on the mantle of a vortex tube but does not encircle the tube, the circulation must be zero*. This follows from (6) since a surface $\mathcal{A}$ spanning this circuit can be taken on the mantle, where the normal component of curl $\mathbf{q}$ is everywhere zero. This is no longer true for a circuit which encircles the tube, as $\mathcal{C}$ or $\mathcal{C}_2$ in Fig. 24. Here it follows from (6) that *the circulation must have a common value for all circuits encircling the same tube*. In fact, in computing the circulation Γ_2 along $\mathcal{C}_2$ one can choose for $\mathcal{A}_2$ the surface consisting of the part of the mantle between $\mathcal{C}_2$ and $\mathcal{C}$ together with any surface $\mathcal{A}$ spanning $\mathcal{C}$. As above, the surface integral is zero on the mantle, so that the integral over $\mathcal{A}_2$ has exactly the same value as the integral over $\mathcal{A}$, giving $\Gamma_2 = \Gamma$.* This common value of the circulation is also called the *vorticity of the tube*; it is a scalar quantity, not to be confused with the magnitude of the vortex vector. In the case of a vortex filament, the vorticity $d\Gamma$ is given by the product of the length of the vortex vector by the normal cross section of the tube.[7]

If a vortex tube be divided into several tubes of finite cross section (or into an infinite number of vortex filaments), the vorticity of the whole tube is the sum (integral) of the individual vorticities. This follows from the additivity of circulation [or from Eq. (6)].

A vortex tube *cannot begin or end in the interior of the fluid*, but must either be a closed tube (like a torus or doughnut) or else (provided it does not meet a boundary) must extend indefinitely in either direction. For at an end, if there were one, a continuous transition would be possible along the mantle from curves of type $\mathcal{C}_1$ to those of type $\mathcal{C}_2$, which is inconsistent with the fact that $\Gamma_1 = 0$, while $\Gamma_2 = \text{constant} \neq 0$.

3. Kelvin's theorem

So far in this article only pure kinematics has been discussed: the concepts of circulation and mean rotation, as well as the relation between them, are valid for any type of continuously distributed material. We shall now specialize to the case of an inviscid elastic fluid, so that the basic equations of Arts. 1 and 2 hold.

The fluid particles lying on any closed circuit at some moment will still form a closed circuit at a later time, since for reasons of continuity no separation of particles can occur; a preliminary question, and one which can be given a very simple and decisive answer, is the following: *how does the circulation change during this transition?*

* When there is no surface spanning $\mathcal{C}$ (in the fluid) the result follows from an argument similar to that at the end of the preceding section.

In Fig. 25 the solid line represents a circuit $\mathcal{C}$, and the dotted line the circuit $\mathcal{C}'$ formed by the same material particles after time dt. The circulations along these two circuits are given by

$$(8) \qquad \Gamma = \oint \mathbf{q} \cdot d\mathbf{l} \qquad \text{and} \qquad \Gamma' = \oint \mathbf{q} \cdot d\mathbf{l}',$$

where the integrals are evaluated along $\mathcal{C}$ and $\mathcal{C}'$ respectively. Let P be the position of an arbitrary particle of $\mathcal{C}$ and Q that of a particle of $\mathcal{C}$ at a distance dl away, and let P' and Q' be the positions of these particles after time dt. Then

$$\overline{PP'} = \mathbf{q}\, dt \qquad \text{and} \qquad \overline{QQ'} = \left(\mathbf{q} + \frac{\partial \mathbf{q}}{\partial l}\, dl\right) dt,$$

where $\partial/\partial l$ denotes the directional derivative in the direction of the tangent to $\mathcal{C}$ at P. The corresponding element of arc $d\mathbf{l}'$ on $\mathcal{C}'$ can be computed from the vector equation $\overline{PP'} + \overline{P'Q'} = \overline{PQ} + \overline{QQ'}$, giving

$$(9) \qquad d\mathbf{l}' = \overline{P'Q'} = \overline{PQ} + \overline{QQ'} - \overline{PP'} = d\mathbf{l} + \frac{\partial \mathbf{q}}{\partial l}\, dl\, dt,$$

while the value of $\mathbf{q}$ corresponding to P' is given by

$$(10) \qquad \mathbf{q}' = \mathbf{q} + \frac{d\mathbf{q}}{dt}\, dt.$$

Then, using Eqs. (8), (9), and (10), and omitting a term of higher order, we find

$$(11) \qquad \begin{aligned} \Gamma' - \Gamma &= \oint \left[\mathbf{q} \cdot \frac{\partial \mathbf{q}}{\partial l}\, dl\, dt + \frac{d\mathbf{q}}{dt} \cdot d\mathbf{l}\, dt + \frac{d\mathbf{q}}{dt} \cdot \frac{\partial \mathbf{q}}{\partial l}\, dl (dt)^2 \right] \\ &= dt \oint \left[\frac{\partial}{\partial l}\left(\frac{q^2}{2}\right) dl + \frac{d\mathbf{q}}{dt} \cdot d\mathbf{l} \right], \end{aligned}$$

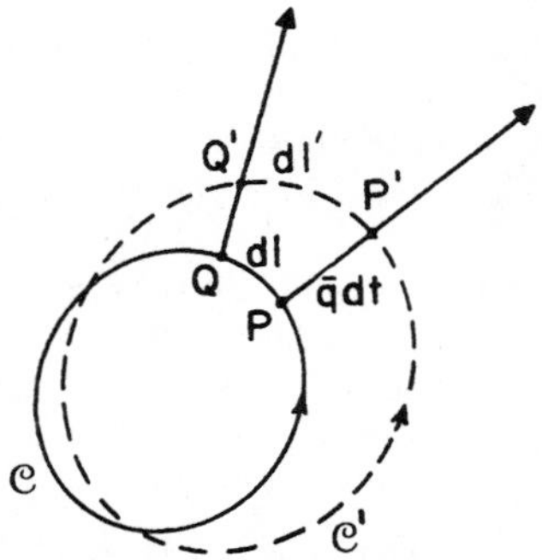

FIG. 25. Two circuits formed by the same particle.

where the integral is to be extended along $\mathcal{C}$.[8] Equation (11) is still true for any continuously distributed mass. For an inviscid fluid, however, the value of the acceleration vector $d\mathbf{q}/dt$ may be taken from the equation of motion (1.I)

$$\frac{d\mathbf{q}}{dt} = \mathbf{g} - \frac{1}{\rho}\,\text{grad}\,p,$$

where in the notation of Sec. 2.1, $\mathbf{g} = -g\,\text{grad}\,h$. Moreover, for an elastic fluid the notion of pressure head P/g (Sec. 2.5) can be used, giving

$$\frac{1}{\rho}\,\text{grad}\,p = \text{grad}\,P.$$

Thus the expression for $d\mathbf{q}/dt$ takes the form

$$\frac{d\mathbf{q}}{dt} = -\text{grad}\,(gh + P), \tag{12}$$

and

$$\frac{d\mathbf{q}}{dt}\cdot d\mathbf{l} = -\text{grad}\,(gh + P)\cdot d\mathbf{l} = -\frac{\partial}{\partial l}\,(gh + P)\,dl.$$

When this is inserted in (11), there results

$$\Gamma' - \Gamma = dt \oint \frac{\partial}{\partial l}\left[\frac{q^2}{2} - gh - P\right] dl. \tag{13}$$

The quantity within the bracket is a single-valued function of position and time. Thus, since at a given time the integral is extended around the closed circuit $\mathcal{C}$ its value is zero. Thus Eq. (13) gives $\Gamma' - \Gamma = 0$ or: *In an inviscid elastic fluid, the circulation around any circuit does not change as the particles forming the circuit move along.* This is Kelvin's theorem.[9] The theorem depends essentially on the fact that the equation of motion can be expressed in the form (12), i.e., on the fact that in an inviscid elastic fluid *the acceleration vector is a gradient*, or (since the curl of a gradient vanishes identically) that *the curl of the acceleration is zero.*[10]

4. Helmholtz' vortex theorems

Starting from Kelvin's theorem, it is easy to derive two theorems on vortex motion which had been proved earlier by Helmholtz (although by a different method; see Sec. 6). Consider a vortex tube $\mathcal{H}$ of infinitesimal cross section (Fig. 26). The particles of $\mathcal{H}$ after time dt still form a tube $\mathcal{H}'$, because no separation of particles can occur; we first wish to determine whether $\mathcal{H}'$ is still a vortex tube. In Sec. 2 we found that the circulation must vanish along any circuit $\mathcal{C}_1$ lying on the mantle of a vortex tube, but not encircling it. From Kelvin's theorem it follows that the circulation along the new position $\mathcal{C}_1'$ of this circuit must also vanish. Also, $\mathcal{C}_1'$ lies on the surface of $\mathcal{H}'$. For an infinitesimal circuit $\mathcal{C}_1'$, according to Stokes' Theorem, the circulation is the product of the area enclosed within the circuit by the component of the vortex vector normal to that area. Since this product is zero, the vortex vector must be tangent to the area element at $\mathcal{C}_1'$, and the circuit $\mathcal{C}_1'$ must lie on the mantle of some vortex tube. This is true for any infinitesimal circuit $\mathcal{C}_1$ on $\mathcal{H}$ and the corresponding $\mathcal{C}_1'$ on $\mathcal{H}'$, so that $\mathcal{H}'$ is also a vortex tube of infinitesimal cross section.[11] Thus we arrive at this statement: Particles lying on a vortex line at some moment move in such a way that they form a vortex line at every moment. A shorter expression of this first vortex theorem is: *The vortex lines are material lines*, in the sense that they always consist of the same particles or material points.

Each vortex tube has a certain vorticity, equal to the circulation along any circuit encircling the tube, such as $\mathcal{C}_2$ in Fig. 26. By Kelvin's theorem, the circulation has the same value along the corresponding circuit $\mathcal{C}_2'$ on $\mathcal{H}'$, so that the vorticity of the vortex tube $\mathcal{H}'$ is the same as that of $\mathcal{H}$. Thus we can state Helmholtz' second theorem: *The vorticity of a vortex tube does not change as its particles move along.*

In Sec. 2 it was seen that vortex tubes cannot come to an end in the interior of the fluid, but must either meet a boundary, extend indefinitely, or be closed. Tubes of the latter type can be observed in air as smoke rings, produced by imparting a rotational motion to the smoke particles. Actually, the smoke rings do not persist indefinitely, in apparent contradiction of the vortex theorems. This is due to the presence of viscosity effects, which are disregarded in the theory of inviscid fluids. The vortex theorems follow from the fact that the acceleration vector is a gradient (and therefore curl-free). To arrive at this statement (12) it was necessary to neglect all stress components other than the pressure p (all shearing stresses),

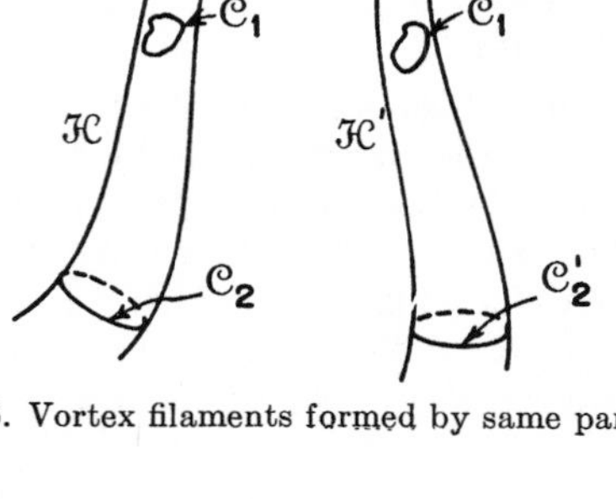

FIG. 26. Vortex filaments formed by same particles.

and to assume the existence of a relation between p and ρ in order to make possible the definition of P.

5. Mean rotation and the Bernoulli function

In Sec. 2.5 we introduced the *total head*

$$H = \frac{q^2}{2g} + h + \frac{P}{g}, \tag{14}$$

which was shown to be constant along each streamline during steady flow (Bernoulli equation). Let us consider the relation of the Bernoulli function H to the mean rotation of the fluid or to curl $\mathbf{q}$.

Starting from the equation of motion for an inviscid elastic fluid in the form (12), we subtract grad $(q^2/2)$ from both sides and use Eq. (14), to obtain

$$\frac{d\mathbf{q}}{dt} - \text{grad}\left(\frac{q^2}{2}\right) = -\text{grad}\,(gH). \tag{15}$$

In order to interpret the vector on the left, we compute its x-component. Using the Euler rule of differentiation and $q^2 = q_x^2 + q_y^2 + q_z^2$, and denoting briefly curl $\mathbf{q}$ by $\boldsymbol{\lambda}$, we see that the x-component of the left-hand side is

$$\begin{aligned} \frac{dq_x}{dt} - \frac{\partial}{\partial x}\left(\frac{q^2}{2}\right) &= \frac{\partial q_x}{\partial t} + q_z\left(\frac{\partial q_x}{\partial z} - \frac{\partial q_z}{\partial x}\right) - q_y\left(\frac{\partial q_y}{\partial x} - \frac{\partial q_x}{\partial y}\right) \\ &= \frac{\partial q_x}{\partial t} + (\lambda_y q_z - \lambda_z q_y) = \frac{\partial q_x}{\partial t} + (\boldsymbol{\lambda} \times \mathbf{q})_x, \end{aligned} \tag{16}$$

and from Eq. (15)

$$\frac{\partial \mathbf{q}}{\partial t} + (\text{curl}\,\mathbf{q} \times \mathbf{q}) = -\text{grad}\,(gH).\text{[12]} \tag{17}$$

This (vector) equation (which is a form of Newton's equation for an inviscid fluid), includes the (scalar) Bernoulli equation and more. In fact, for steady flow $\partial\mathbf{q}/\partial t = 0$, so that

$$\text{grad}\,H = -\frac{1}{g}(\text{curl}\,\mathbf{q} \times \mathbf{q}). \tag{18}$$

Since a vector product is perpendicular to each of its factors, Eq. (18) shows that the vector grad H is perpendicular to $\mathbf{q}$. Hence the directional derivative of H along a streamline is zero, and H must be constant along the streamline. Moreover, grad H has no component in the direction of curl $\mathbf{q}$, the direction of the vortex lines. Thus: *In the steady flow of an inviscid elastic fluid the surfaces on which the Bernoulli function has constant values are composed of streamlines and vortex lines.*

The most important consequence of (18) is the following. If curl $\mathbf{q}$ vanishes at all points, then (18) shows that grad $H \equiv 0$, i.e., the Bernoulli function has one and the same value everywhere and we can state: *In the steady irrotational flow of an inviscid elastic fluid the Bernoulli function, or the total head, has the same value on all streamlines.* The converse is not true in general. It can happen that the streamlines and vortex lines coincide, in which case the vector product curl $\mathbf{q} \times \mathbf{q}$ vanishes and H is constant everywhere, although the motion is not irrotational.[13] This case, however, is a very particular type of motion and cannot occur, for example, in a plane motion: $q_z = 0$, and $\partial/\partial z = 0$ where (5) shows that curl $\mathbf{q}$ is perpendicular to the x,y-plane and therefore cannot coincide anywhere with $\mathbf{q}$.

In the case of a nonsteady irrotational motion, Eq. (17) leads to

$$\frac{\partial \mathbf{q}}{\partial t} = -\text{grad } (gH), \tag{19}$$

which should be noted for later use.

6. Helmholtz' derivation of the vortex theorems[14]

Hermann von Helmholtz derived his two vortex theorems directly from Eq. (17) without using the concept of circulation. An outline of his argument, which, however, does not lead to a rigorous proof, follows.

It is well known that the curl of a gradient is zero. In fact, if for some function ϕ a vector $\mathbf{a}$ satisfies $\mathbf{a} = \text{grad } \phi$ then

$$\frac{\partial a_x}{\partial y} = \frac{\partial}{\partial y}\left(\frac{\partial \phi}{\partial x}\right) = \frac{\partial a_y}{\partial x}$$

etc., and (5) shows that curl $\mathbf{a}$ vanishes identically. Thus, on taking the curl of both sides of Eq. (17), we get

$$\text{curl } \frac{\partial \mathbf{q}}{\partial t} + \text{curl } (\text{curl } \mathbf{q} \times \mathbf{q}) = 0.$$

If we interchange the order of differentiation in the first term, and write $\boldsymbol{\lambda}$ for the vector curl $\mathbf{q}$, this becomes

$$\frac{\partial \boldsymbol{\lambda}}{\partial t} + \text{curl } (\boldsymbol{\lambda} \times \mathbf{q}) = 0. \tag{20}$$

Using the definition (5) of curl and that of vector product, we write the x-component of the second term on the left as

$$\frac{\partial}{\partial y}(\lambda_x q_y - \lambda_y q_x) - \frac{\partial}{\partial z}(\lambda_z q_x - \lambda_x q_z). \tag{21}$$

The terms resulting from the differentiation of the products are listed in the first two columns below:

$$
(21')\qquad
\begin{aligned}
&q_y \frac{\partial \lambda_x}{\partial y} + q_z \frac{\partial \lambda_x}{\partial z} + q_x \frac{\partial \lambda_x}{\partial x} \\
&- q_x \frac{\partial \lambda_y}{\partial y} - q_x \frac{\partial \lambda_z}{\partial z} - q_x \frac{\partial \lambda_x}{\partial x} \\
&+ \lambda_x \frac{\partial q_y}{\partial y} + \lambda_x \frac{\partial q_z}{\partial z} + \lambda_x \frac{\partial q_x}{\partial x} \\
&- \lambda_y \frac{\partial q_x}{\partial y} - \lambda_z \frac{\partial q_x}{\partial z} - \lambda_x \frac{\partial q_x}{\partial x}
\end{aligned}
$$

while the added terms in the third column cancel each other. The terms in each line combine to a simple expression, and (21) can be replaced by

$$
(21'')\qquad q \frac{\partial \lambda_x}{\partial s} - q_x \operatorname{div} \boldsymbol{\lambda} + \lambda_x \operatorname{div} \mathbf{q} - \lambda \frac{\partial q_x}{\partial \sigma},
$$

where $\partial/\partial\sigma$ denotes differentiation in the direction of $\boldsymbol{\lambda}$ (along the vortex line). The second of the terms (21″) is zero, since the divergence of a curl always vanishes, as may be seen from Eq. (5). In the third term, the factor div $\mathbf{q}$ can be replaced by $-(1/\rho)d\rho/dt$, from the equation of continuity (1.II′). Thus (21″) becomes

$$
q \frac{\partial \lambda_x}{\partial s} - \frac{\lambda_x}{\rho} \frac{d\rho}{dt} - \lambda \frac{\partial q_x}{\partial \sigma},
$$

which is exactly the x-component of

$$
(22)\qquad q \frac{\partial \boldsymbol{\lambda}}{\partial s} - \frac{\boldsymbol{\lambda}}{\rho} \frac{d\rho}{dt} - \lambda \frac{\partial \mathbf{q}}{\partial \sigma}.
$$

When (22) is substituted for the second term of Eq. (20), and the terms collected according to the Euler rule of differentiation (1.4), we obtain

$$
(23)\qquad \frac{d\boldsymbol{\lambda}}{dt} - \frac{1}{\rho} \boldsymbol{\lambda} \frac{d\rho}{dt} = \lambda \frac{\partial \mathbf{q}}{\partial \sigma},
$$

or, dividing through by ρ and setting $\mathbf{\Lambda} = \boldsymbol{\lambda}/\rho$,

$$
(23')\qquad \frac{d\mathbf{\Lambda}}{dt} = \Lambda \frac{\partial \mathbf{q}}{\partial \sigma}.
$$

This is *Helmholtz' equation*, which can be interpreted so as to give the two vortex theorems.

Let P and Q (Fig. 27) be two neighboring points lying on the same vortex

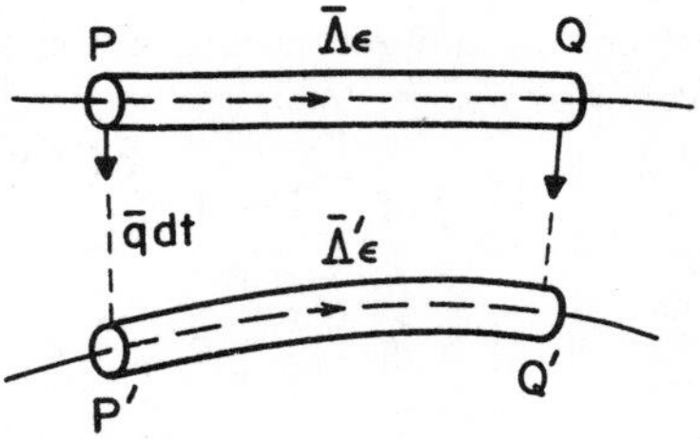

FIG. 27. Interpretation of Helmholtz' equation.

line. Then $\overline{PQ} = \mathbf{\Lambda}\epsilon$ for a suitable small ϵ. After time dt, the particle initially at P has moved to P' and the particle at Q to Q', where

$$\overline{PP'} = \mathbf{q}\, dt, \qquad \overline{QQ'} = \left(\mathbf{q} + \frac{\partial \mathbf{q}}{\partial \sigma} \Lambda\epsilon\right) dt.$$

Then $\overline{P'Q'}$ is given by

$$\overline{P'Q'} = -\overline{PP'} + \overline{PQ} + \overline{QQ'} = \left(\mathbf{\Lambda} + \frac{\partial \mathbf{q}}{\partial \sigma} \Lambda\, dt\right) \epsilon.$$

But then Eq. (23′) gives

$$\overline{P'Q'} = \left(\mathbf{\Lambda} + \frac{d\mathbf{\Lambda}}{dt}\, dt\right) \epsilon = \mathbf{\Lambda}'\epsilon, \tag{24}$$

since the parenthesis is exactly the value $\mathbf{\Lambda}'$ of $\mathbf{\Lambda}$ obtaining at the position P' after time dt. From (24) follows, first, that P' and Q' again lie on a vortex line, up to terms of first order, since $\mathbf{\Lambda}'$ gives the direction of the vortex line at P'; this is in accord with the first vortex theorem. Secondly, Eq. (24) shows that the change in the distance PQ is proportional to the change in Λ. Now, during the transition from PQ to $P'Q'$, the mass of the particle does not change; therefore, if $d\mathcal{A}$ and $d\mathcal{A}'$ are the normal cross sections of the vortex filament before and after the displacement, and ρ and ρ' the corresponding densities, we must have

$$(\rho\, d\mathcal{A})\cdot(PQ) = (\rho'\, d\mathcal{A}')\cdot(P'Q')$$

from which

$$\rho\Lambda\, d\mathcal{A} = \rho'\Lambda'\, d\mathcal{A}'. \tag{25}$$

Now $\rho\Lambda$ is, according to the definition of Λ, the length of the vector curl $\mathbf{q}$, so that Eq. (25) expresses the second vortex theorem for a vortex filament: the product of cross section and vortex magnitude remains constant. This leads to the analogous theorem for a vortex tube of finite cross section.

Article 7

Irrotational Motion[15]

1. Potential

From the vortex theory of the preceding article, it follows that in an inviscid elastic fluid, a particle that at one time has no mean rotation cannot subsequently acquire rotation.[16] If at some time $t = 0$ the entire fluid mass under consideration is irrotational, then it remains so. Thus flow patterns can exist with

$$\text{curl } \mathbf{q} = 0 \qquad \text{for all } t. \tag{1}$$

A transition to rotational flow can occur only if viscosity becomes effective, or if the fluid ceases to be elastic, etc. In particular, if a flow originates from a region where $\mathbf{q}$, p, and ρ are constant (for example, from a state of rest) the flow problem is irrotational everywhere and at all time, whether it is steady or nonsteady.

Mathematically, the condition (1) is equivalent to the statement that $\mathbf{q}$ is a gradient, i.e., that there exists a function $\Phi_1(x,y,z,t)$ whose gradient is $\mathbf{q}$:

$$\mathbf{q} = \text{grad } \Phi_1 \,; \qquad q_x = \frac{\partial \Phi_1}{\partial x}, \qquad q_y = \frac{\partial \Phi_1}{\partial y}, \qquad q_z = \frac{\partial \Phi_1}{\partial z}. \tag{2}$$

In addition to some given (p,ρ)-relation, the flow is subject to (a) the equation of continuity, which we take in the form (1.II′):

$$\text{div } \mathbf{q} = -\frac{1}{\rho}\frac{d\rho}{dt}, \tag{3}$$

and (b) the Newton equation of motion, which we take in the form (6.19) holding for an inviscid elastic fluid when the flow is irrotational:

$$\frac{\partial \mathbf{q}}{\partial t} = -\text{grad } (gH) = -\text{grad}\left(\frac{q^2}{2} + gh + P\right). \tag{4}$$

With the use of $\mathbf{q} = \text{grad } \Phi_1$ and

$$\frac{\partial}{\partial t}\text{grad } \Phi_1 = \text{grad } \frac{\partial \Phi_1}{\partial t},$$

Eq. (4) may be written

$$\text{grad}\left(\frac{\partial \Phi_1}{\partial t} + gH\right) = 0. \tag{5}$$

Consequently the sum $\partial\Phi_1/\partial t + gH$ must be a function of t only, say $f(t)$. If $F(t)$ is the indefinite integral of $f(t)$, then

$$(5') \qquad \frac{\partial\Phi_1}{\partial t} + gH = f = \frac{dF}{dt} \qquad \text{or} \qquad \frac{\partial}{\partial t}(\Phi_1 - F) = -gH.$$

Since F is independent of x, y, and z, the function $\Phi = \Phi_1 - F$ has the same space derivatives as Φ_1 and may, therefore, be used in place of Φ_1 in (2). Then Φ satisfies the four conditions

$$(6) \quad \frac{\partial\Phi}{\partial x} = q_x, \qquad \frac{\partial\Phi}{\partial y} = q_y, \qquad \frac{\partial\Phi}{\partial z} = q_z, \qquad \frac{\partial\Phi}{\partial t} = -\left(\frac{q^2}{2} + gh + P\right).$$

Here the six integrability conditions

$$\frac{\partial^2\Phi}{\partial x\,\partial y} = \frac{\partial^2\Phi}{\partial y\,\partial x}, \ldots, \frac{\partial^2\Phi}{\partial z\,\partial t} = \frac{\partial^2\Phi}{\partial t\,\partial z},$$

are satisfied on account of Eqs. (1) and (4), and Φ is thus determined, for a given flow pattern, to within an additive constant. The function Φ is called the *potential* of the irrotational fluid flow. The reader is familiar with the fact that grad Φ is normal to the surfaces Φ = constant; thus, the velocity vector $\mathbf{q}$ is perpendicular to these *equipotential surfaces*, or more briefly *potential surfaces*. The magnitude of the component of $\mathbf{q}$ in any direction equals the directional derivative of Φ in that direction and, in particular q equals $\partial\Phi/\partial s$ where $\partial/\partial s$ means differentiation in direction of the streamlines. Once $\Phi(x,y,z,t)$ is known, the flow is completely determined, since the first three equations (6) give $\mathbf{q}$, and then the last equation determines P, which, together with the (p,ρ)-relation, determines p and ρ as functions of x,y,z, and t. Obviously an additive constant in Φ has no significance.

In the case of steady motion, $\mathbf{q}$ and P independent of t, it follows from the first three equations (6) that $\partial\Phi/\partial t$ is independent of x, y, and z, because $\partial(\partial\Phi/\partial t)/\partial x = \partial(\partial\Phi/\partial x)/\partial t = 0$, etc.; and from the last equation it follows that $\partial\Phi/\partial t$ is not a function of t either. Thus $\partial\Phi/\partial t$ is constant, everywhere and at all times, and is, in fact, equal to $-gH$ by Eq. (6); this is in agreement with the conclusion in Sec. 6.5 that H is constant in steady irrotational flow.

2. Equation for the potential

In the derivation of Eqs. (6) only Eqs. (1) and (4) were used. Thus, an arbitrary function $\Phi(x,y,z,t)$, together with a (p,ρ)-relation and (6), determines a distribution of values of $\mathbf{q}$, p, and ρ which satisfy the Newton equation, (4), but which will not, in general, satisfy the equation of continuity, (3). This condition will be fulfilled if Φ is a solution of the differential equation which results when Φ is substituted in Eq. (3).

The left-hand member of Eq. (3) is

$$\text{(7)} \qquad \operatorname{div} \mathbf{q} = \operatorname{div} \operatorname{grad} \Phi = \frac{\partial^2 \Phi}{\partial x^2} + \frac{\partial^2 \Phi}{\partial y^2} + \frac{\partial^2 \Phi}{\partial z^2} = \Delta\Phi,$$

where the symbol Δ (Laplace operator) is used exactly as in Art. 4. The sound velocity may be defined by

$$\text{(8)} \qquad a^2 = \frac{dp}{d\rho},$$

as in Sec. 4.1, whenever there exists a (p,ρ)-relation; then the right-hand member of Eq. (3) is

$$-\frac{1}{\rho}\frac{d\rho}{dt} = -\frac{1}{\rho}\frac{d\rho}{dp}\frac{dp}{dt} = -\frac{1}{a^2\rho}\frac{dp}{dt} = -\frac{1}{a^2}\frac{dP}{dt},$$

since by definition $P = \int dp/\rho$ (Sec. 2.5). Thus Eq. (3) becomes

$$\text{(9)} \qquad \Delta\Phi = -\frac{1}{a^2}\frac{dP}{dt}.$$

We shall find [Eqs. (10) and (16)] that both a^2 and dP/dt are expressible in terms of derivatives of Φ.

In the case of an incompressible fluid, when $a = \infty$, the right-hand side of Eq. (9) is zero; the potential Φ must therefore be a solution of Laplace's equation, $\Delta\Phi = 0$, and all the classical results on Laplace's equation are applicable. Further, the equation does not involve t, which means that Φ (and therefore the whole pattern of flow) is determined at each moment only by the boundary conditions holding at that moment: in the irrotational flow of an incompressible inviscid fluid, there is no "after-effect". The situation is much more complex when the fluid is compressible.

From the last equation (6), $-P = \partial\Phi/\partial t + q^2/2 + gh$, so that by Euler's rule of differentiation

$$\text{(10)} \qquad -\frac{dP}{dt} = \frac{\partial^2\Phi}{\partial t^2} + \frac{\partial}{\partial t}\left(\frac{q^2}{2}\right) + q\frac{\partial}{\partial s}\left(\frac{\partial\Phi}{\partial t} + \frac{q^2}{2}\right) + g\frac{\partial\Phi}{\partial h}.$$

In differentiating the function h, use has been made of $\partial h/\partial t = 0$ and of the fact that $\partial h/\partial s$ is the cosine of the angle between the directions of $\mathbf{q}$ (=grad Φ) and grad h. Hence $q\,\partial h/\partial s$ is the component of grad Φ in the direction normal to h = constant; or $q\,\partial h/\partial s = \partial\Phi/\partial h$. Using here Φ_x,

Φ_y, Φ_z, and Φ_t, for the first partial derivatives of Φ, we write out the second and third terms on the right of Eq. (10):

$$\frac{\partial}{\partial t}\left(\frac{q^2}{2}\right) = \tfrac{1}{2}\frac{\partial}{\partial t}(\Phi_x^2 + \Phi_y^2 + \Phi_z^2) = \Phi_x \frac{\partial \Phi_x}{\partial t} + \Phi_y \frac{\partial \Phi_y}{\partial t} + \Phi_z \frac{\partial \Phi_z}{\partial t},$$

$$\text{(11)} \qquad q\frac{\partial}{\partial s}\left(\frac{\partial \Phi}{\partial t}\right) = q_x \frac{\partial \Phi_t}{\partial x} + q_y \frac{\partial \Phi_t}{\partial y} + q_z \frac{\partial \Phi_t}{\partial z} = \Phi_x \frac{\partial \Phi_t}{\partial x} + \Phi_y \frac{\partial \Phi_t}{\partial y} + \Phi_z \frac{\partial \Phi_t}{\partial z}.$$

Hence these two terms are equal. By the use of Eq. (10), Eq. (9) can now be written as

$$\text{(12)} \qquad \Delta\Phi - \frac{1}{a^2}\frac{\partial^2 \Phi}{\partial t^2} = \frac{1}{a^2} q \frac{\partial}{\partial s}\left(2\frac{\partial \Phi}{\partial t} + \frac{q^2}{2}\right) + \frac{1}{a^2} g \frac{\partial \Phi}{\partial h}.$$

Since gravity is comparatively unimportant in problems of gas flow, the last term in (12) will be dropped from now on; retaining this term would not greatly increase the complexity of the equation.

To express Eq. (12) in rectangular coordinates, we use

$$\text{(13)} \qquad q\frac{\partial}{\partial s}\left(\frac{q_x^2}{2}\right) = q_x\left(q_x \frac{\partial q_x}{\partial x} + q_y \frac{\partial q_x}{\partial y} + q_z \frac{\partial q_x}{\partial z}\right) = q_x^2 \frac{\partial^2 \Phi}{\partial x^2} + q_x q_y \frac{\partial^2 \Phi}{\partial x\, \partial y} + q_x q_z \frac{\partial^2 \Phi}{\partial x\, \partial z},$$

and corresponding formulas for q_y and q_z. Applying Eqs. (11) and (13), writing out the terms of $\Delta\Phi$ as in (7), and omitting the gravity term from Eq. (12) we get

$$\text{(14)} \qquad \begin{aligned} &\frac{\partial^2 \Phi}{\partial x^2}\left(1 - \frac{q_x^2}{a^2}\right) + \frac{\partial^2 \Phi}{\partial y^2}\left(1 - \frac{q_y^2}{a^2}\right) + \frac{\partial^2 \Phi}{\partial z^2}\left(1 - \frac{q_z^2}{a^2}\right) - \frac{1}{a^2}\frac{\partial^2 \Phi}{\partial t^2} \\ &\quad - 2\frac{q_x q_y}{a^2}\frac{\partial^2 \Phi}{\partial x\, \partial y} - 2\frac{q_y q_z}{a^2}\frac{\partial^2 \Phi}{\partial y\, \partial z} - 2\frac{q_z q_x}{a^2}\frac{\partial^2 \Phi}{\partial z\, \partial x} \\ &\quad - 2\frac{q_x}{a^2}\frac{\partial^2 \Phi}{\partial x\, \partial t} - 2\frac{q_y}{a^2}\frac{\partial^2 \Phi}{\partial y\, \partial t} - 2\frac{q_z}{a^2}\frac{\partial^2 \Phi}{\partial z\, \partial t} = 0. \end{aligned}$$

This is the general *potential equation for compressible fluid flow*, as compared to $\Delta\Phi = 0$ in the incompressible case. As was mentioned above, a^2 is a function of the first derivatives of Φ: by way of the (p,ρ)-relation, a^2 can be expressed in terms of ρ or p or P, and according to Eq. (6), when gravity is neglected,

$$\text{(15)} \qquad P = -\frac{\partial \Phi}{\partial t} - \frac{q^2}{2} = -\frac{\partial \Phi}{\partial t} - \frac{1}{2}(\text{grad } \Phi)^2.$$

If the flow is steady and the velocity is sufficiently small, so that all second-degree terms in velocity components may be neglected, the potential equation again reduces to the classical one. Without this approximation however, Eq. (14) is *nonlinear*, and therefore the sum of two solutions need not satisfy the equation: solutions cannot be superposed as in the case of linear differential equations.

In the case of the polytropic (p,ρ)-relation $p/p_0 = (\rho/\rho_0)^\kappa$, Eq. (8) gives $a^2 = \kappa p/\rho$, while from Eq. (2.22c), $P = \kappa p/(\kappa - 1)\rho$; thus $a^2 = (\kappa - 1)P$, so that Eq. (15) gives

$$(16)\qquad a^2 = -(\kappa - 1)\left(\frac{\partial \Phi}{\partial t} + \frac{q^2}{2}\right) = -(\kappa - 1)\left[\frac{\partial \Phi}{\partial t} + \frac{1}{2}(\text{grad } \Phi)^2\right].$$

We may also verify from Eq. (2.22d) that the same result holds for $\kappa = -1$, corresponding to the linearized (p,ρ)-relation (see Secs. 1.4 and 2.5). For isothermal flow, p/ρ = constant, we have $a^2 = p_0/\rho_0$ = constant. In any other case, the relation between a^2 and P is computed by means of the (p,ρ)-relation, and then Eq. (15) is used to eliminate P.

The nonlinear partial differential equation (14), combined with (16) [or some other formula for a^2] and with $\mathbf{q} = \text{grad } \Phi$, gives the condition that a function $\Phi(x,y,z,t)$ be the potential of a physically possible irrotational flow of an inviscid elastic fluid. Very few examples are known of solutions in terms of elementary functions. In the remainder of this article we consider some particular cases of Eq. (14).

3. Steady radial flow[17]

The simplest example, other than uniform flow, which needs no further explanation, is that of a steady motion along rays emanating from a fixed point O. Since $\mathbf{q}$ is always directed along the radius from O, the potential surfaces must be concentric spheres with center O; thus Φ is a function of t and $r = (x^2 + y^2 + z^2)^{\frac{1}{2}}$ only. If we denote by q_r the velocity in the direction of increasing r, then

$$q_r = \frac{\partial \Phi}{\partial r} \qquad \text{and} \qquad q_r^2 = q^2$$

are functions of r only. In Sec. 1, it was seen that steady irrotational motion requires that $\partial\Phi/\partial t$ be an absolute constant; then (16) yields the Bernoulli equation

$$(17)\qquad a^2 + \frac{\kappa - 1}{2}q^2 = \text{constant} = a_s^2,$$

where a_s is the value of a corresponding to $q = 0$, the so-called *stagnation value* of the sound velocity. (It is assumed in this section that a polytropic (p,ρ)-relation holds.)

Taking the x-axis along the ray from O to an arbitrary point P, we have $q_y = q_z = 0$, $q_x = q_r$ at P, and since all derivatives of $\partial\Phi/\partial t$ vanish, Eq. (14) reduces to

$$\frac{\partial^2\Phi}{\partial x^2}\left(1 - \frac{q_r^2}{a^2}\right) + \frac{\partial^2\Phi}{\partial y^2} + \frac{\partial^2\Phi}{\partial z^2} = 0. \tag{18}$$

In the first term, $\partial^2\Phi/\partial x^2 = \partial q_x/\partial x = dq_r/dr$. The second term is $\partial q_y/\partial y$, and Fig. 28 shows that $\partial q_y/\partial y = q_r/r$. The same is true of $\partial q_z/\partial z$, and Eq. (18) may be written as

$$\frac{dq_r}{dr}\left(1 - \frac{q_r^2}{a^2}\right) + 2\frac{q_r}{r} = 0. \tag{19}$$

When the value of a^2 is substituted from Eq. (17), the final differential equation for q_r as a function of r reads

$$\frac{dq_r}{dr} \cdot \frac{2a_s^2 - (\kappa + 1)q_r^2}{2a_s^2 - (\kappa - 1)q_r^2} + 2\frac{q_r}{r} = 0. \tag{19'}$$

In this equation the variables can be separated and the integration carried out directly. It is convenient, however, to make a change of variables from r and q_r to the dimensionless quantities ξ and η defined by

$$\xi = \frac{r^2}{r_0^2} \quad \text{and} \quad \eta = \frac{q_r}{a_s},$$

where r_0 is an arbitrary constant. In terms of the new variables, Eq. (19′) is

$$\frac{d\eta}{d\xi} = -\frac{\eta}{\xi} \cdot \frac{2 - (\kappa - 1)\eta^2}{2 - (\kappa + 1)\eta^2}, \tag{19''}$$

the solution of which is

$$\pm\frac{1}{\xi} = \eta\left(1 - \frac{\kappa - 1}{2}\eta^2\right)^{1/(\kappa-1)}, \tag{20}$$

as can be verified by differentiation. Any constant factor could be inserted in one member and Eq. (20) would still satisfy Eq. (19″), but this is un-

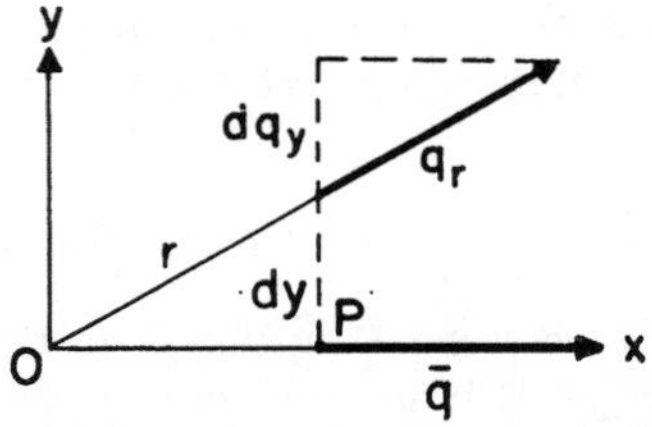

FIG. 28. Auxiliary relation for radial flow.

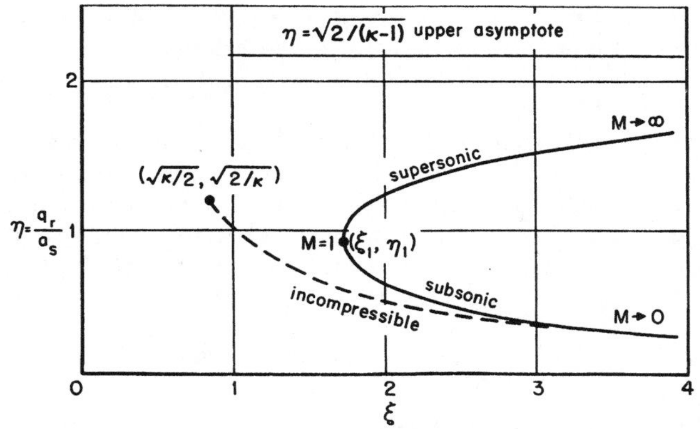

FIG. 29. Dimensionless velocity $\eta = \dfrac{q_r}{a_s}$ versus dimensionless distance ξ in steady radial flow where $\xi = \dfrac{r^2}{r_0^2}$ or $= \dfrac{r}{r_0}$ for spatial or plane flow respectively.

necessary since an arbitrary (positive) factor r_0 has already been included in the definition of ξ.

Before discussing this result let us find the relation between η and the local Mach number q/a; from Eq. (17) it follows that

$$\frac{1}{M^2} = \frac{a^2}{q^2} = \frac{1}{\eta^2} - \frac{\kappa - 1}{2}. \tag{21}$$

In the solution, (20) for $\kappa > 1$, two values of η correspond to $\xi = \infty$: $\eta = 0$ and $\eta = [2/(\kappa - 1)]^{\frac{1}{2}}$, this last value being $\sqrt{5}$ for $\kappa = 1.4$. It follows from (21) that the corresponding values of M are 0 and ∞. Figure 29 (solid line) shows η as a function of ξ [using the positive sign in (20)] when $\kappa = 1.4$. There are two horizontal asymptotes: $\eta = 0$ and $\eta = \sqrt{5}$. The point (ξ_1, η_1) corresponding to the minimum value of ξ may be found by differentiation of ξ as a function of η, giving

$$\eta_1 = \left(\frac{2}{\kappa + 1}\right)^{\frac{1}{2}}, \qquad \xi_1^{\,2} = \left(\frac{\kappa + 1}{2}\right)^{(\kappa+1)/(\kappa-1)}.$$

When the value η_1 is substituted in (21), the corresponding Mach number is seen to be $M = 1$.

If this solution is considered for all rays in a certain solid angle, we have the result: *In a conically divergent channel two radial flows are possible, one subsonic with velocity zero at* ∞, *and the other supersonic with infinite* M *at infinity*. This statement has been derived for $\eta \geqslant 0$, corresponding to an outward flow; if the minus sign is used in the solution (20), the graph is exactly the reflection of Fig. 29 in the ξ-axis, and the same statement is true for an inward flow, where $\eta \leqslant 0$.

The mass flux through unit solid angle in this cone is given by $Q = \rho q r^2 = a_s r_0^2 \rho \xi \eta$. Since mass is conserved, Q must be the same for all r, and therefore the same for all points on the ξ,η-curve; thus Q may be computed on the lower branch of the curve, with ξ tending to ∞ and η to 0. From Eq. (20) it follows that $\xi\eta \to 1$; the limit of ρ may be called ρ_s, the stagnation value, since it corresponds to $\eta = q = 0$. Then

$$Q = a_s r_0^2 \rho \xi \eta = a_s r_0^2 \lim_{\xi \to \infty} \rho \xi \eta = a_s \rho_s r_0^2 . \tag{22}$$

When Q is given, (22) determines r_0.

The fact that the flow does not extend to $r = 0$ is not surprising; this is true even in the case of an incompressible fluid. The broken line in Fig. 29 gives the velocity distribution in the case of an incompressible fluid for the flow with the same stagnation values p_s, ρ_s, and the same flux Q. Here $\rho = \rho_s$, so that $Q = \rho_s q r^2 = \rho_s a_s r_0^2 \xi \eta$. Compared with (22), this implies that the equation of the curve is $\xi\eta = 1$. The left-hand endpoint is determined by the condition that p may not become negative. The Bernoulli equation (2.20), taken for an incompressible fluid and with gravity omitted, reads

$$\frac{q^2}{2} + \frac{p}{\rho_s} = \frac{p_s}{\rho_s} \qquad \text{or} \qquad p = p_s - \rho_s \frac{q^2}{2},$$

from which it can be concluded that p goes through zero when $q^2 = 2p_s/\rho_s$ and $\eta^2 = q^2/a_s^2 = 2p_s/\rho_s a_s^2$. Since a_s^2 is related to p_s, ρ_s by $a_s^2 = \kappa p_s/\rho_s$, the value of η^2 at the critical point for the incompressible flow is, therefore, $2/\kappa$ and, from $\xi\eta = 1$, the value of ξ^2 is $\kappa/2$. As M decreases, the curve for the compressible case is more and more nearly the same as the curve for incompressible flow.

From (22) we obtain $\rho\xi\eta = \rho_s$, which with Eq. (20) gives

$$\rho = \rho_s \left(1 - \frac{\kappa - 1}{2} \eta^2\right)^{1/(\kappa-1)} \tag{23}$$

and also, because $p/\rho^\kappa = \text{constant}$,

$$p = p_s \left(1 - \frac{\kappa - 1}{2} \eta^2\right)^{\kappa/(\kappa-1)} . \tag{23'}$$

Alternatively, Eq. (23) could be derived from the Bernoulli equation in the form

$$\frac{q^2}{2} + \frac{\kappa}{\kappa - 1}\frac{p}{\rho} = \frac{\kappa}{\kappa - 1}\frac{p_s}{\rho_s},$$

together with the (p,ρ)-relation. The main result (20) would then follow from the continuity condition (22), with no reference to the potential.

If a *plane radial flow*,[18] where $r = (x^2 + y^2)^{\frac{1}{2}}$, is studied, the only change is that $\partial^2\Phi/\partial z^2$ is to be omitted in (18); then the factor 2 is missing from the second term in Eq. (19). However, if ξ is taken to be r/r_0, rather than r^2/r_0^2, the same differential equation (19″) results. Thus all conclusions, including Fig. 29, hold as before, provided ξ is given the new interpretation.

4. Nonsteady parallel flow

If all particles move parallel to the x-axis, the equipotential surfaces are planes perpendicular to this axis. Thus, Φ depends only on x and t, while

$$q_x = \frac{\partial\Phi}{\partial x} \qquad \text{and} \qquad q_x^2 = q^2.$$

Since $q_y = q_z = 0$, Eq. (14) reduces to

$$\frac{\partial^2\Phi}{\partial x^2}\left(1 - \frac{q_x^2}{a^2}\right) - \frac{1}{a^2}\frac{\partial^2\Phi}{\partial t^2} - 2\frac{q_x}{a^2}\frac{\partial^2\Phi}{\partial x\,\partial t} = 0. \tag{24}$$

As before, a^2 also involves derivatives of Φ; if the polytropic (p,ρ)-relation is adopted, then Eq. (16) gives

$$a^2 = -(\kappa - 1)\left(\frac{\partial\Phi}{\partial t} + \frac{q_x^2}{2}\right). \tag{25}$$

The problem characterized by Eqs. (24) and (25) will be studied in detail in Chapter III, but a fairly general type of particular integral of these equations will be indicated here.

Assuming

$$\Phi(x, t) = \alpha\frac{x^2}{2} + \beta x + \gamma, \quad q_x = \frac{\partial\Phi}{\partial x} = \alpha x + \beta, \tag{26}$$

where α, β, and γ are functions of t only, can we choose α, β, and γ so that Eqs. (24) and (25) are satisfied? When (26) is substituted in (25), a^2 is given as a quadratic function in x; also, $\partial^2\Phi/\partial x^2$ does not involve x. Thus, when (24) is multiplied through by a^2 and (26) substituted in it, the left-hand member of the resulting equation is also a quadratic function of

x; the differential equation is satisfied if the coefficients of x^2 and x and the constant term, all functions of t, vanish identically. If these coefficients are set equal to zero, we have the conditions

$$(27)\qquad \begin{aligned} \alpha'' + (\kappa + 3)\,\alpha\alpha' + (\kappa + 1)\,\alpha^3 &= 0, \\ \beta'' + (\kappa + 1)\,\alpha\beta' + \beta(2\alpha' + (\kappa + 1)\,\alpha^2) &= 0, \\ \gamma'' + (\kappa - 1)\,\alpha\gamma' + \beta\left(\frac{\kappa + 1}{2}\alpha\beta + 2\beta'\right) &= 0. \end{aligned}$$

These are three ordinary differential equations which can be solved successively for α, for β, and finally for γ. Two examples of solutions follow:

$$(28)\qquad \begin{aligned} &\text{(a)}\quad \alpha = \frac{2}{\kappa + 1}\frac{1}{t}, \qquad \beta = \text{constant} = c_1, \\ &\gamma = -\frac{\kappa + 1}{\kappa - 1}\frac{c_1^2}{2}\,t + c_2 t^{(3-\kappa)/(\kappa+1)}, \\ &q_x = \frac{2}{\kappa + 1}\frac{x}{t} + c_1, \\ &a^2 = \left(\frac{\kappa - 1}{\kappa + 1}\frac{x}{t} - c_1\right)^2 - \frac{3 - \kappa}{\kappa + 1}(\kappa - 1)c_2 t^{-2(\kappa-1)/(\kappa+1)}. \end{aligned}$$

In the particular case $c_2 = 0$, both q_x and a^2 are functions of x/t only; this motion plays an important role in the theory of nonsteady one-dimensional flow (see Art. 13, "centered simple waves"). Including a c_2-term has no effect on the particle lines (see Sec. 1.2), which are the lines in the x,t-plane defined by $dx/dt = q_x$ and which trace the history of each particle; however, the pressure distribution along these lines does depend on c_2. For this example the particle lines are given by

$$x = c_1 \frac{\kappa + 1}{\kappa - 1}\,t + A t^{2/(\kappa+1)},$$

where A is the parameter; this may be verified by differentiation. Along each particle line, a^2 is $t^{-2(\kappa-1)/(\kappa+1)}$ times a constant that is a simple function of c_2 and A. Now $p = \text{constant}\cdot\rho^\kappa$, so that $a^2 = dp/d\rho = \text{constant}\cdot \rho^{\kappa-1}$. Thus, along each particle line, ρ is proportional to $t^{-2/(\kappa+1)}$ and p to $t^{-2\kappa/(\kappa+1)}$; the constants of proportionality are again functions of c_2 and A.

$$(29)\qquad \begin{aligned} &\text{(b)}\quad \alpha = \frac{1}{t}, \qquad \beta = ct^{1-\kappa}, \qquad \gamma = \frac{5 - 3\kappa}{3 - 2\kappa}\cdot\frac{c^2}{2(\kappa - 1)}\,t^{3-2\kappa} \\ &q_x = \frac{x}{t} + ct^{1-\kappa}, \qquad a^2 = (\kappa - 2)ct^{-\kappa}[(\kappa - 1)x + ct^{2-\kappa}]. \end{aligned}$$

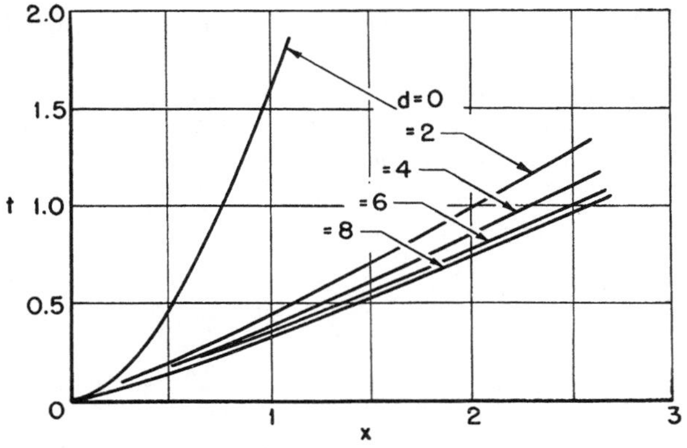

FIG. 30. Particle lines $\frac{x}{t} + \frac{c}{\kappa - 1} t^{1-\kappa}$ = constant = k, for $c = -0.3$, $\kappa = 1.4$, for equidistant values of $d = k^{\kappa/(\kappa-1)}$.

Here the particle lines are given by

$$\frac{x}{t} + \frac{c}{\kappa - 1} t^{1-\kappa} = \text{constant} = k; \tag{30}$$

some of these curves are shown in Fig. 30, with $c = -0.3$ and $\kappa = 1.4$. From Eqs. (29) and (30) we have $a^2 = ck(\kappa - 1)(\kappa - 2)t^{1-\kappa}$ along any particle line, so that ρ is here proportional to $1/t$. For constant t and variable k, a^2 is a multiple of k, and ρ is proportional to $k^{1/(\kappa-1)}$. Also when t is constant, $dx = t\,dk$ from Eq. (30), so that on computing the fluid mass included between two positions at a given time, we obtain

$$\int_{x_1}^{x_2} \rho\,dx = \text{constant} \cdot \int_{k_1}^{k_2} k^{1/(\kappa-1)}\,dk = \text{constant} \cdot [k_2^{\kappa/(\kappa-1)} - k_1^{\kappa/(\kappa-1)}].$$

The particle lines in Fig. 30 are drawn for equidistant values of $k^{\kappa/(\kappa-1)} = d$, namely for $d = 0, 2, 4, 6, 8$. Hence, the mass between any two successive lines is the same.

5. Steady plane motion

Since a substantial part of this book will be devoted to problems of steady plane potential flow, only a preliminary discussion will be given here.

By hypothesis $q_z = 0$ and $\partial\Phi/\partial t$ = constant, so that Eq. (14) reduces to

$$\frac{\partial^2\Phi}{\partial x^2}\left(1 - \frac{q_x^2}{a^2}\right) - 2\,\frac{q_x q_y}{a^2}\,\frac{\partial^2\Phi}{\partial x\,\partial y} + \frac{\partial^2\Phi}{\partial y^2}\left(1 - \frac{q_y^2}{a^2}\right) = 0. \tag{31}$$

With the polytropic (p, ρ)-relation, the formula for a^2 again takes the form

(17), where $a_s^2 = (1 - \kappa)\partial\Phi/\partial t$ is the square of the stagnation value of the sound velocity.

It is often advantageous to use polar coordinates r, θ; then $\mathbf{q}$ is decomposed into a radial component q_r and a circumferential component q_θ. Taking the x-axis along a radius, we have at any point on this radius

$$\frac{\partial\Phi}{\partial x} = q_x = q_r = \frac{\partial\Phi}{\partial r}, \qquad \frac{\partial\Phi}{\partial y} = q_y = q_\theta = \frac{1}{r}\frac{\partial\Phi}{\partial\theta},$$

(32) $$\frac{\partial^2\Phi}{\partial x^2} = \frac{\partial q_x}{\partial x} = \frac{\partial q_r}{\partial r} = \frac{\partial^2\Phi}{\partial r^2}, \qquad \frac{\partial^2\Phi}{\partial y^2} = \frac{\partial q_y}{\partial y} = \frac{1}{r}\frac{\partial q_\theta}{\partial\theta} + \frac{q_r}{r} = \frac{1}{r^2}\frac{\partial^2\Phi}{\partial\theta^2} + \frac{1}{r}\frac{\partial\Phi}{\partial r},$$

$$\frac{\partial^2\Phi}{\partial x\,\partial y} = \frac{\partial q_y}{\partial x} = \frac{\partial q_\theta}{\partial r} = \frac{1}{r}\frac{\partial^2\Phi}{\partial r\,\partial\theta} - \frac{1}{r^2}\frac{\partial\Phi}{\partial\theta} = \frac{\partial^2\Phi}{\partial y\,\partial x} = \frac{\partial q_x}{\partial y} = \frac{\partial q_r}{r\,\partial\theta} - \frac{q_\theta}{r}.$$

To derive the last of these formulas, a figure generalizing Fig. 28 can be used. That is, we note that the y-derivative of the y-component of q_r is q_r/r, the same as in the discussion in Sec. 3. Here, however, we must also take account of the y-component of q_θ, whose y-derivative is $(1/r)\,\partial q_\theta/\partial\theta$. Equations (32) are the usual formulas for the derivatives of Φ in polar coordinates, written for $\theta = 0$.

Equation (31) can now be written as

(33) $$\frac{\partial^2\Phi}{\partial r^2}\left(1 - \frac{q_r^2}{a^2}\right) - 2\,\frac{q_r q_\theta}{a^2}\left(\frac{1}{r}\frac{\partial^2\Phi}{\partial r\,\partial\theta} - \frac{1}{r^2}\frac{\partial\Phi}{\partial\theta}\right) + \left(\frac{1}{r^2}\frac{\partial^2\Phi}{\partial\theta^2} + \frac{1}{r}\frac{\partial\Phi}{\partial r}\right)\left(1 - \frac{q_\theta^2}{a^2}\right) = 0,$$

or in terms of velocity components as

(34) $$\frac{\partial q_r}{\partial r} + \frac{\partial q_\theta}{r\,\partial\theta} + \frac{q_r}{r} = \frac{1}{a^2}\left[q_r^2\frac{\partial q_r}{\partial r} + q_r q_\theta\left(\frac{\partial q_\theta}{\partial r} + \frac{\partial q_r}{r\,\partial\theta}\right) + q_\theta^2\frac{\partial q_\theta}{r\,\partial\theta}\right].$$

The latter equation is actually the polar form of

$$\frac{\partial q_x}{\partial x}\left(1 - \frac{q_x^2}{a^2}\right) - \frac{q_x q_y}{a^2}\left(\frac{\partial q_x}{\partial y} + \frac{\partial q_y}{\partial x}\right) + \frac{\partial q_y}{\partial y}\left(1 - \frac{q_y^2}{a^2}\right) = 0$$

rather than of Eq. (31). Thus, if Eq. (34) is used, the condition (1) for the existence of a potential function must be added; for plane motion Eq. (1) reduces to

$$\frac{\partial q_y}{\partial x} - \frac{\partial q_x}{\partial y} = 0$$

in rectangular coordinates, or

(35) $$\frac{\partial q_\theta}{\partial r} - \frac{\partial q_r}{r\,\partial\theta} + \frac{q_\theta}{r} = 0$$

in polar coordinates, as is seen from (32). Equations (34) and (35) combined are equivalent to Eq. (33).

The case of radial motion, $q_\theta = 0$ and q_r a function of r only, has already been discussed at the end of Sec. 3. Now we consider a more general case of what may be called an *axially symmetric flow*:[19] q_r and q_θ are independent of θ, but q_θ does not vanish. The condition (35) then reads

$$\frac{dq_\theta}{dr} + \frac{q_\theta}{r} = 0, \qquad \text{or} \qquad rq_\theta = \text{constant}. \tag{36}$$

Now $2\pi r q_\theta$ is the value of the circulation Γ on the circle of radius r about the origin, so we have here an illustration of the case mentioned previously (Sec. 6.1), where Γ is constant (but different from zero) on circuits surrounding an infinite cylindrical obstacle. The "obstacle" here is a certain circular cylinder, $r = r_l$ (see Sec. 6). The differential equations apply only to the region outside the corresponding circle with center at O. In this (doubly connected) region a regular potential flow with $2\pi r q_\theta = \text{constant} = \Gamma$ exists. It will be seen presently that the immediate neighborhood of $r = 0$ is without interest for us.

Once q_θ has been found, Eq. (34) can serve to determine q_r.[20] This equation, with the θ-derivatives omitted, reads after multiplication by r

$$\frac{d}{dr}(rq_r) = \frac{r}{a^2} q_r \left(q_r \frac{dq_r}{dr} + q_\theta \frac{dq_\theta}{dr}\right) = \frac{r}{a^2} q_r \frac{d}{dr}\left(\frac{q^2}{2}\right). \tag{37}$$

Here q^2 is written for $q_r^2 + q_\theta^2$. Now, considering a general elastic fluid, we set $dH/dr = d(\frac{1}{2} q^2 + P)/dr = 0$ (see end of Sec. 1), or

$$\frac{d}{dr}\left(\frac{q^2}{2}\right) + \frac{1}{\rho}\frac{dp}{dr} = \frac{d}{dr}\left(\frac{q^2}{2}\right) + \frac{a^2}{\rho}\frac{d\rho}{dr} = 0, \tag{38}$$

and see that the right member of Eq. (37) equals*

$$-rq_r \frac{1}{\rho}\frac{d\rho}{dr}.$$

Thus, after dividing by the factor rq_r, we can integrate both sides to obtain

$$\log(rq_r) = -\log\rho + \text{constant}, \qquad rq_r = \frac{\text{constant}}{\rho}. \tag{39}$$

This, in conjunction with (36), solves the problem. For if we call the two constants in Eqs. (36) and (39) $C(= \Gamma/2\pi)$ and k respectively, the equations squared and added show that

$$r^2q^2 = C^2 + \frac{k^2}{\rho^2}, \qquad \text{or} \qquad r^2 = \frac{C^2}{q^2} + \frac{k^2}{(\rho q)^2}. \tag{40}$$

* This result also follows directly from the equation of continuity (3).

The Bernoulli equation in its integrated form establishes, as we know, a relation between q^2 and ρ. If this relation is used to eliminate ρ from (40) the latter equation links r^2 to q^2, and since q_θ^2 is already known as a function of r, we have finally a relation between q_r^2 and r. An examination of these relationships reveals two important phenomena, which will be discussed in the following section. (See also end of Sec. 17.4.)

6. Transition between subsonic and supersonic flow. Limit line

The relation between ρ and q established by the Bernoulli equation will be discussed in some detail in Art. 8. Here it will suffice to know that ρ decreases monotonically when q increases, while the product ρq first increases from zero at $q = 0$, reaches a maximum at the sonic point $q = a, M = 1$, and then decreases towards zero as q increases through supersonic values. The function $k^2/(\rho q)^2$ is plotted against q^2 in Fig. 31; the abscissa OE_0 of the minimum point is, as just stated, $q^2 = a^2$. The graph shows also the hyperbola with the ordinates C^2/q^2. The ordinates of the heavily drawn curve in Fig. 31 are the sums of the ordinates of the other two curves, and these sums must equal r^2 according to Eq.(40). Since one curve has a minimum at the sonic point and then increases without bound, while the other decreases monotonically, it is apparent that whatever the (positive) constants C^2 and k^2 are, the resultant curve must have a minimum F with an ordinate different from zero and an abscissa $OF_0 > OE_0$, that is, lying in the supersonic region. Denote this minimum ordinate F_0F by r_l^2.

If some value $r^2 = OA$ is given, the graph shows that two different values

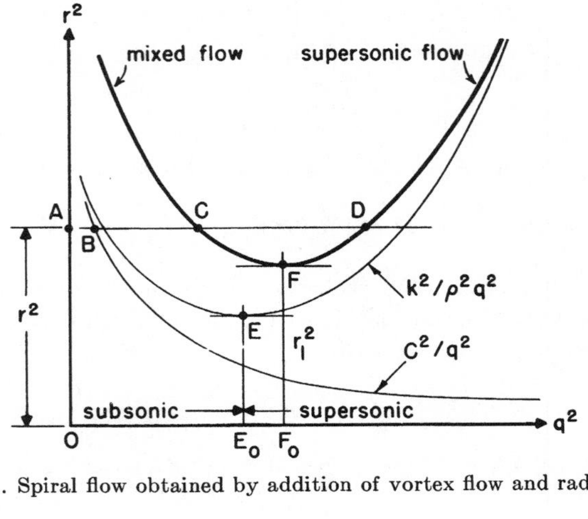

FIG. 31. Spiral flow obtained by addition of vortex flow and radial flow.

of q^2 correspond to it when $r^2 > r_l^2$; if $r^2 < r_l^2$ no q^2 can be found that will satisfy Eq. (40). This means: There exist for each pair of constants C and k two different axially symmetric potential flows, both extending over the region from $r = r_l$ to $r = \infty$, with the same velocity at $r = r_l$. One flow, corresponding to the branch of the curve to the right of F, is entirely supersonic, while the other one includes subsonic as well as supersonic velocities. At the circle $r = r_l$ the flow ends; it has here a "natural limit". Something similar was found in Sec. 3 in the discussion of radial flow. But in the radial flow, the limit line coincided with the line on which $M = 1$ and thus could have been attributed to the fact that the sound velocity had been reached. We now learn that the natural flow limit has nothing to do with the border line between a subsonic and a supersonic region. On the contrary, the present example shows that a "mixed" potential flow is possible without any singularity or irregularity occurring at the border between the regions where $M < 1$ and $M > 1$.

In Figs. 32 and 33 are indicated the streamlines corresponding to the two solutions determined by Fig. 31. To find these streamlines, one has only

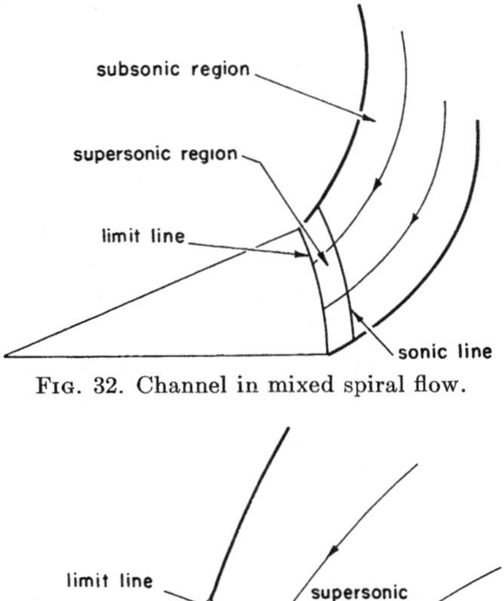

FIG. 32. Channel in mixed spiral flow.

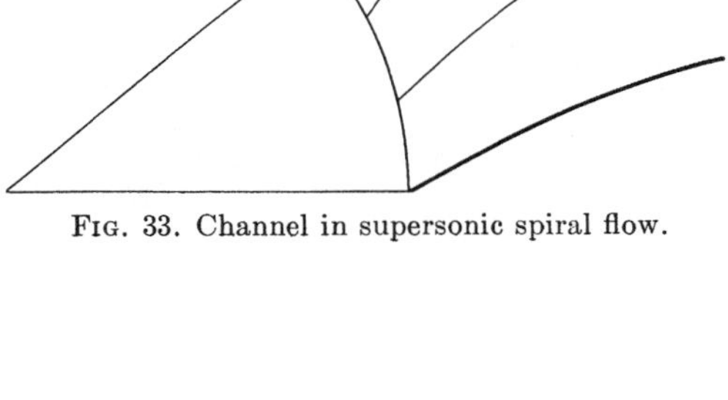

FIG. 33. Channel in supersonic spiral flow.

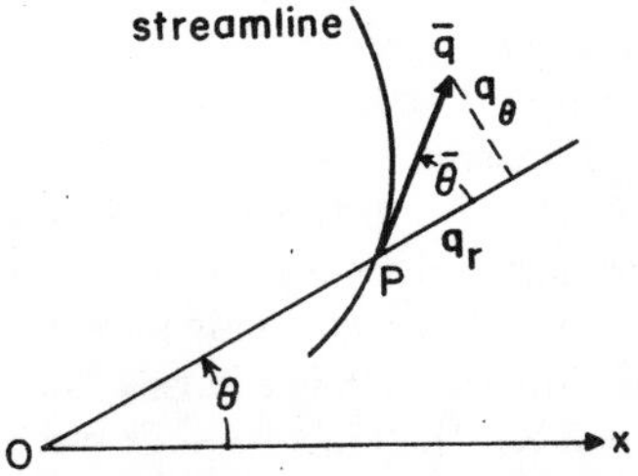

FIG. 34. Showing angle between velocity and radius vector in spiral flow.

to take from Fig. 31, for various positions of the point A on the vertical axis, the magnitudes $AB = q_\theta^2$ and BC or BD for q_r^2. Then the ratio $q_\theta : q_r$ gives the slope of the streamline, namely, tan $\bar{\theta}$, with respect to the radius vector (see Fig. 34). Graphical or numerical integration supplies the streamlines.

Figure 32 refers to the mixed flow, with the smaller values of q^2. In this case the velocity is zero at infinity, it increases monotonically up to the sound velocity at the sonic circle, and then to supersonic values in the annular region between the sonic and the limit circles. The limit at $r = \infty$ of the slope of the streamline with respect to the radius vector is, from Eqs. (36) and (39),

$$\lim_{r\to\infty} \frac{q_\theta}{q_r} = \lim_{q\to 0} \frac{q_\theta}{q_r} = \lim_{q\to 0} \frac{C\rho}{k} = \frac{C\rho_s}{k} = \text{constant.}$$

In this case, then, the streamlines approach logarithmic spirals as $r \to \infty$. If the two heavily drawn curves in Fig. 32 represent the walls of a channel, the flow inside the channel can follow the pattern shown in the sketch up to the limit circle.

Figure 33 shows the streamlines in the case of the completely supersonic motion. Here the velocity at the limit line is the same as in the case of Fig. 32. But now the velocity increases as we go outward. Since q_θ tends to zero with increasing r, the flow becomes more and more radial. The direction of the flow can be inward, as indicated in the figures, or outward in both cases.

7. Other particular cases of the general potential equation

In the case of steady motion the general equation (14) becomes

$$\begin{aligned}\frac{\partial^2\Phi}{\partial x^2}\left(1-\frac{q_x^2}{a^2}\right)+\frac{\partial^2\Phi}{\partial y^2}\left(1-\frac{q_y^2}{a^2}\right)+\frac{\partial^2\Phi}{\partial z^2}\left(1-\frac{q_z^2}{a^2}\right)\\ -2\frac{q_xq_y}{a^2}\frac{\partial^2\Phi}{\partial x\,\partial y}-2\frac{q_yq_z}{a^2}\frac{\partial^2\Phi}{\partial y\,\partial z}-2\frac{q_zq_x}{a^2}\frac{\partial^2\Phi}{\partial z\,\partial x}=0\end{aligned} \tag{41}$$

where from Eq. (16)

$$a^2 = a_s^2 - \frac{\kappa - 1}{2}(q_x^2 + q_y^2 + q_z^2).$$

This same equation in cylindrical coordinates r, θ, and z, with $q_r = \partial\Phi/\partial r$, $q_\theta = (1/r)\,\partial\Phi/\partial\theta$ takes the form

$$(42)\qquad \begin{aligned} &\frac{\partial^2\Phi}{\partial r^2}\left(1 - \frac{q_r^2}{a^2}\right) + \frac{1}{r^2}\frac{\partial^2\Phi}{\partial\theta^2}\left(1 - \frac{q_\theta^2}{a^2}\right) + \frac{\partial^2\Phi}{\partial z^2}\left(1 - \frac{q_z^2}{a^2}\right) \\ &\quad - 2\frac{q_r q_\theta}{ra^2}\frac{\partial^2\Phi}{\partial r\,\partial\theta} - 2\frac{q_\theta q_z}{ra^2}\frac{\partial^2\Phi}{\partial\theta\,\partial z} - 2\frac{q_z q_r}{ra^2}\frac{\partial^2\Phi}{\partial z\,\partial r} + \frac{q_r}{r}\left(1 + \frac{q_\theta^2}{a^2}\right) = 0. \end{aligned}$$

Note that from (42) for $a^2 \to \infty$ we obtain the polar form of the Laplace equation, namely,

$$\frac{\partial^2\Phi}{\partial r^2} + \frac{1}{r^2}\frac{\partial^2\Phi}{\partial\theta^2} + \frac{\partial^2\Phi}{\partial z^2} + \frac{1}{r}\frac{\partial\Phi}{\partial r} = 0.$$

For $q_z = 0$ in Eq. (42) we find Eq. (33) again. Another case with only two independent variables is that of an *axial symmetry* where Φ depends only on z and r and not on θ, and Eq. (42) reduces to

$$\frac{\partial^2\Phi}{\partial r^2}\left(1 - \frac{q_r^2}{a^2}\right) + \frac{\partial^2\Phi}{\partial z^2}\left(1 - \frac{q_z^2}{a^2}\right) - 2\frac{q_r q_z}{a^2}\frac{\partial^2\Phi}{\partial r\,\partial z} + \frac{q_r}{r} = 0.$$

If here z and r are replaced by x and y respectively, so that the x-axis is the axis of revolution, this equation differs from Eq. (31) only by the term q_y/y. One may introduce the equation

$$(43)\qquad \frac{\partial^2\Phi}{\partial x^2}\left(1 - \frac{q_x^2}{a^2}\right) + \frac{\partial^2\Phi}{\partial y^2}\left(1 - \frac{q_y^2}{a^2}\right) - 2\frac{q_x q_y}{a^2}\frac{\partial^2\Phi}{\partial x\,\partial y} + \frac{\nu}{y}q_y = 0$$

where $\nu = 0$ for the case of plane flow, $\nu = 1$ for that of rotational symmetry.[21]

Next we consider nonsteady motions. We start with the equation for plane nonsteady motion derived from Eq. (14):

$$(44)\qquad \begin{aligned} &\frac{\partial^2\Phi}{\partial x^2}\left(1 - \frac{q_x^2}{a^2}\right) + \frac{\partial^2\Phi}{\partial y^2}\left(1 - \frac{q_y^2}{a^2}\right) - 2\frac{q_x q_y}{a^2}\frac{\partial^2\Phi}{\partial x\,\partial y} \\ &\qquad - \frac{1}{a^2}\frac{\partial^2\Phi}{\partial t^2} - 2\frac{q_x}{a^2}\frac{\partial^2\Phi}{\partial x\,\partial t} - 2\frac{q_y}{a^2}\frac{\partial^2\Phi}{\partial y\,\partial t} = 0. \end{aligned}$$

Again introducing polar coordinates [by means of Eqs. (32)] we obtain, if Φ is independent of θ:

$$\frac{\partial^2\Phi}{\partial r^2}\left(1 - \frac{q_r^2}{a^2}\right) - \frac{1}{a^2}\frac{\partial^2\Phi}{\partial t^2} - 2\frac{q_r}{a^2}\frac{\partial^2\Phi}{\partial r\,\partial t} + \frac{1}{r}q_r = 0,$$

the case of nonsteady flow with *cylindrical symmetry*. Also generalizing Eq. (19) to nonsteady flow or working directly from Eq. (14), we obtain for *spherical symmetry*:

$$\frac{\partial^2\Phi}{\partial r^2}\left(1 - \frac{q_r^{\,2}}{a^2}\right) - \frac{1}{a^2}\frac{\partial^2\Phi}{\partial t^2} - 2\frac{q_r}{a^2}\frac{\partial^2\Phi}{\partial r\,\partial t} + \frac{2}{r}\,q_r = 0.$$

Comparing these last two equations with Eq. (24), we see that we may write

$$\frac{\partial^2\Phi}{\partial r^2}\left(1 - \frac{q_r^{\,2}}{a^2}\right) - \frac{1}{a^2}\frac{\partial^2\Phi}{\partial t^2} - 2\frac{q_r}{a^2}\frac{\partial^2\Phi}{\partial r\,\partial t} + \frac{\nu}{r}\,q_r = 0, \tag{43'}$$

where $\nu = 0$, 1, or 2 stands for nonsteady parallel flow, nonsteady flow with cylindrical symmetry, or with spherical symmetry, respectively. Again there are only two independent variables, namely, r and t. We find from Eq. (16) that in each case

$$a^2 = -(\kappa - 1)\left(\frac{\partial\Phi}{\partial t} + \frac{q_r^{\,2}}{2}\right).$$

If in the sense of Art. 4 we replace a^2 by $a_0^{\,2}$, the sound velocity of the fluid at rest, omit the terms of higher order, and replace r by x, we obtain the generalized one-dimensional wave equation

$$\frac{\partial^2\Phi}{\partial t^2} - a_0^{\,2}\frac{\partial^2\Phi}{\partial x^2} - a_0^{\,2}\frac{\nu}{x}\frac{\partial\Phi}{\partial x} = 0, \tag{45}$$

with the above meaning of ν. The general solution for $\nu = 2$ is

$$\Phi = \frac{1}{x}\,[f_1(x - a_0t) + f_2(x + a_0t)],$$

where f_1 and f_2 are arbitrary functions of one variable; this may be compared to d'Alembert's solution for $\nu = 0$. The theory of cylindrical waves, $\nu = 1$, is more difficult than, and essentially different from, that for $\nu = 0$ or $\nu = 2$. This has an analogy in the general theory of one-, two-, and three-dimensional waves (see end of Art. 4, where we found that the second case differs essentially from the first and third ones).

Article 8

Steady Flow Relations

1. General relations among q, p, ρ, and T

When the flow of a compressible fluid is steady, i.e., independent of time, and a (p, ρ)-relation holds on each streamline, then each of the four quantities q, p, ρ, and T (velocity magnitude, pressure, density, and absolute temperature) can be expressed, on a given streamline, as a function of any single one of them. In fact, the equation of state, the (p, ρ)-relation, and the Bernoulli equation supply three relations among these four variables. If the same (p, ρ)-relation holds on all streamlines (elastic fluid) and if the flow is steady and irrotational (steady potential flow), then the relationships among q, p, ρ, and T are the same for the entire flow, as will be shown presently.

When gravity is neglected, the Bernoulli equation in differential form is (2.21′):

$$q\,dq + \frac{dp}{\rho} = 0, \tag{1}$$

where the differentials refer to changes in q and p along a streamline. On integrating, and introducing the *stagnation pressure* p_s, the value which p assumes at a point of the streamline where $q = 0$, we find

$$\frac{q^2}{2} + \int_{p_s}^{p} \frac{dp}{\rho} = 0, \qquad q = \sqrt{2\int_{p}^{p_s} \frac{dp}{\rho}}. \tag{2}$$

This equation gives q as a function of p for all points of the streamline, the function depending on the parameter p_s. Since p, ρ, and q^2 are nonnegative quantities it follows that $p \leqq p_s$ everywhere along the streamline, the maximum value being attainable only at stagnation points.

In the case of an elastic fluid we may consider the function P of p (Sec. 2.5) whose derivative is $1/\rho$, say $P = \int_{p_0}^{p} dp/\rho$ where p_0 is some reference pressure. Then P is a monotonically increasing function of p. In this case Eq. (2) may be written

$$\frac{q^2}{2} = P(p_s) - P(p), \tag{2′}$$

which gives again $p \leqq p_s$.

When gravity is neglected, the Bernoulli function H (Sec. 2.5) is defined by

$$\frac{q^2}{2} + P = gH. \tag{3}$$

For steady irrotational flow this quantity H was shown in Secs. 6.5 and 7.1 to be constant throughout the whole fluid mass. From Eq. (3), this constant value is also the value of P/g at all stagnation points; thus P, and consequently p, has the same value at all stagnation points. When p is known, the value of ρ may be determined by means of the (p, ρ)-relation, and the value of T follows from the values of p and ρ by means of the equation of state. Thus we conclude: *In the steady irrotational flow of an elastic fluid, the* *stagnation values p_s, ρ_s, T_s of p, ρ, T are the same on all streamlines.* Under these conditions, Eq. (2), giving q as a function of p, is also the same for all streamlines. We now study this relationship.

Whether the (p, ρ)-relation holds for a single streamline only or for the whole fluid mass, it is assumed that $p = 0$ corresponds to $\rho = 0$ and that p increases monotonically with ρ so that from $p \leqq p_s$ follows $\rho \leqq \rho_s$ on the streamline. We even assume strictly monotonical increase, i.e., $dp/d\rho = a^2$ does not vanish, except possibly at $\rho = 0$. Then a graph of p versus $1/\rho$ will have the form shown in the right half of Fig. 35. To the left, the integral of $1/\rho$ from p to p_s is plotted as the abscissa corresponding to the ordinate p; each curve corresponds to a particular value of the parameter p_s, which appears as the p-intercept of the curve. From Eq. (2) it follows that the horizontal axis to the left is also the (positive) axis of $q^2/2$. In Fig. 35 it has been assumed that the integral $\int^p dp/\rho$ converges for $\rho = 0$, which

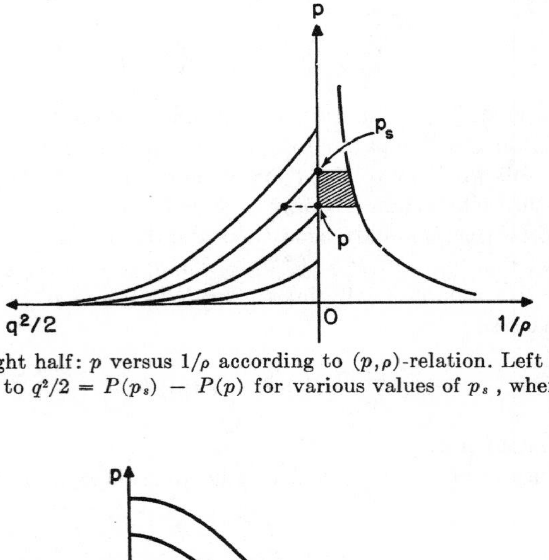

FIG. 35. Right half: p versus $1/\rho$ according to (p,ρ)-relation. Left half: p versus $q^2/2$ according to $q^2/2 = P(p_s) - P(p)$ for various values of p_s, where $P = \int^p dp/\rho$.

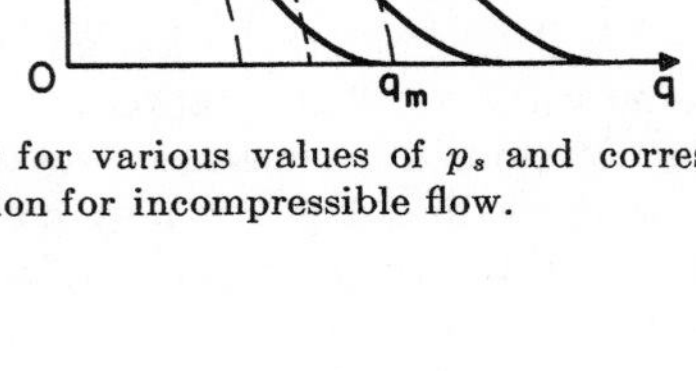

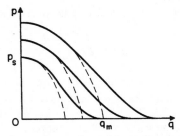

FIG. 36. p versus q for various values of p_s and corresponding values of q_m. Dotted line shows relation for incompressible flow.

implies in particular that $dp/d\rho = a^2$ vanishes for $\rho = 0$. With this assumption, the curves at the left meet the horizontal axis at finite values of $q^2/2$; in the contrary case, all curves would have this axis as an asymptote.

Considering q, rather than $q^2/2$, we obtain graphs of p versus q for various values of the parameter p_s (Fig. 36). From Eq. (1), we find

$$\frac{dp}{dq} = -\rho q; \tag{4}$$

thus each curve has a horizontal tangent at $q = 0$, $p = p_s$ and another at $p = 0$ ($\rho = 0$), with q finite or infinite. Each curve must therefore have an inflection point for some intermediate point (q, p).* Differentiating Eq. (4), and using $dp/d\rho = a^2$, we have

$$\frac{d^2p}{dq^2} = -\rho - q\frac{d\rho}{dq} = -\rho - q\frac{d\rho}{dp}\frac{dp}{dq} = -\rho + \rho\frac{q^2}{a^2}$$

or, in terms of the Mach number, $M = q/a$,

$$\frac{d^2p}{dq^2} = -\frac{d}{dq}(\rho q) = \rho(M^2 - 1). \tag{5}$$

The product ρq is the flux per unit area and can be called the *flow intensity*. Then the conclusions that can be drawn from Eq. (5) may be expressed by this statement: *The curve of p versus q has an inflection point when the Mach number equals* 1 *(at the sonic point); the flow intensity ρq increases with the velocity q in subsonic flow, reaches a maximum at the sonic point, and decreases as q increases in supersonic flow. To each value of the parameter p_s (or of ρ_s or of T_s) there corresponds a certain transition velocity q_t (abscissa of the inflection point) where $M = 1$, and a certain maximum velocity q_m where* $p = \rho = 0$. If the value of q_m is finite then the Macł number $M \to \infty$ as $q \to q_m$, since $a^2 \to 0$. For exceptional (p, ρ)-relations where the above mentioned $\int dp/\rho$, does not converge as $\rho \to 0$ and q_m is infinite (which then occurs for all values of the parameter), it can happen that M remains finite as $q \to \infty$.

In the case of an incompressible fluid, when $\rho = \rho_s$, we have $\int_{p_s}^{p} dp/\rho = (p - p_s)/\rho_s$, and the curves of p versus q are the parabolas

$$q^2 = 2\frac{p_s - p}{\rho_s}.$$

These curves for various values of p_s are the dotted lines in Fig. 36. It is clear from the figure that the compressible fluid behaves in the subsonic

* We shall assume that there is only one such inflection point. It can be shown that this will certainly be the case if $d^2p/d\rho^2 > 0$.

range, very much like an incompressible fluid while the flow has an entirely different character when the velocities are supersonic.

The above statement concerning flow intensity may be given another interpretation. In any infinitesimal stream tube the fact that mass is preserved and the flow is steady means that the flux (which equals ρq times the cross section) is the same across any cross section of the tube; thus the flow intensity ρq must be inversely proportional to the cross section of the tube. Then decreasing cross section corresponds to increasing q—and vice versa—in subsonic flow, exactly as in the incompressible case. For supersonic flow, however, increasing cross section corresponds to increasing velocity of flow: the minimum cross section corresponds to $M = 1$. This behavior is illustrated in the radial flow studied in Sec. 7.3: in the subsonic flow the velocity decreases from $q = q_t$ to $q = 0$, while the cross section of the channel increases; in the supersonic flow the velocity increases from q_t to q_m with increasing cross section.[22]

If an overall (p,ρ)-relation is given, each possible steady irrotational flow pattern is characterized, so far as the relationship between p and q (or ρ and q, or T and q) is concerned, by the value of a *single parameter*. Any one of the following may be used as parameter: the stagnation values p_s, ρ_s, or T_s; the value of the transition velocity q_t; the value of the maximum velocity q_m (provided this value is finite); or the Bernoulli constant H, which is equal to the stagnation value of P/g or to the square of q_m divided by $2g$, if in the definition of P the reference pressure is zero.

2. Hodograph representation[23]

In steady flow of any type there is a definite velocity vector $\mathbf{q}$, independent of t, for each point P. If these vectors are plotted with a fixed origin O', so that $\mathbf{q} = \overline{O'P'}$, then to each point P of the flow pattern there will correspond some point P' (see Fig. 37). Then, for example, to all points of one streamline there will correspond a curve formed by the respective points P'. This correspondence, or *mapping*, is known as the *hodograph transformation*. We shall also speak of the *hodograph space* in which the points P' lie, as contrasted with the *physical space* in which the flow actually occurs. Note that each point P is mapped into exactly one point P', although the converse is not true: one point P' can represent several points P of the physical space—namely, points with the same $\mathbf{q}$.

If a streamline $\mathcal{L}$ of physical space is mapped into a curve $\mathcal{L}'$ of the hodograph space, the tangent to $\mathcal{L}'$ at P' has the direction of the instantaneous rate of change of $\mathbf{q}$, i.e., the direction of the acceleration vector. Now, in the absence of gravity and viscosity, the acceleration vector has, by Eq. (1.I), the direction of $-\text{grad } p$; thus the *isobar* (surface of constant pressure p) through any point P of the physical space is perpendicular to the tangent to $\mathfrak{L}'$ at P'.

The results of Sec. 1 show that, in the case of a steady irrotational flow, all points P' must lie on or within the sphere with center O' and radius q_m (maximum velocity): all of physical space is mapped into the interior and boundary of this sphere in the hodograph space. For exceptional (p, ρ)-relations, this sphere may be infinite. All stagnation points (where $q = 0$ and $M = 0$) map into the center O'; points with $q = q_m$ and hence (in the nonexceptional case) $M = \infty$ correspond to points on the boundary of the sphere; and the sonic points of physical space (where $M = 1$) map onto the *sonic sphere* with center O' and radius q_t. The subsonic region of flow maps into the interior of the sonic sphere and the supersonic domain into the shell between the spherical surfaces $q = q_t$ and $q = q_m$.

The hodograph transformation is used extensively in the study of plane motion. In this case, all points P, as well as all points P', lie in a plane, and so we speak of the *physical plane* and of the *hodograph plane*. The spheres are now replaced by concentric circles C_t and C_m with center O' and radii q_t and q_m, respectively (Fig. 37); the subsonic part of the flow maps into the interior of the *sonic circle* C_t, the supersonic part into the annular region between C_t and the *maximum circle* C_m, stagnation points map into O', sonic points into points on the circle C_t, and points with $q = q_m$ into points of C_m.

Corresponding to each point P' within C_m, there is a value of the pressure p, determined by the distance $O'P' = q$ and Eq. (2). If these p-values

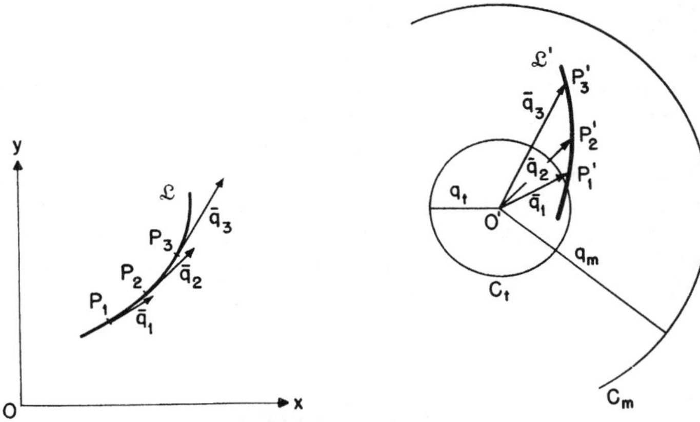

FIG. 37. Physical plane and hodograph plane. Streamline $\mathfrak{L}$ in physical plane and $\mathfrak{L}'$ in hodograph. Sonic circle and maximum circle.

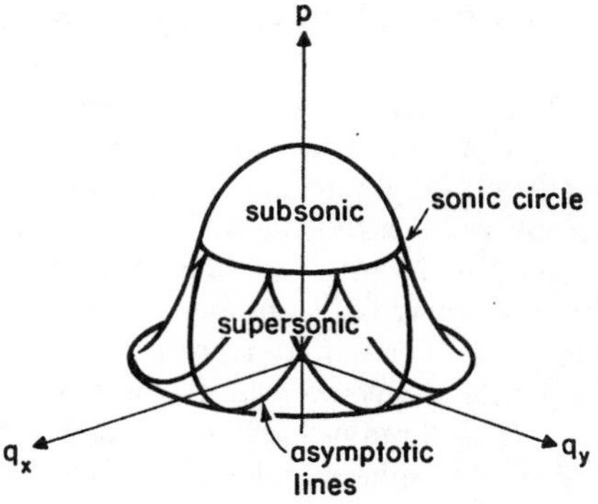

FIG. 38. Pressure hill with subsonic and supersonic region.

are plotted above the hodograph plane as points Q, such that the perpendicular distance to the plane is $P'Q = p$, then the points Q corresponding to a given streamline lie on the bell-shaped surface of revolution obtained by rotating one of the p,q-curves of Fig. 36 about the p-axis. In the case of an elastic fluid in irrotational motion the same surface holds for all streamlines. A sketch of this surface, sometimes called the *pressure hill*, is given in Fig. 38. Each position of the generating curve is called a meridian of the surface of revolution; the path of any point of the generating curve is a parallel. The surface is divided into an upper and a lower part by the parallel circle whose projection is C_t ; the upper part resembles an ellipsoid of revolution in the neighborhood of one vertex, while the lower portion has the character of a hyperboloid (of revolution) of one sheet.

In the terminology of differential geometry* the upper portion of the pressure hill consists of elliptic points, while the points below the critical circle are hyperbolic points. This means that if the surface is cut by a plane parallel to the tangent plane at Q and sufficiently close to it, the curve of intersection approximates an ellipse in the first case, and a hyperbola in the second. The axes of this conic (called Dupin's indicatrix), are parallel to the "principal directions" on the surface at Q; in the case of a surface of revolution these are the directions of the tangent to the meridian through Q and the tangent to the parallel circle. If coordinates in these two directions are called x and y respectively, the equation of the indicatrix has the form

$$\frac{x^2}{R_1} + \frac{y^2}{R_2} = \text{constant}. \tag{6}$$

* A reader who is unfamiliar with the elements of differential geometry may omit the remainder of this section.

Here, $|R_1|$ and $|R_2|$ are the radii of curvature of the two plane curves cut off on the surface by normal sections through the two principal directions at Q. The signs of R_1 and R_2 are to be taken the same if both centers of curvature lie on the same side of the tangent plane, and opposite in the contrary case.

For a surface of revolution, one principal section gives the meridian through Q. Thus $|R_1|$ is the radius of curvature at Q of the generating curve defined by Eq. (2). Using Eq. (5) and

$$(7) \qquad \frac{dp}{dq} = \tan\,(90° + \theta) = -\cot\theta,$$

where θ is the angle between the normal to the p,q-curve and the q-axis, we obtain from the formula for the radius of curvature

$$(8) \qquad \frac{1}{R_1} = \frac{\dfrac{d^2p}{dq^2}}{\left[1 + \left(\dfrac{dp}{dq}\right)^2\right]^{3/2}} = \rho \sin^3\theta\,(M^2 - 1)$$

and K_1 (Fig. 39) is the center of curvature. The second principal section is the normal section through the tangent to the parallel circle. The theorem of Meunier (proved in elementary differential geometry), when applied to this case, shows that the center of curvature K_2 of this section lies on the axis of revolution. Hence

$$(9) \qquad |R_2| = \frac{q}{\cos\theta},$$

since θ is also the angle between the horizontal plane of the parallel circle and the normal section, while q is the radius (of curvature) of the parallel circle.

For points Q on the lower portion of the pressure hill, R_1 and R_2 have opposite signs, as in Fig. 39a; also, θ is acute and $M^2 > 1$. Thus, with the use of Eqs. (8) and (9), Eq. (6) becomes

$$(6') \qquad x^2\rho\sin^3\theta\,(M^2 - 1) - \frac{y^2}{q}\cos\theta = \text{constant}.$$

The angle β between the x-axis and the asymptotes of the hyperbola (6′) is determined by

$$\tan^2\beta = \frac{q}{\cos\theta}\,\rho\sin^3\theta\,(M^2 - 1).$$

If the axes and asymptotes of (6′), all in the tangent plane at Q, are projected onto the horizontal plane (Fig. 39b), then Q projects onto P', the

x-axis (first principal direction) onto the radial direction $O'P'$, and the y-axis onto the (dotted) line normal to $O'P'$ at P'. The projections of the asymptotes make an angle β' with $O'P'$, where

$$\tan^2 \beta' = \frac{\tan^2 \beta}{\cos^2 (90^\circ - \theta)} = \rho q \tan \theta \, (M^2 - 1).$$

From Eqs. (4) and (7), $\rho q = \cot \theta$, and therefore $\tan^2 \beta' = M^2 - 1$. Now, the sine of the Mach angle α (Sec. 5.2) is $1/M$, so that

$$(10) \qquad \tan^2 \beta' = M^2 - 1 = \cot^2 \alpha, \qquad \beta' = 90^\circ - \alpha.$$

On a surface with hyperbolic points, the lines which at each point have the direction of the asymptotes of the indicatrix are called the asymptotic lines of the surface; these lines are indicated in Fig. 38. Equation (10) thus

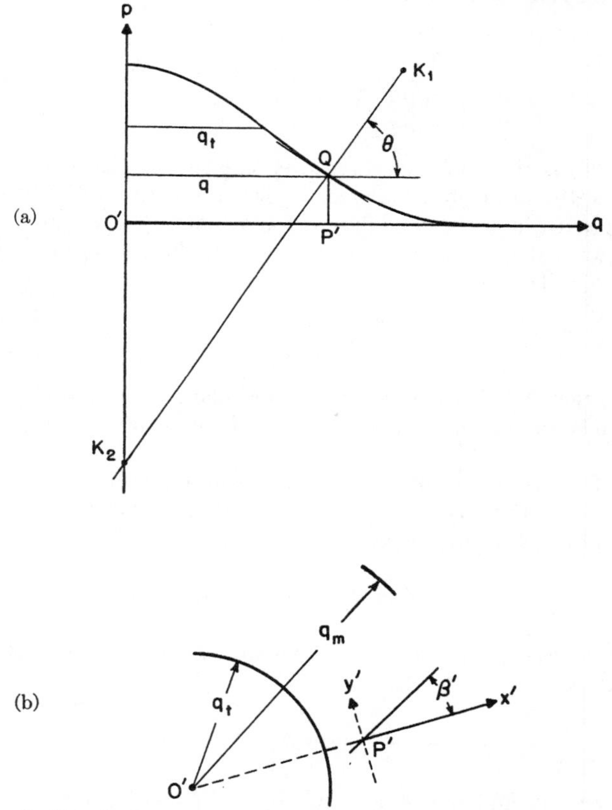

FIG. 39. Principal centers of curvature K_1 and K_2 for hyperbolic point. Projection of an asymptotic direction on pressure hill.

shows that the asymptotic lines on the pressure hill are projected onto the lines in the hodograph plane making the angle $\pm(90° - \alpha)$ with the direction of $\mathbf{q}$.

It will be shown later (Sec. 16.6) that, in any steady plane irrotational flow, the hodograph transformation maps the Mach lines (Sec. 5.2) of the physical plane into curves in the hodograph plane which meet the rays through O' at the angles $\pm(90° - \alpha)$. Thus, the result expressed by Eq. (10) can also be stated as follows: *The Mach lines of a steady plane irrotational flow of an elastic fluid are mapped into curves in the hodograph plane which are the projections of the asymptotic lines on the pressure hill.* This result was first given by L. Prandtl and A. Busemann.[24]

3. Case of polytropic (p,ρ)-relation

The relation between p and q (or between ρ and q or T and q) is particularly simple if the (p, ρ)-relation has the polytropic form

$$\frac{p}{\rho^\kappa} = \text{constant} = \frac{p_s}{\rho_s{}^\kappa}. \tag{11}$$

For given κ—and we shall usually take $\kappa = \gamma$, where γ is the adiabatic exponent (Sec. 1.5)—Eq. (11) depends on one parameter: the value of the constant. As seen in Sec. 1, the function p of q depends, once the (p, ρ)-relation is given, on the value of the parameter p_s. Therefore, the function p of q involves the *two independent parameters* p_s and ρ_s. Actually it turns out that, if the (p, ρ)-relation is given by Eq. (11), and in this case alone, these parameters appear as *scale factors* only, i.e., for the dimensionless variables p/p_s or ρ/ρ_s (rather than p or ρ) in terms of the Mach number M (rather than q), a *single function* is to be computed for each variable.

On carrying out the integration indicated in Eq. (2), or using the value of P from (2.22c) with $p_0 = p_s$, we find

$$q^2 = \frac{2\kappa}{\kappa - 1}\frac{p_s}{\rho_s}\left[1 - \left(\frac{p}{p_s}\right)^{(\kappa-1)/\kappa}\right]. \tag{12}$$

On the other hand, differentiation of Eq. (11) yields

$$\frac{dp}{d\rho} = a^2 = \kappa\frac{p_s}{\rho_s{}^\kappa}\rho^{\kappa-1} = \kappa\frac{p_s}{\rho_s}\left(\frac{p}{p_s}\right)^{(\kappa-1)/\kappa}. \tag{13}$$

Dividing (12) by (13), we get

$$M^2 = \frac{q^2}{a^2} = \frac{2}{\kappa - 1}\left[\left(\frac{p}{p_s}\right)^{-(\kappa-1)/\kappa} - 1\right] \tag{14}$$

or, solving for p/p_s in Eq. (14),

$$\frac{p}{p_s} = \left(\frac{\kappa - 1}{2}M^2 + 1\right)^{-\kappa/(\kappa-1)}. \tag{15}$$

From Eqs. (11) and (15) we obtain also

$$\text{(16)} \qquad \frac{\rho}{\rho_s} = \left(\frac{p}{p_s}\right)^{1/\kappa} = \left(\frac{\kappa - 1}{2} M^2 + 1\right)^{-1/(\kappa-1)},$$

and finally, from the equation of state (1.6), assuming the fluid to be a perfect gas,

$$\text{(17)} \qquad \frac{T}{T_s} = \frac{p\rho_s}{p_s\rho} = \left(\frac{\kappa - 1}{2} M^2 + 1\right)^{-1}.$$

Equations (15), (16), and (17) express the result mentioned at the end of the preceding paragraph.

In expressing the velocity q in dimensionless form, we take as scale factor the stagnation value a_s of the sound velocity; by means of Eqs. (13), (15), and (17), a_s is given by

$$\text{(18)} \qquad a_s^2 = \kappa \frac{p_s}{\rho_s} = a^2 \left(\frac{p}{p_s}\right)^{-(\kappa-1)/\kappa} = a^2 \frac{T_s}{T}.$$

We note that the equality between the first and last members of (18) is consistent with the result $a^2 = \kappa gRT$ following from the definitions of a^2 and of a perfect gas. Then Eqs. (18) and (15) give

$$\text{(19)} \qquad \frac{q^2}{a_s^2} = M^2 \frac{a^2}{a_s^2} = M^2 \left(\frac{p}{p_s}\right)^{(\kappa-1)/\kappa} = M^2 \left(\frac{\kappa - 1}{2} M^2 + 1\right)^{-1}.$$

Next, from Eqs. (19) and (16), the flow intensity ρq satisfies

$$\text{(20)} \qquad \frac{\rho q}{\rho_s a_s} = M \left(\frac{\kappa - 1}{2} M^2 + 1\right)^{-(\kappa+1)/2(\kappa-1)},$$

and the so-called *dynamic pressure* $\rho q^2/2$ is given by

$$\text{(21)} \qquad \frac{\rho q^2}{\rho_s a_s^2} = M^2 \left(\frac{\kappa - 1}{2} M^2 + 1\right)^{-\kappa/(\kappa-1)}.$$

If we denote the values of all quantities at a sonic point, $M = 1$, by the subscript t, Eqs. (15) to (17) and (19) to (21) give the relations

$$\text{(22)} \qquad p_t = p_s \left(\frac{\kappa + 1}{2}\right)^{-\kappa/(\kappa-1)}, \qquad \rho_t = \rho_s \left(\frac{\kappa + 1}{2}\right)^{-1/(\kappa-1)}, \qquad T_t = T_s \frac{2}{\kappa + 1}$$

and

$$\text{(23)} \qquad \begin{gathered} q_t^2 = a_s^2 \frac{2}{\kappa + 1}, \qquad \rho_t q_t = \rho_s a_s \left(\frac{\kappa + 1}{2}\right)^{-(\kappa+1)/2(\kappa-1)}, \\ \rho_t q_t^2 = \rho_s a_s^2 \left(\frac{\kappa + 1}{2}\right)^{-\kappa/(\kappa-1)}. \end{gathered}$$

It was seen in Sec. 1 that $\rho_t q_t$ is the maximum value of ρq.

The maximum velocity q_m, corresponding to $p = 0$, may be determined from Eqs. (12) and (18):

$$q_m^2 = \frac{2\kappa}{\kappa - 1}\frac{p_s}{\rho_s} = \frac{2}{\kappa - 1}a_s^2. \tag{24}$$

For all $\kappa > 1$ the maximum value q_m is finite. All the preceding equations hold also in the isothermal case, $\kappa = 1$, if limits are taken as $\kappa \to 1$; in particular, q_m is then infinite for all values of the parameters p_s and ρ_s.

Combining Eq. (24) and the first equation (23), we derive the important relation

$$q_m^2 = \frac{\kappa + 1}{\kappa - 1} q_t^2. \tag{25}$$

It can be verified, from Eqs. (15) or (19), that for $p = 0$, $q = q_m$, the Mach number is infinite if $\kappa > 1$. From (17) the corresponding temperature T_m is seen to be zero, from which it is clear that the limit $q = q_m$ can never be reached in an actual flow.

It is to be emphasized, again, that all formulas derived in this section are valid for:

(a) Steady flow along a single streamline (stream tube) if Eq. (11) holds at all points of the streamline;

(b) Steady *irrotational* flow in one, two, or three dimensions, if Eq. (11) holds throughout the fluid.

4. Adiabatic (irrotational) airflow

Dry air can be considered as approximating to a diatomic perfect gas, for which the theoretical value of the adiabatic exponent $\gamma = c_p/c_v$ is $7/5 = 1.4$. Experiments lead to a slightly higher value, not above 1.405. The choice of one of these values rather than the other makes little difference in the results obtained, in view of the fact that the whole theory is approximate. For example, the ratio of the pressures at sonic and stagnation points, given by (22), is

$$\frac{p_t}{p_s} = \left(\frac{\gamma + 1}{2}\right)^{-\gamma/(\gamma-1)} \begin{array}{ll} = 0.5283 & \text{for } \gamma = 1.400, \\ = 0.5274 & \text{for } \gamma = 1.405. \end{array}$$

For the sake of simplicity, all numerical data in this book concerning adiabatic airflow will be based on the assumption that the value of κ in the preceding formulas is 1.4. In particular, this means that the ratio of the areas of the two circles C_m and C_t in the hodograph plane, using Eq. (25), is $(1.4 + 1)/(1.4 - 1) = 6$, while the corresponding velocity ratios, from Eqs. (25) and (24), are

$$\frac{q_m}{q_t} = \sqrt{6} = 2.45, \qquad \frac{q_m}{a_s} = \sqrt{5} = 2.24, \qquad \frac{q_t}{a_s} = \sqrt{\frac{5}{6}} = 0.91. \tag{26}$$

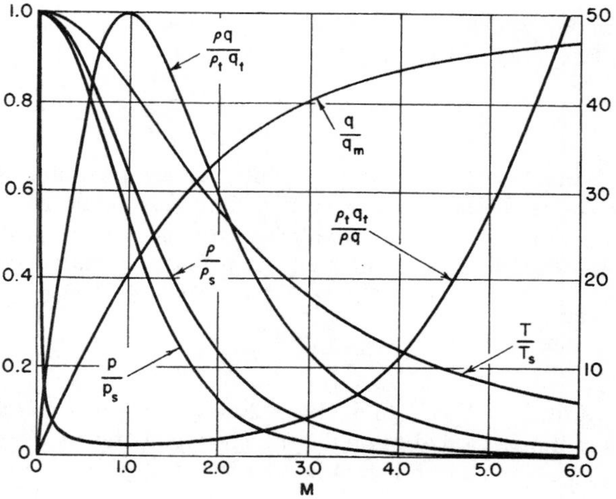

FIG. 40. $\frac{p}{p_s}$, $\frac{\rho}{\rho_s}$, $\frac{T}{T_s}$, $\frac{q}{q_m}$, $\frac{\rho q}{\rho_t q_t}$, $\frac{\rho_t q_t}{\rho q}$ versus Mach number. Use scale at right for last curve.

In Fig. 40, the ratios p/p_s, ρ/ρ_s, T/T_s, and q/q_m are plotted against the Mach number M, with the use of Eqs. (15) to (17) and (19). A further curve represents the ratio $\rho_t q_t/\rho q$, which may be computed from Eq. (20) and the second equation (23) as

$$\frac{\rho_t q_t}{\rho q} = \frac{1}{M}\left(\frac{\kappa - 1}{\kappa + 1} M^2 + \frac{2}{\kappa + 1}\right)^{(\kappa+1)/2(\kappa-1)}. \tag{27}$$

Since cross sectional area is inversely proportional to flow intensity, the ratio (27) is also that of the cross section at any point of the stream tube to the cross section at a sonic point, i.e., to the minimum cross section of the tube. The remaining curve shows the flow intensity $\rho q/\rho_t q_t$, which has its maximum value at $M = 1$. Numerical values for the first five functions are given in the Table on the following page (Table I).

The (p,q)-relation of Fig. 36 may now be graphed exactly, since the (p,ρ)-relation is given explicitly in Eq. (11), and this curve is given in Fig. 41. The formula expressing this relation is[25]

$$\frac{p}{p_s} = \left(1 - \frac{\kappa - 1}{2}\frac{q^2}{a_s^2}\right)^{\kappa/(\kappa-1)}. \tag{28}$$

This formula was obtained previously, in the particular case of purely radial flow, as Eq. (7.23′). Except for scale factors, this curve represents the meridian of the pressure hill for all values of the parameters p_s and ρ_s.[26]

TABLE I

DEPENDENCE OF FLOW VARIABLES ON MACH NUMBER M FOR ADIABATIC AIRFLOW

M	p/p_s	ρ/ρ_s	T/T_s	q/q_m	$\rho_t q_t/\rho q$
0.00	1.0	1.0	1.0	0.0	∞
0.01	0.99993	0.99995	0.99998	0.00447	57.874
0.02	0.99972	0.99980	0.99992	0.00894	28.942
0.03	0.99937	0.99955	0.99982	0.01342	19.300
0.04	0.99888	0.99920	0.99968	0.01789	14.481
0.05	0.99825	0.99875	0.99950	0.02236	11.591
0.1	0.99303	0.99502	0.99800	0.04468	5.8218
0.2	0.97250	0.98028	0.99206	0.08909	2.9635
0.4	0.89561	0.92427	0.96899	0.17609	1.5901
0.6	0.78400	0.84045	0.93284	0.25916	1.1882
0.8	0.65602	0.73999	0.88652	0.33686	1.0382
1.0	0.52828	0.63394	0.83333	0.40825	1.0
1.2	0.41238	0.53114	0.77640	0.47287	1.0304
1.5	0.27240	0.39498	0.68966	0.55709	1.1762
2.0	0.12780	0.23005	0.55556	0.66667	1.6875
3.0	0.02722	0.07623	0.35714	0.80178	4.2346
4.0	0.00659	0.02766	0.23810	0.87287	10.719
6.0	0.000633	0.00519	0.12195	0.93704	53.180
8.0	0.000102	0.00141	0.07246	0.96309	190.11
10.0	0.0000236	0.000493	0.04762	0.97590	535.94
15.0	—	—	0.02173	0.98907	3755.2
20.0	—	—	0.01235	0.99381	15377
25.0	—	—	0.00794	0.99602	46305

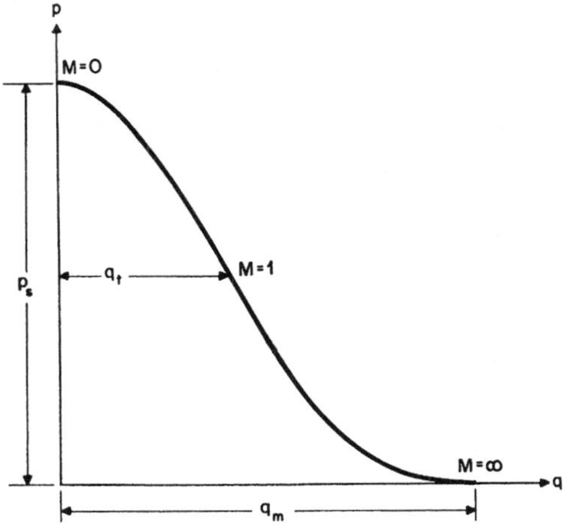

FIG. 41. p versus q for adiabatic airflow.

Article 9

Theory of Characteristics

1. Introduction

In order to explain the concept of *characteristics of partial differential equations*, to the extent that is needed in the theory of compressible fluid flow, we start with a preliminary examination of the relatively simple case of steady potential flow in two dimensions. It has been seen (Sec. 7.5) that in this case the potential function $\Phi(x,y)$ must satisfy the second-order partial differential equation

$$(1) \qquad \left(1 - \frac{q_x^2}{a^2}\right)\frac{\partial^2\Phi}{\partial x^2} - 2\,\frac{q_xq_y}{a^2}\,\frac{\partial^2\Phi}{\partial x\,\partial y} + \left(1 - \frac{q_y^2}{a^2}\right)\frac{\partial^2\Phi}{\partial y^2} = 0.$$

Here q_x and q_y are the first partial derivatives of Φ with respect to x and y respectively, and a^2 the square of the sound velocity; a^2 can also be expressed in terms of first-order derivatives of Φ, e.g., by Eq. (7.17) in the case of a polytropic (p,ρ)-relation. Equation (1) therefore falls within the general class of equations of the form

$$(2) \qquad A\,\frac{\partial^2\Phi}{\partial x^2} + 2B\,\frac{\partial^2\Phi}{\partial x\,\partial y} + C\,\frac{\partial^2\Phi}{\partial y^2} = F,$$

where A, B, C, and F are functions of Φ and its first-order derivatives, and possibly also of x and y.[27] Now we ask: Of what significance is the fact that Φ satisfies an equation of type (2)?

Suppose that the values of Φ and $\partial\Phi/\partial x$ are known for all points of a certain straight line parallel to the y-axis, say $x = x_0$. These values then determine for all points of the line $x = x_0$, all derivatives with respect to y of both these quantities, in particular $\partial\Phi/\partial y$, $\partial^2\Phi/\partial x\,\partial y$, and $\partial^2\Phi/\partial y^2$. The only second-order derivative of Φ not determined on $x = x_0$ by the given values is $\partial^2\Phi/\partial x^2$, and this derivative can be computed from Eq. (2), in the form

$$(2') \qquad \frac{\partial^2\Phi}{\partial x^2} = -\frac{2B}{A}\,\frac{\partial^2\Phi}{\partial x\,\partial y} - \frac{C}{A}\,\frac{\partial^2\Phi}{\partial y^2} + \frac{F}{A},$$

whenever A is different from zero. The given values also determine Φ, except for terms of higher order, on any nearby parallel line, say $x = x_1 = x_0 + dx$:

$$(3) \qquad \Phi(x_1,y) = \Phi(x_0,y) + \frac{\partial\Phi}{\partial x}(x_0,y)\,dx\;;$$

and we can even obtain

$$\frac{\partial \Phi}{\partial x}(x_1, y) = \frac{\partial \Phi}{\partial x}(x_0, y) + \frac{\partial^2 \Phi}{\partial x^2}(x_0, y)\, dx,$$

provided the coefficient of dx can be determined by Eq. (2′). Thus, if A does not vanish anywhere on $x = x_0$, the data can be extended to the line $x = x_1$. Then one can proceed to compute the values of the derivatives with respect to y on $x = x_1$ and, if A does not vanish on $x = x_1$, the values of $\partial^2\Phi/\partial x^2$ on $x = x_1$ and the values of Φ and $\partial\Phi/\partial x$ on $x = x_2 = x_1 + dx$. Thus, so long as A is different from zero, one can carry out an approximate integration of the differential equation in the half-plane $x \geqq x_0$ (and, of course, in the same way in the other half-plane), and the smaller the increment dx the better the approximation. It may be noted that the transition from (x_0, y_0) to $(x_0 + dx, y_0)$ does not require a knowledge of the values on the entire line $x = x_0$, but only on a small interval of this line containing the point (x_0, y_0), or even on an arc of some curve tangent to $x = x_0$ at (x_0, y_0).

For the particular case of Eq. (1) we have

$$A = 1 - \frac{q_x^2}{a^2}. \tag{4}$$

Now, if the motion under consideration is subsonic ($q < a$), then *a fortiori* $q_x^2/a^2 < 1$ and A remains positive everywhere. In this case no difficulty can arise in extending the solution. The situation is different in a region of supersonic flow ($q > a$). In this case it can happen that q_x, the x-component of $\mathbf{q}$, has exactly the value of the local sound velocity a. At such a point P, the angle α between the y-direction, and the velocity vector $\mathbf{q}$ (Fig. 42) satisfies

$$\sin \alpha = \frac{q_x}{q} = \frac{a}{q} = \frac{1}{M}; \tag{5}$$

thus α is the Mach angle defined in Sec. 5.2, and the y-direction coincides

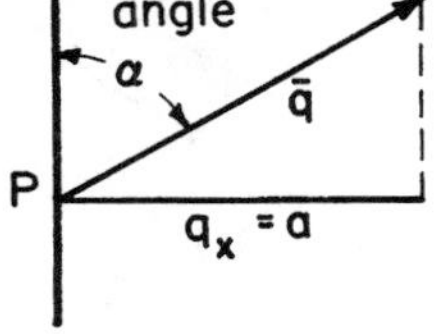

FIG. 42. y-direction as characteristic direction.

with one of the Mach directions at P. It appears that the integration process cannot be extended across an element dy if the angle between dy and $\mathbf{q}$ is the Mach angle.

Since the x- and y-directions in (1) are entirely arbitrary, this means that the process of step-by-step integration fails whenever one must cross a line element which makes the angle α with the velocity vector, i.e., whenever the component q_ν normal to the line element is a. The curves which cross streamlines at the angle α have been called *Mach lines* (Sec. 5.4); through each point P pass two Mach lines C^+ and C^- (Fig. 43), with $q_\nu = a$ along each line. The significance of the Mach lines in connection with the differential equation (1) may now be seen. Even if Φ is known everywhere in the region R between C^+ and C^-, the differential equation leads to no information about the values of Φ on the other sides of the Mach lines. If a solution of Eq. (1) can be found in the region R' with the same values of $\partial\Phi/\partial x$ and $\partial\Phi/\partial y$ on C^+ as the solution in R, then the two solutions combined determine a flow pattern in $R + R'$ with continuous velocity (and pressure) values, which satisfies the differential equation everywhere, even on the

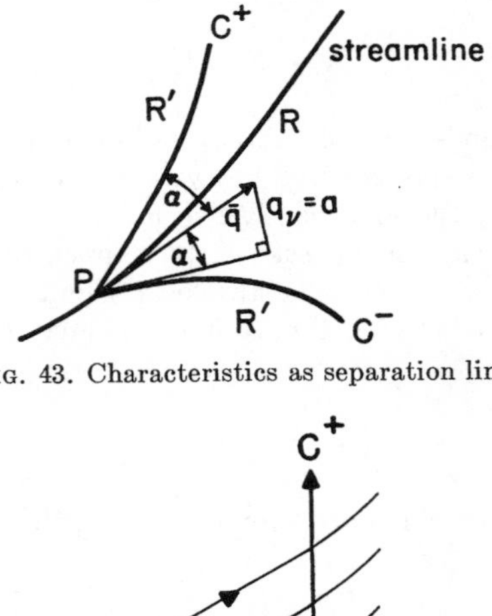

FIG. 43. Characteristics as separation lines.

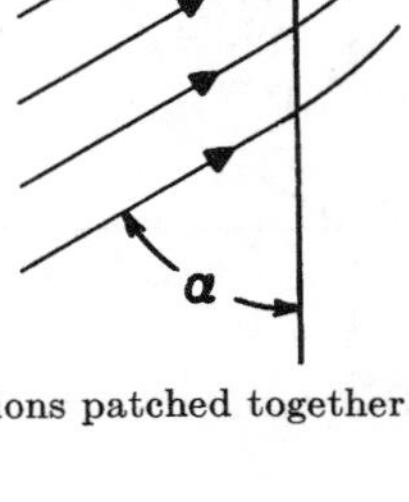

FIG. 44. Two different solutions patched together along straight characteristic.

boundary line C^+. (If, e.g., the line element on C^+ is dy, then the possibly discontinuous quantity $\partial^2\Phi/\partial x^2$ occurs with coefficient $A = 0$.) These two solutions can be quite different *analytically*, in the sense that there is no Taylor series which converges to the combined solution in any region which is partly in R and partly in R'. For example, it can happen that a uniform parallel flow changes into a curvilinear flow along a Mach line (which in this case must be a straight line, as shown in Fig. 44). Such a change is not possible in subsonic flow; there the solutions are everywhere analytic, and there are no curves beyond which a solution cannot be extended.

Curves (or surfaces, in the case of more dimensions) *along which analytically different solutions of a differential equation or a system of such equations can be patched together* (with certain restrictions following from the equations themselves) *are called characteristics of the equation or system.* Insofar as the preceding argument is accepted as rigorous, it has been shown that the Mach lines are characteristics of the two-dimensional potential equation (1).

2. General theory[28]

In most branches of physics the mathematical problem can be formulated as follows: A set of k unknowns $u_1, u_2, \cdots, u_k$, which are functions of n independent variables $x_1, x_2, \cdots, x_n$, is subject to k first-order differential equations which are linear in the derivatives $\partial u_\kappa/\partial x_\nu$. Since the coefficients may depend on the unknowns u_κ, as well as on the independent variables, the differential equations are, in general, nonlinear. We shall use the term *planar* to indicate that they are nevertheless linear in the derivatives. In the case of fluid flow, there are $k = 5$ unknowns: q_x, q_y, q_z, p, and ρ; and $n = 4$ independent variables: x, y, z, and t; the system of equations consists of Newton's equation (I), the equation of continuity (II), and the specifying equation (III).

An expression of the form

$$a_1 \frac{\partial u}{\partial x_1} + a_2 \frac{\partial u}{\partial x_2} + \cdots + a_n \frac{\partial u}{\partial x_n}$$

can always be considered as the (n-dimensional) scalar product of a vector $\mathbf{a}$, with components $a_1, \cdots, a_n$, and the vector grad u, which also has n components $\partial u/\partial x_\nu$. Since there are k unknowns u_κ and k equations, it requires k^2 n-dimensional vectors $\mathbf{a}_{\iota\kappa}$, where both ι and κ run from 1 to k, to express the system of planar differential equations in the form

$$\sum_{\kappa=1}^{k} \mathbf{a}_{\iota\kappa}\cdot\operatorname{grad} u_\kappa = b_\iota \qquad (\iota = 1, 2, \cdots, k). \tag{6}$$

Writing the system in vector form has the advantage of its being independent of the choice of coordinate axes. In discussing any particular physical fact in connection with Eqs. (6), one may choose the coordinate axes so as to make the components of the vectors describing the situation as simple as possible.[27]

The k gradients have nk components. Suppose that for some choice of the axes, the components of the gradients are known in $n - 1$ of the coordinate directions at P, i.e., $(n - 1)k$ values are given arbitrarily; then the k equations (6) serve, in general, to determine the remaining k components, in the direction normal to the "plane" of the $n - 1$ given directions at P. In the case $n = 3$, there are given $2k$ gradient components, parallel to some plane (a plane in the usual sense of the word), and Eqs. (6) serve to compute the remaining k components. It can happen, however, that for some plane E and $(n - 1)k$ given components of grad u_κ parallel to E, Eqs. (6) fail to determine the remaining k components. A necessary and sufficient condition for this is that the determinant of coefficients of these k components in Eqs. (6) vanishes or, what is equivalent, that there exists a linear combination of the Eqs. (6) that does not contain any of these components. A plane E for which this happens will be termed *exceptional* for the system (6) at P. Our problem is to find at each point which planes, if any, are exceptional, or, equivalently, to find the direction of the normal $\boldsymbol{\lambda}$ to any such exceptional plane.

We consider the linear combination of the equations of the system (6) obtained by multiplying by factors $\alpha_1, \alpha_2, \cdots, \alpha_k$ respectively and summing:

$$\sum_{\iota=1}^{k} \alpha_\iota \sum_{\kappa=1}^{k} \mathbf{a}_{\iota\kappa}\cdot\text{grad } u_\kappa = \sum_{\iota=1}^{k} \alpha_\iota b_\iota, \tag{7}$$

or

$$\sum_{\kappa=1}^{k} \mathbf{A}_\kappa\cdot\text{grad } u_\kappa = B, \tag{7'}$$

where

$$\begin{aligned} \mathbf{A}_\kappa &= \sum_{\iota=1}^{k} \alpha_\iota \mathbf{a}_{\iota\kappa} \qquad (\kappa = 1, \cdots, k), \\ B &= \sum_{\iota=1}^{k} \alpha_\iota b_\iota . \end{aligned} \tag{8}$$

Equation (7′) is a necessary consequence of (6) and must be satisfied by solutions u_κ of (6) for arbitrary choices of the factors α_ι. The left-hand member of Eq. (7′) is a sum of scalar products, each of the form $\mathbf{A}_\kappa\cdot$ grad u_κ, with $\mathbf{A}_\kappa$ a linear combination of the vectors $\mathbf{a}_{\iota\kappa}$, $\iota = 1, \cdots, k$; each scalar product is also equal to the product of A_κ by the component of grad u_κ in the direction of $\mathbf{A}_\kappa$. Suppose that for some choice of the factors α_ι (not all zero) the vectors $\mathbf{A}_\kappa$ are all parallel to a common plane E; then this plane is surely exceptional since there exists a linear combination of Eqs. (6) that does not contain gradient components normal to E. For this same reason the $(n - 1)k$ gradient components parallel to E are not arbitrary since there exists a linear relation between them. Now if the vectors $\mathbf{A}_\kappa$ are parallel to some plane E and if $\boldsymbol{\lambda}$ is the normal to E, then $\boldsymbol{\lambda}$ satisfies

$$(9) \qquad \boldsymbol{\lambda}\cdot\mathbf{A}_1 = 0, \qquad \boldsymbol{\lambda}\cdot\mathbf{A}_2 = 0, \qquad \cdots, \qquad \boldsymbol{\lambda}\cdot\mathbf{A}_k = 0.$$

When the $\mathbf{A}_\kappa$ are replaced by the expressions from (8), we obtain from Eqs. (9) k linear homogeneous equations to be satisfied by the k factors α_ι

$$(9') \qquad \begin{aligned} \alpha_1(\boldsymbol{\lambda}\cdot\mathbf{a}_{11}) + \alpha_2(\boldsymbol{\lambda}\cdot\mathbf{a}_{21}) + \cdots + \alpha_k(\boldsymbol{\lambda}\cdot\mathbf{a}_{k1}) &= 0, \\ \alpha_1(\boldsymbol{\lambda}\cdot\mathbf{a}_{12}) + \alpha_2(\boldsymbol{\lambda}\cdot\mathbf{a}_{22}) + \cdots + \alpha_k(\boldsymbol{\lambda}\cdot\mathbf{a}_{k2}) &= 0, \\ \cdot \quad \cdot \quad \cdot \quad \cdot \quad \cdot \quad & \cdot \\ \alpha_1(\boldsymbol{\lambda}\cdot\mathbf{a}_{1k}) + \alpha_2(\boldsymbol{\lambda}\cdot\mathbf{a}_{2k}) + \cdots + \alpha_k(\boldsymbol{\lambda}\cdot\mathbf{a}_{kk}) &= 0. \end{aligned}$$

A nontrivial solution $\alpha_1, \cdots, \alpha_k$ of these equations exists if the determinant of coefficients vanishes, i.e., if

$$(10) \qquad \begin{vmatrix} \boldsymbol{\lambda}\cdot\mathbf{a}_{11} & \boldsymbol{\lambda}\cdot\mathbf{a}_{12} & \cdots & \boldsymbol{\lambda}\cdot\mathbf{a}_{1k} \\ \boldsymbol{\lambda}\cdot\mathbf{a}_{21} & \boldsymbol{\lambda}\cdot\mathbf{a}_{22} & \cdots & \boldsymbol{\lambda}\cdot\mathbf{a}_{2k} \\ \cdot & \cdot & \cdot & \cdot \\ \boldsymbol{\lambda}\cdot\mathbf{a}_{k1} & \boldsymbol{\lambda}\cdot\mathbf{a}_{k2} & \cdots & \boldsymbol{\lambda}\cdot\mathbf{a}_{kk} \end{vmatrix} = 0$$

Eq. (10) is therefore a *sufficient* condition that $\boldsymbol{\lambda}$ be normal to an exceptional plane; by reversing the steps in the proof it is easily verified that Eq. (10) is also a *necessary* condition on $\boldsymbol{\lambda}$.

To write the determinant (10) one has simply to arrange the original k equations (6) in such a way that the terms involving u_1 form the first column, the terms with u_2 the second, etc., and to replace each x_1-derivative by λ_1, each x_2-derivative by λ_2, etc. If, using the so-called summation convention, we write the Eqs. (6) as

$$(6') \qquad a_{\iota\kappa\mu}\frac{\partial u_\kappa}{\partial x_\mu} = b_\iota,$$

the determinant equation (10) reads

$$(10') \qquad \| a_{\iota\kappa\mu}\lambda_\mu \| = 0 \qquad \begin{pmatrix} \mu = 1, \cdots, n \\ \iota, \kappa = 1, \cdots, k \end{pmatrix}.$$

The fact that condition (10) is *necessary*, can also be seen as follows. We will show that if $\boldsymbol{\lambda}$ does not satisfy Eq. (10), then $\boldsymbol{\lambda}$ cannot be the normal to an exceptional plane. Thus suppose λ is such that the determinant in (10) does not vanish, and that λ has length 1. Take the x_1-axis parallel to λ; then λ has components $(1, 0, \cdots, 0)$ and the elements of the determinant in (10) are the x_1-components of the k^2 vectors $\mathbf{a}_{\iota\kappa}$. If the components of the k gradients are given in the plane normal to λ, the unknowns in Eqs. (6) are the x_1-components of the gradients, viz., $\partial u_1/\partial x_1, \partial u_2/\partial x_1, \cdots, \partial u_k/\partial x_1$. Then (6) is a system of k linear equations for the k unknowns $\partial u_i/\partial x_1$ whose determinant consists of the x_1-components of the vectors $\mathbf{a}_{\iota\kappa}$ and does not vanish, by hypothesis. Thus it is possible to compute the unknown components from the given ones, which means, we recall, that the plane normal to λ is not exceptional.

If Eq. (10) is expressed in terms of the components of λ, the result is a homogeneous algebraic equation of degree k in these components, with coefficients depending on the coefficients of the given equations at P. Thus the endpoints of the vectors λ satisfying condition (10) lie on a "cone" of order k in n-space with vertex at P. The cone need not be real and may be degenerate, but in general Eq. (10) defines a cone of directions at P. Each plane through P normal to a direction λ is exceptional, and any (real) surface tangent to such a plane at each of its points is a characteristic surface, according to the definition given at the end of Sec. 1.[30]

3. Compatibility relations

We have not quite completed the basic theory. Equation (10) defines the exceptional directions λ^*. When such a direction $\lambda = \lambda^*$ has been found, we have still to determine the previously mentioned combinations (7′) of the original equations, where the α_ι^* are such that all $\mathbf{A}_\kappa^*$ are parallel to one and the same exceptional "plane" E^* which is perpendicular to λ^*. In other words, once a λ^* with components $\lambda_1^*, \lambda_2^*, \cdots, \lambda_n^*$ has been found as a solution of (10)—a generator of the cone mentioned above—we must substitute these $\lambda_1^*, \lambda_2^*, \cdots, \lambda_n^*$ in (9′) and determine a corresponding set of multipliers $\alpha_1^*, \alpha_2^*, \cdots, \alpha_k^*$; the set of vectors $\mathbf{A}_1^*, \mathbf{A}_2^*, \cdots, \mathbf{A}_k^*$ defined by (8) and formed with these α_ι^* is then parallel to the exceptional plane E^*, and substitution of these $\mathbf{A}_\kappa^*$ into (7′) yields the desired combination:

$$(11) \qquad \sum_{\kappa=1}^{k} \mathbf{A}_\kappa^* \text{ grad } u_\kappa = \sum_{\iota=1}^{k} \alpha_\iota^* b_\iota = B^*.$$

On the left side of (11) there are only differentiations in directions $\mathbf{A}_\kappa^*$, i.e. *parallel to* E^* and perpendicular to λ^*. We call (11) a *compatibility relation*. This name is justified by the fact that (11) restricts the arbitrariness of the derivatives of the u_κ "along E^*", or parallel to E^*. In the particular case $n = 2$, an exceptional plane E is simply a line of characteristic direction (since $n - 1 = 1$) and differentiation along E becomes the well-known directional differentiation. Corresponding to one λ^* there may be more than one compatibility relation, i.e., combinations of the original equations which do not contain any differentiation in the λ^*-direction. (See Sec. 6.)

Thus we can state that with respect to any λ^* the original set of k equations can be so combined as to split into two parts: one part, consisting of r equations ($k - r$ being the rank of the coefficient matrix in (9′) for $\lambda = \lambda^*$), includes derivatives of the u_κ only in directions perpendicular to λ^*. These equations are the compatibility relations. Each of the additional $(k - r)$ equations, forming the other part, includes at least one derivative of an unknown in the λ^*-direction.[31]

These rather brief indications must suffice. They will become more concrete when we consider various examples in the next sections and, particularly, the case of general fluid motion where $n = 4$, $k = 5$ in Sec. 6. In Art. 10 the case $n = k = 2$ will be dealt with in detail, and we shall see that even in this case the general and detailed discussion of the compatibility relations (Sec. 10.2) is not too simple. In Chapters III, IV, and V we shall deal with the problems of nonsteady parallel flow and steady plane flow respectively. In both cases the basic equations present comparatively simple examples of the Eqs. (10.1). As far as the compatibility relations are concerned, the main simplification consists in the fact that the right sides of the basic equations are zero.

4. First examples

Suppose that a function $\Phi(x,y)$ is subject to a *second-order* differential equation of the form (2) where A, B, C, and F depend on x, y, $\partial\Phi/\partial x$, $\partial\Phi/\partial y$, but not explicitly on Φ. To express this problem in the form of system (6), we take the first derivatives of Φ as unknowns:

$$\frac{\partial \Phi}{\partial x} = u_1\,, \qquad \frac{\partial \Phi}{\partial y} = u_2\,.$$

Then Eq. (2) and the condition that u_1 and u_2 are derivatives of the same function may be written as

$$(12)\qquad \begin{gathered} A\,\frac{\partial u_1}{\partial x} + B\left(\frac{\partial u_1}{\partial y} + \frac{\partial u_2}{\partial x}\right) + C\,\frac{\partial u_2}{\partial y} = F, \\ \frac{\partial u_1}{\partial y} - \frac{\partial u_2}{\partial x} = 0.^* \end{gathered}$$

* If the coefficients A, B, C, and F also depend on Φ, a third equation, $\partial\Phi/\partial x = u_1$, is added to Eqs. (12) so as to form a system of three planar equations for u_1, u_2, and Φ. The characteristics are still given by (13).

Here $k = n = 2$. When this system is put in the form (6), the four vectors $\mathbf{a}_{\iota\kappa}$ are

$$\mathbf{a}_{11} : \ (A,B) \qquad \mathbf{a}_{12} : \ (B,C)$$

$$\mathbf{a}_{21} : \ (0,1) \qquad \mathbf{a}_{22} : \ (-1,0).$$

Thus, condition (10), determining $\boldsymbol{\lambda}$, becomes

$$\begin{vmatrix} A\lambda_1 + B\lambda_2 & B\lambda_1 + C\lambda_2 \\ \lambda_2 & -\lambda_1 \end{vmatrix} = -[A\lambda_1^2 + 2B\lambda_1\lambda_2 + C\lambda_2^2] = 0. \tag{13}$$

In this two-dimensional case, the "cone" of directions at P consists of two directions $\boldsymbol{\lambda}$ in the x,y-plane with components satisfying Eq. (13); the "exceptional planes" are straight lines each normal to a $\boldsymbol{\lambda}$ and in the x,y-plane and the "surface elements" normal to these directions $\boldsymbol{\lambda}$ are elements of arc. Now λ_2/λ_1, determined from (13), gives the slopes of the two $\boldsymbol{\lambda}$-directions; hence the slopes of the two *characteristic directions* are given by

$$-\frac{\lambda_1}{\lambda_2} = \frac{1}{A}[B \pm \sqrt{B^2 - AC}]. \tag{13$'$}$$

For the particular case of steady potential flow, governed by Eq. (1), the coefficients A, B, and C may be given explicitly. Choosing the x-axis parallel to $\mathbf{q}$ at P, we have $A = 1 - M^2$, $B = 0$, $C = 1$; then

$$-\frac{\lambda_1}{\lambda_2} = \pm\frac{1}{\sqrt{M^2 - 1}} = \pm \tan\alpha,$$

where α is the Mach angle. Thus, if $M < 1$, no real characteristic directions exist; for $M \geqq 1$, the characteristic directions are the Mach directions at P, so that *the characteristic lines turn out to be the Mach lines*. Moreover, since there are no other solutions of Eq. (13), the Mach lines are the only characteristic lines in the case of steady plane potential flow, and, more generally, of Eq. (7.43).

Let us now consider the compatibility relations corresponding to Eqs. (12). Call ϕ the angle a characteristic direction makes with the x-axis. Then Eq. (13) shows that there are two values of ϕ which we denote by ϕ^+ and ϕ^-. With $\lambda_1 = -\sin\phi$, $\lambda_2 = \cos\phi$, Eq. (13) gives $A\sin^2\phi - 2B\sin\phi\cos\phi + C\cos^2\phi = 0$, or

$$2B - A\tan\phi - \frac{C}{\tan\phi} = 0.$$

Denote by $\partial/\partial\sigma$ differentiation along a characteristic. Then, multiplying the first Eq. (12) by $\cos\phi$, the second by $A\sin\phi - B\cos\phi$ (these being appropriate multipliers α_ι), adding, and using the above relation, we obtain immediately

$$A \frac{\partial u_1}{\partial \sigma} + C \cot \phi \frac{\partial u_2}{\partial \sigma} = F \cos \phi \tag{14}$$

as the desired compatibility relation. It consists of two equations, one for ϕ^+ and $d\sigma^+$, the other for ϕ^- and $d\sigma^-$, or, in other words, one along each characteristic.

Equation (14) allows a useful modification: from the above quadratic equation for $\tan \phi$ it is seen that

$$\tan \phi^+ \tan \phi^- = \frac{C}{A}.$$

Hence the two equations contained in (14) can be written

$$A \left(\frac{\partial u_1}{\partial \sigma^\pm} + \tan \phi^\mp \frac{\partial u_2}{\partial \sigma^\pm} \right) = F \cos \phi^\pm. \tag{14$'$}$$

In differential form, and using the abbreviation $dx = \cos \phi \, d\sigma$, we obtain

$$du_1^\pm + \tan \phi^\mp \, du_2^\pm = \frac{F}{A} \, dx^\pm \tag{14$''$}$$

as the required compatibility relation.

If our equations have come from one of the flow problems governed by Eq. (7.43) we know that the characteristic curves are the Mach lines so that

$$\phi^+ = \theta + \alpha, \qquad \phi^- = \theta - \alpha,$$

where θ is the angle between the x-axis and the flow direction and α the Mach angle. In Eq. (7.43) with $\nu = 0$ for plane potential flow we have $F = 0$ and (14′) reduces to

$$\left(\frac{du_2}{du_1} \right)^\pm = -\cot \phi^\mp. \tag{15}$$

This relation between the derivatives of the dependent variables along the characteristics, may be compared to $(dy/dx)^\pm = \tan \phi^\pm$.

For (7.43), with $\nu = 1$, $q_x = u_1$, $q_y = u_2$, $F = -u_2 a^2/y$, and $A = a^2 - u_1^2$ we obtain with $\phi^\pm = \theta \pm \alpha$:

$$du_1^\pm + \tan (\theta \mp \alpha) \, du_2^\pm = -\frac{u_2 a^2}{y(a^2 - u_1^2)} \, dx^\pm. \tag{16}$$

This equation simplifies if polar variables q, θ are used, since α is a function of q alone. Introducing into (16)

$$du_1 = dq \cos\theta - d\theta\, q \sin\theta, \qquad du_2 = dq \sin\theta + d\theta\, q \cos\theta,$$

we obtain after some simple manipulation

$$\text{(17)} \qquad \begin{aligned} \frac{dq}{q\tan\alpha} - d\theta &= \frac{\sin\alpha\sin\theta}{\cos(\theta+\alpha)}\frac{dx}{y}, && \text{along a } C^+, \\ \frac{dq}{q\tan\alpha} + d\theta &= \frac{\sin\alpha\sin\theta}{\cos(\theta-\alpha)}\frac{dx}{y}, && \text{along a } C^-. \end{aligned}$$

These relations, as well as Eqs. (15), will be better understood after the study of Secs. 10.6,7. If the same polar variables are introduced into (15) rather than (16) we obtain Eqs. (17) with their right sides replaced by zeros (see Sec. 16.3).

5. Further examples

Some caution is needed in considering the *three-dimensional problem of steady potential flow*. Suppose the differential equation for Φ is of the form

$$\text{(18)} \quad A_1\frac{\partial^2\Phi}{\partial x^2} + A_2\frac{\partial^2\Phi}{\partial y^2} + A_3\frac{\partial^2\Phi}{\partial z^2} + 2B_1\frac{\partial^2\Phi}{\partial y\,\partial z} + 2B_2\frac{\partial^2\Phi}{\partial z\,\partial x} + 2B_3\frac{\partial^2\Phi}{\partial x\,\partial y} = 0,$$

where, as before, the coefficients do not depend on Φ itself. If the first-order derivatives of Φ are taken as the unknowns u_1, u_2, u_3, then $n = k = 3$, and Eq. (18) may be replaced by the system of three equations

$$\text{(19)} \qquad \begin{aligned} &A_1\frac{\partial u_1}{\partial x} + A_2\frac{\partial u_2}{\partial y} + A_3\frac{\partial u_3}{\partial z} + B_1\left(\frac{\partial u_3}{\partial y} + \frac{\partial u_2}{\partial z}\right) \\ &\qquad + B_2\left(\frac{\partial u_1}{\partial z} + \frac{\partial u_3}{\partial x}\right) + B_3\left(\frac{\partial u_2}{\partial x} + \frac{\partial u_1}{\partial y}\right) = 0, \\ &\frac{\partial u_1}{\partial z} - \frac{\partial u_3}{\partial x} = 0, \qquad \frac{\partial u_2}{\partial x} - \frac{\partial u_1}{\partial y} = 0. \end{aligned}$$

As above, the additional equations result from the condition that the unknowns are derivatives of the same function Φ. This condition can also be expressed by curl $\mathbf{u} = 0$; in (19) only the y- and z-components of this last equation are written. The x-component, namely

$$\text{(19}'\text{)} \qquad \frac{\partial u_3}{\partial y} - \frac{\partial u_2}{\partial z} = 0,$$

is not a consequence of the other two. In fact, from the vanishing of the y- and z-components of curl $\mathbf{u}$ we can conclude that the x-component is independent of x, but not that it vanishes entirely. [That $(\text{curl}\,\mathbf{u})_x \equiv \partial u_3/\partial y -$

$\partial u_2/\partial z$ is independent of x is seen by straightforward differentiation with respect to x and then substitution from the last two of Eqs. (19) to show that $\partial(\text{curl } \mathbf{u})_x/\partial x$ vanishes.] Thus the problem is not completely stated by Eqs. (19), and eventually we shall have to take into account Eq. (19′) also. Let us first determine the characteristic surfaces of the system (19).

For this set of equations, the components of the coefficient vectors $\mathbf{a}_{\iota\kappa}$ are

$$\begin{array}{llll} \iota = 1: & (A_1, B_3, B_2), & (B_3, A_2, B_1), & (B_2, B_1, A_3); \\ \iota = 2: & (0, \; 0, \; 1), & (0, \; 0, \; 0), & (-1, 0, \; 0); \\ \iota = 3: & (0, -1, \; 0), & (1, \; 0, \; 0), & (0, \; 0, \; 0). \end{array}$$

Thus the $\boldsymbol{\lambda}$-equation (10) becomes

$$(20) \qquad \begin{vmatrix} A_1\lambda_1 + B_3\lambda_2 + B_2\lambda_3 & B_3\lambda_1 + A_2\lambda_2 + B_1\lambda_3 & B_2\lambda_1 + B_1\lambda_2 + A_3\lambda_3 \\ \lambda_3 & 0 & -\lambda_1 \\ -\lambda_2 & \lambda_1 & 0 \end{vmatrix}$$

$$= \lambda_1[A_1\lambda_1^2 + A_2\lambda_2^2 + A_3\lambda_3^2 + 2B_1\lambda_2\lambda_3 + 2B_2\lambda_3\lambda_1 + 2B_3\lambda_1\lambda_2] = 0.$$

The expression in square brackets determines a cone of second order; again at a particular point P the cone may be real, possibly degenerate, or imaginary, depending upon the values of the coefficients A_i and B_i at P. In particular, if we consider the case of steady three-dimensional potential flow, i.e., Eq. (7.14) with $\partial/\partial t = 0$, and choose the x-direction parallel to $\mathbf{q}$ at P, the coefficients in (19) are $A_1 = 1 - M^2$, $A_2 = A_3 = 1$, $B_1 = B_2 = B_3 = 0$. Then the cone is given by

$$(21) \qquad (M^2 - 1)\lambda_1^2 - \lambda_2^2 - \lambda_3^2 = 0.$$

This represents a cone of revolution about the x-axis, with semivertex angle $\bar{\alpha}$ satisfying

$$\tan \bar{\alpha} = \frac{\sqrt{\lambda_2^2 + \lambda_3^2}}{\lambda_1} = \sqrt{M^2 - 1} = \cot \alpha,$$

the angle (and cone) being real only for $M \geqq 1$. Thus a characteristic surface, normal to $\boldsymbol{\lambda}$ at P, makes the Mach angle α with the x-axis, or direction of $\mathbf{q}$, and is tangent to the Mach cone at P, as defined in Sec. 5.2.[32]

We see that Eq. (20) has the additional solution $\lambda_1 = 0$ for all points P, so that any plane parallel to the x-axis is exceptional for the system (19). These planes are, however, not exceptional for the original Eq. (18), as follows from considering Eq. (19′) in addition to (19). Take, for instance, the x,z-plane, and try to compute the derivatives with respect to y from the known x- and z-derivatives. This cannot be done from (19) alone since the second of these equations does not involve any of the unknowns

$\partial u_1/\partial y$, $\partial u_2/\partial y$, $\partial u_3/\partial y$. From (19′), however, $\partial u_3/\partial y$ is obtained at once, and then $\partial u_1/\partial y$ and $\partial u_2/\partial y$ are determined by the first and last of Eqs. (19). Since the y-derivatives can be determined, the x,z-plane is not exceptional.* Thus, the cone (21) includes in fact all exceptional directions.†

As a further example, let us briefly consider one-dimensional nonsteady motion; the potential in this case satisfies Eq. (7.24), which is of the type (2) in the independent variables x and t. Here $A = a^2 - q^2$, $B = -q$, $C = -1$, so that Eq. (13′) gives

$$-\frac{\lambda_1}{\lambda_2} = \frac{-q \pm \sqrt{a^2}}{a^2 - q^2} = \frac{1}{q \pm a} = \frac{1}{q_x \pm a} \tag{22}$$

for the slope of the characteristic directions in the x,t-plane. The characteristic lines make with the t-axis an angle whose tangent is $q_x \pm a$. This result plays an important role in the theory of one-dimensional motion (see Chapter III), where the compatibility relations will be taken up for Eq. (7.24), or Eq. (7.43′) with $\nu = 0$. The compatibility relations corresponding to this latter equation with $\nu = 0, 1, 2$ may be derived directly from Eq. (14″). This is left as an exercise for the reader. The characteristics are, in appropriate variables, the same for all three cases.

6. General case of fluid motion

In considering the general differential equations of nonsteady nonpotential flow in three dimensions at a point P, we again take the x-axis in the direction of $\mathbf{q}$. Then the particle derivative d/dt of (1.4) is expressed by $\partial/\partial t + q\,\partial/\partial x$. We use Eq. (I) of Art. 1, the equation of motion for an inviscid fluid, but neglect gravitational forces.[34] A somewhat more general case than that of an elastic fluid may be included by supposing the existence of a function $\mathcal{S}(p,\rho)$ (such as entropy), which has a constant value for each particle, although not necessarily the same constant everywhere, as in the case of an elastic fluid. The specifying condition is then $d\mathcal{S}/dt = 0$. As pointed out in Sec. 5.2, the local sound velocity a can still be defined, using

$$a^2 = \frac{dp}{d\rho} = -\frac{\partial\mathcal{S}/\partial\rho}{\partial\mathcal{S}/\partial p}. \tag{23}$$

* If in (19) we had used the x- and y-components of curl $\mathbf{u} = 0$, the factor λ_3 would have appeared in the analogue of (20), rather than the factor λ_1, and similar conclusions would apply.

† If the general argument of Sec. 2 is applied to a single second-order planar differential equation for $\Phi(x_1, x_2, \cdots, x_n)$, it is seen that the λ-equation is immediately obtained by replacing each factor $\partial^2\Phi/\partial x_\iota\,\partial x_\kappa$ in the original equation by $\lambda_\iota\lambda_\kappa$. This can be proved in exactly the same way as Eq. (20) is derived from Eq. (18). Actually the theory of characteristics for second-order equations can be arrived at directly by means of a reasoning analogous to that of Sec. 1, and the result here explained is then obtained.[33]

Then

$$\frac{dS}{dt} = \frac{\partial S}{\partial p}\frac{dp}{dt} + \frac{\partial S}{\partial \rho}\frac{d\rho}{dt} = \frac{\partial S}{\partial p}\left[\frac{dp}{dt} - a^2\frac{d\rho}{dt}\right], \tag{24}$$

so that, when $\partial S/\partial p$ is different from zero, the specifying condition may be expressed by setting the bracket in (24) equal to zero, with d/dt expanded as above. Then the system of five partial differential equations in the five unknowns q_x, q_y, q_z, p, and ρ is

$$\begin{gathered} q\frac{\partial q_x}{\partial x} + \frac{\partial q_x}{\partial t} + \frac{1}{\rho}\frac{\partial p}{\partial x} = 0, \\ q\frac{\partial q_y}{\partial x} + \frac{\partial q_y}{\partial t} + \frac{1}{\rho}\frac{\partial p}{\partial y} = 0, \\ q\frac{\partial q_z}{\partial x} + \frac{\partial q_z}{\partial t} + \frac{1}{\rho}\frac{\partial p}{\partial z} = 0, \\ \frac{\partial q_x}{\partial x} + \frac{\partial q_y}{\partial y} + \frac{\partial q_z}{\partial z} + \frac{q}{\rho}\frac{\partial \rho}{\partial x} + \frac{1}{\rho}\frac{\partial \rho}{\partial t} = 0, \\ q\frac{\partial p}{\partial x} + \frac{\partial p}{\partial t} - a^2\left(q\frac{\partial \rho}{\partial x} + \frac{\partial \rho}{\partial t}\right) = 0. \end{gathered} \tag{25}$$

This is a system of the form (6), with $k = 5$, $n = 4$. There are $k^2 = 25$ vector coefficients $\mathbf{a}_{\iota\kappa}$, each having four components (in the directions x, y, z, and t), though many of these vanish because of the choice of the coordinate axes. For example, the components of $\mathbf{a}_{11}$ are the coefficients of the derivatives of q_x in the first equation: q, 0, 0, 1. For $\mathbf{a}_{12}$, $\mathbf{a}_{13}$, and $\mathbf{a}_{15}$, the components are all zero, since derivatives of the unknowns q_y, q_z, and ρ do not appear in the first equation, while $\mathbf{a}_{14}$ has components $1/\rho$, 0, 0, 0. The entries in the first line of the $\boldsymbol{\lambda}$-determinant are the scalar products of these vectors with $\boldsymbol{\lambda} = (\lambda_1, \lambda_2, \lambda_3, \lambda_4)$, namely, $q\lambda_1 + \lambda_4$, 0, 0, λ_1/ρ, 0. The remaining terms can be found similarly, and the equation for $\boldsymbol{\lambda}$ is

$$\begin{vmatrix} q\lambda_1 + \lambda_4 & 0 & 0 & \lambda_1/\rho & 0 \\ 0 & q\lambda_1 + \lambda_4 & 0 & \lambda_2/\rho & 0 \\ 0 & 0 & q\lambda_1 + \lambda_4 & \lambda_3/\rho & 0 \\ \lambda_1 & \lambda_2 & \lambda_3 & 0 & (q\lambda_1 + \lambda_4)/\rho \\ 0 & 0 & 0 & q\lambda_1 + \lambda_4 & -a^2(q\lambda_1 + \lambda_4) \end{vmatrix} \tag{26}$$

$$= -\frac{(q\lambda_1 + \lambda_4)^3}{\rho}\left[(q\lambda_1 + \lambda_4)^2 - a^2(\lambda_1^2 + \lambda_2^2 + \lambda_3^2)\right] = 0.$$

There are two types of characteristics, corresponding to the two factors in Eq. (26). Before discussing the general case, let us consider the particular case of steady motion, when the t-component λ_4 drops out. The second factor in Eq. (26) then supplies

$$a^2[(M^2 - 1)\lambda_1^2 - \lambda_2^2 - \lambda_3^2] = 0, \tag{27}$$

and the corresponding characteristic surfaces are again tangent to the Mach cones (for $M \geqq 1$, of course). The first factor supplies $\lambda_1^3 = 0$, so that any plane passing through the x-axis (or velocity direction) is exceptional. It follows that, in the general case of steady motion under the specifying condition $dS/dt = 0$, *all surfaces composed of streamlines are also characteristics*. These stream surfaces are real characteristics even for subsonic flow whenever the flow is not assumed to be irrotational, and in fact even in rotational *incompressible* flow. For this case, however, the system (25) must be modified, but the formula corresponding to (26) is

$$-\frac{(q\lambda_1 + \lambda_4)^2}{\rho}(\lambda_1^2 + \lambda_2^2 + \lambda_3^2) = 0,$$

so that the same factor is present.[35]

In the case of a nonsteady flow, the second factor in (26) represents a "cone" of second-order in x,y,z,t-space. This "cone" intersects the x,y,z-space in the cone given by Eq. (27). For the intersection with the x,t-plane—i.e., when λ_2 and λ_3 drop out—the bracket reduces to

$$\lambda_1^2(q^2 - a^2) + 2q\lambda_1\lambda_4 + \lambda_4^2,$$

and this last expression equated to zero leads to the same two directions in the x,t-plane as in the one-dimensional nonsteady case: Eq. (22), with λ_4 in place of λ_2. The first factor in Eq. (26) leads to the surfaces in the x,y,z,t-space consisting of the world lines of the particles (see Sec. 1.2).[36] Thus *all surfaces composed of world lines are characteristic*. (See also Sec. 15.2).

It is seen that the Mach cone keeps its role as the envelope of exceptional planes irrespective of whether or not the flow is irrotational.[37]

Consider finally the compatibility relations, remembering in particular the end of Sec. 3. Firstly, corresponding to the triple root, i.e., to the first factor in Eq. (26), there are $r = 3$ compatibility relations.[38] They consist of the specifying equation, the last of Eqs. (25), and two components of the Newton equation which are perpendicular to the $\boldsymbol{\lambda}$-direction.

With respect to the second factor in Eq. (26) there is, for each direction of the cone, one compatibility relation. Consider first the steady case:

Eq. (27) supplies $(M^2 - 1)\lambda_1^{*2} - \lambda_2^{*2} - \lambda_3^{*2} = 0$, where the λ_i^* are only determined up to a common factor.

Next compute the $\alpha_1^*, \alpha_2^*, \cdots, \alpha_5^*$ from (9′). Omitting for the moment the * in α and λ, we have the equations

$$\alpha_1 q\lambda_1 + \alpha_4\lambda_1 = 0,$$

$$\alpha_2 q\lambda_1 + \alpha_4\lambda_2 = 0,$$

$$\alpha_3 q\lambda_1 + \alpha_4\lambda_3 = 0,$$

$$\alpha_1\lambda_1 + \alpha_2\lambda_2 + \alpha_3\lambda_3 + \alpha_5\rho q\lambda_1 = 0,$$

$$\alpha_4 q\lambda_1 - \alpha_5 a^2\rho q\lambda_1 = 0.$$

First we note that $\lambda_1 = \lambda_1^* \neq 0$ since $\boldsymbol{\lambda}^*$ is perpendicular to a direction that makes the Mach angle $\alpha \neq 0$ with the flow direction, which is here the x-direction. Hence the first and last of these equations give

$$\alpha_4 = a^2\rho\alpha_5, \qquad \alpha_1 = -\frac{\alpha_4}{q} = -\frac{a^2\rho}{q}\alpha_5$$

independent of $\boldsymbol{\lambda}$. Further, we have

$$\alpha_2 = -\frac{\alpha_4}{q}\frac{\lambda_2}{\lambda_1} = -\frac{a^2\rho}{q}\alpha_5\frac{\lambda_2}{\lambda_1},$$

$$\alpha_3 = -\frac{\alpha_4}{q}\frac{\lambda_3}{\lambda_1} = -\frac{a^2\rho}{q}\alpha_5\frac{\lambda_3}{\lambda_1}.$$

Since the α are likewise determined up to a common factor only, we choose $\alpha_5 = -1/a\rho$ and obtain, again using stars,

$$\text{(28)} \quad \alpha_1^* = \frac{1}{M}, \quad \alpha_2^* = \frac{1}{M}\frac{\lambda_2^*}{\lambda_1^*}, \quad \alpha_3^* = \frac{1}{M}\frac{\lambda_3^*}{\lambda_1^*}, \quad \alpha_4^* = -a, \quad \alpha_5^* = -\frac{1}{a\rho}.$$

We now choose $\lambda_1^* = 1/M$ and have

$$\text{(29)} \qquad \lambda_2^{*2} + \lambda_3^{*2} = 1 - 1/M^2, \quad \text{or} \quad \lambda_1^{*2} + \lambda_2^{*2} + \lambda_3^{*2} = 1,$$

thus making $\boldsymbol{\lambda}^*$ a unit vector. Using $\lambda_1^* = 1/M$, Eqs. (28) simplify to

$$\text{(28')} \quad \alpha_1^* = \frac{1}{M}, \quad \alpha_2^* = \lambda_2^*, \quad \alpha_3^* = \lambda_3^*, \quad \alpha_4^* = -a, \quad \alpha_5^* = -\frac{1}{a\rho},$$

where (29) holds. Note that the fourth of the equations for the α_ι has not been used; it is seen that it reduces identically to (29) if the values of (28′) are introduced.

These α_ι^* determine a compatibility relation. Using Eqs. (25) and (28′) we find according to (11):

$$(30) \qquad \boldsymbol{\lambda}^* \cdot \left(\frac{d\mathbf{q}}{dt} + \frac{1}{\rho}\operatorname{grad} p\right) - \left(a \operatorname{div} \mathbf{q} + \frac{1}{\rho a}\frac{dp}{dt}\right) = 0$$

or, using $\boldsymbol{\lambda}^* \cdot \mathbf{q} = a$:

$$(30') \qquad \boldsymbol{\lambda}^* \cdot \left(\frac{d\mathbf{q}}{dt} - \mathbf{q} \operatorname{div} \mathbf{q} + \frac{1}{\rho}\operatorname{grad} p - \frac{1}{\rho a^2}\mathbf{q}\frac{dp}{dt}\right) = 0.$$

Here $\boldsymbol{\lambda}^*$ may be any one of the single infinity of unit vectors perpendicular to the tangent planes of the Mach cone.

In the general, i.e., nonsteady case, the compatibility condition is quite similar to Eq. (30), except that $\boldsymbol{\lambda}^*$ must now be a vector which makes the second factor of Eq. (26) vanish.

Article 10

The Characteristics in the Case of Two Independent Variables

1. Characteristic directions

In many problems of aerodynamics the number of independent variables reduces to two: $n = 2$. Almost all that is known at present about compressible fluid flow refers to such cases. Examples (some of which will be treated in more detail in later chapters) include: (a) plane steady flow, in which the independent variables are x and y (Chapters IV, V); (b) nonsteady parallel flow, variables x and t (Chapter III); (c) nonsteady radial flow, variables r and t; and (d) steady axially symmetric flow, variables r and z, where r is the distance from the axis of symmetry and z is measured parallel to the axis, as in cylindrical coordinates. In all these cases the characteristics are curves in the plane of the independent variables, as was seen in Art. 9. A study of these curves will be of great use in the task of integrating the differential equations.

If the specifying equation is a (p,ρ)-relation and if, in examples (a) and (d), the motion is further assumed to be irrotational, then the number of unknowns is also two: $k = 2$. In (b) and (c) the unknowns are ρ (or p) and a velocity component; in cases (a) and (d) two velocity components may be taken as the unknowns, since we have seen (Art. 8) that in a steady irrotational flow ρ (or p or a^2) may be expressed as a function of the magnitude of the velocity vector.

If we let $x_1 = x$, $x_2 = y$ for the independent variables, and $u_1 = u$, $u_2 = v$ for the dependent variables, the system (9.6) of planar differential equations can be written out as

$$
(1)\qquad
\begin{aligned}
a_1 \frac{\partial u}{\partial x} + a_2 \frac{\partial u}{\partial y} + a_3 \frac{\partial v}{\partial x} + a_4 \frac{\partial v}{\partial y} &= a, \\
b_1 \frac{\partial u}{\partial x} + b_2 \frac{\partial u}{\partial y} + b_3 \frac{\partial v}{\partial x} + b_4 \frac{\partial v}{\partial y} &= b,
\end{aligned}
$$

where the vectors of Eq. (9.6) are $\mathbf{a}_{11} = (a_1, a_2)$, $\mathbf{a}_{12} = (a_3, a_4)$, $\mathbf{a}_{21} = (b_1, b_2)$, $\mathbf{a}_{22} = (b_3, b_4)$. All the coefficients and the right-hand members are, in general, functions of x, y, u, and v;[39] in the case of a *linear problem* the a_i and b_i depend on x and y only, while a and b are linear functions of u and v, with coefficients depending on x,y, if indeed they depend on u and v at all.

Next we consider the vector $\boldsymbol{\lambda}$ (Sec. 9.2) which is normal to a characteristic direction; the components λ_1 and λ_2 of $\boldsymbol{\lambda}$ in the x- and y-directions are determined by Eq. (9.10), which here becomes

$$
(2)\qquad
\begin{vmatrix}
\lambda_1 a_1 + \lambda_2 a_2 & \lambda_1 a_3 + \lambda_2 a_4 \\
\lambda_1 b_1 + \lambda_2 b_2 & \lambda_1 b_3 + \lambda_2 b_4
\end{vmatrix} = 0.
$$

Since the normal directions are in the same plane as the characteristic directions, this condition may be so expressed as to determine the characteristic directions themselves, rather than the normals to these directions. If ϕ is the angle between the x-axis and a characteristic direction, then the slope of the characteristic at any point is

$$
\frac{dy}{dx} = \tan \phi = -\frac{\lambda_1}{\lambda_2}.
$$

If we take $\boldsymbol{\lambda}$ of length 1 and suitably directed, we may write $\lambda_1 = -\sin \phi$, $\lambda_2 = \cos \phi$, and Eq. (2) becomes

$$
(2')\qquad
\begin{aligned}
0 &= \begin{vmatrix}
a_2 \cos \phi - a_1 \sin \phi & a_4 \cos \phi - a_3 \sin \phi \\
b_2 \cos \phi - b_1 \sin \phi & b_4 \cos \phi - b_3 \sin \phi
\end{vmatrix} \\
&= \Delta_{24} \cos^2 \phi - (\Delta_{14} + \Delta_{23}) \cos \phi \sin \phi + \Delta_{13} \sin^2 \phi,
\end{aligned}
$$

where Δ_{ik} is used as an abbreviation for $a_i b_k - a_k b_i$. All three coefficients of this quadratic equation vanish in the following two cases: (a) if the left-hand member of one of the equations (1) is a multiple of the left-hand member of the other; (b) if the derivatives of u and v appear only as linear combinations $\mu\, \partial u/\partial x + \nu\, \partial v/\partial x$ and $\mu\, \partial u/\partial y + \nu\, \partial v/\partial y$. We assume that neither of these occurs for any set of values x, y, u and v considered.

In all other cases*, Eq. (2′) has two, one, or no real solutions for $\tan \phi$; the system (1) is then called *hyperbolic*, *parabolic*, or *elliptic*, respectively. Thus, for example, we saw in Sec. 9.3 that the equations governing steady potential flow in two dimensions are elliptic in a subsonic region, parabolic on a sonic line, and hyperbolic in a supersonic region. When the a_i and b_i are independent of u and v, the coefficients of (2′) depend only on x and y and it is possible to state that the system (1) is hyperbolic in a certain region of the x,y-plane. In this case the two roots $\tan \phi = dy/dx$ of Eq. (2′) are functions of x and y alone. By integration of the resulting ordinary differential equations, the two sets of characteristics can be found independently of u and v; thus, *in the case of linear equations in two dimensions the characteristics form a network of curves determined once and for all, before any boundary conditions are given to single out a particular solution u, v.*

In the general case, however, where the coefficients a_i and b_i are functions of u and v also, the system may be described as hyperbolic or elliptic at a point only in connection with a particular solution u, v of the system. Here the characteristics are different for different solutions u, v and must be determined progressively along with the solutions themselves, which depend on the boundary conditions. Actual use of the characteristics will be made only in a hyperbolic region and at its boundary, which is parabolic.

As a simple example, consider the case of a small perturbation of a fluid at rest, discussed in Sec. 4.2. The equations corresponding to (1) are Eqs. (4.5), where u and v are q_x and ρ, while x and t stand for x and y; in the present notation these equations are given by

$$\rho_0 \frac{\partial u}{\partial x} + \frac{\partial v}{\partial y} = 0, \qquad \rho_0 \frac{\partial u}{\partial y} + a_0^2 \frac{\partial v}{\partial x} = 0. \tag{3}$$

Here all coefficients are constants: $a_1 = \rho_0$, $a_4 = 1$, $b_2 = \rho_0$, $b_3 = a_0^2$, $a_2 = a_3 = b_1 = b_4 = 0$. Eq. (2′) therefore reads

$$-\rho_0 \cos^2 \phi + \rho_0 a_0^2 \sin^2 \phi = 0, \qquad \tan \phi = \pm \frac{1}{a_0}.$$

The differential equations

$$\frac{dy}{dx} = \frac{1}{a_0}, \qquad \frac{dy}{dx} = -\frac{1}{a_0},$$

are satisfied by $y = x/a_0 +$ constant and $y = -x/a_0 +$ constant, respectively, so that the characteristics consist of two sets of parallel straight

* $\Delta_{24} = 0$ and $\Delta_{13} = 0$ imply that there are linear combinations of equations (1) with left-hand sides $\mu_1\, \partial u/\partial x + \nu_1\, \partial v/\partial x$ and $\mu_2\, \partial u/\partial y + \nu_2\, \partial v/\partial y$, respectively. If in addition $\Delta_{14} + \Delta_{23} = 0$, then $\mu_1\nu_2 - \mu_2\nu_1 = 0$. From this it follows that (a) and (b) are the only exceptional cases.

lines meeting the x-axis at angles arc cot a_0 and arc cot $(-a_0)$ respectively. If, however, the one-dimensional flow were not treated as a small perturbation, i.e., with terms of higher order neglected, the constant ρ_0 in (3) would be replaced by $\rho = v$, and other terms would be added. Then the left-hand members of (3) would be nonlinear, and the characteristics would change with the boundary conditions to which the flow is subjected.

2. Compatibility relations

From the discussion in Sec. 9.3, it follows that there is for each characteristic a suitable linear combination (9.11) of the left-hand members of the system (1), involving only the components of grad u and grad v parallel to that characteristic, that is, $\partial u/\partial\sigma$ and $\partial v/\partial\sigma$, if $\partial/\partial\sigma$ means differentiation in that characteristic direction. We called the equation so derived, which links the changes of u and v along a characteristic, the compatibility relation.

The compatibility relation in the present case, where $n = k = 2$, will now be derived directly and discussed in some detail. We first introduce the abbreviations A_1, A_2, B_1, and B_2 for the four elements of the determinant in (2′), so that Eq. (2′) becomes $A_1B_2 - A_2B_1 = 0$. Then, with the aid of the relation (2′) between $\tan\phi$ and the Δ_{ij}'s occurring there, it can be seen by direct computation that the following identities hold:

$$\text{(4a)}\qquad \frac{a_1B_1 - b_1A_1}{\cos\phi} = \frac{a_2B_1 - b_2A_1}{\sin\phi} = \Delta_{12},$$

$$\text{(4b)}\qquad \frac{a_3B_2 - b_3A_2}{\cos\phi} = \frac{a_4B_2 - b_4A_2}{\sin\phi} = \Delta_{34},$$

$$\text{(4c)}\qquad \frac{a_3B_1 - b_3A_1}{\cos\phi} = \frac{a_4B_1 - b_4A_1}{\sin\phi} = \Delta_{32} + \Delta_{13}\tan\phi$$
$$= \Delta_{14} + \Delta_{42}\cot\phi = K,$$

$$\text{(4d)}\qquad \frac{a_1B_2 - b_1A_2}{\cos\phi} = \frac{a_2B_2 - b_2A_2}{\sin\phi} = \Delta_{14} + \Delta_{31}\tan\phi$$
$$= \Delta_{32} + \Delta_{24}\cot\phi = L.$$

Equations (4c) and (4d) are definitions of K and L.

If the first and second of Eqs. (1) are multiplied by B_1 and A_1 (or B_2 and A_2) respectively, and subtracted, the results, in view of the identities (4), are

$$\text{(5)}\quad \Delta_{12}\frac{\partial u}{\partial\sigma} + K\frac{\partial v}{\partial\sigma} = B_1a - A_1b, \qquad L\frac{\partial u}{\partial\sigma} + \Delta_{34}\frac{\partial v}{\partial\sigma} = B_2a - A_2b,$$

where for K and L either one of the expressions in Eqs. (4c) and (4d) may be taken.[40] In general, any one of the four alternatives can be used as the compatibility relation since they are equivalent. In certain special cases, however, some of the alternatives may fail. For example, if in (1) all coefficients except a_1 and b_4 vanish, the only nonvanishing Δ is $\Delta_{14} = a_1b_4$, and it follows from Eq. (2′) that the characteristic directions are $\phi = 0°$ and $\phi = 90°$. For $\phi = 0°$ the second expression for K and the same for L become indefinite $(0 \cdot \infty)$. If the alternative expressions for K and L are used, the first Eq. (5) has only zero terms, but the second gives $\Delta_{14}\,\partial u/\partial\sigma = B_2a = b_4a$ or $a_1\,\partial u/\partial\sigma = a$, which is identical with the first Eq. (1). Hence one of the given equations is the compatibility relation in this case. In all cases in which Eq. (2′) has two distinct real solutions, at least one of the four alternative forms of (5) supplies the compatibility relation.

We may now return to the compatibility relations derived in Sec. 9.4 for particular equations and check the results obtained against the present theory.

3. Two important theorems

We are now in a position to formulate, for the hyperbolic case, two principal theorems concerning the existence and the uniqueness of solutions of the system (1) which satisfy certain boundary (or initial) conditions. The real characteristic curves, which first appeared as curves *across* which a solution could not be extended by means of the system (1), can now be used as curves *along* which a solution can be extended by means of the compatibility relations (5).

Theorem A. Let the values of u and v be given along a curve AB in the x,y-plane (see Fig. 45a), in such a way that at no point does the direction of AB coincide with either of the two characteristic directions determined by x, y, u, and v (Cauchy's problem). *Then a solution, assuming the given values at the points of AB, exists in the neighborhood of AB on both sides, lying within a region bounded (partly) by the four curves AC_1, AD_1, BC_2*

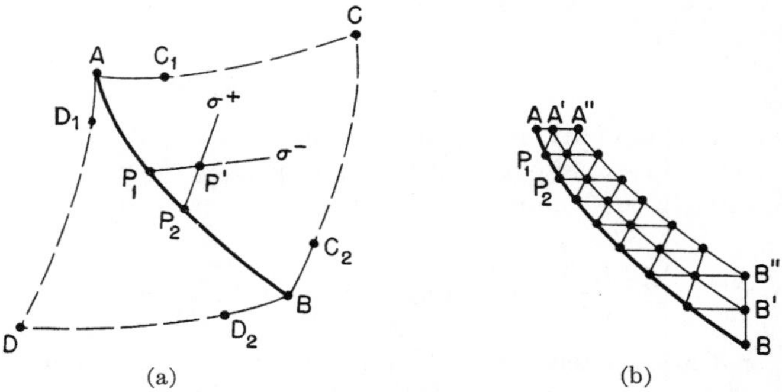

FIG. 45. Illustration of the Cauchy problem.

and BD_2 that are characteristics for this solution. As far out as the solution exists, that is, at most in a characteristic quadrangle $ACBD$ (Fig. 45a), it is uniquely determined by the given values on AB. In the linear case (see Sec. 4) the existence (and the uniqueness) of the solution can be proved in the whole characteristic quadrangle $ACBD$, whose boundaries are known independently of the given values of u and v.[41]

To understand this theorem let us consider on the curve AB two points P_1 and P_2 an infinitesimal distance apart. Since at each point of AB there exist two characteristic directions σ^+ and σ^- that are distinct from each other and from the direction of the tangent to AB, it follows that a straight line of direction σ^- through P_1 and another line of direction σ^+ through P_2 must have an intersection P' at a distance from AB of the same order of magnitude as P_1P_2. We shall see how u' and v', the values of u and v at P', can be computed by using the compatibility relation (5).[42]

Neglecting terms of higher order one can substitute in (5) for the differential quotient $\partial u/\partial\sigma$ the quotient $(u' - u_1)/(P_1P')$ when the transition from P_1 to P' is considered, and the quotient $(u' - u_2)/(P_2P')$ for the transition from P_2 to P' and similarly for $\partial v/\partial\sigma$. Using the first form of (5) and the third member of (4c) for K, one gets for u' and v' the two linear relations

$$(5')\quad \begin{aligned} \Delta_{12}(u' - u_1) + (\Delta_{32} + \Delta_{13}\tan\phi^-)(v' - v_1) &= (B_1a - A_1b)(P_1P'), \\ \Delta_{12}(u' - u_2) + (\Delta_{32} + \Delta_{13}\tan\phi^+)(v' - v_2) &= (B_1a - A_1b)(P_2P'), \end{aligned}$$

with the determinant $\Delta_{12}\Delta_{13}(\tan\phi^+ - \tan\phi^-)$. Here all coefficients are to be evaluated at the same point, say P_1. Without loss of generality it may be assumed that both ϕ-values are different from 0° and 90° at P_1. Then according to (2′), Δ_{13} and Δ_{24} are not zero, and also the factor $(\tan\phi^+ - \tan\phi^-)$ in the expression for the determinant has a finite value. Thus, in the case $\Delta_{12} \neq 0$, the system (5′) can be solved for u' and v'. The same argument holds in the case $\Delta_{34} \neq 0$ if the second form of the compatibility relation (5) is used. Finally, if $\Delta_{12} = \Delta_{34} = 0$, it is seen from Eq. (2′) that one value of $\tan\phi$, say $\tan\phi^+$, equals $a_2/a_1 = b_2/b_1$ and the other equals $a_4/a_3 = b_4/b_3$. Then, using once the first and once the second form of (5), one finds the equations

$$(u' - u_1)L = (B_2a - A_2b)(P_1P'), \qquad (v' - v_2)K = (B_1a - A_1b)(P_2P'),$$

where

$$K = \Delta_{13}(\tan\phi^- - \tan\phi^+), \qquad L = \Delta_{24}(\cot\phi^+ - \cot\phi^-),$$

which can also be solved for u' and v'.

If the whole arc AB is subdivided (Fig. 45b) into n small segments by a sequence of points $P_1, P_2, \cdots$, one can find, in the manner just described, for each couple P_i, P_{i+1} a new point P_i' and the values of u and v at this point. These points may be joined to form a new curve $A'B'$, where AA' (BB') has the direction of one of the characteristics passing through A (B). Starting from the cross curve $A'B'$ and the u,v-values computed for its points P', one can continue in the same way and obtain a new cross curve $A''B''$, etc. The procedure stops automatically when a point is reached where a characteristic direction coincides with the direction of the respective cross curve. If all functions involved and their derivatives are continuous, this can happen only at a finite distance from AB. Up to this distance a network of characteristics with one set of diagonal curves such as $A'B'$, $A''B''$, $\cdots$, together with the values of u and v at each crossing point, can be developed.

If as $n \to \infty$ all distances P_iP_{i+1} tend to zero, the values of u and v tend to solutions of the given system (1) in a certain neighborhood of AB (of the type described in the theorem), satisfying the given boundary conditions on AB. In such a neighborhood existence (and uniqueness) of the solution is guaranteed. If, in addition, the existence of a regular bounded solution is assured beyond this neighborhood of AB, our procedure (provided it does not break down) will furnish an approximation to this solution. The accuracy of the procedure will increase if, as n gets larger, all the distances P_iP_{i+1} are made smaller.[43] On the other hand, even if the mechanics of our method does not meet any obstacle, we cannot be sure that the approximation converges to a solution in the region covered unless we know from elsewhere that such a solution exists.[44]

The construction can be extended to the other side of AB by interchanging the roles of σ^+ and σ^-. The special circumstances prevailing in the case of linear equations will be discussed in the following sections.

Theorem B. Suppose two arcs AB and AC in the x,y-plane and the values of u and v along them are given in such a way that at each point of AB (AC) the direction of the curve coincides with the σ^+ (σ^-) characteristic direction determined by x, y, u, and v and that, likewise at each point, u, v, and their derivatives satisfy the appropriate compatibility relation. *Then a solution, assuming the given values along AB and AC, exists in a neighborhood of the point A in a region bounded by segments AB_1, AC_1 of the characteristics AB, AC and the two other characteristics through B_1 and C_1 respectively.* As far out as the solution exists—that is, at most in a characteristic quadrangle $ACDB$—it is uniquely determined by the given values along AB and AC. (See Fig. 46). Note that if a curve is given in the form $y = y(x)$ and the values of u and v are given by $u = u(x)$ and $v = v(x)$, only one of the three functions may be assumed arbitrarily. In fact, if the curve is to be a characteristic, the slope of the curve must be a given function of x, y, u, and v according to (2′), and a second relation between the three functions is the compatibility relation. In the linear case, $y(x)$ by itself is determined by a differential equation, namely Eq. (2′), and thus either u or v can be chosen, the other variable being determined by the compatibility relation.[45]

We again show how an approximate solution can be built up from the given "compatible" values of u and v along AB and AC (compatible with each other and with the curves). Let P_1P_2 and P_1P_3 (Fig. 46) be two infinitesimal elements in the characteristic directions. If u and v are known at P_2 and P_3, then the same argument as for theorem A shows that the values of u and v at the point P_4 (intersection of a σ^+ through P_2 and σ^- through P_3) can be computed. The point P_4 combined with a subsequent point on AB, represents the same situation and allows us to compute u and v at a subsequent point on the characteristic $A'B'$, etc. Thus, at most in a whole quadrangle determined by AB and AC, the network of characteristics can be constructed and for each cross point the values of u and v can be computed.

Often, the actual boundary problems occurring in fluid dynamics are more complicated than the cases envisaged in the two theorems above. A typical example is indicated in Fig. 47. Here the values of both variables, u and v, are prescribed along a noncharacteristic arc AB, exactly as in theorem A, but, in addition, either u or v or a linear combination $c_1u + c_2v$ is prescribed along lines AA_1 and BB_1. In many cases the process explained above leads to a solution.[46] First, the values interior to ABC may be found as above, where AC and BC are the characteristics through A and B respectively. Next, let P_1 be a point of AC near A, and suppose that the other characteristic segment through P_1 meets AA_1 at P_2. Then the values u_2 and v_2 at P_2 satisfy the compatibility relation along P_1P_2; and this equation, together with the value of $c_1u + c_2v$ prescribed at P_2, is sufficient to determine u_2 and v_2. Next, one computes the values of u and v at P_4, from the two relations along the characteristic segments P_2P_4 and P_3P_4, then at P_5, etc., until the entire wedge adjacent to AA_1 is filled up.

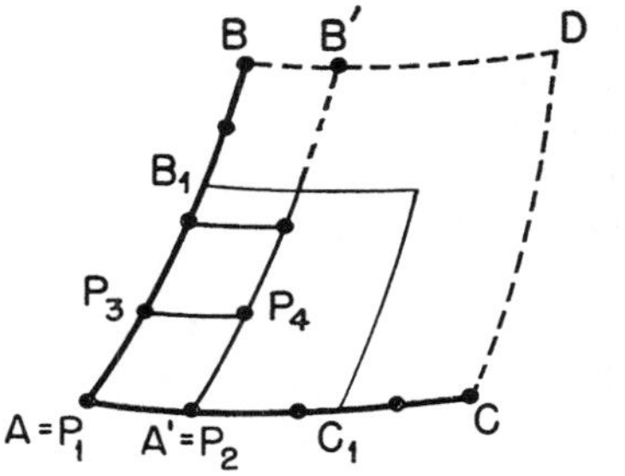

FIG. 46. Illustration of the characteristic boundary-value problem.

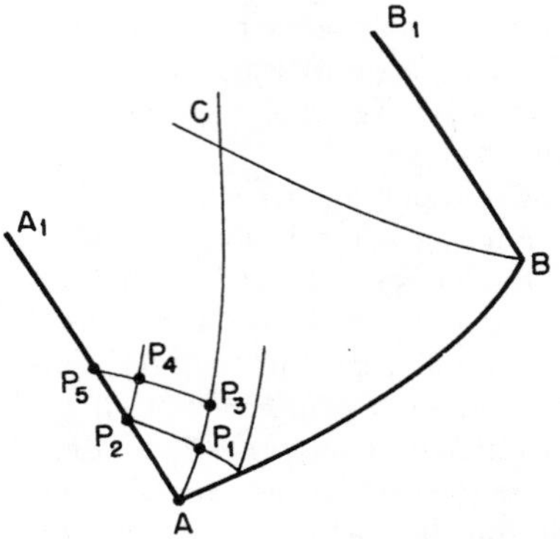

FIG. 47. Illustration of a more general boundary-value problem.

Obviously this procedure fails if AA_1 or any part of it, or BB_1, lies inside the triangle ABC, or if the characteristic segments through P_1, through P_4, etc., fail to meet AA_1. It will be shown later that for quite simple and physically admissible boundary conditions of this kind, no solution exists (see Secs. 14.1 and 22.1).

4. The linear case

The two theorems A and B of the preceding section have been discussed for the general case of a planar system (1). The step-by-step procedure outlined above can be considered as a method of approximate integration. A rigorous proof of the two theorems requires convergence considerations, which can be supplied in the case of A for a sufficiently small neighborhood of the arc AB and in the case of B for a sufficiently small neighborhood of the point A. In the particular case that the equations (1) are *linear* differential equations, and not merely planar, an existence proof can be given for the region consisting of the whole characteristic quadrangle. For convenience we shall develop the argument for equations which are *homogeneous* also, i.e., linear differential equations in which the right-hand members are of the form $\alpha(x,y)u + \beta(x,y)v$, but it will easily be seen that this restriction is not necessary.

If the left-hand members of Eqs. (1) are linear, i.e., if the a_i and b_i do not depend on u and v, and if the system is hyperbolic, Eq. (2′) determines the two values of $\tan\phi$ as explicit functions of x and y. These two ordinary differential equations, $dy/dx = \tan\phi^+$ and $dy/dx = \tan\phi^-$, can be integrated, giving the equations of the two sets of characteristics in the form

$$f(x,y) = \text{constant} \qquad \text{and} \qquad g(x,y) = \text{constant},$$

which form a nonsingular curvilinear mesh, at least in a bounded region of regularity.

Let ξ and η be new variables defined by

$$\xi = f(x,y), \qquad \eta = g(x,y). \tag{6}$$

In terms of the new variables the system (1) is still linear, e.g., the coefficient of $\partial u/\partial \xi$ in the first equation is $a_1\, \partial \xi/\partial x + a_2\, \partial \xi/\partial y$, etc. The characteristics of the transformed system in the ξ,η-plane are the lines $\xi =$ constant and $\eta =$ constant. The linear relation between the derivatives in a characteristic direction will, in this case, involve derivatives with respect to ξ only or derivatives with respect to η only. These relations will therefore be of the form

$$a_1' \frac{\partial u}{\partial \xi} + a_3' \frac{\partial v}{\partial \xi} = a', \qquad b_2' \frac{\partial u}{\partial \eta} + b_4' \frac{\partial v}{\partial \eta} = b', \tag{7}$$

where the coefficients are functions of ξ and η only. Since the equations (7) are merely suitable linear combinations of the equations (1), they will also be homogeneous in the same sense as equations (1). The equations (7) may be further simplified by introducing new unknowns U and V, taking

$$U = a_1'u + a_3'v, \qquad V = b_2'u + b_4'v. \tag{8}$$

Then Eqs. (7) transform into

$$\begin{aligned} \frac{\partial U}{\partial \xi} &= \alpha_1 U + \alpha_2 V, \\ \frac{\partial V}{\partial \eta} &= \beta_1 U + \beta_2 V, \end{aligned} \tag{9}$$

where α_1, α_2, β_1, and β_2 are known functions of ξ and η. Either U or V can be eliminated from this system. If we differentiate the first equation with respect to η, and substitute for $\partial V/\partial \eta$ its value from the second equation and for V its value from the first equation, we obtain

$$\begin{aligned} \frac{\partial^2 U}{\partial \eta\, \partial \xi} = \left(\beta_2 + \frac{1}{\alpha_2}\frac{\partial \alpha_2}{\partial \eta}\right)\frac{\partial U}{\partial \xi} + \alpha_1 \frac{\partial U}{\partial \eta} + \\ \left(\alpha_2\beta_1 - \alpha_1\beta_2 + \frac{\partial \alpha_1}{\partial \eta} - \frac{\alpha_1}{\alpha_2}\frac{\partial \alpha_2}{\partial \eta}\right) U, \end{aligned} \tag{10}$$

if α_2 does not vanish. If $\alpha_2 = 0$, the first equation (9) is already an equation for U, and U will also satisfy

$$\frac{\partial^2 U}{\partial \eta\, \partial \xi} = \alpha_1 \frac{\partial U}{\partial \eta} + \frac{\partial \alpha_1}{\partial \eta} U. \tag{10a}$$

Similarly one can obtain a linear second-order partial differential equation of this type for V.

Under the hypotheses of theorem A the values of u and v are given along an arc AB in the x,y-plane which is nowhere tangent to a characteristic. Under the transformation (6) the arc AB is mapped into an arc $\bar{A}\bar{B}$ in the ξ,η-plane, which can never have a horizontal or vertical tangent (see Fig. 48). The values of u and v given on AB are the values of u and v on $\bar{A}\bar{B}$, and these determine the values of U and V along $\bar{A}\bar{B}$, by (8). The values of U along $\bar{A}\bar{B}$ determine the derivative of U in the direction tangent to the curve, which, by hypothesis, is never the ξ-direction; the first of Eqs. (9) gives $\partial U/\partial\xi$ along $\bar{A}\bar{B}$. From the derivatives in two different directions, the derivative in any direction can be found, and in particular $\partial U/\partial\eta$. Thus the problem of finding U may be stated as follows: U satisfies a differential equation of the form

$$\mathcal{L}(U) \equiv \frac{\partial^2 U}{\partial\xi\,\partial\eta} + a\frac{\partial U}{\partial\xi} + b\frac{\partial U}{\partial\eta} + cU = 0, \tag{11}$$

where a, b, and c are functions of ξ and η [not to be confused with the a, b in Eqs. (1)], and the values of U, $\partial U/\partial\xi$, and $\partial U/\partial\eta$ are given at all points of an arc $\bar{A}\bar{B}$ which at no point has its tangent parallel to either of the axes. The problem is the same for V. The problem is solved if it can be shown how to find the value of U at a point $\bar{C}$ which has the same abscissa as $\bar{B}$ and the same ordinate as $\bar{A}$ (or vice versa). The rectangle $\bar{A}\bar{C}\bar{B}\bar{D}$ in the ξ,η-plane corresponds to the curvilinear quadrangle $ACBD$ in the x,y-plane. It then follows that U may be computed at any point $\bar{P}$ interior to this rectangle, using the given data on an appropriate segment of $\bar{A}\bar{B}$.

The solution of this problem was given in 1860 by Bernhard Riemann, who laid the foundation for the analytic theory of supersonic flow.

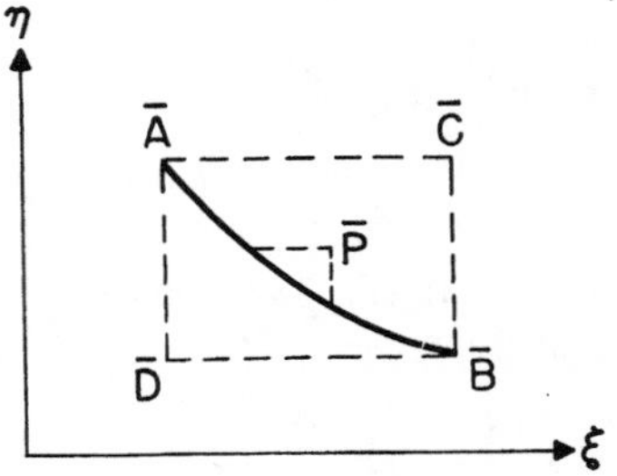

FIG. 48. Cauchy problem in characteristic plane.

5. Riemann's solution[47]

The solution given by Riemann for Eq. (11), subject to the conditions given along $\bar{A}\bar{B}$, is based on the use of a second function $\Omega(\xi,\eta)$, which satisfies another differential equation, the so-called *adjoint equation*, namely,

$$\text{(12)} \qquad \mathfrak{M}(\Omega) \equiv \frac{\partial^2\Omega}{\partial\xi\,\partial\eta} - a\frac{\partial\Omega}{\partial\xi} - b\frac{\partial\Omega}{\partial\eta} + \left(c - \frac{\partial a}{\partial\xi} - \frac{\partial b}{\partial\eta}\right)\Omega = 0,$$

where a, b, and c are the same as in Eq. (11). It may be noted that the adjoint of (12) is exactly (11).

For any two differentiable functions $U(\xi,\eta)$ and $\Omega(\xi,\eta)$ we consider the following expressions:

$$\text{(13)} \qquad \begin{aligned} X &= \frac{1}{2}\frac{\partial}{\partial\eta}(U\Omega) - U\left(\frac{\partial\Omega}{\partial\eta} - a\Omega\right), \\ Y &= \frac{1}{2}\frac{\partial}{\partial\xi}(U\Omega) - U\left(\frac{\partial\Omega}{\partial\xi} - b\Omega\right). \end{aligned}$$

Differentiating these expressions, adding and subtracting the product $cU\Omega$. and collecting terms, we find that

$$\text{(14)} \qquad \frac{\partial X}{\partial\xi} + \frac{\partial Y}{\partial\eta} = \Omega\mathfrak{L}(U) - U\mathfrak{M}(\Omega),$$

where $\mathfrak{L}(U)$ has been written for the left-hand member of Eq. (11) and $\mathfrak{M}(\Omega)$ for the left-hand member of Eq. (12). Now if U satisfies (11) and Ω satisfies (12), the right-hand member of Eq. (14) vanishes. Let a curve $\mathcal{C}$ bound an area A in the ξ,η-plane in which both equations are satisfied. If Stokes' Theorem (Sec. 6.1) is applied to this area for the vector $(-Y,X,0)$, we find

$$\text{(15)} \qquad \oint (X\,d\eta - Y\,d\xi) = \iint \left(\frac{\partial X}{\partial\xi} + \frac{\partial Y}{\partial\eta}\right) dA = 0$$

with the line integral taken around $\mathcal{C}$. The same result can be obtained by specializing the Divergence Theorem (2.27) to two dimensions and applying it to the vector (X,Y).

This formula (15) will be applied to the circuit $\bar{A}\bar{B}\bar{C}$ (Fig. 48). First, however, we shall prescribe boundary conditions for Ω, which so far is subject only to the differential equation (12). If (ξ_1,η_1) are the coordinates of $\bar{C}$, we shall require that

$$\text{(16)} \qquad \Omega(\bar{C}) = \Omega(\xi_1,\eta_1) = 1, \qquad \begin{aligned} \frac{\partial\Omega}{\partial\xi} &= b\Omega \quad \text{along } \bar{A}\bar{C}, \\ \frac{\partial\Omega}{\partial\eta} &= a\Omega \quad \text{along } \bar{C}\bar{B}. \end{aligned}$$

The function Ω depending on the four variables ξ, η, ξ_1, η_1, which satisfies Eqs. (12) and (16), is known as the *Riemann function* of the problem (11). It depends only on the coefficients of the differential equation (11) and is independent of prescribed boundary data for U. It may therefore be determined once and for all for a given equation $\mathfrak{L}(U) = 0$. On account of (16) the expressions for X and Y are considerably simpler on the straight line parts of $\mathfrak{C}$; along $\bar{C}\bar{A}$, for example, we have $Y = \frac{1}{2}\partial(U\Omega)/\partial\xi$. For this same line, $d\eta = 0$ so that the part of the integral in (15) along $\bar{C}\bar{A}$ is

$$2\int_{\bar{C}}^{\bar{A}} (-Y)\, d\xi = \int_{\bar{A}}^{\bar{C}} \frac{\partial}{\partial\xi}(U\Omega)\, d\xi = U(\bar{C}) - U(\bar{A})\Omega(\bar{A}).$$

Similarly, the part of the integral along $\bar{B}\bar{C}$ is

$$2\int_{\bar{B}}^{\bar{C}} X\, d\eta = \int_{\bar{B}}^{\bar{C}} \frac{\partial}{\partial\eta}(U\Omega)\, d\eta = U(\bar{C}) - U(\bar{B})\Omega(\bar{B}).$$

When these values are inserted in (15) and the equation solved for $U(\bar{C})$, there results

$$(17)\quad U(\bar{C}) = \frac{1}{2}[U(\bar{A})\Omega(\bar{A}) + U(\bar{B})\Omega(\bar{B})] - \int_{\bar{A}}^{\bar{B}} (X\, d\eta - Y\, d\xi),$$[48]

where the last integral is a line integral along the curve $\bar{A}\bar{B}$. If the solution Ω of Eq. (12), under the boundary conditions (16), has been found, then formula (17) gives an expression for U, using the values of Ω and its derivatives as well as the prescribed values of U and its derivates along $\bar{A}\bar{B}$.[49] Assuming the existence of Ω, it is possible (though lengthy) to verify that the formula (17) actually does satisfy Eq. (11) and takes the required values along $\bar{A}\bar{B}$.

The proof is now complete except for the construction of the Riemann function Ω, corresponding to Eq. (11). The boundary conditions for Ω are relatively simple.[50] In some simple cases Ω can be given explicitly, and in general, Ω can be constructed by the method of successive approximations. In the same way the function V can also be found anywhere in the rectangle $\bar{A}\bar{B}\bar{C}\bar{D}$. From U and V the solutions u and v are found in terms of ξ and η, by means of (8), everywhere in $\bar{A}\bar{B}\bar{C}\bar{D}$ or, in terms of x and y, everywhere in the quadrangle bounded by the characteristics through A and B. Thus Riemann's formula (17) leads to a complete proof of theorem A [which establishes the existence and uniqueness of the solution of the system (1) in the characteristic quadrangle for given values of u and v along AB] provided that the system (1) is linear.

6. Interchange of variables

If the right-hand members of the two equations (1) are zero, a very efficient transformation may be applied, one which will be used extensively later on. The essential idea of the transformation is interchanging the roles of the dependent and independent variables.

If two variables u and v depend on two independent variables x and y, e.g., by Eqs. (1) and appropriate boundary conditions, the relationship may be considered as a correspondence between a point $P(x,y)$ in the x,y-plane and a point $Q(u,v)$ in the u,v-plane. (See Fig. 49.) A displacement of P in the x,y-plane causes a corresponding displacement of Q. The same relationship can be considered in reverse, as a correspondence between Q and P; then x and y are considered functions of u and v. Any function Φ of x and y may equally well be considered a function of u and v. The differential of Φ, for corresponding displacements of P and Q, may be written in the two forms

$$d\Phi = \frac{\partial\Phi}{\partial x}\,dx + \frac{\partial\Phi}{\partial y}\,dy = \frac{\partial\Phi}{\partial u}\,du + \frac{\partial\Phi}{\partial v}\,dv. \tag{18}$$

In addition

$$dx = \frac{\partial x}{\partial u}\,du + \frac{\partial x}{\partial v}\,dv, \qquad dy = \frac{\partial y}{\partial u}\,du + \frac{\partial y}{\partial v}\,dv.$$

Next, these expressions for dx and dy are substituted in (18), and two particular displacements, one with $dv = 0$ and the other with $du = 0$, are considered. We see that this leads to the equations

$$\frac{\partial\Phi}{\partial x}\frac{\partial x}{\partial u} + \frac{\partial\Phi}{\partial y}\frac{\partial y}{\partial u} = \frac{\partial\Phi}{\partial u}, \qquad \frac{\partial\Phi}{\partial x}\frac{\partial x}{\partial v} + \frac{\partial\Phi}{\partial y}\frac{\partial y}{\partial v} = \frac{\partial\Phi}{\partial v}. \tag{19}$$

The pair of equations (19) can be solved for $\partial\Phi/\partial x$ and $\partial\Phi/\partial y$ whenever the determinant of coefficients

$$J = \frac{\partial x}{\partial u}\frac{\partial y}{\partial v} - \frac{\partial x}{\partial v}\frac{\partial y}{\partial u}$$

does not vanish, yielding

$$\frac{\partial\Phi}{\partial x} = \frac{1}{J}\left(\frac{\partial\Phi}{\partial u}\frac{\partial y}{\partial v} - \frac{\partial\Phi}{\partial v}\frac{\partial y}{\partial u}\right), \qquad \frac{\partial\Phi}{\partial y} = \frac{1}{J}\left(\frac{\partial\Phi}{\partial v}\frac{\partial x}{\partial u} - \frac{\partial\Phi}{\partial u}\frac{\partial x}{\partial v}\right). \tag{20}$$

It should be noted that J is independent of Φ.

In the particular cases $\Phi = u$ and $\Phi = v$, respectively, these solutions (20) lead to the formulas

$$(21) \quad \frac{\partial u}{\partial x} = \frac{1}{J}\frac{\partial y}{\partial v}, \qquad \frac{\partial u}{\partial y} = -\frac{1}{J}\frac{\partial x}{\partial v}, \qquad \frac{\partial v}{\partial x} = -\frac{1}{J}\frac{\partial y}{\partial u}, \qquad \frac{\partial v}{\partial y} = \frac{1}{J}\frac{\partial x}{\partial u}.$$

When these substitutions are made in Eqs. (1) for the case where the right-hand members are zero, the factor $1/J$ may be divided out, and we obtain

$$(22) \quad \begin{aligned} a_1 \frac{\partial y}{\partial v} - a_2 \frac{\partial x}{\partial v} - a_3 \frac{\partial y}{\partial u} + a_4 \frac{\partial x}{\partial u} &= 0, \\ b_1 \frac{\partial y}{\partial v} - b_2 \frac{\partial x}{\partial v} - b_3 \frac{\partial y}{\partial u} + b_4 \frac{\partial x}{\partial u} &= 0. \end{aligned}$$

This is again a planar system, with x,y as unknowns and u,v as independent variables. This interchange of variables in (1) is possible only if the right-hand members of Eqs. (1) vanish.

The importance of this transformation in fluid mechanics lies in the fact that in actual applications the coefficients $a_1, \cdots, b_4$ in Eqs. (1) are often functions of u and v only, and do not involve x and y. These same coefficients occur in Eqs. (22), which is then a *linear system in the new independent variables u and v*. The advantages of this situation are obvious.

In obtaining Eqs. (20) to (22) it was assumed that the determinant:

$$(23) \quad J = \frac{\partial x}{\partial u}\frac{\partial y}{\partial v} - \frac{\partial x}{\partial v}\frac{\partial y}{\partial u} \equiv \frac{\partial(x,y)}{\partial(u,v)}$$

was different from zero.[51] From (21) and (23) it follows also that:

$$(24) \quad j = \frac{\partial u}{\partial x}\frac{\partial v}{\partial y} - \frac{\partial u}{\partial y}\frac{\partial v}{\partial x} = \frac{\partial(u,v)}{\partial(x,y)} = \frac{1}{J}.$$

The geometrical significance of these *Jacobian determinants*, or *Jacobians*, is well known: when P is displaced so as to cover a rectangle of area $dx\,dy$, the corresponding point Q covers a curvilinear "parallelogram" of area $j\,dx\,dy$; conversely, a rectangular area $du\,dv$ about Q corresponds to an area $J\,du\,dv$ about P. The case $J = \infty$ (i.e., $j = 0$) means that an area around P is mapped into a single arc through Q, while if J vanishes, an area about Q corresponds to an arc at P. (In either case the arc may degenerate to a single point.) In both cases, therefore, the transformation degenerates and the two systems (1) and (22) are no longer equivalent. The geometrical significance of these exceptional cases will be discussed in the following section, while the physical meaning of the transformation in general will appear later (e.g., Sec. 12.3).

7. Geometrical interpretation

When the correspondence between the x,y-plane and the u,v-plane is defined by differentiable functions $u(x,y)$, $v(x,y)$, or $x(u,v)$, $y(u,v)$, then in the infinitesimal neighborhoods of a pair of corresponding points P and Q the mapping can be considered a linear (affine) transformation:

$$(25)\qquad \begin{aligned} du &= \alpha_{11}\,dx + \alpha_{12}\,dy \\ dv &= \alpha_{21}\,dx + \alpha_{22}\,dy \end{aligned} \qquad \text{or} \qquad \begin{aligned} dx &= \beta_{11}\,du + \beta_{12}\,dv \\ dy &= \beta_{21}\,du + \beta_{22}\,dv. \end{aligned}$$

Here the coefficients α and β are, of course, partial derivatives:

$$(26)\qquad \alpha_{11} = \frac{\partial u}{\partial x}, \quad \alpha_{12} = \frac{\partial u}{\partial y}, \quad \cdots, \quad \beta_{11} = \frac{\partial x}{\partial u}, \quad \beta_{12} = \frac{\partial x}{\partial v}, \quad \cdots,$$

and the determinant J of (23) becomes

$$(27)\qquad J = \beta_{11}\beta_{22} - \beta_{12}\beta_{21} = \frac{1}{\alpha_{11}\alpha_{22} - \alpha_{12}\alpha_{21}}.$$

Either set α or β determines the other set, from the formulas (21). Thus, the specifying of a set of four independent parameters determines the affine transformation. The system (1) of two planar differential equations, which can now be written

$$(28)\qquad \begin{aligned} a_1\alpha_{11} + a_2\alpha_{12} + a_3\alpha_{21} + a_4\alpha_{22} &= a, \\ b_1\alpha_{11} + b_2\alpha_{12} + b_3\alpha_{21} + b_4\alpha_{22} &= b, \end{aligned}$$

serves to reduce the number of independent parameters α by two; i.e., the differential equations single out a subset of ∞^2 transformations from the set of ∞^4 possible transformations. The equations (28) are therefore not sufficient to determine completely the mapping of the neighborhood of P onto the neighborhood of Q. (It is different in the case of one-dimensional problems, where there is only one parameter to be determined and one differential equation to determine it: $du/dx = f(x,u)$ determines α in $du = \alpha\, dx$ as $\alpha = f$.) Two out of the four parameters can still be chosen. This may be done by supposing that two neighboring pairs of corresponding points $P(x,y)$, $Q(u,v)$ and $P_1(x_1, y_1)$, $Q_1(u_1, v_1)$ are known (see Fig. 49); this is equivalent to supposing that for one special set of values dx_1, dy_1 (not both zero), the corresponding values du_1, dv_1 are given, or

$$(29)\qquad \alpha_{11}\,dx_1 + \alpha_{12}\,dy_1 = du_1, \qquad \alpha_{21}\,dx_1 + \alpha_{22}\,dy_1 = dv_1.$$

Together, Eqs. (28) and (29) give four linear equations for the four parameters α. One must then consider whether this system determines the parameters in question. A unique solution is obtained if and only if the determinant of coefficients is different from zero. Taking ϕ as the angle between the x-direction and PP_1, we set $dx_1 = ds \cos\phi$ and $dy_1 = ds \sin\phi$. Then the determinant of coefficients of our system (28), (29), namely,

$$\begin{vmatrix} a_1 & a_2 & a_3 & a_4 \\ b_1 & b_2 & b_3 & b_4 \\ dx_1 & dy_1 & 0 & 0 \\ 0 & 0 & dx_1 & dy_1 \end{vmatrix}$$

is exactly the right-hand member of Eq. (2′) except for a factor $-(ds)^2$, with the a's and b's evaluated for the given values of the coordinates of $P(x,y)$ and $Q(u,v)$.

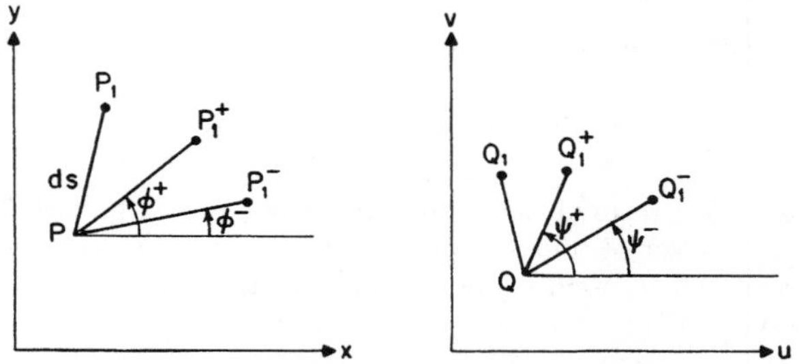

FIG. 49. Increments in corresponding planes.

Thus, in the elliptic case, when Eq. (2′) has no real roots, the equations (28) and (29) always determine the parameters α for arbitrary values of du_1 and dv_1, i.e., PP_1 and QQ_1 may be chosen arbitrarily. In the hyperbolic case, however, there are two characteristic directions through P, determined by Eq. (2′), for the given correspondence P, Q. If PP_1 is not one of these directions, the equations can be solved, as above, for arbitrary Q_1. But if PP_1 is a characteristic, say a σ^+, the four equations (28) and (29), whose determinant vanishes, are in general (i.e., for arbitrary right sides) not consistent. They will be consistent only for particular right sides such that the four equations become linearly dependent. In this case a definite relation must exist between du_1 and dv_1. It is obvious that this relation must be the compatibility condition. To show it formally write in Eqs. (28) and (29), for brevity, x_1, x_2, x_3, x_4 instead of α_{11}, α_{12}, α_{21}, α_{22}. Then, exactly as on p. 119, multiply the first Eq. (28) by B_1 (with $\phi = \phi^+$), the second by $-A_1$, and add. In view of the identities (4a) and (4c) this gives immediately

$$\Delta_{12}(x_1 \cos \phi^+ + x_2 \sin \phi^+) + K(x_3 \cos \phi^+ + x_4 \sin \phi^+) = B_1 a - A_1 b.$$

Then multiply the first Eq. (29) by Δ_{12}, the second by K, and add; the result is:

$$\Delta_{12}(x_1 \cos \phi^+ + x_2 \sin \phi^+) + K(x_3 \cos \phi^+ + x_4 \sin \phi^+) = \frac{du_1}{d\sigma^+}\Delta_{12} + \frac{dv_1}{d\sigma^+} K.$$

These equations are consistent only if

$$\frac{du_1}{d\sigma^+}\Delta_{12} + \frac{dv_1}{d\sigma^+}K = B_1 a - A_1 b, \tag{30}$$

and this is the first Eq. (5) for the σ^+-direction. Hence, if PP_1 has the σ^+-direction, Eq. (30) must hold for the increments du_1 and dv_1 of QQ_1, in order to make the equations consistent. In this case there exists a one-parameter family of solutions α_{ik}.

The geometric interpretation can be followed up further in the case where the right sides of the given equations vanish: $a = b = 0$. Then the condition (30) leads to

$$\tan \psi = \frac{dv_1}{du_1} = -\frac{\Delta_{12}}{K}, \tag{31}$$

where ψ is the angle between the u-direction and QQ_1. On using the second of the expressions in (4c) for K, Eq. (31) may be written

$$\Delta_{24} \cot \phi - \Delta_{12} \cot \psi = \Delta_{14}, \tag{32}$$

where $\phi^\pm$ corresponds to $\psi^\pm$ (see Fig. 49).

Furthermore, in this case ($a = b = 0$) the interchange of variables transforms (1) into the system (22), which may be written as

$$a_4\beta_{11} - a_2\beta_{12} - a_3\beta_{21} + a_1\beta_{22} = 0,$$

$$b_4\beta_{11} - b_2\beta_{12} - b_3\beta_{21} + b_1\beta_{22} = 0,$$

with the β_{ik} as in (26). For this system the equation analogous to (2′) is

$$\Delta_{12} \cos^2 \psi + (\Delta_{14} - \Delta_{23}) \cos \psi \sin \psi + \Delta_{34} \sin^2 \psi = 0, \tag{33}$$

which is satisfied by $\cot \psi$ as given in Eq. (32) whenever $\cot \phi$ satisfies Eq. (2′). Therefore, if PP_1 is a characteristic direction for the system (1), with $a = b = 0$, the equations (28) and (29) are consistent only when the corresponding direction QQ_1 is also characteristic for the transformed system; to each characteristic direction ϕ^+ or ϕ^- in the x,y-plane, there corresponds a definite characteristic direction, ψ^+ or ψ^-, in the u,v-plane, given by (32). Thus we can state: A system of two planar equations between x,y and u,v with vanishing right sides determines for each pair of corresponding points $P(x,y)$ and $Q(u,v)$ a set of ∞^2 affine transformations, *each of which maps two directions ϕ^+ and ϕ^- through P into two directions ψ^+ and ψ^- through Q. These two pair of directions (real only in the hyperbolic case) are the characteristic directions of the system* (1) *and of the transformed system* (22), *respectively.*

The directions ϕ^+ and ϕ^- are determined from Eq. (2′) as soon as P and Q are given; the directions ψ^+ and ψ^- are then determined from Eq. (32). A particular affine transformation can be singled out by giving the two quotients QQ_1^+/PP_1^+ and QQ_1^-/PP_1^-, where P_1^+, P_1^-, Q_1^+, and Q_1^- are points on the corresponding characteristics. (That these quotients, and not the lengths themselves, fix the transformation follows from the fact that multiplying dx_1, dy_1, du_1, and dv_1 by the same constant leaves Eqs. (29) unaltered.) The transformation is determined, except for a scale factor, by the ratio of these quotients. In particular, if the first quotient is zero, Eqs. (25) show that $1/D$ must vanish [see Eq. (27)]; the area around P is mapped into a segment of the line QQ_2, at least if only first-order terms are considered. Thus, if $D = \infty$ (or $D = 0$), the area around P (or around Q) is mapped into a curve element in the other plane *whose direction is one of the two characteristic directions in that plane.* This phenomenon, known as that of the *limit line* of the x,y-plane in the case $D = 0$, and that of the *edge* in the u,v-plane in the case $D = \infty$, appears repeatedly in connection with applications. A general discussion will be given in Art. 19.

CHAPTER III

ONE-DIMENSIONAL FLOW

Article 11

Steady Flow with Viscosity and Heat Conduction

1. General equations for parallel nonsteady flow

A one-dimensional, or parallel flow is one in which (a) the velocity vector **q** is always parallel to a fixed direction, and (b) all derivatives in directions perpendicular to this direction vanish. Taking the x-axis in the direction of motion, we have

$$(1) \qquad q_x = u, \qquad q_y = q_z = 0, \qquad \frac{\partial}{\partial y} = \frac{\partial}{\partial z} = 0.$$

In this case the divergence of **q** reduces to $\partial u/\partial x$, and the equation of continuity (1.II) becomes

$$(2) \qquad \frac{\partial}{\partial x}(\rho u) + \frac{\partial \rho}{\partial t} = 0.$$

In Newton's equation (1.I), when gravity is omitted, only the first component remains, giving

$$\rho \frac{du}{dt} = -\frac{\partial p}{\partial x}$$

in the case of inviscid flow.

In the more general case of a viscous fluid it is clear that for any volume element $dx\,dy\,dz$ (Fig. 6) no shearing stresses due to viscosity (friction) can exist on the four faces parallel to the x-direction, since the particles adjacent to those faces move along with the same speed. Consequently, all shearing components vanish, and the only viscosity effect in the x-direction is the normal tensile stress σ_x' (positive when directed outward) on the two faces perpendicular to this direction. Thus the quantity p used in the inviscid case is to be replaced by $p - \sigma_x'$, leading to

$$(3) \qquad \rho \frac{du}{dt} = -\frac{\partial}{\partial x}(p - \sigma_x'),$$

in agreement with Eq. (3.8). The relation of σ_x' to the velocity u will be discussed presently.

As *specifying condition* we suppose that the motion is adiabatic except for heat conduction, i.e., simply adiabatic (see Sec. 1.5). The equation specifying simply adiabatic conditions, Eq. (3.23), was discussed at the end of Sec. 3.5; when Q' is replaced by the left-hand member of the energy equation (3.19), with gravity omitted, we have [see Eq. (3.24)]

$$\frac{d}{dt}\left(\frac{q^2}{2} + U\right) + \frac{w + w'}{\rho} = \frac{1}{\rho} \operatorname{div} (k \operatorname{grad} T)$$

in the general case of simply adiabatic flow, and

$$\frac{d}{dt}\left(\frac{u^2}{2} + U\right) + \frac{w + w'}{\rho} = \frac{1}{\rho}\frac{\partial}{\partial x}\left(k\frac{\partial T}{\partial x}\right) \tag{4}$$

in the one-dimensional case. Here w is defined by Eq. (2.5) and w' by Eq. (3.10); in the one-dimensional case these equations give

$$w = \frac{\partial(pu)}{\partial x}, \qquad w' = -\frac{\partial(u\sigma_x')}{\partial x}, \qquad w + w' = \frac{\partial[u(p - \sigma_x')]}{\partial x}.$$

If we assume, further, that the fluid is a perfect gas, then the relations (1.6) and (1.9) hold:

$$p = gR\rho T \qquad \text{and} \qquad c_v = \frac{gR}{\gamma - 1}. \tag{5}$$

Then U [see Eq. (2.13)] is given by $c_vT = gRT/(\gamma - 1)$. Thus the specifying condition is

$$\frac{d}{dt}\left(\frac{u^2}{2} + \frac{gR}{\gamma - 1}T\right) + \frac{1}{\rho}\frac{\partial}{\partial x}[u(p - \sigma_x')] = \frac{1}{\rho}\frac{\partial}{\partial x}\left(k\frac{\partial T}{\partial x}\right). \tag{4$'$}$$

Equations (2) through (5) are four equations for the four unknowns u, p, ρ, and T, provided that σ_x' is expressed in terms of these variables.[1] The usual assumption, made in the Navier-Stokes theory, is that σ_x' is proportional to the rate of expansion $\partial u/\partial x$, as mentioned at the end of Sec. 3.1:

$$\sigma_x' = \mu_0 \frac{\partial u}{\partial x}. \tag{6}$$

Here μ_0 is a function, presumed to be known, of some or all of the variables u, p, ρ, and T. The value

$$\mu = \tfrac{3}{4}\mu_0,$$

however, rather than μ_0 itself, is usually called the (physical) viscosity, or coefficient of viscosity, for the following reason.

Under the assumption of one-dimensional flow, the y- and z- directions are interchangeable. Then σ_y' and σ_z' must be equal, and the condition (3.7), $\sigma_x' + \sigma_y' + \sigma_z' = 0$, gives

$$\sigma_y' = \sigma_z' = -\tfrac{1}{2}\,\sigma_x' . \tag{7}$$

Now consider an element of quadratic cross section $dx = dy$. Equilibrium considerations of the half-cell (Fig. 50) with cross section ABC require

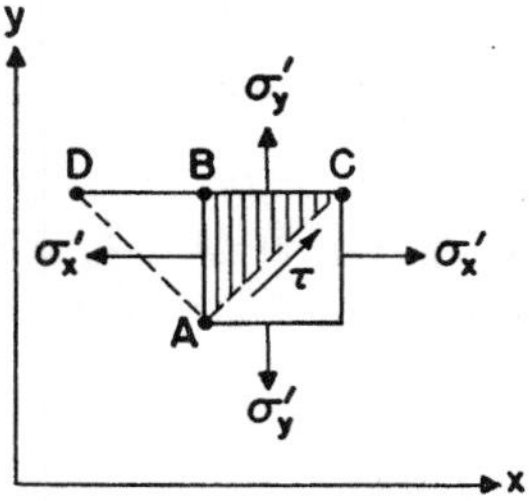

FIG. 50. The viscous stresses on a rectangular fluid element in one-dimensional flow.

that on the face represented by AC a shearing force must result in the direction AC of magnitude

$$-\frac{1}{\sqrt{2}}\,(\sigma_y'\,dx\,dz) + \frac{1}{\sqrt{2}}\,(\sigma_x'\,dy\,dz) = \frac{1}{\sqrt{2}}\,(\sigma_x' - \sigma_y')\,dx\,dz,$$

and since the area of this diagonal surface element is $(dx\,\sqrt{2})\,dz$, the shearing stress is

$$\tau = \tfrac{1}{2}(\sigma_x' - \sigma_y') = \tfrac{3}{4}\sigma_x' .$$

On the other hand, the rate of shear η, i.e., the time rate of change of the angle CAD, where DA is perpendicular to AC in the x,y-plane, is determined by the differences between the velocities at A, C, and D. Point C advances against A by $(\partial u/\partial x)\,dx$ units of length per second, and D is left behind by the same amount. Thus

$$\eta = \frac{1}{AC}\left[\left(\frac{\partial u}{\partial x}\,dx\,\frac{1}{\sqrt{2}}\right) - \left(-\frac{\partial u}{\partial x}\,dx\,\frac{1}{\sqrt{2}}\right)\right] = \frac{\partial u}{\partial x}$$

and, using (6), the ratio of shearing stress to rate of shear is $\tfrac{3}{4}\mu_0 = \mu$. For its numerical value, see Sec. 5.

2. Equations for steady motion

The partial differential equations (2), (3), and (5), in which the independent variables are x and t, become ordinary differential equations in x when a state of steady motion, $\partial/\partial t = 0$, is considered. In the Euler rule of differentiation, (1.4), d/dt reduces to $u(d/dx)$ and the three differential equations are now

$$\frac{d}{dx}(\rho u) = 0, \qquad \rho u \frac{du}{dx} + \frac{d(p - \sigma_x')}{dx} = 0, \tag{8}$$

$$u \frac{d}{dx}\left(\frac{u^2}{2} + \frac{gR}{\gamma - 1} T\right) + \frac{1}{\rho}\frac{d}{dx}[u(p - \sigma_x') - K] = 0;$$

here K represents the heat flux $k(dT/dx)$.

The first equation shows that the mass flux

$$m = \rho u, \tag{9}$$

or rate of flow of mass across a unit cross section normal to the x-axis is a constant. Using this, the second equation gives

$$mu + p - \sigma_x' = \text{constant} = C_1 m, \tag{10}$$

say, and the third equation yields

$$m\left(\frac{u^2}{2} + \frac{gR}{\gamma - 1} T\right) + u(p - \sigma_x') - K = \text{constant} = C_2 m,$$

say. From (10) we have $p - \sigma_x' = m(C_1 - u)$, so that the last equation may be written

$$m\left(-\frac{u^2}{2} + \frac{gR}{\gamma - 1} T\right) - K = m(C_2 - C_1 u). \tag{11}$$

Finally, p may be replaced in (10) by use of the equation of state (4): $p = gR\rho T = mgRT/u$. When Eqs. (10) and (11) are solved for $\sigma_x' = \mu_0\, du/dx$ and $K = k\, dT/dx$ respectively, there results

$$\frac{\mu_0}{m}\frac{du}{dx} = u + gR\frac{T}{u} - C_1, \tag{12}$$

$$\frac{k}{m}\frac{dT}{dx} = C_1 u - C_2 - \frac{u^2}{2} + \frac{gR}{\gamma - 1} T.$$

These are two simultaneous ordinary differential equations for u and T. The solutions of this system, depending upon the constants C_1, C_2, and m already introduced, and upon two additional constants of integration, represent all possible patterns of one-dimensional steady flow of a perfect gas with viscosity and heat conduction.[2]

It is convenient to replace u and T by dimensionless variables v and Θ; since C_1 has the dimensions of velocity [see Eq. (10)] and C_2 the dimensions of velocity squared, we introduce

$$c = \frac{C_2}{C_1^2}, \qquad v = \frac{u^2}{2C_1^2}, \qquad \Theta = \frac{gRT}{C_1^2} = \frac{p}{C_1^2\rho}. \tag{13}$$

Then the system (12) takes the form

$$\frac{\mu_0}{m}\frac{dv}{dx} = 2v - \sqrt{2v} + \Theta$$

(14)

$$\frac{1}{gR}\frac{k}{m}\frac{d\Theta}{dx} = \frac{\Theta}{\gamma - 1} - v + \sqrt{2v} - c.$$

The solutions v and Θ of this system will depend on four arbitrary constants: m, c, and two constants of integration. Of these, m occurs in the equations only as a factor of dx and therefore in the general solution only as a factor of x, and one constant of integration can be absorbed by translating the origin $x = 0$ (since Eqs. (14) are unchanged in form if, instead of x, $x' = x + C'$ is used as the independent variable). Thus, except for similarity transformations: $x'' = mx + C''$, the solution depends on only two parameters.

Before discussing the general equations (14), we turn to a special case.

3. Steady flow without heat conduction

G. I. Taylor has shown[3] that the system (14) can be integrated in closed form when k is set equal to zero.* This is not a realistic assumption, since it is known that the ratio μ_0/k varies over a small finite range (see Sec. 5). It will be seen later, however, that some principle features of the flow can be found in the solution of (14) under the assumption $k = 0$.

Eliminating Θ from Eqs. (14), with $k = 0$, we find

$$\frac{\mu_0}{m}\frac{dv}{dx} = (\gamma + 1)v - \gamma\sqrt{2v} + c(\gamma - 1). \tag{15}$$

Except for the scale factor m and translation of x, the solution of this equation depends on only one parameter, c.

Suppose that c lies within the limits

$$0 < c < \frac{\gamma^2}{2(\gamma^2 - 1)} = \frac{49}{48} \qquad (\gamma = 1.4). \tag{16}$$

Then there exist two real positive values v_1 and v_2 ($v_1 > v_2$, say) for which the right-hand member of (15) vanishes:

$$(\gamma + 1)v - \gamma\sqrt{2v} + (\gamma - 1)c = 0, \tag{17}$$

and it can be written as

$$(\gamma + 1)v - \gamma\sqrt{2v} + c(\gamma - 1) = (\gamma + 1)(\sqrt{v} - \sqrt{v_1})(\sqrt{v} - \sqrt{v_2}). \tag{17'}$$

For constant μ_0 and for v between v_1 and v_2, the solution of (15) for x as

* Equation (3.23) shows that this amounts to strictly adiabatic flow.

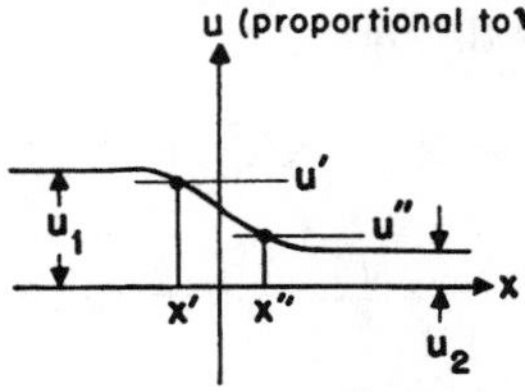

FIG. 51. Variation of velocity u (or $\sqrt{v}$) with position x, for $k = 0$.

a function of v takes the form

$$x = \frac{2}{\gamma + 1} \frac{\mu_0}{m} \frac{\sqrt{v_1} \log (\sqrt{v_1} - \sqrt{v}) - \sqrt{v_2} \log (\sqrt{v} - \sqrt{v_2})}{\sqrt{v_1} - \sqrt{v_2}} + \text{constant.} \tag{18}$$

Here* x decreases from $+\infty$ to $-\infty$ as v increases from v_2 to v_1. For $v > v_1$ the argument of the first logarithm in (18) must be changed in sign, and then x increases from $-\infty$ to $+\infty$ as v increases from v_1 to $+\infty$. For $0 < v < v_2$ the argument of the second logarithm must be changed, and then x increases from a finite value to $+\infty$ as v increases from 0 to v_2. If we now restrict our attention to flows for which the state variables tend to finite limits as $x \to \pm\infty$, then these last two branches of the solution may be neglected and attention focused on the function $v(x)$ defined by (18). Figure 51 shows $\sqrt{v}$ as a function of x; since $\sqrt{v}$ is proportional to the velocity u, we also obtain the graph of u as a function of x merely by taking a different scale on the vertical axis. Since ρu = constant, $\sqrt{v}$ is also inversely proportional to ρ. Two considerations are of major interest to us: the relation between the initial and final values of u, and the steepness of the descent from u_1 to u_2.

Since $\sqrt{v_1}$ and $\sqrt{v_2}$ are roots of (17), considered as a quadratic equation in $\sqrt{v}$, we have

$$\sqrt{v_1} + \sqrt{v_2} = \frac{\sqrt{2j}}{\gamma + 1} \tag{19}$$

or, in terms of u, using (13),

$$\frac{u_1 + u_2}{2} = \frac{\gamma}{\gamma + 1} C_1, \tag{19'}$$

which gives an interpretation of the constant C_1. When $k = 0$, the second

* For the remainder of this article it is assumed that m is positive, i.e., x is chosen to increase along the direction of the flow.

of Eqs. (14) shows that v and Θ satisfy

$$\frac{\Theta}{\gamma - 1} - v + \sqrt{2v} - c = 0. \tag{20}$$

Eliminating c from (20) and (17), which is valid however only for $v = v_1$ or v_2, and then eliminating $\sqrt{2v}$ from the same equations, we find

$$\Theta_i + 2v_i - \sqrt{2v_i} = 0 \quad \text{and} \quad \frac{\gamma}{\gamma - 1}\Theta_i + v_i = c \qquad (i = 1, 2), \tag{21}$$

so that

$$\begin{aligned} \frac{\Theta_1}{\sqrt{2v_1}} + \sqrt{2v_1} &= \frac{\Theta_2}{\sqrt{2v_2}} + \sqrt{2v_2}, \\ \frac{\gamma}{\gamma - 1}\Theta_1 + v_1 &= \frac{\gamma}{\gamma - 1}\Theta_2 + v_2. \end{aligned} \tag{21$'$}$$

When v and Θ are replaced by their values from (13), $v = u^2/2C_1^2$ and $\Theta = p/C_1^2\rho$, the first equation multiplied by $m = \rho u$ and C_1 gives

$$p_1 + mu_1 = p_2 + mu_2, \tag{22}$$

and the second, multiplied by C_1^2, gives

$$\frac{u_1^2}{2} + \frac{\gamma}{\gamma - 1}\frac{p_1}{\rho_1} = \frac{u_2^2}{2} + \frac{\gamma}{\gamma - 1}\frac{p_2}{\rho_2}. \tag{23}$$

Equation (23) is the same as the Bernoulli equation (see Eq. (2.20′) with gravity omitted) found in the case of a steady, strictly adiabatic, inviscid flow, and implies conservation of energy. Equation (22) may be interpreted as expressing the conservation of momentum. Finally, the continuity equation (9) yields

$$\rho_1 u_1 = \rho_2 u_2, \tag{24}$$

or conservation of mass. Equations (22), (23), and (24) also follow directly from (9), (10), and (11) on setting $\sigma_x = K = 0$ for $x = \pm\infty$.

In studying the transition from u_1 to u_2, we use Eq. (18), in which $\sqrt{v}$, $\sqrt{v_1}$, and $\sqrt{v_2}$ can be replaced by u, u_1, and u_2 without further change, except in the additive constant. Let ϵ be any number satisfying $0 < \epsilon < \frac{1}{2}$ and consider the two intermediate values u' and u'' (Fig. 51) satisfying

$$u_1 - u' = \epsilon(u_1 - u_2) = u'' - u_2 \qquad (0 < \epsilon < \tfrac{1}{2}). \tag{25}$$

Then also $u_1 - u'' = (1 - \epsilon)(u_1 - u_2)$, and the change from u' to u'' is $(1 - 2\epsilon)$ times the total velocity change from u_1 to u_2. The difference between the abscissas x corresponding to these values u is

$$L_0 = x'' - x' = \frac{2}{\gamma + 1} \log \left(\frac{1}{\epsilon} - 1 \right) \frac{\mu_0}{m} \frac{u_1 + u_2}{u_1 - u_2}. \tag{26}$$

For given values of the flux m and the ratio u_2/u_1, the right-hand side tends to zero as μ_0 decreases, no matter how small ϵ may be. Let ρ^* denote the value of the density at that point of flow where $u = (u_1 + u_2)/2$ and u^* the velocity at the point where $\rho = (\rho_1 + \rho_2)/2$; then the last factor in (26) may be written in two ways:

$$\frac{\mu_0}{m} \frac{u_1 + u_2}{u_1 - u_2} = \frac{\mu_0}{m} \frac{\rho_2 + \rho_1}{\rho_2 - \rho_1} = \frac{2\mu_0}{\rho^*(u_1 - u_2)} = \frac{2\mu_0}{u^*(\rho_2 - \rho_1)}. \tag{27}$$

If, for example, we take $\epsilon = 0.05$ and use the standard values (see Sec. 5) $\mu_0 \sim 5 \times 10^{-7}$ slug/ft sec and $\rho^* \sim 0.0025$ slug/ft^3, then (26) and (27) give $x'' - x' \sim 0.001/(u_1 - u_2)$. Thus, if the total velocity drop $u_1 - u_2$ amounts to 10 ft/sec, then 90 per cent of this drop is effected within a distance of 0.0001 ft, or about 0.03 mm. This is a significant result: *the thickness of the layer within which occurs the major part of the transition from u_1, p_1, ρ_1 to u_2, p_2, ρ_2 tends toward zero with μ, and is actually extremely small in air under normal conditions.*

If we use $a^2 = \gamma p/\rho$ as the expression for the sound velocity, as is usual in discussing the adiabatic flow of an inviscid fluid, the Mach number is given in terms of our dimensionless variables by

$$M^2 = \frac{u^2}{a^2} = \frac{u^2}{\gamma p/\rho} = \frac{u^2}{\gamma g R T} = \frac{2v}{\gamma \Theta}. \tag{28}$$

Using the first equation (21), we obtain

$$\frac{1}{\gamma M_1^2} = \frac{\Theta_1}{2v_1} = \frac{1}{\sqrt{2v_1}} - 1, \qquad \frac{1}{\gamma M_2^2} = \frac{1}{\sqrt{2v_2}} - 1. \tag{29}$$

Since $v_1 > v_2$, Eqs. (29) imply $M_1 > M_2$. Moreover, v_1 and v_2 satisfy (17). If the right-hand member of (15), $(\gamma + 1)v - \gamma\sqrt{2v} + (\gamma - 1)c$, is considered as a function of $\sqrt{2v}$ its derivative is $(\gamma + 1)\sqrt{2v} - \gamma$, which vanishes only for $\sqrt{2v} = \gamma/(\gamma + 1)$. Since the zero of the derivative must lie between the zeros $\sqrt{2v_1}$ and $\sqrt{2v_2}$ of the function, $\sqrt{2v_2} < \gamma/(\gamma + 1) < \sqrt{2v_1}$ follows. If these inequalities are introduced into (29), we find that $M_1^2 > 1$ and $M_2^2 < 1$. *The transition flow represented by* (18) *begins in a supersonic and ends in a subsonic state.* The inflection point of the curve of v against x occurs when $\sqrt{2v} = \gamma/(\gamma + 1)$. This value of $\sqrt{2v}$ would correspond to $M = 1$ if (29) held for all the flow; for v between v_1 and v_2, however, the left-hand member of (17) is negative [see (17′)] and we find $M > 1$ at the inflection point.

An interpretation of c may be found by combining Eqs. (28) and (29) with the second of Eqs. (21) for v_1, Θ_1, or for v_2, Θ_2; one finds

$$c = \frac{M_1^2}{2} \frac{M_1^2 + 2/(\gamma - 1)}{[M_1^2 + 1/\gamma]^2} = \frac{M_2^2}{2} \frac{M_2^2 + 2/(\gamma - 1)}{[M_2^2 + 1/\gamma]^2}. \tag{30}$$

From the last equality we find also

$$2\gamma M_1^2 M_2^2 = 2 + (\gamma - 1)(M_1^2 + M_2^2), \tag{30'}$$

as a relation between the Mach numbers M_1 and M_2 before and after the transition, which will be of importance later on.

4. The complete problem[4]

If we do not neglect heat conduction, as in Sec. 3, a steady one-dimensional flow is determined by the two first-order differential equations (14) in the unknowns v and Θ. One could eliminate either v or Θ to obtain one second-order differential equation for Θ or v, respectively. The resulting equation, however, would not be easy to handle, and a better procedure is to eliminate x by dividing the two equations (14), obtaining

$$\frac{d\Theta}{dv} = \frac{\mu_0 g R}{k} \frac{\Theta/(\gamma - 1) - v + \sqrt{2v} - c}{\Theta + 2v - \sqrt{2v}}. \tag{31}$$

For each value of c the solutions of the first-order differential equation (31) consist of a set of ∞^1 curves in the v,Θ-plane; the solutions of (31) represent, therefore, ∞^2 different v,Θ-relations. For each such solution the relations between x and v and between x and Θ may then be found by quadratures, using the original equations (14).

The coefficient of the right-hand member in (31) is, of course, a dimensionless quantity. It is customary to use the name *Prandtl number*[5] for the quotient

$$P = \frac{\gamma c_v \mu}{k} = \frac{\gamma}{\gamma - 1} gR \frac{\mu}{k}. \tag{32}$$

For dry air under normal conditions, the value of P varies only slightly, roughly between 0.68 and 0.77 (see Sec. 5). We then write (31) in the form

$$\frac{d\Theta}{dv} = \frac{4}{3} \frac{\gamma - 1}{\gamma} P\lambda, \tag{33}$$

where

$$\lambda = \frac{\Theta/(\gamma - 1) - v + \sqrt{2v} - c}{\Theta + 2v - \sqrt{2v}}. \tag{34}$$

We first consider the case P = constant; then a curve λ = constant in the v,Θ-plane is an isocline of Eq. (31), i.e., the locus of all points at which the solutions of (31) have a given slope, namely, $4P\lambda(\gamma - 1)/3\gamma$. The isoclines have equations of the form:

$$(av + b\Theta + c)^2 = 2(\lambda + 1)^2 v$$

where $a = 2\lambda + 1$, $b = \lambda - 1/(\gamma - 1)$, and all pass through the singular points which make both the numerator N and the denominator D of λ vanish. The points of intersection of the isoclines $D = 0$, $N = 0$ are the same as those of the isoclines $D = 0$, $N + D = 0$. Hence consider the two isoclines:

(35)
$$D = 0, \qquad \lambda = \pm\infty$$
$$\Theta + 2v - \sqrt{2v} = 0 \qquad \text{or} \qquad (\Theta + 2v)^2 = 2v$$

and

(36)
$$N + D = 0, \qquad \lambda = -1$$
$$\frac{\gamma}{\gamma - 1}\Theta + v - c = 0 \qquad \text{or} \qquad \Theta = \frac{\gamma - 1}{\gamma}(c - v).$$

Eliminating Θ between (35) and (36), which hold simultaneously at a point of intersection (Θ_i, v_i), we find

(37)
$$(\gamma + 1)v_i - \gamma\sqrt{2v_i} + (\gamma - 1)c = 0 \qquad (i = 1, 2).$$

Note that (35), (36), and (37) are exactly the same equations as (21) and (17), holding at the beginning and the end of the special flow considered in Sec. 3.* It follows then, as in Sec. 3, that the u-, p-, ρ-, and T-values for the points 1 and 2 satisfy Eqs. (22), (23), and (24). Also for M_1, M_2, Eqs. (30) and (30′) hold.

As seen previously, (37) will have two distinct positive roots in v if and only if c satisfies (16). For $c < \frac{1}{2}$, however, one or both of the corresponding values of Θ will be negative. Thus if

(38)
$$\frac{1}{2} < c < \frac{\gamma^2}{2(\gamma^2 - 1)} = \frac{49}{48},$$

there will be two distinct points of intersection, 1 and 2, with both v and Θ positive, and they are the same points as in the preceding section. For $c = \frac{1}{2}$, one point of intersection is $v = \frac{1}{2}$, $\Theta = 0$; from (28) it is seen that this point corresponds to $M = \infty$. When $c = 49/48$, the points 1 and 2 coincide at $v = \gamma^2/2(\gamma + 1)^2$, $\Theta = \gamma/(\gamma + 1)^2$, corresponding to $M = 1$. From the result at the end of Sec. 3 it is clear that the straight line $2v/\gamma\Theta = 1$, on

* This can be easily understood since in the second Eq. (14) the result is the same whether k or $d\Theta/dx$ is put equal to zero.

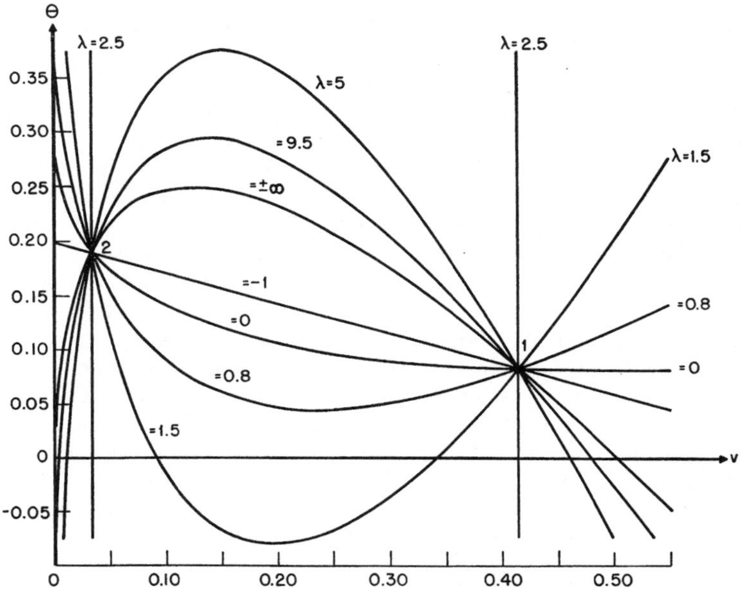

FIG. 52. Set of parabolas λ = constant in the velocity-temperature plane, for $c = 0.7$.

which $M = 1$, separates the points 1 and 2. In what follows we assume that (38) holds.

Figure 52 shows a full set of λ-curves; except for $\lambda = -1$ and $\lambda = 1/(\gamma - 1)$, all these curves are parabolas passing through the points 1 and 2 and tangent to the Θ-axis. The parabola $\lambda = \pm\infty$, Eq. (35), is independent of c, has a vertical tangent at $v = \Theta = 0$, and passes through the point $v = \frac{1}{2}$, $\Theta = 0$. It intersects the curve

$$\text{(39)} \qquad \lambda = 0,\ N = 0: \qquad \left(\frac{\Theta}{\gamma - 1} - v - c\right)^2 = 2v$$

at the points 1 and 2, through which passes also the straight line $\lambda = -1$, Eq. (36). The slope of the latter, $-(\gamma - 1)/\gamma$, does not depend on c, but its location does. For $\lambda = 1/(\gamma - 1)$, the λ-curve degenerates into a pair of vertical lines through 1 and 2, respectively. Using this set of λ-curves, for P constant, the integral curves for (31) may be found graphically (see Fig. 54 and following text).

For P not necessarily constant, the λ-curves are no longer isoclines of (31). The geometrical construction of the integral curves, by means of the so-called isocline method, fails. Further information about the solutions must come from the general theory of first-order differential equations. We now assume that P is, at least, a continuous function of v and Θ, not vanishing at either of the points 1 and 2. The points 1 and 2 are then singular points for the differential equation (31). If we analyze each of these singular points by considering the linear terms in the Taylor expansions of the numerator and denominator of (31) about that point, we find that 1 (supposing $v_1 > v_2$) is a nodal point, and 2 a saddle point. Thus exactly two integral curves pass through point 2, while an infinite number of solutions pass through point 1, all of them except one having a common tangent at 1. The two pairs of directions at 1 and 2 can be computed easily, as will be seen in Sec. 6. One and only one integral curve of (30) passes through both singular points, and this solution represents a transition between the two states v_1,Θ_1 and v_2,Θ_2. This corresponds to the solution which was considered in Sec. 3; there $k = 0$, so that $P = \infty$, and the complete solution is represented in the v,Θ-plane by the parabola $\lambda = 0$ (depending on c). Furthermore, in the shaded region of Fig. 53, between the parabolas (35) and (39), we have $N > 0$ and $D < 0$. Consequently $d\Theta/dx$ and dv/dx as determined by (14) are positive and negative, respectively, in this region, the solutions of (31) have negative slopes there, and the direction of the negative v-axis corresponds to increasing x. It will be shown in Sec. 6 that the transition curve lies in this region.

The integral curves in Fig. 54 have been found under the assumption that P = constant = $\frac{3}{4}$. Thus, from (33), the slope of an integral curve is $(\gamma - 1)\lambda/\gamma$ as it crosses any λ-curve. In particular, the slope of an integral

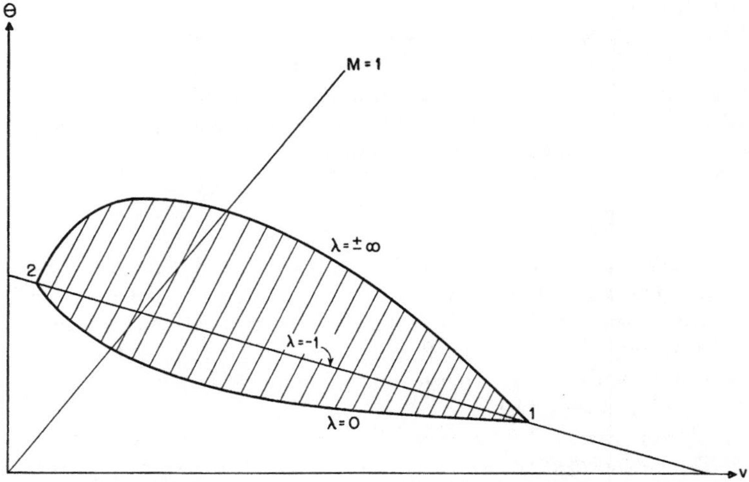

FIG. 53. The curves $\lambda = 0, -1, \pm\infty$ in the v,Θ-plane for $c = 0.7$.

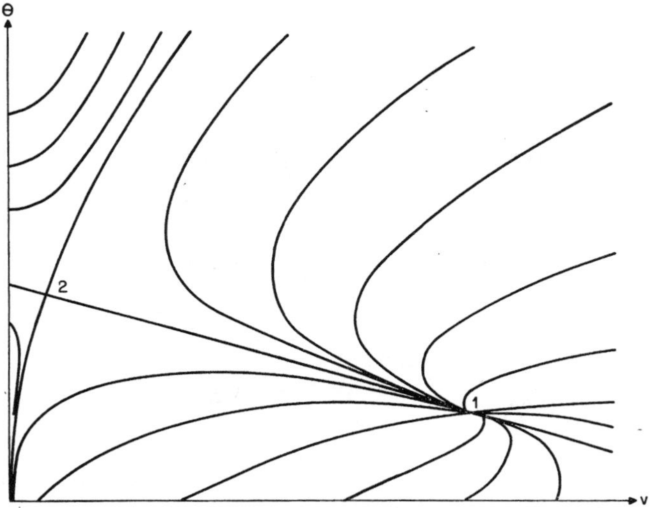

FIG. 54. The set of integral curves corresponding to Fig. 52 for Prandtl number $P = 3/4$. The transition curve from 1 to 2 is a straight line.

curve at any point of the straight line $\lambda = -1$ is $-(\gamma - 1)/\gamma$, which is also the slope of the straight line $\lambda = -1$, Eq. (36). In the case $P = \frac{3}{4}$, therefore, the transition from 1 to 2 occurs along the straight line (36). When this solution is expressed, by means of (13), in terms of the given variables u and T, one finds

$$\frac{u^2}{2} + \frac{\gamma}{\gamma - 1} gRT = \frac{u^2}{2} + c_pT = C_2 = \text{constant.} \tag{40}$$

This relation, valid only for $P = \frac{3}{4}$, was found by R. Becker in 1922.[6]

Finally, Fig. 55 shows the solutions* determining the transition from 1 to 2 for the three values $P = \frac{1}{2}$, $P = \frac{3}{4}$, and $P = 1$. These are found by computing the directions at the singular points 1 and 2, and then by graphical integration from a diagram such as Fig. 52. The interpretation of these results will be given in Sec. 6.

5. Numerical data

The conclusions to be drawn in the following section depend, to a certain extent, upon the numerical values of the physical constants which appear in the equations. Therefore we begin by establishing a set of standard values of constants, referring to the conditions of air at room temperature under

* The ordinates have been magnified 5:1.

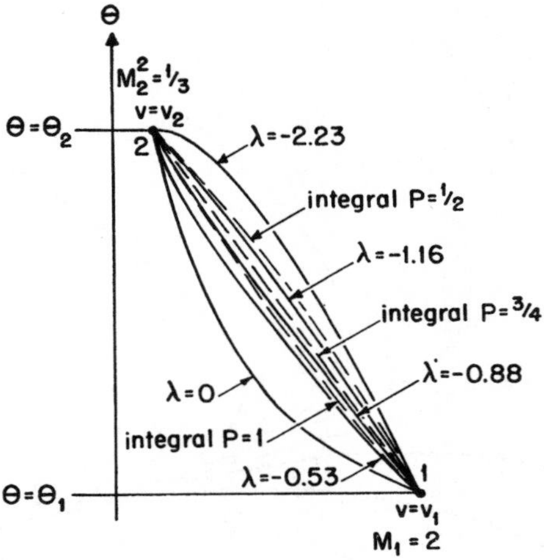

FIG. 55. The transition curves for $P = \frac{1}{2}, \frac{3}{4}$, and 1, with bounding parabolas.

a pressure of 1 atmosphere. For the convenience of the reader there follows a list of the numerical values of constants employed in this book.

(a) General constant:

$$g = 32.17 \text{ ft/sec}^2 = 980.7 \text{ cm/sec}^2.$$

(b) Dry air constants (diatomic, perfect gas):

$$\gamma = \tfrac{7}{5} = 1.4; \qquad R = 53.33 \text{ ft/°F} = 29.26 \text{ m/°C};$$

$$c_v = \frac{gR}{\gamma - 1} = 4289 \text{ ft}^2/\text{sec}^2 \text{ °F} = 0.717 \times 10^7 \text{ cm}^2/\text{sec}^2 \text{ °C};$$

$$c_p = \gamma c_v = 6005 \text{ ft}^2/\text{sec}^2 \text{ °F} = 1.004 \times 10^7 \text{ cm}^2/\text{sec}^2 \text{ °C}.$$

(c) Assumed standard conditions:

Temperature = 59°F = 15°C, corresponding to the absolute temperature

$$T = 518.4 \text{ in Fahrenheit degrees} = 288 \text{ in Celsius degrees};$$

$$\rho = 0.00238 \text{ slug/ft}^3 = 0.001226 \text{ g/cm}^3;$$

$$p = gR\rho T = 2116 \text{ lb/ft}^2 = 1.013 \times 10^6 \text{ dynes/cm}^2;$$

$$a = \sqrt{\frac{\gamma p}{\rho}} = 1116 \text{ ft/sec} = 340.1 \text{ m/sec}.$$

(d) Viscosity (experimental values):

$$\mu = 3.73 \times 10^{-7} \text{ slug/ft sec} = 0.000179 \text{ g/cm sec};$$

$$\nu = \frac{\mu}{\rho} = 0.000157 \text{ ft}^2\text{/sec} = 0.146 \text{ cm}^2\text{/sec}.$$

(e) Heat conductivity (experimental values):

$$k = 0.00314 \text{ lb/sec °F} = 2.52 \times 10^3 \text{ dynes/sec °C}.$$

(f) Prandtl number:

$$P = \frac{c_p\mu}{k} = 0.713.$$

The coefficients of viscosity and heat conductivity depend, in general, on the state variables p, ρ, T. Experimental evidence indicates that μ and k (but not ν) can be considered as functions of temperature alone. An older formula of Lord Rayleigh[7] gives μ proportional to the $\frac{3}{4}$ power of T in the range between the freezing and boiling points of water:

$$\frac{\mu'}{\mu''} = \left(\frac{T'}{T''}\right)^{0.75}, \tag{41}$$

from which one derives

$$\frac{\nu'}{\nu''} = \left(\frac{T'}{T''}\right)^{0.75} \cdot \frac{\rho''}{\rho'} = \left(\frac{T'}{T''}\right)^{1.75} \cdot \frac{p''}{p'} = \left(\frac{p'}{p''}\right)^{0.75}\left(\frac{\rho'}{\rho''}\right)^{-1.75}. \tag{42}$$

A more recent investigation[8] gives, within the limits 32°F to 950°F,

$$\frac{\mu'}{\mu''} = \left(\frac{T'}{T''}\right)^{1.5} \times \left(\frac{T'' + 223.2}{T' + 223.2}\right), \tag{43}$$

when T' and T'' are absolute temperatures in Fahrenheit degrees. For example, if one uses the value given above for μ at $T = 518.4$, the first formula (41), gives $10^7\mu = 4.07$ slug/ft sec, while the second formula (43) yields 4.08 slug/ft sec at $T = 581.4$ (i.e., 122°F). Table II for ν is computed from Eq. (43).

A formula very similar to (43) has been found in experiments on heat conduction:[9]

$$\frac{k'}{k''} = \left(\frac{T'}{T''}\right)^{1.5} \times \left(\frac{T'' + 225}{T' + 225}\right). \tag{44}$$

From (43) and (44) it would follow that the ratio μ/k, and consequently the Prandtl number also, is essentially constant, giving $P = 0.713$ in general.

TABLE II—KINEMATIC VISCOSITY ν IN FT²/SEC
$p_0 = 2116.2 \text{ lb/ft}^2 = 29.92$ in Hg
Values of $\nu \times 10^4$

	Temperature (deg. F)						
p/p_0	14	32	50	68	86	104	122
1	1.33	1.42	1.52	1.62	1.72	1.82	1.92
10	0.133	0.142	0.152	0.162	0.172	0.182	0.192
20	0.066	0.071	0.076	0.081	0.086	0.091	0.096

On the other hand a theoretical value of P, based on the kinetic theory of gases, is[10]

$$P = \frac{4\gamma}{9\gamma - 5} = 0.737 \qquad \text{for} \qquad \gamma = 1.4. \tag{45}$$

In any case, the value 0.75 is a close approximation to the actual value of P.

6. Conclusions

We consider first the case when the Prandtl number P is 0.75. Then, as seen in Sec. 4, the solution passing through the singular points 1 and 2 in the v,Θ-plane, the transition curve, is the straight line (36). If the value of Θ from the solution (36) is substituted into the first of Eqs. (14), we find

$$\frac{\mu_0}{m}\frac{dv}{dx} = \frac{1}{\gamma}\left[(\gamma + 1)v - \gamma\sqrt{2v} + (\gamma - 1)c\right], \tag{46}$$

which differs only by the factor $1/\gamma$ from Eq. (15), which was derived under the assumption $k = 0$. Consequently the former discussion (Sec. 3) holds unchanged, provided that x is replaced by x/γ. In particular, the thickness of the transition layer defined there is now

$$L = \gamma L_0 \qquad (P = \tfrac{3}{4}), \tag{47}$$

where L_0 is the expression (26).

There remains the case of a finite P different from $\frac{3}{4}$, but constant, and the case in which P is not constant.

For general P the integral curves of (30) are qualitatively similar to those shown in Fig. 54 for the case $P = \frac{3}{4}$, except that the singular integral curve S joining the points 1 and 2 is no longer a straight line. In particular, the sign of $d\Theta/dv$ is unchanged in the various regions bounded by the isoclines $\lambda = \pm\infty$ and $\lambda = 0$ (these two curves being isoclines even for non-constant P). Although the solution S cannot be given in closed form, it is possible to find bounds for $\Theta(v)$ on S, and these are sufficient for giving significant estimates concerning the thickness of the transition layer.

First we compute the slopes of the integral curves at 1 and 2, in the case P = constant. Since $N = D = 0$ at these points, the slopes are found by applying l'Hospital's rule to the right-hand side of (31). Thus the slope $\Theta' = d\Theta/dv$ of any solution at 1 or 2 must satisfy the quadratic equation

$$\Theta' = \frac{4P}{3\gamma}\frac{\Theta' - (\gamma - 1)(1 - 1/\sqrt{2v_i})}{\Theta' + 2 - 1/\sqrt{2v_i}} \qquad (i = 1, 2), \tag{48}$$

or, using $1/\sqrt{2v_i} = 1 + 1/\gamma M_i^2$ from (29),

$$\Theta'^2 + \Theta'\left(1 - \frac{1}{\gamma M_i^2} - \frac{4P}{3\gamma}\right) - \frac{4P}{3\gamma}(\gamma - 1)\frac{1}{\gamma M_i^2} = 0. \tag{48'}$$

Thus, of the two values of Θ' at each of the two points, one is positive and one negative. As in Fig. 54, the two negative roots, say Θ_1' at 1 and Θ_2' at 2, must give the slopes of the transition curve S at the singular points 1 and 2, respectively. A more detailed study of (48') shows that $\Theta_1' < -(\gamma - 1)/\gamma < \Theta_2'$ when $P < \frac{3}{4}$, and that the inequalities are reversed if $P > \frac{3}{4}$. For $P = \frac{3}{4}$, Eq. (48') is satisfied by $\Theta_1' = \Theta_2' = -(\gamma - 1)/\gamma$, which agrees with (40).

Equation (48) also expresses the fact that any solution passing through 1 or 2 in one of the possible directions determined by (48) is actually tangent to the appropriate isocline at the point (i.e., to the isocline whose λ follows by (33) from the slope Θ'). In particular, the singular solution curve S which passes through 1 and 2 is tangent to a certain isocline parabola A_1 at 1 and tangent to another parabola A_2 at 2. The bounds on the solution S follow from the fact that S *must lie on the concave side of* A_1 *between* 1 *and* 2 *and on the convex side of* A_2.

For example, suppose $P < \frac{3}{4}$. Then, since $\Theta_1' < -(\gamma - 1)/\gamma$, the isocline A_1 lies above its chord 12, as shown in Fig. 56a. At any point of A_1, a solution of (30) has the same slope as the tangent T, at points below A_1 a less negative slope, and at points above A_1 a more negative slope (more vertical) or positive slope (beyond $\lambda = \pm\infty$). No integral curve can leave the point 1 in the area between A_1 and T, for, geometrically, such a curve must have a slope less negative than that of T near 1, whereas a solution of (31) must have a more negative slope than T. Along an integral curve leaving 1 above T, the tangent to the curve must rotate in a clockwise direction and it is clear that such a curve cannot reach the point 2. Thus the integral curve S runs below A_1 as it leaves the point 1. It cannot meet A_1 at a point P between 1 and 2 (this assumption is illustrated in Fig. 56a). For at P a solution has the same slope as T, but no longer the same slope

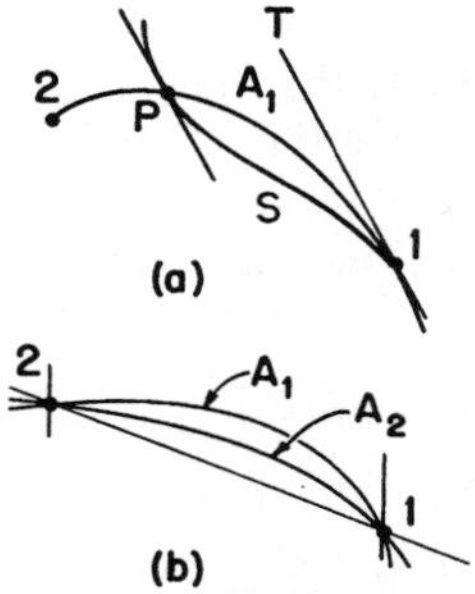

FIG. 56. The two isoclines bounding a transition curve when P is constant; shown for $P < 3/4$.

as A_1, and can only continue beyond P in a manner similar to the solutions leaving 1 above T. Thus S runs below A_1 between 1 and 2.

Since $\Theta_2' > -(\gamma - 1)/\gamma$ (which is the slope of the chord joining 1 and 2), the isocline A_2 lies above this chord (but of course below A_1, since $\Theta_1' < \Theta_2'$), as in Fig. 56b. An analogous argument shows that S must run above A_2 between 1 and 2, since a solution at a point of A_2, different from 1, has the slope Θ_2' and continues below A_2 with less negative slope, so that it cannot reach the point 2. Thus the two isoclines A_1 and A_2 bound a crescent-shaped area within which S must lie between 1 and 2. For $P > \frac{3}{4}$, these two isoclines lie below the chord joining the points 1 and 2. If $P = \frac{3}{4}$, then A_1, A_2, and S coincide with the chord, while if $k = 0$ (so that $P = \infty$) all coincide with the isocline $\lambda = 0$.

Now let α and β be the λ-values corresponding to the isoclines A_1 and A_2 with $\alpha \geqq \beta$, e.g., if $P < \frac{3}{4}$, then α corresponds to A_2 and β to A_1, and let $\Theta_\alpha(v)$ and $\Theta_\beta(v)$ denote the ordinates on these two isoclines and $\Theta(v)$ the ordinate on S. Then the fact that S runs between A_1 and A_2 is expressed by the inequalities

$$\Theta_\alpha(v) \leqq \Theta(v) \leqq \Theta_\beta(v), \qquad v_2 \leqq v \leqq v_1. \tag{49}$$

In the particular cases $P = \frac{3}{4}$ and $P = \infty$, equality holds throughout in (49).

Next we introduce the abbreviations

$$\begin{aligned} \Theta_\infty &= \sqrt{2v} - 2v \\ \Theta_0 &= (\gamma - 1)(v - \sqrt{2v} + c); \end{aligned} \tag{50}$$

i.e., Θ_∞ and Θ_0 are ordinates on the two isoclines $\lambda = \pm\infty$ and $\lambda = 0$ respectively. Then the first differential equation (14) may be written as

$$\frac{4}{3}\frac{\mu}{m}\frac{dv}{dx} = \Theta - \Theta_\infty,$$

so that (49) leads to

$$\Theta_\alpha - \Theta_\infty \leqq \frac{4}{3}\frac{\mu}{m}\frac{dv}{dx} \leqq \Theta_\beta - \Theta_\infty . \tag{51}$$

Also, Eq. (34) may be written as

$$\lambda = \frac{\Theta - \Theta_0}{(\Theta - \Theta_\infty)(\gamma - 1)},$$

from which it follows that the ordinate Θ_λ on the isocline corresponding to any value λ satisfies

$$\Theta_\lambda - \Theta_\infty = -\frac{\Theta_\infty - \Theta_0}{1 - (\gamma - 1)\lambda}.$$

Consequently, (51) may be written as

$$-\frac{\Theta_\infty - \Theta_0}{1 - (\gamma - 1)\alpha} \leqq \frac{4}{3}\frac{\mu}{m}\frac{dv}{dx} \leqq -\frac{\Theta_\infty - \Theta_0}{1 - (\gamma - 1)\beta}. \tag{52}$$

This result may be compared with the differential equation (15) determining v in the case $k = 0$, $P = \infty$, discussed in Sec. 3, namely,

$$\frac{4}{3}\frac{\mu}{m}\frac{dv}{dx} = -(\Theta_\infty - \Theta_0). \tag{53}$$

It is seen that the upper and lower limits for dv/dx, as determined by the equality signs in (52), lead to the same differential equation as (53), except that x is to be replaced by $x/[1 - (\gamma - 1)\alpha]$ and $x/[1 - (\gamma - 1)\beta]$, respectively. Consequently, if we assume μ constant, the thickness L of the transition layer must satisfy[11]

$$L_0[1 - (\gamma - 1)\alpha] \leqq L \leqq L_0[1 - (\gamma - 1)\beta], \tag{54}$$

where L_0 is the expression (26). Here α and β are the λ-values for the two isoclines A_1 and A_2 tangent to S at 1 and 2 respectively, with $\alpha \geqq \beta$. To determine the exact limits in (54), it is merely necessary to compute the slopes Θ_1' and Θ_2' of the solution $\Theta(v)$ at the points 1 and 2, i.e., the negative root of (48′) at each of these two points. Then α and β are the (negative) values

$$\frac{3}{4}\frac{\gamma}{\gamma - 1}\frac{\Theta_1'}{P} \quad \text{and} \quad \frac{3}{4}\frac{\gamma}{\gamma - 1}\frac{\Theta_2'}{P}; \tag{55}$$

here α is the second expression if $P < \frac{3}{4}$ and the first if $P > \frac{3}{4}$. When $P = \frac{3}{4}$, we find $\Theta_1' = \Theta_2' = -(\gamma - 1)/\gamma$, as mentioned above; in this case both brackets in (54) reduce to γ, in agreement with (47).

A similar argument can be applied to the case where μ is not constant, but a given increasing function of T. It follows from (52) that the estimate

(54) still holds, provided that L_0 on the right is computed for the maximum value of μ (the value at point 2) while L_0 on the left is computed for the minimum value of μ (the value at point 1).

Finally there remains the case of P not constant, to which the preceding theory can be adapted. The result is that if the expressions in (55) are now computed for the maximum and minimum values of P in the shaded region of Fig. 53, and the two extreme values taken for α and β, then the estimate (54) still holds. The transition curve lies between the parabolas $\lambda = \alpha$ and $\lambda = \beta$. Any variation in μ is treated as in the last paragraph.

Example

Suppose the initial Mach number is $M_1 = 2$. Then from (30′) we have $M_2^2 = \frac{1}{3}$ for $\gamma = 1.4$, while from (29) we find

$$\frac{1}{\sqrt{2v_1}} - 1 = \frac{1}{\gamma M_1^2} = 0.179, \qquad \frac{1}{\sqrt{2v_2}} - 1 = \frac{1}{\gamma M_2^2} = 2.143.$$

Thus the equations (48) for the slopes at the two endpoints are

$$\Theta_1' = \frac{4P}{3 \times 1.4}\,\frac{\Theta_1' + 0.071}{\Theta_1' + 0.821} \qquad \text{and} \qquad \Theta_2' = \frac{4P}{4.2}\,\frac{\Theta_2' + 0.857}{\Theta_2' - 1.143}.$$

The negative roots are

$$\Theta_1' = -0.43, \quad \Theta_2' = -0.22 \qquad \text{for } P = \tfrac{1}{2}$$

$$\Theta_1' = -0.20, \quad \Theta_2' = -0.34 \qquad \text{for } P = 1.$$

The limits for the thickness of the transition layer are then

$$1.46\, L_0 \leqq L \leqq 1.88\, L_0 \qquad \text{for } P = \tfrac{1}{2}$$

$$1.21\, L_0 \leqq L \leqq 1.35\, L_0 \qquad \text{for } P = 1.$$

The quantity L_0 was found, for a numerical example in Sec. 3, to be of the order of less than one-tenth mm.

The situation in the v,Θ-plane is sketched in Fig. 55 (see end of Sec. 4) with the ordinates Θ magnified in the ratio 5:1. The broken lines are the integral curves for $P = \frac{1}{2}$, $P = \frac{3}{4}$, and $P = 1$; the solid lines are the isoclines for the four slopes Θ' found above (and for $\lambda = 0$).

Article 12

Nonsteady Flow of an Ideal Fluid[12]

1. General equations

We consider an ideal fluid: a perfect gas with $\mu_0 = k = 0$; that is, viscosity and heat conduction are neglected. The Eqs. (11.12) for steady adiabatic one-dimensional flow reduce to a pair of relations between u and T, not involving x. The only possible solutions are therefore of the form u = constant and T = constant, from which follow also ρ = constant and p = constant. In other words, the only steady, strictly adiabatic one-dimensional flow of an ideal fluid is a uniform flow with constant values of u, p, ρ, T, etc. Thus, only nonsteady flow is of interest in this case.[13]

Returning to Eqs. (11.2) and (11.3), and omitting the viscosity term, we obtain

$$\rho\frac{\partial u}{\partial x} + u\frac{\partial \rho}{\partial x} + \frac{\partial \rho}{\partial t} = 0, \tag{1}$$

$$\rho\frac{\partial u}{\partial t} + \rho u\frac{\partial u}{\partial x} + \frac{\partial p}{\partial x} = 0. \tag{2}$$

The specifying equation which expresses the fact that the motion is adiabatic may be taken from Sec. 1.5. In the case of an ideal fluid it reads: p/ρ^{γ} = constant for each particle, and this last qualification may be omitted if initially all particles have the same entropy. The mathematical problem remains unchanged if γ is replaced by any other constant greater than one, so we shall, in general, use as specifying equation the polytropic relation

$$\frac{p}{\rho^{\kappa}} = \text{constant} = C, \tag{3}$$

where κ is any constant greater than 1.

If (3) holds, or more generally an arbitrary (p,ρ)-relation with $dp/d\rho > 0$, we can introduce the square of the sound velocity

$$a^2 = \frac{dp}{d\rho}$$

as a known function of ρ (see Sec. 5.2) and use this function to eliminate p from (2). Since $\partial p/\partial x$ equals $dp/d\rho$ times $\partial\rho/\partial x$, Eq. (2) takes the form

$$\rho\frac{\partial u}{\partial t} + \rho u\frac{\partial u}{\partial x} + a^2\frac{\partial \rho}{\partial x} = 0. \tag{4}$$

Equations (1) and (4) may be slightly rearranged to yield

$$\tag{5} \begin{gathered} \frac{\partial}{\partial x}(\rho u) + \frac{\partial \rho}{\partial t} = 0, \\ \frac{\partial}{\partial t}(\rho u) + u\frac{\partial}{\partial x}(\rho u) - u\frac{\partial \rho}{\partial t} + (a^2 - u^2)\frac{\partial \rho}{\partial x} = 0, \end{gathered}$$

a system of two equations for the unknowns ρu and ρ.

Another convenient form of the equations may be obtained by using the pressure head P, defined in Sec. 2.5:

$$\tag{6} P = \int \frac{dp}{\rho} = \int a^2 \frac{d\rho}{\rho}, \qquad \frac{\partial P}{\partial x} = \frac{a^2}{\rho}\frac{\partial \rho}{\partial x}; \qquad \frac{\partial P}{\partial t} = \frac{a^2}{\rho}\frac{\partial \rho}{\partial t}.$$

Then it is easily verified, by carrying out the indicated differentiations, that (4) and (1) are equivalent, respectively, to

$$\tag{7} \begin{gathered} \frac{\partial}{\partial x}\left(\frac{u^2}{2} + P\right) + \frac{\partial u}{\partial t} = 0, \\ \frac{\partial}{\partial t}\left(\frac{u^2}{2} + P\right) + u\frac{\partial}{\partial x}\left(\frac{u^2}{2} + P\right) - u\frac{\partial u}{\partial t} + (a^2 - u^2)\frac{\partial u}{\partial x} = 0. \end{gathered}$$

Alternatively, the first equations of (5) and (7) form a system equivalent to (1) and (2). It may be noted that (5) and (7) are very similar: the variables ρu and ρ in (5) parallel the variables $(u^2/2 + P)$ and u in (7).

In the particular case of the polytropic condition (3), we have

$$\tag{8} a^2 = \kappa C \rho^{\kappa-1}, \qquad P = \frac{\kappa}{\kappa - 1} C\rho^{\kappa-1} = \frac{a^2}{\kappa - 1}.$$

With the velocities u and a as variables in place of u and ρ, the first equations of (5) and (7) become

$$\tag{9} \begin{gathered} \frac{\partial}{\partial x}(u a^{2/(\kappa-1)}) + \frac{\partial}{\partial t}(a^{2/(\kappa-1)}) = 0, \\ \frac{\partial}{\partial x}\left(\frac{u^2}{2} + \frac{a^2}{\kappa - 1}\right) + \frac{\partial u}{\partial t} = 0. \end{gathered}$$

For adiabatic flow of a diatomic perfect gas (as assumed for air), we have $\kappa = \gamma = 1.4 = 7/5$, and therefore

$$\tag{9'} \begin{gathered} \frac{\partial}{\partial x}(u a^5) + \frac{\partial}{\partial t}(a^5) = 0, \\ \frac{\partial}{\partial x}\left(\frac{u^2}{2} + \frac{5}{2}a^2\right) + \frac{\partial u}{\partial t} = 0. \end{gathered}$$

For a monatomic gas, $\kappa = 5/3$, we have 3 instead of 5 in the above equations.

The two equations (9) are homogeneous, planar (but not linear) differential equations for the unknowns u and a with independent variables x and t, namely,

$$\text{(10)} \qquad \begin{aligned} \frac{\kappa - 1}{2} a \frac{\partial u}{\partial x} + u \frac{\partial a}{\partial x} + \frac{\partial a}{\partial t} &= 0, \\ \frac{2}{\kappa - 1} a \frac{\partial a}{\partial x} + u \frac{\partial u}{\partial x} + \frac{\partial u}{\partial t} &= 0. \end{aligned}$$

Of course, Eq. (10) could have been derived directly from (1) and (4) by introducing the variable a in place of ρ, using (8).

2. Potential and particle function

Either one of the two parallel sets of equations (5) and (7) can be reduced to a single differential equation of second order. Indeed, the first Eq. (5) expresses the fact that ρ and $-\rho u$ are, respectively, the x- and t-derivatives of one and the same function $\psi(x,t)$:

$$\text{(11)} \qquad \rho = \frac{\partial \psi}{\partial x}, \qquad \rho u = -\frac{\partial \psi}{\partial t}.$$

The second Eq. (5) then supplies the following condition on ψ:

$$\text{(11')} \qquad \frac{\partial^2 \psi}{\partial t^2} + 2u \frac{\partial^2 \psi}{\partial x\, \partial t} + (u^2 - a^2) \frac{\partial^2 \psi}{\partial x^2} = 0.$$

In the same way, the system (7) is satisfied if a function $\Phi(x,t)$ is introduced for which

$$\text{(12)} \qquad u = \frac{\partial \Phi}{\partial x}, \qquad \frac{u^2}{2} + P = -\frac{\partial \Phi}{\partial t}$$

and which satisfies

$$\text{(12')} \qquad \frac{\partial^2 \Phi}{\partial t^2} + 2u \frac{\partial^2 \Phi}{\partial x\, \partial t} + (u^2 - a^2) \frac{\partial^2 \Phi}{\partial x^2} = 0.$$

Equations (11′) and (12′) appear to be the same; the coefficients $2u$ and $u^2 - a^2$, however, do not bear the same relation to the unknown function, Φ or ψ, in the two cases.

The function $\Phi(x,t)$ is none other than the potential introduced in Sec. 7.1. Thus, one-dimensional isentropic flow of an ideal fluid is always a potential flow and irrotational. Equation (12′) is exactly the same as Eq. (7.24), discussed in Sec. 7.4. Each function $\Phi(x,t)$ satisfying (12′) determines a particular one-dimensional flow of an ideal fluid and, conversely, for each continuous one-dimensional flow pattern there exists a potential function satisfying (12′).

The function $\psi(x,t)$, which is known as the *particle function*, may be interpreted in the following way. The rectilinear motion of any material particle is completely determined when its position x is given as a function of t. The curve in the x,t-plane representing this function for any particle has been called a particle line (Sec. 1.2). The slope of the particle line, measured by the tangent of its angle with the t-axis, is dx/dt or the instantaneous velocity u of the particle. Now, along the lines ψ = constant we have, using (11),

$$(13) \qquad 0 = \frac{\partial\psi}{\partial x}\,dx + \frac{\partial\psi}{\partial t}\,dt = \rho\,dx - \rho u\,dt, \qquad \frac{dx}{dt} = u.$$

Thus the family of curves ψ = constant (Fig. 57) consists of the particle lines for all elements of the fluid mass under consideration. The difference between the ψ-values on two distinct particle lines may be found by integrating the first Eq. (11) along any parallel to the x-axis:

$$(14) \qquad \psi_B - \psi_A = \int_A^B \frac{\partial\psi}{\partial x}\,dx = \int_A^B \rho\,dx,$$

and is therefore equal to the quantity of mass enclosed in the flow by a cylinder of unit cross section extending between the two planes $x = x_A$, and $x = x_B$.

Both (11′) and (12′) are second-order differential equations of the type (9.2), with

$$A = 1, \qquad B = u, \qquad C = u^2 - a^2, \qquad F = 0,$$

where the present independent variables x and t are to be identified with the former y and x, respectively. From (9.13′) [see end of Sec. 9.5] the slopes of the characteristics are given by

$$(15) \qquad \frac{dx}{dt} = \frac{1}{A}\left(B \pm \sqrt{B^2 - AC}\right) = u \pm a.$$

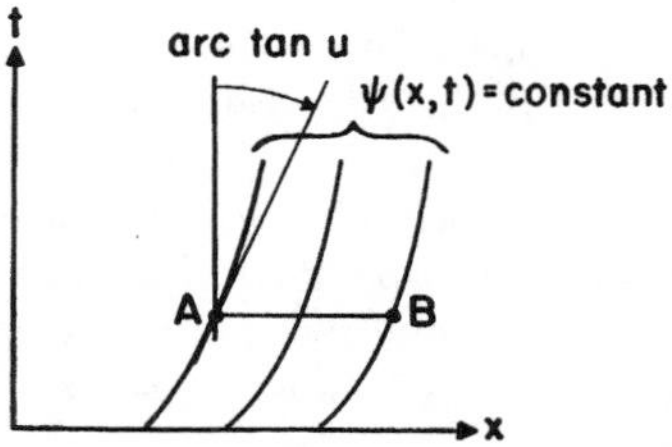

FIG. 57. Particle lines ψ = constant.

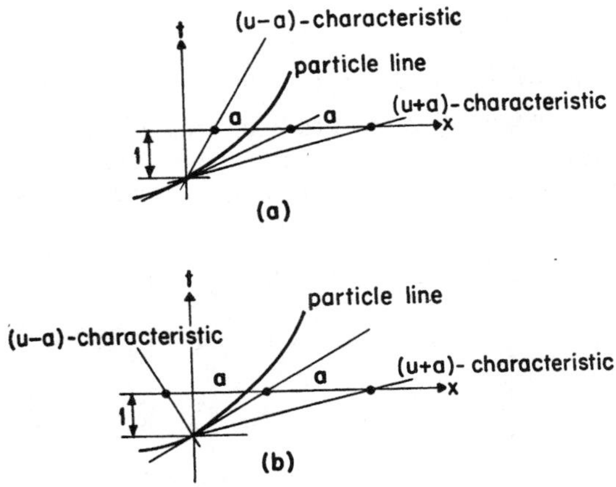

FIG. 58. Orientation of the characteristics with respect to the t-axis for velocity $u > 0$.
(a) u supersonic, (b) u subsonic.

Since both u and a depend on derivatives of the solution, Φ or ψ, these slopes vary with the solution being considered, in agreement with the fact that the equations (11′) and (12′) are nonlinear. The slopes are real in all cases: *whether the flow is supersonic or subsonic, the problem is hyperbolic.* In the first case $u + a$ and $u - a$ have the same sign, so that the characteristic directions lie on the same side of the vertical in the x,t-plane (Fig. 58a, if $u > 0$), while in subsonic flow the characteristic directions fall on opposite sides of the vertical line (Fig. 58b, if $u > 0$).

The same conclusions can be obtained by applying to (5) or (7) the general theory of characteristics (Sec. 9.2) or the discussion of the two-dimensional case (Art. 10).

Some examples of one-dimensional flow were given in Sec. 7.4. These were all such that Φ was a quadratic, and u a linear function in x with coefficients depending on t. We arrive at essentially the same solutions if we start from the assumption

$$(16) \qquad \psi(x,t) = (\alpha x + \beta)^n,$$

where α and β depend only on t, and n is constant. From (11) we have then

$$(16') \qquad \rho = \frac{\partial \psi}{\partial x} = n\alpha(\alpha x + \beta)^{n-1}, \qquad u = -\frac{\partial \psi}{\partial t} \bigg/ \frac{\partial \psi}{\partial x} = -\frac{\alpha' x + \beta'}{\alpha},$$

and, with the polytropic relation (3),

$$(16'') \qquad a^2 = \kappa C \rho^{\kappa-1} = \kappa C (n\alpha)^{\kappa-1} (\alpha x + \beta)^{(n-1)(\kappa-1)}.$$

When these expressions are substituted into (11′) or (4), the equation reduces to

$$(2\alpha'^2 - \alpha\alpha'')x + (2\alpha'\beta' - \alpha\beta'') + \kappa(n-1)\alpha^{\kappa+2}Cn^{\kappa-1}(\alpha x + \beta)^{(n-1)(\kappa-1)-1} = 0.$$

This equation can be satisfied identically in x only if all coefficients vanish, or if the equation is independent of x, or if the last term is a suitable linear function of x. The three values of n making these results possible, and the corresponding conditions on the coefficients, are:

$$\text{(a) } n = 1;\quad 2\alpha'^2 - \alpha\alpha'' = 0,\quad 2\alpha'\beta' - \alpha\beta'' = 0;$$

$$\text{(17)}\quad \text{(b) } n = \frac{\kappa}{\kappa - 1};\quad 2\alpha'^2 - \alpha\alpha'' = 0,\quad 2\alpha'\beta' - \alpha\beta'' = K_1\alpha^{\kappa+2};$$

$$\text{(c) } n = \frac{\kappa+1}{\kappa-1};\quad 2\alpha'^2 - \alpha\alpha'' = K_2\alpha^{\kappa+3},\quad 2\alpha'\beta' - \alpha\beta'' = K_2\alpha^{\kappa+2}\beta,$$

where $K_1 = -C[\kappa/(\kappa-1)]^{\kappa}$, $K_2 = -2C\kappa(\kappa+1)^{\kappa-1}(\kappa-1)^{-\kappa}$.

A particular solution of (17c) is

$$\alpha = \text{constant}\cdot t^{-2/(\kappa+1)},\qquad \beta = \text{constant}\cdot t^{(\kappa-1)/(\kappa+1)},$$

which leads to

$$\psi(x,t) = At^{-2/(\kappa-1)}(x + ct)^{(\kappa+1)/(\kappa-1)},$$

where A is a simple function of C and κ, and c is arbitrary. This is the particle function* corresponding to the example (a), Eqs. (7.28), of Sec. 7.4 with $c_2 = 0$, except for different meaning of the constant A. In the same way a simple solution of (17b) supplies

$$\psi(x,t) = B\left(\frac{x}{t} + c_1 t^{1-\kappa} + c_2\right)^{\kappa/(\kappa-1)},$$

where B and c_2 are arbitrary and c_1 is a simple function of B, C, and κ. For $c_2 = 0$ this is the particle function for example (b) of the same section [see Eqs. (7.29)]. The detailed discussion of the differential equations (17) and of the corresponding flow patterns is left to the reader.

3. Interchange of variables. Speedgraph

As has been seen in Sec. 1, Eqs. (1) and (4) may serve as the equations governing the one-dimensional nonsteady flow of an inviscid elastic fluid:

$$\text{(18)}\quad \begin{aligned} \rho\frac{\partial u}{\partial x} + u\frac{\partial \rho}{\partial x} + \frac{\partial \rho}{\partial t} &= 0, \\ u\frac{\partial u}{\partial x} + \frac{a^2}{\rho}\frac{\partial \rho}{\partial x} + \frac{\partial u}{\partial t} &= 0, \end{aligned}$$

where $a^2 = dp/d\rho$ is a known function of ρ.

* This flow is the centered simple wave of Sec. 13.2 [see Eq. (13.13)].

The further argument is greatly simplified if the variable ρ is replaced by a certain function of ρ, namely,[14]

(19) $$v = \int_{\rho_1}^{\rho} \frac{a}{\rho}\, d\rho, \qquad \frac{dv}{d\rho} = \frac{a}{\rho}, \qquad \frac{\partial}{\partial \rho} = \frac{a}{\rho}\frac{\partial}{\partial v},$$

with an appropriately chosen lower limit ρ_1 in the integral;* there is a one-to-one correspondence between v and ρ. The new variable v has the dimensions of velocity. For example, in the case of the polytropic relation (3), we choose $\rho_1 = 0$ and find

(19′) $$v = \sqrt{\kappa C}\,\frac{2}{\kappa - 1}\,\rho^{(\kappa-1)/2} = \frac{2a}{\kappa - 1}, \qquad a = \frac{\kappa - 1}{2}\, v,$$

and, in particular, if $\kappa = \gamma = 1.4$,

(19″) $$v = 5a.$$

If an isothermal condition $p = \text{constant}\cdot\rho$ holds, we take $\rho_1 = 1$ and have

(19‴) $$a = \text{constant}, \qquad v = a \log \rho.$$

When the substitution (19) is made, Eqs. (18) read

(20) $$\begin{aligned} a\,\frac{\partial u}{\partial x} + u\,\frac{\partial v}{\partial x} + \frac{\partial v}{\partial t} &= 0, \\ u\,\frac{\partial u}{\partial x} + a\,\frac{\partial v}{\partial x} + \frac{\partial u}{\partial t} &= 0. \end{aligned}$$

To each couple x,t there corresponds a pair of values of u and v determined by (20). This mapping of the x,t-plane onto the u,v-plane will be called the *speedgraph* transformation of the one-dimensional flow. The speedgraph is analogous in many respects to the hodograph of a two-dimensional steady flow (see Sec. 8.2). In the hodograph the lines of constant speed q, and consequently of constant Mach number M, are concentric circles. Here, in the polytropic case, the straight lines through the origin in the u,v-plane are the lines on which the Mach number is constant, since $M = |u|/a$ equals $2/(\kappa - 1)$ times $|u|/v$. In particular, the rays $v/u = \pm 2/(\kappa - 1)$ separate the subsonic region in the middle from supersonic regions on either side (Fig. 59).

By adding and subtracting Eqs. (20) we find that

(21) $$(u \pm a)\,\frac{\partial u}{\partial x} + \frac{\partial u}{\partial t} = \mp\left[(u \pm a)\,\frac{\partial v}{\partial x} + \frac{\partial v}{\partial t}\right].$$

Now the left-hand member is the rate of change of u in the respective characteristic direction $dx/dt = (u \pm a)$, while the bracket on the right is

* Note that v is defined also for a particlewise (p,ρ)-relation.

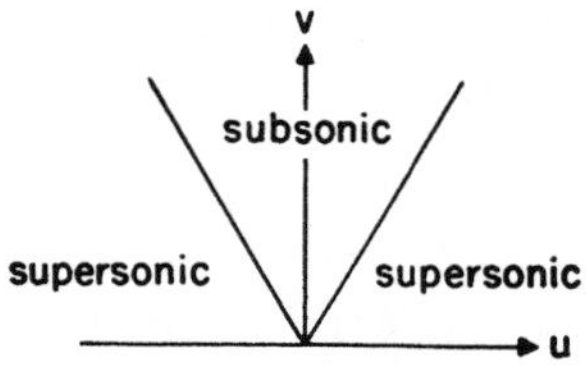

FIG. 59. Regions in which $M \gtrless 1$.

the same for v. Thus, along the $(u + a)$-characteristic in the x,t-plane we have $du = -dv$, and along the other characteristic $du = dv$. This shows that *the two sets of characteristics $dx/dt = u \pm a$ in the x,t-plane are mapped into the $\mp 45°$ lines*

$$v + u = \text{constant} \qquad \text{and} \qquad v - u = \text{constant}, \tag{22}$$

respectively, in the speedgraph plane.

In the two planar first-order equations (20), the coefficients depend only on the dependent variables u and v; hence we proceed as described in Sec. 10.6 and interchange the dependent and independent variables. The Eqs. (20) may be identified with Eqs. (10.1); if the present t replaces y then the coefficients of (10.1) are

$$a_1 = b_3 = a, \qquad a_3 = b_1 = u, \qquad a_4 = b_2 = 1, \qquad a_2 = b_4 = 0,$$

and the transformed equations (10.22) read

$$\begin{aligned} a\frac{\partial t}{\partial v} - u\frac{\partial t}{\partial u} + \frac{\partial x}{\partial u} &= 0, \\ u\frac{\partial t}{\partial v} - a\frac{\partial t}{\partial u} - \frac{\partial x}{\partial v} &= 0. \end{aligned} \tag{23}$$

This system, in contrast to (18) and (20), is linear, so that the characteristics are independent of the unknowns. The characteristic directions can be computed as in Sec. 10.1 and are found to be $dv/du = \pm 1$ in agreement with (22).

To obtain from (23) a single second-order equation for one unknown, we can eliminate x by differentiating the first equation with respect to v and the second with respect to u. Remembering that a is a function of v only, we obtain

$$\frac{\partial^2 t}{\partial v^2} - \frac{\partial^2 t}{\partial u^2} = -\frac{1}{a}\left(1 + \frac{da}{dv}\right)\frac{\partial t}{\partial v}. \tag{24}$$

The equation for x, obtained by an analogous elimination of t, is similar, except that the first partial derivative of the unknown, on the right, is replaced by a more complicated first-order expression. The second degree terms of (24) again show that the characteristics of the present problem are $dv/du = \pm 1$.

Alternatively, by the use of $a\,d\rho = \rho\,dv$, Eqs. (23) may be rewritten in the form

$$\begin{aligned} &\frac{\partial}{\partial u}(x - ut) + \frac{a}{\rho}\frac{\partial}{\partial v}(\rho t) = 0, \\ (25)\qquad &\frac{\partial}{\partial v}(x - ut) + \frac{a}{\rho}\frac{\partial}{\partial u}(\rho t) = 0. \end{aligned}$$

To integrate the first equation, we introduce a function U of u and v by setting

$$(26)\qquad x - ut = \frac{a}{\rho}\frac{\partial U}{\partial v}, \qquad t = -\frac{1}{\rho}\frac{\partial U}{\partial u};$$

the second equation then shows that U must satisfy

$$(26')\qquad \frac{\partial^2 U}{\partial v^2} - \frac{\partial^2 U}{\partial u^2} = \frac{1}{a}\left(1 - \frac{da}{dv}\right)\frac{\partial U}{\partial v}.$$

Similarly, the second equation (25) is satisfied if we set

$$(27)\qquad x - ut = \frac{\partial V}{\partial u}, \qquad t = -\frac{1}{a}\frac{\partial V}{\partial v},$$

and then the first equation supplies the condition[15]

$$(27')\qquad \frac{\partial^2 V}{\partial v^2} - \frac{\partial^2 V}{\partial u^2} = -\frac{1}{a}\left(1 - \frac{da}{dv}\right)\frac{\partial V}{\partial v}.$$

Also, either $x - ut$ or ρt may be eliminated from the system (25), and it is seen that ρt and $x - ut$ satisfy the same second-order equations as U and V, respectively.

It is also of interest to study the relations between the functions U and V on the one hand, and the previously introduced potential and particle functions Φ and ψ on the other. From the definition (11) it follows that

$$\begin{aligned} &\frac{\partial \psi}{\partial u} = \rho\frac{\partial x}{\partial u} - \rho u\frac{\partial t}{\partial u} = \rho\frac{\partial}{\partial u}(x - ut) + \rho t, \\ (28)\qquad &\frac{\partial \psi}{\partial v} = \rho\frac{\partial x}{\partial v} - \rho u\frac{\partial t}{\partial v} = \frac{\partial}{\partial v}[\rho(x - ut)] - \frac{\rho}{a}(x - ut). \end{aligned}$$

It is easily seen that the two expressions on the right are the derivatives with respect to u and v, respectively, of $a\,\partial U/\partial v - U$. Thus, except for an additive constant, which has no significance,

$$\psi = a\frac{\partial U}{\partial v} - U. \tag{29}$$

In exactly the same way we find that

$$\Phi = u\frac{\partial V}{\partial u} + \frac{1}{a}\left(P - \frac{u^2}{2}\right)\frac{\partial V}{\partial v} - V. \tag{30}$$

Equations analogous to the first parts of (28) can be written for $\partial\Phi/\partial u$ and $\partial\Phi/\partial v$. When these equations are compared with (28) we can verify, using (25), that

$$\begin{aligned} \frac{\partial \Phi}{\partial u} &= A_1\frac{\partial\psi}{\partial u} + B_1\frac{\partial\psi}{\partial v}, \\ \frac{\partial \Phi}{\partial v} &= A_2\frac{\partial\psi}{\partial u} + B_2\frac{\partial\psi}{\partial v}, \end{aligned} \tag{31}$$

where

$$A_1 = B_2 = \frac{u}{\rho}, \qquad A_2 = B_1 = \frac{1}{a\rho}\left(P - \frac{u^2}{2}\right). \tag{31$'$}$$

Also, using $a\,d\rho = \rho\,dv$ and $dP = a\,dv$, we have

$$\begin{aligned} \frac{\partial A_1}{\partial v} - \frac{\partial A_2}{\partial u} = 0, \qquad \frac{\partial B_1}{\partial v} - \frac{\partial B_2}{\partial u} &= -\frac{1}{a^2\rho}\left(P - \frac{u^2}{2}\right)\left(1 + \frac{da}{dv}\right) \\ &= -\frac{A_2}{a}\left(1 + \frac{da}{dv}\right). \end{aligned}$$

Thus, when Φ is eliminated from (31), the result is

$$\frac{\partial^2\psi}{\partial v^2} - \frac{\partial^2\psi}{\partial u^2} = \frac{1}{a}\left(1 + \frac{da}{dv}\right)\frac{\partial\psi}{\partial v}. \tag{32}$$

Elimination of ψ leads to a similar equation for Φ, but with a more complicated first-order expression on the right.

All these equations, (24), (26′), (27′), and (32), for t, U, V, and ψ, respectively, are of similar form, differing only in the factors $\pm\,(1 \pm da/dv)$ on the right. In the polytropic case these factors are all constants, since (19′) gives $da/dv = (\kappa - 1)/2$. The right-hand members consist then of the derivative of the unknown with respect to v, multiplied, respectively, by the factors

$$-\frac{\kappa+1}{\kappa-1}\frac{1}{v}, \qquad \frac{3-\kappa}{\kappa-1}\frac{1}{v}, \qquad -\frac{3-\kappa}{\kappa-1}\frac{1}{v}, \qquad \frac{\kappa+1}{\kappa-1}\frac{1}{v}, \tag{33}$$

which in the particular case of $\kappa = 1.4$ are

$$-\frac{6}{v}, \qquad \frac{4}{v}, \qquad -\frac{4}{v}, \qquad \frac{6}{v}. \tag{33$'$}$$

It will be shown in the next section that the general integral of the equation can be given in simple form in the cases (33′) and in certain other cases also.

4. General integral in the adiabatic case

In the preceding section we have seen that, if the (p,ρ)-relation[16] has the form (3), and the polytropic exponent κ the value 1.4, the second-order differential equations for t, U, V, ψ, $x - ut$, and ρt have the common form[17]

$$\frac{\partial^2 z_n}{\partial v^2} - \frac{\partial^2 z_n}{\partial u^2} = \frac{2n}{v}\frac{\partial z_n}{\partial v}, \tag{34}$$

with $n = 2$ for $z_n = U$ or ρt, $n = -2$ for V or $x - ut$, $n = 3$ for ψ, and $n = -3$ for t. It can be shown that, if n in (34) is any integer, the general integral of (34) can be given in a simple form. This result applies, more generally, whenever κ has any value which leads to an even integer for the quotients $(3 - \kappa)/(\kappa - 1)$ and $(\kappa + 1)/(\kappa - 1)$.[18] Moreover, the formulas can be extended in a certain way to cover the case of any real n.

First, if $n = 0$, Eq. (34) is the one-dimensional wave equation (4.6), so that from (4.7) the general solution is

$$z_0(u,v) = f(v + u) + g(v - u), \tag{35}$$

where f and g are arbitrary, sufficiently differentiable functions of a single variable. The successive derivatives of these functions will be denoted by $f', f'', \cdots, g', g'', \cdots$.

For the sake of abbreviation we introduce the notation

$$z_0' = f' + g', \qquad z_0'' = f'' + g'', \qquad \cdots, \qquad z_0^{(n)} = f^{(n)} + g^{(n)}, \tag{36}$$

where the arguments parallel those in (35). Then each $z_0^{(\nu)}$ is the sum of a function of $v + u$ and a function of $v - u$. It may be noted that $z_0^{(\nu+1)}$ is the derivative of $z_0^{(\nu)}$ with respect to v, while $z_0^{(\nu+2)}$ is the result of differentiating $z_0^{(\nu)}$ twice with respect to either v or u.

With an arbitrary z_0 of the form (35), the general solution of (34) for n a positive integer can be written in the form[19]

$$z_n(u,v) = z_0 + \alpha_1 v z_0' + \alpha_2 v^2 z_0'' + \cdots + \alpha_n v^n z_0^{(n)}, \tag{37}$$

where the numerical coefficients α_ν, depending on n, are found by substituting (37) into (34):

$$\alpha_1 = -1; \qquad \alpha_\nu = (-1)^\nu \frac{2^{\nu-1}}{\nu!} \frac{(n-1)(n-2)\cdots(n-\nu+1)}{(2n-1)(2n-2)\cdots(2n-\nu+1)} \qquad (\nu = 2, 3, \cdots, n). \tag{37'}$$

If n is a negative integer, it is more convenient to set $m = -n > 0$ and write

$$(38)\quad z_{-m} = z_n = v^{1-2m}[z_0 + \beta_1 v z_0' + \beta_2 v^2 z_0'' + \cdots + \beta_{m-1} v^{m-1} z_0^{(m-1)}],$$

where

$$(38')\quad \beta_1 = -1; \qquad \beta_\nu = (-1)^\nu \frac{2^{\nu-1}}{\nu!} \frac{(m-2)(m-3)\cdots(m-\nu)}{(2m-3)(2m-4)\cdots(2m-\nu-1)} \qquad (\nu = 2, 3, \cdots, m-1).$$

The first examples are

$$(39)\quad z_1 = z_0 - vz_0', \qquad z_2 = z_0 - vz_0' + \tfrac{1}{3}v^2 z_0'', \qquad z_3 = z_0 - vz_0' + \tfrac{2}{5}v^2 z_0'' - \tfrac{1}{15}v^3 z_0''',$$

$$(40)\quad z_{-1} = \frac{z_0}{v}, \qquad z_{-2} = \frac{1}{v^3}(z_0 - vz_0'), \qquad z_{-3} = \frac{1}{v^5}(z_0 - vz_0' + \tfrac{1}{3}v^2 z_0'').$$

In order to prove formulas (37′) and (38′), we start by working out the result of substituting a single term of (37) into the differential equation:

$$\left(\frac{\partial^2}{\partial v^2} - \frac{\partial^2}{\partial u^2} - \frac{2n}{v}\frac{\partial}{\partial v}\right)(\alpha_\nu v^\nu z_0^{(\nu)}) = \alpha_\nu[\nu(\nu - 1 - 2n)v^{\nu-2} z_0^{(\nu)} + 2(\nu - n)v^{\nu-1} z_0^{(\nu+1)}].$$

Thus, when the whole of (37) is substituted into (34) and the coefficient of $v^{\nu-2} z_0^{(\nu)}$ equated to zero, we find

$$\alpha_1 = -1; \qquad \alpha_\nu \nu(\nu - 1 - 2n) + \alpha_{\nu-1} 2(\nu - 1 - n) = 0 \qquad (\nu = 2, 3, \cdots, n),$$

and this recursion formula gives (37′). Formulas (38′) can be proved in the same way. In either case no terms of higher degree than those given in (37) or (38) are necessary, for the coefficients of such terms would vanish since n is an integer. This suggests that when n is not an integer, the solution can be expanded in an infinite series beginning with the terms (37) or (38), the coefficients being determined by recursion.[20] Even in this case the series reduces to a finite sum if f and g are polynomials.

It seems that the function best suited for the study of most problems is V. We repeat Eq. (27′) for adiabatic flow with $\kappa = \gamma = 1.4$, and its solution from (40):

$$(41)\quad \frac{\partial^2 V}{\partial v^2} - \frac{\partial^2 V}{\partial u^2} = -\frac{4}{v}\frac{\partial V}{\partial v},$$

$$(42)\quad V = \frac{f(\xi) + g(\eta)}{v^3} - \frac{f'(\xi) + g'(\eta)}{v^2}; \qquad \xi = v + u, \qquad \eta = v - u.$$

It still remains to determine the functions f and g for particular problems.

If ξ and η are used as independent variables in place of u and v, we have

$$\frac{\partial}{\partial v} = \frac{\partial}{\partial \xi} + \frac{\partial}{\partial \eta}, \qquad \frac{\partial}{\partial u} = \frac{\partial}{\partial \xi} - \frac{\partial}{\partial \eta},$$

so that the differential equation for V becomes[21]

$$\frac{\partial^2 V}{\partial \xi\, \partial \eta} + \frac{2}{\xi + \eta}\left(\frac{\partial V}{\partial \xi} + \frac{\partial V}{\partial \eta}\right) = 0. \tag{43}$$

Equation (43) is of the form (10.11) with

$$a = b = \frac{2}{\xi + \eta}, \qquad c = 0,$$

and the solution of (10.11) was found in terms of a particular solution of the adjoint equation (10.12). The adjoint equation for (43) is

$$\frac{\partial^2 \Omega}{\partial \xi\, \partial \eta} - \frac{2}{\xi + \eta}\left(\frac{\partial \Omega}{\partial \xi} + \frac{\partial \Omega}{\partial \eta}\right) + \frac{4}{(\xi + \eta)^2}\,\Omega = 0. \tag{44}$$

Now the general solution of (44) is

$$\Omega = (\xi + \eta)[F(\xi) + G(\eta)] - \tfrac{1}{2}(\xi + \eta)^2[F'(\xi) + G'(\eta)], \tag{45}$$

where F and G are arbitrary functions of a single variable.* The particular solution, or Riemann function, used in Sec. 10.5 was the one satisfying the boundary conditions

$$\Omega(\xi_1, \eta_1) = 1, \qquad \frac{\partial \Omega}{\partial \xi} = b\Omega \text{ for } \eta = \eta_1, \qquad \frac{\partial \Omega}{\partial \eta} = a\Omega \text{ for } \xi = \xi_1,$$

where ξ_1, η_1 were a pair of parameters. It can be verified by differentiation that the Riemann function corresponding to Eq. (43) is[22]

$$\begin{aligned} \Omega(\xi, \eta; \xi_1, \eta_1) &= \frac{\xi + \eta}{(\xi_1 + \eta_1)^3}[2\xi_1\eta_1 + 2\xi\eta + (\xi - \eta)(\xi_1 - \eta_1)] \\ &= \frac{v}{v_1^3}\left[\frac{1}{2}(v_1^2 - u_1^2 + v^2 - u^2) + uu_1\right]. \end{aligned} \tag{46}$$

The general initial value problem can be solved explicitly once the Riemann function is known, as shown in Sec. 10.5 (see also Sec. 7).

* By setting $\Omega = (\xi + \eta)\omega$, we obtain

$$\frac{\partial^2 \omega}{\partial \xi \partial \eta} - \frac{1}{\xi + \eta}\left(\frac{\partial \omega}{\partial \xi} + \frac{\partial \omega}{\partial \eta}\right) = 0,$$

whose general solution is given by z_1 in (39).

5. Application of the speedgraph. Initial-value problem

The two variables u,ρ or u,v are determined as functions of x and t by the pair of planar differential equations (18) or (20), respectively. We have seen (Sec. 2) that the problem is hyperbolic in all cases; thus there are real characteristics and the theorems of Sec. 10.3 can be used. The first of these states that if both u and v are given at all points of an arc AB in the x,t-plane nowhere tangent to a characteristic, then the solution is determined (at most) in the characteristic quadrangle having A and B as opposite corners. A natural problem is to ask for the solution when the initial values, i.e., the values of u and v at $t = 0$, are given on a certain interval of the x-axis, say from $x = 0$ to $x = l$. In this case the arc AB is a segment of the x-axis; now the characteristics have slopes $u \pm a$ as measured from the t-axis, so that no characteristic can have the same direction as AB as long as u and a have finite values.

The numerical approximation method described in detail in Sec. 10.3 consisted of a step-by-step construction of a network of characteristic arcs together with the determination of the values of the unknowns at the nodal points of the network. Both parts of this program take a very simple form

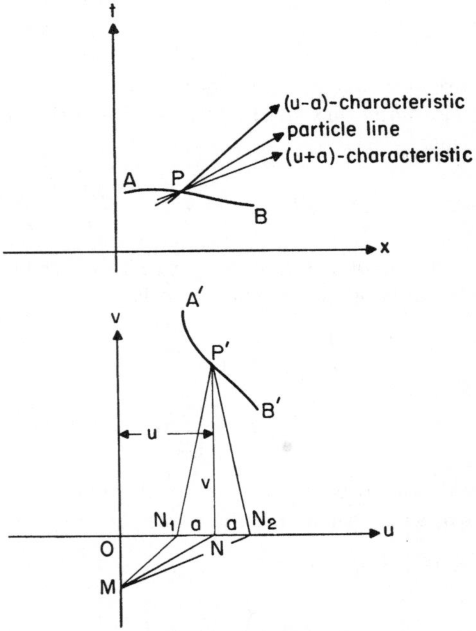

FIG. 60. Construction of the characteristic directions at a point P of the x,t-plane, for $v = 5a$. These directions are given by MN_1, MN_2.

in the present problem if the speedgraph is used.[23] The pairs of values given for the points of the segment AB map this segment into an arc $A'B'$ in the u,v-plane (Fig. 60). Let P' with coordinates u,v correspond to a point P on AB. The characteristic directions at P make the angles arc tan $(u \pm a)$ with the t-axis. Suppose that $\kappa = 1.4$, so that $v = 5a$. Then lines from P' with slope ± 5 meet the u-axis in the points N_1 and N_2 with abscissas $u - a$ and $u + a$ respectively. If $OM = 1$, then the tangents of the angles between the vertical direction and the lines MN_1 and MN_2 are $u - a$ and $u + a$, respectively. Therefore lines through P parallel to MN_1 and MN_2 give the characteristic directions at P.

Also we have seen (Sec. 3) that the $(u \pm a)$-characteristics in the x,t-plane correspond to the $\mp 45°$ lines respectively in the u,v-plane. Thus the images of the characteristics are very easily constructed and form a rectangular network. Furthermore, the u,v-coordinates of a point of intersection give the values of the unknowns at the corresponding intersection in the x,t-plane.

To carry out the computation, we 1) choose a sequence of points P on AB (see Fig. 61), 2) locate the corresponding points P' on $A'B'$, 3) con-

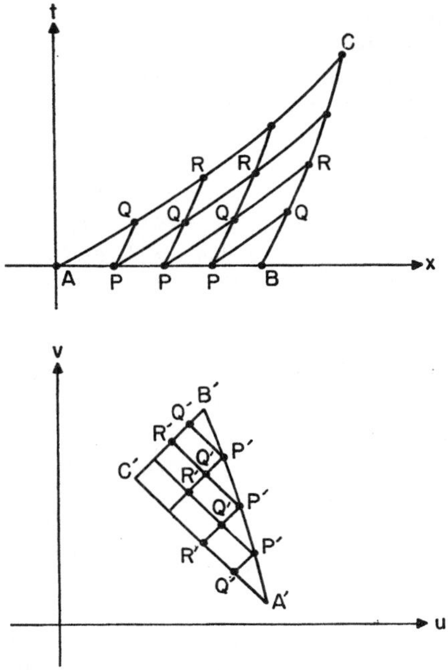

FIG. 61. Step-by-step construction of an approximate solution to an initial-value problem using the speedgraph.

struct the characteristic directions at the points P by means of the parallel method described above, leading to a series of intersections Q in the x,t-plane, 4) find the image points Q'—and thus the values of the unknowns at Q—by drawing $\pm 45°$ lines through the points P'. The process may then be repeated, starting from the points Q and finding a new set of intersections R and corresponding points R' whose u,v-coordinates supply the values of the unknowns at the points R, until the complete mapping in the x,t-plane of the network in $A'B'C'$ has been determined. Then the values of the two flow variables u,v will have been found in the interior of the characteristic triangle ABC, that is, in the whole domain (for $t > 0$) in which these values are determined by the initial conditions. In addition, the direction of the particle line, $dx/dt = u$, at any point P may be found by drawing a parallel to the line MN in the u,v-plane (Fig. 60), where N is the projection of the corresponding point P' onto the u-axis.

This method works rapidly and with sufficient accuracy for all practical cases. Three remarks may be added. First, it is not necessary to take $OM = 1$ in Fig. 60, provided a suitable adjustment is made in the scale of t. For example, if it is convenient to choose $OM = c$ ft/sec, then drawing parallels still gives the correct lines in the x,t-plane provided that 1 sec on the t-axis has the same length as c feet on the x-axis.

Second, in obtaining the solution for $t > 0$ only half the characteristic quadrangle is used, and it is necessary to choose the appropriate part of the image rectangle in the u,v-plane. For $t > 0$ the $(u + a)$-characteristic through a point P(whose image is the $-45°$ line through the corresponding P') must meet the $(u - a)$-characteristic (whose image is the $+45°$ line) through the next point to the right on AB within the half-quadrangle. Therefore, *as one goes along $A'B'$ in the sense corresponding to increasing x, the right side of $A'B'$ is to be used as long as dv/du lies between* -1 *and* 1, *while the left side is chosen when dv/du lies in the interval from* $+1$ *through* $\pm\infty$ *to* -1 (as in Fig. 61).

Finally, it can happen that some domain in the u,v-plane is covered more than once.[24] For example, suppose that the u,v-values given along the segment ABC (Fig. 62) lead to a curve $A'B'C'$ in the speedgraph with $dv/du = 1$ at B'. Then the right side is used from A' to B', giving the solution in the triangle ABD corresponding to $A'B'D'$, and the left side is used from B' to C', giving the solution in BCE. The solution in $DBEF$ is determined, according to the second theorem of Sec. 10.3, from the (compatible) values of u and v along the two characteristics BD and BE. The speedgraph of $DBEF$ is the rectangle $D'B'E'F'$. Thus the region $A'B'D'$ is covered twice. To complete the picture in the x,t-plane, it is best to transfer the lines $D'B'$ and $B'E'$, together with the points at which they are intersected by characteristics already plotted, to a new speedgraph figure. The $\pm 45°$ lines may

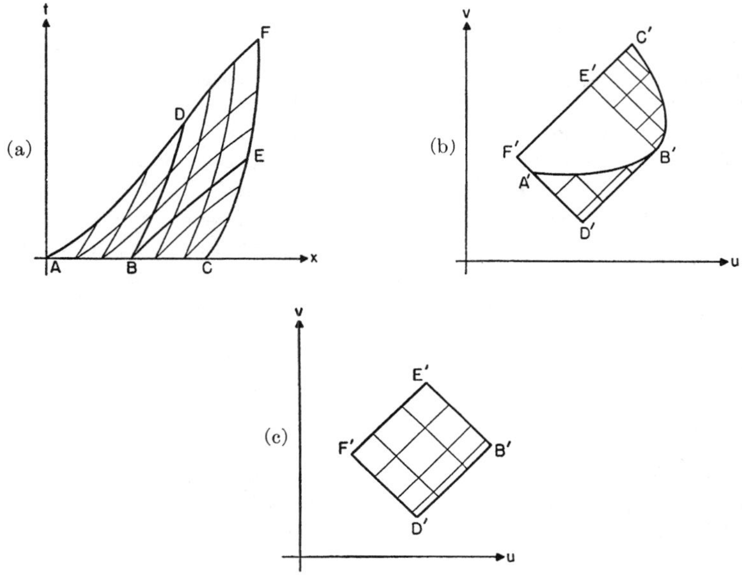

FIG. 62. Double covering of the speedgraph.
(a) The characteristic net in the x,t-plane.
(b) Image regions in the speedgraph with $A'B'D'$ covered twice.
(c) Image region $B'D'F'E'$ shown separately.

be drawn in the rectangle, and the network mapped onto the x,t-plane as described above. Each $(u + a)$-characteristic has an inflection point as it crosses BD, corresponding to a reversal of direction along the image line $B'D'$ in the speedgraph plane.*

6. Analytic solution: values given on two characteristics

A one-dimensional flow problem is completely solved when the values of the state variables u,ρ (or u,v) are known for each point x,t in the x,t-plane. Our methods, however, which are based on the interchange of the original independent and dependent variables, will supply x and t as functions of u and v. To invert this solution, i.e., to solve for u,v in terms of x,t is not possible, in general, by the use of known operators. When this situation occurs, it is final: no other method of integration can supply u and v as explicit functions of x and t.

Giving x and t as functions of u and v, however, does supply a parametric

* See Sec. 19.6 for the general study of a "branch line", which appears here (namely BD) for the first time.

representation for the characteristics in the x,t-plane (by setting $v + u$ or $v - u$ equal to a constant), and in most cases this is sufficient for practical problems. To find an analytic expression for the particle lines it is still necessary to integrate a first-order differential equation.

In the remainder of this article we shall deal with the two important boundary-value problems introduced in Art. 10, as they appear in one-dimensional flow. We shall consider here the linearized form of these problems where the speedgraph variables u,v are the independent variables and shall obtain complete and explicit solutions. In this investigation we shall work with the function $V(u,v)$ defined in (27) and the general solution for V given in (42). From these equations we find, for the case $\kappa = 1.4$,

$$\text{(47)}\quad \begin{aligned} x - ut &= \frac{\partial V}{\partial u} = \frac{f' - vf''}{v^3} - \frac{g' - vg''}{v^3}, \\ at = \frac{v}{5}t &= -\frac{\partial V}{\partial v} = \frac{3f - 3vf' + v^2 f''}{v^4} + \frac{3g - 3vg' + v^2 g''}{v^4}, \end{aligned}$$

where f is some function of $\xi = v + u$ and g is a function of $\eta = v - u$. It then remains to choose $f(\xi)$ and $g(\eta)$ so that given initial conditions are satisfied. Of course, f and g are not uniquely determined; certain arbitrary expressions in the one function may be compensated for by terms in the other.

Consider now first the *characteristic boundary-value problem*,[25] wherein we are given compatible values of u and ρ along two characteristics starting from

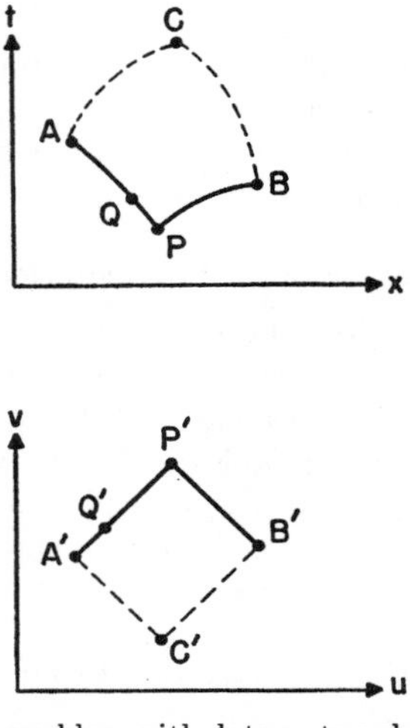

FIG. 63. Boundary-value problem with data on two characteristics PA, PB, showing a quadrilateral in which solution is determined in physical plane and in speedgraph.

a common point P in the x,t-plane (as in the second of the theorems of Sec. 10.3). This problem corresponds to "wave penetration" (see Art. 13). Of course, u and v and the shape of the curves cannot be given independently, as it is necessary that $v - u$ be constant along one curve, say PA (Fig. 63), and $v + u$ be constant along the other, PB. Let us use the subscript 1 to denote the value of any variable at the point P. Then P corresponds to $P'(u_1, v_1)$ in the speedgraph (Fig. 63), and the characteristics are mapped into the $\pm 45°$ lines through P' as shown. Each point Q on PA (or PB) corresponds to a point Q' on $P'A'$ (or $P'B'$) determined by the values of u and v at Q. A single one of these, say v, is sufficient to identify Q'. Of course, the values of x and t are known for each point Q' on either characteristic (since they are the coordinates of Q in the x,t-plane), and they are uniquely determined if the given values of v change monotonically along PA and PB. Then we have

$$
\text{(48)}\qquad
\begin{aligned}
&\text{along } PA\text{:} \quad \eta = v - u = \eta_1, \qquad \xi = v + u = 2v - \eta_1, \\
&\qquad\qquad\qquad\qquad\qquad\qquad\quad v = \tfrac{1}{2}(\xi + \eta_1); \\
&\text{along } PB\text{:} \quad \xi = v + u = \xi_1, \qquad \eta = v - u = 2v - \xi_1, \\
&\qquad\qquad\qquad\qquad\qquad\qquad\quad v = \tfrac{1}{2}(\xi_1 + \eta).
\end{aligned}
$$

The boundary conditions determining the flow pattern must be rewritten with v (or u) as the independent variable, and $V(u,v)$ as the unknown function. Thus we may consider either $\partial V/\partial u = x - ut$ or $-\partial V/\partial v = at = vt/5$ as given along $P'A'$ and $P'B'$.

(a) *Given $x - ut$ along the characteristics.* The data may be assumed in the form

$$
\text{(49)}\qquad
\begin{aligned}
x - ut &= \alpha(v), \qquad v = \tfrac{1}{2}(\xi + \eta_1) \qquad \text{along } P'A', \\
x - ut &= \beta(v), \qquad v = \tfrac{1}{2}(\xi_1 + \eta) \qquad \text{along } P'B',
\end{aligned}
$$

where $\alpha(v)$ and $\beta(v)$ are given functions. Of course, $\alpha(v_1) = \beta(v_1)$, and this value may be taken to be zero by locating the origin suitably on the x-axis.

When expressions (49) are substituted in (47), and $g'(\eta_1)$ is abbreviated to g'_1, etc., we find that f and g must be such that

$$
\text{(50)}\qquad
\begin{aligned}
\frac{f'(\xi) - vf''(\xi)}{v^3} &= \alpha(v) + \frac{g'_1}{v^3} - \frac{g''_1}{v^2}, \qquad \text{where } v = \frac{1}{2}(\xi + \eta_1), \\
\frac{g'(\eta) - vg''(\eta)}{v^3} &= -\beta(v) + \frac{f'_1}{v^3} - \frac{f''_1}{v^2}, \qquad \text{where } v = \frac{1}{2}(\xi_1 + \eta).
\end{aligned}
$$

The first of Eqs. (50) is an ordinary first-order differential equation determining f', while the second determines g'. Both are easily integrated. If we introduce the functions

$$A(v) = \int_{v_1}^{v} \alpha(v)\, dv, \qquad B(v) = \int_{v_1}^{v} \beta(v)\, dv, \tag{51}$$

and if c and k are constants of integration, the solutions of (50) are

$$\begin{aligned} f'(\xi) &= -2v^2 A(v) + g_1' - 2vg_1'' + cv^2, & v &= \tfrac{1}{2}(\xi + \eta_1), \\ g'(\eta) &= 2v^2 B(v) + f_1' - 2vf_1'' + kv^2, & v &= \tfrac{1}{2}(\xi_1 + \eta), \end{aligned} \tag{52}$$

as can be verified by differentiation.

Of the six constants f_1', f_1'', g_1', g_1'', c, and k, which appear in (52), only two can be chosen arbitrarily. We get two conditions restricting the constants if we evaluate (52) for $\xi = \xi_1$ and $\eta = \eta_1$, to obtain

$$f_1' = g_1' - 2v_1 g_1'' + cv_1^2, \qquad g_1' = f_1' - 2v_1 f_1'' + kv_1^2. \tag{53$'$}$$

Furthermore, differentiating Eqs. (52) and then setting $\xi = \xi_1$, $\eta = \eta_1$ we obtain

$$f_1'' = -g_1'' + cv_1, \qquad g_1'' = -f_1'' + kv_1, \tag{53$''$}$$

where we have used $\alpha_1 = \beta_1 = 0$. The four conditions (53) still leave some freedom of choice.[26] For example, all conditions are satisfied if we choose

$$f_1' = g_1' = 0, \qquad f_1'' = g_1'' = 2Cv_1, \qquad c = k = 4C,$$

where C is an arbitrary constant. Then Eqs. (52) take the form

$$\begin{aligned} f'(\xi) &= -2v^2 A(v) + 4Cv(v - v_1), & v &= \tfrac{1}{2}(\xi + \eta_1), \\ g'(\eta) &= 2v^2 B(v) + 4Cv(v - v_1), & v &= \tfrac{1}{2}(\xi_1 + \eta), \end{aligned} \tag{54}$$

where the integrals A and B are known functions of v; when v is expressed in terms of ξ and η respectively, we have

$$\begin{aligned} f'(\xi) &= -\frac{A}{2}(\xi + \eta_1)^2 + C(\xi + \eta_1)(\xi - \xi_1), \\ g'(\eta) &= \frac{B}{2}(\xi_1 + \eta)^2 + C(\xi_1 + \eta)(\eta - \eta_1). \end{aligned} \tag{54$'$}$$

In terms of ξ and η the integrals A and B have upper limits $(\xi + \eta_1)/2$ and $(\xi_1 + \eta)/2$, respectively. Equations (54′) solve the problem. By differentiation, we find f'' and g'' and thus $x - ut$ is given explicitly as a function of u and v, using the first equation (47); this expression is independent of C. To evaluate vt by means of the second Eq. (47), it is necessary to integrate the expressions (54) to find f and g, the constants of integration being chosen so that the second equation (47) is correct at P'. An example will be studied in Sec. 13.5.

(b) *Given t along the characteristics.* Here we assume that the values of at (equal, in our case, to $vt/5$) are given functions of v on two characteristics intersecting at $P'(u_1,v_1)$. These data may be written in the form

$$(55) \quad \begin{aligned} t &= \alpha(v), \qquad v = \tfrac{1}{2}(\xi + \eta_1) \qquad \text{along } P'A', \\ t &= \beta(v), \qquad v = \tfrac{1}{2}(\xi_1 + \eta) \qquad \text{along } P'B'. \end{aligned}$$

By choosing the x-axis through P we can suppose that $t = 0$ at $v = v_1$, i.e., that $\alpha(v_1) = \beta(v_1) = 0$. If again g_1, g_1', etc., are written as abbreviations for $g(\eta_1)$, $g'(\eta_1)$, etc., the conditions (55) may be written, using (47), as

$$(56) \qquad 3f - 3vf' + v^2f'' = \tfrac{1}{5}v^5\,\alpha(v) - 3g_1 + 3vg_1' - v^2g_1''$$

along $P'A'$, and an analogous equation, with f, g, and α replaced by g, f, and β, respectively, along $P'B'$. If we now define

$$(57) \quad A(v) = \frac{1}{5}\int_{v_1}^{v}\left(1 - \frac{z}{v}\right)\alpha(z)\,dz, \qquad B(v) = \frac{1}{5}\int_{v_1}^{v}\left(1 - \frac{z}{v}\right)\beta(z)\,dz,$$

the general solution of (56) is

$$(58) \qquad f(\xi) = 4v^4A(v) - g_1 + 2g_1'v - 2g_1''v^2 + c_1v^3 + c_2v^4,$$

where $v = \frac{1}{2}(\xi + \eta_1)$, and c_1, c_2 are constants of integration. An analogous equation gives $g(\eta)$, introducing two more constants of integration k_1, k_2.

As in the preceding case, the constants are not independent, and six conditions may be found restricting the ten constants f_1, f_1', f_1'', g_1, g_1', g_1'', c_1, c_2, k_1, and k_2, by computing the values f, f', f'' at $v = v_1$ from (58) and equating them to f_1, f_1', f_1'', etc.[27] Since these conditions are homogeneous (for the first and second derivatives of A and B vanish at $v = v_1$), a possible choice is to have all ten constants vanish. Then the solution takes the form

$$(59) \quad \begin{aligned} f(\xi) &= 4v^4A(v) = \frac{4}{5}v^3\int_{v_1}^{v}(v - z)\,\alpha(z)\,dz, \qquad v = \tfrac{1}{2}(\xi + \eta_1), \\ g(\eta) &= 4v^4B(v) = \frac{4}{5}v^3\int_{v_1}^{v}(v - z)\,\beta(z)\,dz, \qquad v = \tfrac{1}{2}(\xi_1 + \eta). \end{aligned}$$

From the formulas (59) and their derivatives, x and t are expressed in terms of u and v by means of (47).

These solutions hold for (u,v) in the rectangle $A'P'B'C'$ in the speedgraph plane, corresponding to a curvilinear quadrangle in the x,t-plane bounded by characteristics.

7. Analytic solution: given u and v at $t = 0$[28]

This is the initial-value problem discussed in Sec. 5; the solution may be given analytically in simple form in the case of a polytropic gas with $\kappa = 1.4$. There are given two functions

$$u = u(x), \qquad v = v(x), \qquad a \leqq x \leqq b, \tag{60}$$

representing the velocity and density distributions at $t = 0$, along an interval of the x-axis from $x = a$ to $x = b$. We assume that Eqs. (60) give a one-to-one correspondence between the segment AB of the x-axis and an arc $A'B'$ (Fig. 64) in the speedgraph plane which at no inner point has a tangent in the $\pm 45°$ directions. In a contrary case such as the one explained at the end of Sec. 5 (see Fig. 62), the solution is obtained in several sections, each part satisfying either boundary conditions of the type here considered or those of the characteristic initial-value problem, for which the analytic solution has been given in the preceding section.

Under the above hypotheses, to each point P' interior to the speedgraph triangle $A'B'C'$ there correspond exactly two points on $A'B'$: P_1', having the same value of $\xi = v + u$ as P', and P_2', having the same $\eta = v - u$. The points P_1' and P_2' correspond to two distinct points on AB, whose x-coordinates we designate as $x_1(\xi)$ and $x_2(\eta)$, respectively. Analytically, $x_1(\xi)$ is the inverse of the function $\xi = v(x) + u(x)$, and $x_2(\eta)$ the inverse of $\eta = v(x) - u(x)$. Each point P' in $A'B'C'$ can thus be characterized by the two quantities or "coordinates" $x_1(\xi)$ and $x_2(\eta)$. The points on the boundary $A'B'$, and these points only (which correspond to $t = 0$), are characterized by $x_1(\xi) = x_2(\eta)$. This suggests using $x_1(\xi)$ and $x_2(\eta)$ to determine the form of the functions $f(\xi)$ and $g(\eta)$.

When $t = 0$ Eqs. (47) have the form

$$\begin{aligned} f' - g' - v(f'' - g'') &= v^3 x, \\ 3(f + g) - 3v(f' + g') + v^2(f'' + g'') &= 0, \end{aligned} \tag{61}$$

where v is a given function of x, from (60). The values of u as a given function of x also enter into (61) by way of the arguments ξ and η of f and g.

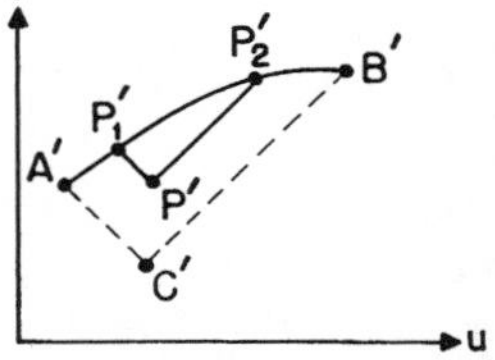

FIG. 64. Location of the points P_1', P_2' in the analytic solution of an initial-value problem.

It is easily verified that the left-hand member of each of these equations is not changed if we add $\mathfrak{A}_0 + \xi\mathfrak{B}_0 + \xi^2\mathfrak{C}_0$ to f and $-\mathfrak{A}_0 + \eta\mathfrak{B}_0 - \eta^2\mathfrak{C}_0$ to g, where $\mathfrak{A}_0$, $\mathfrak{B}_0$, and $\mathfrak{C}_0$ are any constants. This fact suggests that, as in the well-known method of "variation of parameters", $f(\xi)$ and $g(\eta)$ should be set up in the form[29]

$$(62) \qquad \begin{aligned} f(\xi) &= \mathfrak{A}[x(\xi)] + \xi\mathfrak{B}[x(\xi)] + \xi^2\mathfrak{C}[x(\xi)], \\ g(\eta) &= -\mathfrak{A}[x(\eta)] + \eta\mathfrak{B}[x(\eta)] - \eta^2\mathfrak{C}[x(\eta)]. \end{aligned}$$

The new $\mathfrak{A}$, $\mathfrak{B}$, $\mathfrak{C}$ are functions of a single variable, to be chosen so that the conditions (61) are satisfied. These conditions apply only for $t = 0$, and therefore for $x(\xi) = x(\eta)$, so that we may work with $\mathfrak{A}$, $\mathfrak{B}$, $\mathfrak{C}$ simply as functions of x and write

$$\mathfrak{A}' = \frac{d\mathfrak{A}}{dx} = \frac{d\mathfrak{A}}{d\xi}\frac{d\xi}{dx} = \frac{d\mathfrak{A}}{d\xi}(v' + u') = \frac{d\mathfrak{A}}{d\eta}\frac{d\eta}{dx} = \frac{d\mathfrak{A}}{d\eta}(v' - u'),$$

etc. Then for points on the boundary arc $A'B'$ we have

$$(63) \qquad \begin{aligned} f &= \mathfrak{A} + \xi\mathfrak{B} + \xi^2\mathfrak{C}, & g &= -\mathfrak{A} + \eta\mathfrak{B} - \eta^2\mathfrak{C}, \\ f' &= \mathfrak{B} + 2\xi\mathfrak{C} + Y_1, & g' &= \mathfrak{B} - 2\eta\mathfrak{C} + Y_2, \\ f'' &= 2\mathfrak{C} + Z_1, & g'' &= -2\mathfrak{C} + Z_2, \end{aligned}$$

where the Y_i and Z_i are the functions of x defined by

$$(64) \qquad \begin{aligned} (v' + u')Y_1 &= \mathfrak{A}' + \xi\mathfrak{B}' + \xi^2\mathfrak{C}', & (v' - u')Y_2 &= -\mathfrak{A}' + \eta\mathfrak{B}' - \eta^2\mathfrak{C}', \\ (v' + u')Z_1 &= Y_1' + \mathfrak{B}' + 2\xi\mathfrak{C}', & (v' - u')Z_2 &= Y_2' + \mathfrak{B}' - 2\eta\mathfrak{C}'. \end{aligned}$$

When the expressions in (63) are inserted into the conditions (61), these reduce to the conditions

$$(65) \qquad \begin{aligned} Y_1 - Y_2 - v(Z_1 - Z_2) &= v^3x, \\ 3(Y_1 + Y_2) - v(Z_1 + Z_2) &= 0. \end{aligned}$$

Equations (64) and (65) are six linear relations among the seven variables $\mathfrak{A}'$, $\mathfrak{B}'$, $\mathfrak{C}'$, Y_1, Y_2, Z_1, Z_2. To eliminate $\mathfrak{A}'$, $\mathfrak{B}'$, $\mathfrak{C}'$, we multiply the equations in the second line of (64) by $-v$ and add all four equations together, obtaining

$$(66) \qquad \begin{aligned} v(Y_1' + Y_2') = v'\,[v(Z_1 + Z_2) &- (Y_1 + Y_2)] \\ &+ u'\,[v(Z_1 - Z_2) - (Y_1 - Y_2)]. \end{aligned}$$

The values $Z_1 \pm Z_2$ may be taken from (65), yielding

$$Y_1' + Y_2' - \frac{2v'}{v}(Y_1 + Y_2) = -v^2xu'.$$

This is a linear first-order differential equation for the function $Y_1 + Y_2$, and its general integral is

$$Y_1 + Y_2 = -v^2 \int^x xu'\,dx = -v^2 \int^x x\,du = -v^2V(x), \tag{67}$$

where the lower limit of integration is arbitrary. On the boundary $A'B'$, which corresponds to $t = 0$, the expression $\partial V/\partial u = x - ut$ reduces to x, and $\partial V/\partial v = -at$ to zero; thus the integral of $x\,du$, which we have called $V(x)$, is the value of $V(u,v)$ at the point of $A'B'$ corresponding to the point x on AB.

Since only six relations govern our seven variables, one choice is open. We decide to set

$$Y_1 = Y_2 = -\frac{v^2}{2}V(x), \qquad Y_1' = Y_2' = -vv'V(x) - \frac{1}{2}v^2xu'. \tag{68}$$

Then Eqs. (65) determine Z_1 and Z_2, and finally any three Eqs. (64) serve to determine $\mathcal{A}'$, $\mathcal{B}'$, $\mathcal{C}'$. The results are

$$\begin{aligned} \mathcal{A}'(x) &= \frac{V}{4}(v^2 - 3u^2)u' + \frac{v}{4}[x(v^2 - u^2) + 2uV]v', \\ \mathcal{B}'(x) &= \frac{3V}{2}uu' + \frac{v}{2}(xu - V)v', \\ \mathcal{C}'(x) &= -\frac{3V}{4}u' - \frac{v}{4}xv'. \end{aligned} \tag{69}$$

The right-hand members are expressed entirely in terms of the given functions $u(x)$ and $v(x)$, since the function $V(x)$ is defined by (67). Then $f(\xi)$ and $g(\eta)$ are given throughout $A'B'C'$ by (62), with

$$\mathcal{A}[x(\xi)] = \int^{x_1(\xi)} \mathcal{A}'(x)\,dx, \qquad \mathcal{A}[x(\eta)] = \int^{x_2(\eta)} \mathcal{A}'(x)\,dx, \qquad \text{etc;} \tag{70}$$

the lower limits in the integrals are arbitrary, but must be the same in each pair of integrals.

Examples

(a) Let α and β be distinct constants, and suppose that for $t = 0$

$$u = \alpha x, \qquad v = \beta x, \qquad \text{for all } x.$$

Then

$$V(x) = \int^x x\,du = \frac{\alpha}{2}x^2, \qquad \xi = (\beta + \alpha)x, \qquad \eta = (\beta - \alpha)x,$$

$$x(\xi) = \frac{\xi}{\beta + \alpha}, \qquad x(\eta) = \frac{\eta}{\beta - \alpha}.$$

Equations (69) read

$$\mathcal{A}'(x) = \frac{x^4}{8}(2\beta^2 + 3\alpha^2)(\beta^2 - \alpha^2), \qquad \mathcal{B}'(x) = \frac{x^3}{4}(\beta^2 + 3\alpha^2)\alpha,$$

$$\mathcal{C}'(x) = -\frac{x^2}{8}(2\beta^2 + 3\alpha^2).$$

Then, if Eqs. (70) are evaluated, with lower limit 0, and substituted into (62), we find

$$f(\xi) = C_1\xi^5, \qquad g(\eta) = C_2\eta^5$$

with

$$C_1 = -\frac{1}{240}\frac{8\beta^2 + 9\alpha\beta + 3\alpha^2}{(\beta + \alpha)^3}, \qquad C_2 = \frac{1}{240}\frac{8\beta^2 - 9\alpha\beta + 3\alpha^2}{(\beta - \alpha)^3}.$$

Together with (47) these solutions give t, $x - ut$, and therefore x, as functions of u and v.

(b) A particularly simple formula results if $u(x) \equiv 0$, so that the flow starts from rest.[30] Then $V(x)$ in (67) is 0, Eqs. (68) give $Y_1 = Y_2 = 0$, and finally from (69) we get

$$\mathcal{A}'(x) = \frac{x}{4}v^3v', \qquad \mathcal{B}'(x) = 0, \qquad \mathcal{C}'(x) = -\frac{x}{4}vv'.$$

Here we may find f' and g' directly from Eqs. (63), rather than follow all the steps used in the general case:

$$f'(\xi) = 2\xi\mathcal{C} = -\frac{\xi}{2}\int^{x_1(\xi)} xvv'\,dx, \qquad g'(\eta) = -2\eta\mathcal{C} = \frac{\eta}{2}\int^{x_2(\eta)} xvv'\,dx.$$

In this case, $\xi = v(x) + u(x)$ reduces to $\xi = v(x)$, so that if $x = \alpha(v)$ denotes the inverse function of $v = v(x)$, we may write

$$f'(\xi) = -\frac{\xi}{2}\int^{\xi} \alpha(v)\,v\,dv, \qquad g'(\eta) = \frac{\eta}{2}\int^{\eta} \alpha(v)\,v\,dv.$$

It can be verified that this result agrees with that obtained from the formulas that give the solutions in the general case.

Article 13

Simple Waves. Examples

1. Simple waves; definition and basic relations

In the preceding article we treated some problems of one-dimensional nonsteady flow by transforming from the x,t-plane onto the speedgraph plane of the variables u,v. In general, any finite region of the x,t-plane is mapped onto a finite area in the u,v-plane, and any element of area $dx\, dt$ is transformed into an element of area in the speedgraph. The extreme degenerate case is that of uniform flow: u = constant, ρ = constant, where the entire x,t-plane corresponds to a single point of the speedgraph plane. In between lies the case, now to be considered, in which the whole x,t-area is mapped into a single arc Γ in the speedgraph plane and each element of area $dx\, dt$ is transformed into an element of arc on Γ. This type of flow has been given the name *simple wave*, for reasons which will appear below.[31] As was mentioned in Sec. 10.6, this situation can occur only when the Jacobian of the transformation mapping the physical plane into the u,v-plane vanishes at all points; consequently some of the results of the preceding article are not applicable, e.g., Eq. (12.23).

The curve Γ in the speedgraph plane cannot be an arbitrary curve. Indeed, we have seen that the $(u \pm a)$-characteristics in the x,t-plane are always mapped into the $\mp 45°$ straight lines in the u,v-plane. At a point P with coordinates x,t one of the two line elements $dx/dt = u \pm a$ may be mapped into the point P', but certainly not both of them, for then the whole element of area $dx\, dt$ at P would map into a single point. Thus each element of arc on Γ must be the image of an element of arc of a characteristic and have either the $+45°$ or $-45°$ direction. Accordingly we can distinguish two kinds of simple waves, which will be called *forward* and *backward waves*, respectively. In the first case the complete speedgraph consists of a single line in the $+45°$ direction, with equation $v = v_0 + u$; in the whole flow pattern (that is, for all x and all t) the difference $v - u$ has a constant value. In a backward wave the speedgraph is a line $v = v_0 - u$ and the sum $v + u$ is constant for all points of the flow.

In both kinds of simple wave each curve of the second set of characteristics is mapped into a single point of the line Γ, since the image must lie on the line Γ and on a line having the other of the $\pm 45°$ directions. This means that along these characteristics u and v are constant, and with v also ρ and a. Since the slope of these characteristics, measured from the t-axis, is either $u + a$ or $u - a$, it follows that the curves of this second set of characteristics must be straight lines. Thus we arrive at the following defi-

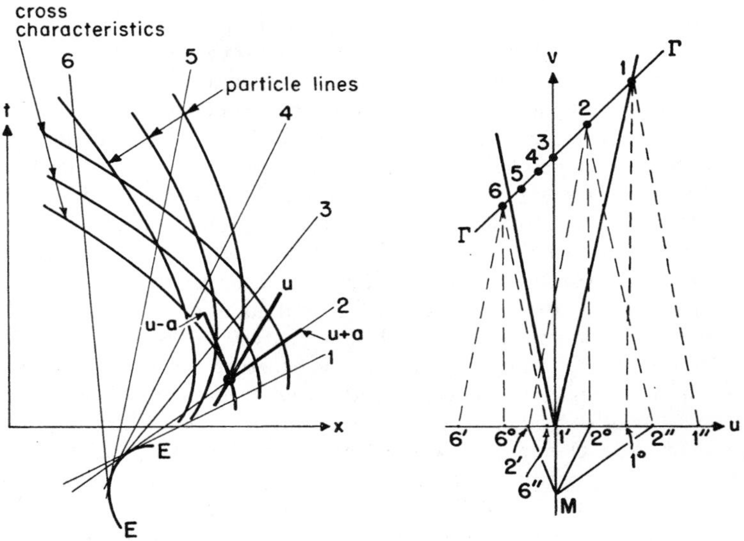

FIG. 65. Construction of a forward rarefaction wave. Assumed given: The envelope E of straight $(u + a)$-characteristics in the x,t-plane, and the $+45°$ line Γ^+: $v = v_0 + u$ forming the speedgraph.

nition:* *A simple wave is a flow pattern in which one set of characteristics in the x,t-plane consists of straight lines along which u and v* (and also ρ and a) *have constant values. Moreover, in a forward wave $v - u$, and in a backward wave $v + u$ is constant for all x and t; in the first case the straight lines are the $(u + a)$-characteristics, while in the second they are the $(u - a)$-characteristics.* The characteristics of the other set, e.g. the $(u - a)$-characteristics in the case of a forward wave, are usually called *cross-characteristics.*

The expression "wave" refers to the fact that the state u, ρ, i.e., the geometrical point at which u and ρ have a given pair of values, "progresses" at constant speed, $u + a$ or $u - a$, while, of course, any material particle moves with velocity u. Thus, a forward wave propagates forwards relative to the moving particles, while a backward wave propagates backwards. Obviously a uniform flow, u = constant, ρ = constant, with two sets of straight (and parallel) characteristics, is a limiting case of a simple wave.

In Fig. 65 is given an example of a forward wave. It is assumed 1) that the 45° line Γ in the speedgraph plane has equation $v = v_0 + u$, where v_0 is arbitrary; and 2) that the straight-line characteristics are an arbitrary

* The proof that such a flow pattern does in fact satisfy the equations of motion follows the same lines as the proof for simple waves in the x,y-plane given in Sec. 18.1.

given set of lines in the x,t-plane, tangent to an envelope E.[32] From these data the flow pattern can be derived in the following way. As in Sec. 12.5, the three directions $u - a$, u, $u + a$ corresponding to any point on Γ (such as point 2 in the figure) may be found by means of the vertical line 2-2^0 and the lines with slopes ± 5, 2-2′ and 2-2″. The straight characteristic that maps into the point 2 on Γ is the tangent to E which is parallel to the line $M2''$. At any point on this characteristic the direction of the particle line is that of $M2^0$ and the direction of the cross-characteristic is that of $M2'$. Thus these two sets of curves, particle lines and cross-characteristics, are determined respectively by two direction fields; analytically, each set is determined by a single ordinary differential equation of first order. The isoclines for either set are the straight-line characteristics. If both parts of Fig. 65 are reflected on a vertical axis, they picture a backward wave. In this example the particles move (in both cases) from points of higher pressure (greater v) to points of lower pressure. This type of flow is called a *rarefaction wave*. One may, however, reverse simultaneously the signs of u and t in any flow pattern, as may be seen from Eq. (12.20). Geometrically, this means that the first of Fig. 65 is reflected in a horizontal axis and the other in a vertical one. Then we have a *compression wave* (backward or forward), and the particles move so that pressure increases. It is seen that the density changes in the same sense as the velocity in a forward wave (both diminish in Fig. 65), and in the opposite sense in a backward wave, i.e., if u increases, then v decreases. The two solid lines with slopes ± 5 in the second part of Fig. 65 are the boundaries between subsonic and supersonic flow, as mentioned in Sec. 12.3. At a sonic point one of the two characteristics in the x,t-plane is vertical.

A physically possible flow is represented only by portions of the x,t-plane in which the straight characteristics do not cross each other. Intersections occur necessarily in the neighborhood of the envelope E. Moreover, the particle lines approach each other, which means that the density increases indefinitely, as they come closer to E. Thus the following difference between rarefaction and compression waves can be stated: *For given initial conditions at $t = 0$, the rarefaction wave can extend to $t = \infty$, while the compression wave is restricted to a finite interval of time.*

Having constructed the flow geometrically, let us now consider the analytic solution, which can be found starting from the same data: the constant v_0 and the chosen set of straight characteristics. Let the latter be given in terms of a parameter β in the form

$$(1) \qquad x = x_0(\beta) + \beta t.$$

Obviously, β is the slope of the characteristic as measured from the t-axis. Then, using the upper signs for a forward (and the lower for a backward) wave, we have

$$\beta = u \pm a, \qquad v = v_0 \pm u, \qquad \pm\beta = a + v - v_0 . \tag{2}$$

Since v is a known function of a (equal to $5a$ in the case of a diatomic gas), the last equation serves to express a in terms of β. We can then define another function F of β by

$$\log F = \mp \int^{\beta} \frac{d\beta}{a}. \tag{3}$$

Differentiation of (1) gives

$$dx = (x_0' + t)d\beta + \beta\, dt. \tag{4}$$

Now particle lines are defined by the condition $dx = u\, dt$. When the differentiation in (4) takes place in the direction of a particle line, i.e., if dx is replaced by $u\, dt$ we find

$$\frac{dt}{d\beta} \pm \frac{t + x_0'}{a} = 0, \tag{5}$$

where $\beta - u$ has been replaced by $\pm a$ from (2). Integration of (5) now yields

$$t = F\left[\mp \int^{\beta} \frac{x_0'}{aF}\, d\beta + C\right]. \tag{6}$$

Equation (6) gives t as a function of β (and a constant of integration C) along a particle line; when (6) is substituted in (1), we obtain x also as a function of β and thus have a parametric representation of the particle lines.

On the cross-characteristics the condition is $dx = (u \mp a)\, dt$, and when this is used to replace $dx/d\beta$ in (4), we obtain Eq. (5) with the denominator a replaced by $2a$, and Eq. (6) is replaced by

$$t = \sqrt{F}\left[\mp \int^{\beta} \frac{x_0'}{2a\sqrt{F}}\, d\beta + C_1\right]. \tag{7}$$

In the polytropic case $v = 2a/(\kappa - 1)$ [Eq. (12.19′)] so that from (2) and (3)

$$a = \frac{v_0 \pm \beta}{h^2}, \qquad F = (v_0 \pm \beta)^{-h^2}, \qquad \text{where } h^2 = \frac{\kappa + 1}{\kappa - 1};$$

and therefore from (6) and (7)

$$t = h^2(v_0 \pm \beta)^{-h^2}\left[\mp \int^{\beta} x_0'(v_0 \pm \beta)^{2/(\kappa-1)}\, d\beta + C\right], \tag{6′}$$

$$t = \frac{1}{2} h^2(v_0 \pm \beta)^{-h^2/2}\left[\mp \int^{\beta} x_0'(v_0 \pm \beta)^{(3-\kappa)/2(\kappa-1)}\, d\beta + C_1\right]. \tag{7′}$$

From Eqs. (6) and (7) we can deduce the following property of either set of curves, particle lines or cross-characteristics. *The integral curves, particle lines or cross-characteristics, corresponding to equidistant values of the constant,* C_0, $C_0 + c$, $C_0 + 2c$, $\cdots$, *intersect any straight characteristic at equidistant points.*

2. Centered waves[33]

A particularly simple case occurs when the envelope E reduces to a single point, i.e., when the set of straight characteristics consists of radii emanating from one point. A flow pattern of this kind is called a *centered wave.*

If we suppose that the center lies at the origin, the quantity x_0 in Eq. (1) vanishes and (1) reduces to $\beta = x/t$. For polytropic flow Eq. (6′) now gives

$$(8) \qquad t = k\left(v_0 \pm \frac{x}{t}\right)^{-h^2} \qquad \text{or} \qquad \pm x = -v_0 t + k\left(\frac{t}{k}\right)^{2/(\kappa+1)}$$

as the equation for the particle lines, and Eq. (7′) gives

$$(9) \qquad t = k_1\left(v_0 \pm \frac{x}{t}\right)^{-h^2/2} \qquad \text{or} \qquad \pm x = -v_0 t + k_1\left(\frac{t}{k_1}\right)^{(3-\kappa)/(\kappa+1)}$$

as the equation for the cross-characteristics. These equations include all four cases of centered waves (forward or backward, compression or rarefaction). The constants k and k_1, which single out individual curves, have positive values in the case of a rarefaction wave and negative values (together with negative t) for a compression wave. The velocity $u = dx/dt$ can be found by differentiating (8) or, alternatively, by solving for u in the relations (2), with $\beta = x/t$:

$$(10) \qquad u = \frac{2}{\kappa + 1}\left(\frac{x}{t} \mp \frac{\kappa - 1}{2} v_0\right) = \frac{2}{\kappa + 1}\left(\frac{x}{t} \mp a_0\right),$$

where $a_0 = (\kappa - 1)v_0/2$ is the sound velocity at a stagnation point (which occurs when the particle line has a vertical tangent). From (10) and $v = v_0 \pm u$ there follows

$$(11) \qquad v = \pm\frac{2}{\kappa + 1}\frac{x}{t} + \frac{2}{\kappa + 1} v_0 = \frac{2}{\kappa + 1}\left(v_0 \pm \frac{x}{t}\right).$$

Since $dx/dt = u$ becomes infinite as we approach the origin along any particle line, as can be noted from (8) and (10), we see that the center of the wave corresponds in a certain sense to the state $|u| = \infty$, $v = \infty$ (and therefore $\rho = \infty$ and $a = \infty$). On the other hand, since on each ray through the center u and v have constant values, the center is mapped into the whole line Γ of the speedgraph.

By means of (11), Eqs. (8) and (9) can be written as*

$$(11') \qquad \begin{aligned} t &= \text{constant}\cdot v^{-h^2} = t_0 \left(\frac{v_0}{v}\right)^{h^2} \quad \text{on a particle line,} \\ t &= \text{constant}\cdot v^{-h^2/2} = t_0 \left(\frac{v_0}{v}\right)^{h^2/2} \quad \text{on a cross-characteristic.} \end{aligned}$$

Then, from (10) we obtain, for later use,

$$(11'') \qquad \begin{aligned} x - ut &= \frac{1}{h^2}(x \pm v_0 t) = \pm\frac{\kappa - 1}{2} vt \\ &= \pm a_0 t_0 \left(\frac{v_0}{v}\right)^{2/(\kappa-1)} \quad \text{on a particle line,} \\ &= \pm a_0 t_0 \left(\frac{v_0}{v}\right)^{(3-\kappa)/2(\kappa-1)} \quad \text{on a cross-characteristic.} \end{aligned}$$

Since v is proportional to a and a^2 is proportional to $\rho^{\kappa-1}$, one concludes from (11′) that, for a given particle, ρ is proportional to the $-2/(\kappa + 1)$ power of $|t|$, where $|t|$ is the time elapsing between the actual state and the state $|u| = \infty$, $\rho = \infty$. Consequently the density ρ changes monotonically, decreasing in rarefaction waves, increasing in compression waves.

Figure 66 shows the particle lines for various values of the constant k and the speedgraph of a forward rarefaction wave, with $v_0 = 5$, $\kappa = 1.4$. All other waves of the same type are obtained by affine transformations, while reflections in the x- or t-axis or both supply waves of the other three kinds (see Fig. 67). As has been mentioned, the "center" point on each particle line corresponds to the state $|u| = \infty$, $\rho = \infty$ (and to the point at infinity in the speedgraph); by setting $t = \infty$ in (11′), we see that the point at infinity on each particle line corresponds to $u = \mp v_0 = \mp 2a_0/(\kappa - 1)$, and $v = 0$ in the speedgraph. A physically possible centered wave can be represented only by a portion of one of the four regions indicated in Fig. 67.

If C is the constant in the polytropic relation, i.e., $C = p/\rho^\kappa$, then Eqs. (12.19′) and (11) give

$$(12) \qquad \rho = \left(\frac{\kappa - 1}{2\sqrt{\kappa C}}\, v\right)^{2/(\kappa-1)} = (h^2\sqrt{\kappa C})^{-2/(\kappa-1)} \left(v_0 \pm \frac{x}{t}\right)^{2/(\kappa-1)}.$$

* For Eqs. (11′) and (11″) t_0 is a parameter which varies with the particle line or cross-characteristic; it is the time at which $u = 0$ on the curve considered. On the other hand, v_0 and a_0 are constant for the whole flow.

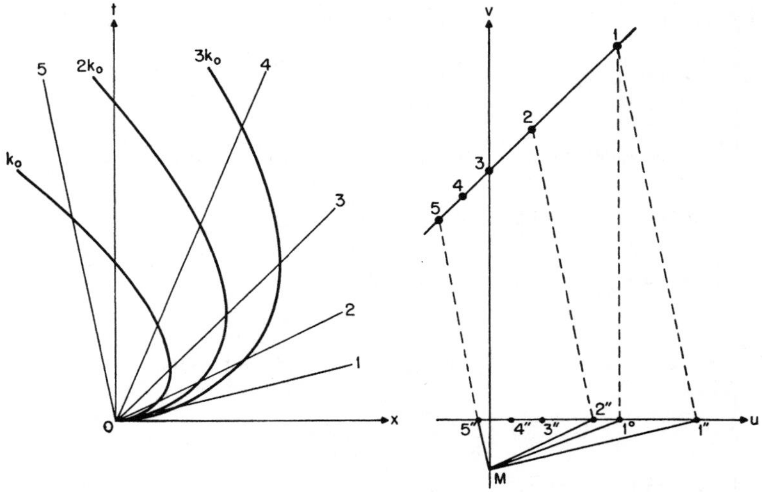

FIG. 66. Example of a centered forward rarefaction wave: $v_0 = 5$. Shown are particle lines for equidistant values of the constant k.

Together, (12) and (10) determine the derivatives $\partial\psi/\partial x = \rho$ and $\partial\psi/\partial t = -\rho u$ of the particle function $\psi(x,t)$ of the centered wave. One can easily verify by differentiation that

$$\psi(x,t) = \pm \frac{t}{h^2} (h^2\sqrt{\kappa C})^{-2/(\kappa-1)} \left(v_0 \pm \frac{x}{t}\right)^{h^2}. \tag{13}$$

On the other hand, the potential function $\Phi(x,t)$ has u as derivative with respect to x and

$$-\frac{u^2}{2} - P = -\frac{u^2}{2} - \frac{a^2}{\kappa - 1} = -\frac{u^2}{2} - \frac{\kappa - 1}{4} v^2$$

as derivative with respect to t. These conditions are fulfilled by the expression

$$\Phi(x,t) = \frac{t}{\kappa + 1}\left(\frac{x}{t} \mp a_0\right)^2 - \frac{t}{\kappa - 1} a_0^2. \tag{14}$$

It is seen that the centered simple wave is one of the examples discussed in Sec. 7.4 [example (a) with $c_2 = 0$] and in Sec. 12.2 [the example of a solution for (12.17c)].

For each point (x,t) the Mach number is given by

$$M = \frac{|u|}{a} = \frac{2}{\kappa - 1}\frac{|u|}{v} = \frac{2}{\kappa - 1}\frac{|x \mp a_0 t|}{v_0 t \pm x}, \tag{15}$$

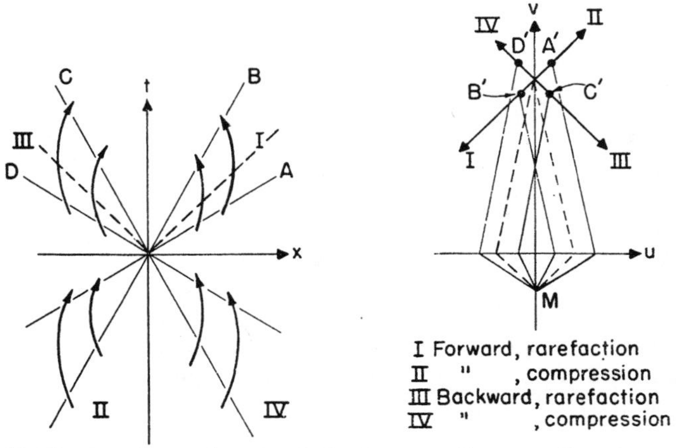

FIG. 67. The four types of centered simple wave. Corresponding changes in the two planes.

with the upper signs for a forward wave and the lower signs for a backward wave.

3. Other examples of simple waves

Let us assume that the envelope E of the straight characteristics consists of two arcs AB and CD (see Fig. 68), both tangent to a straight line BC.

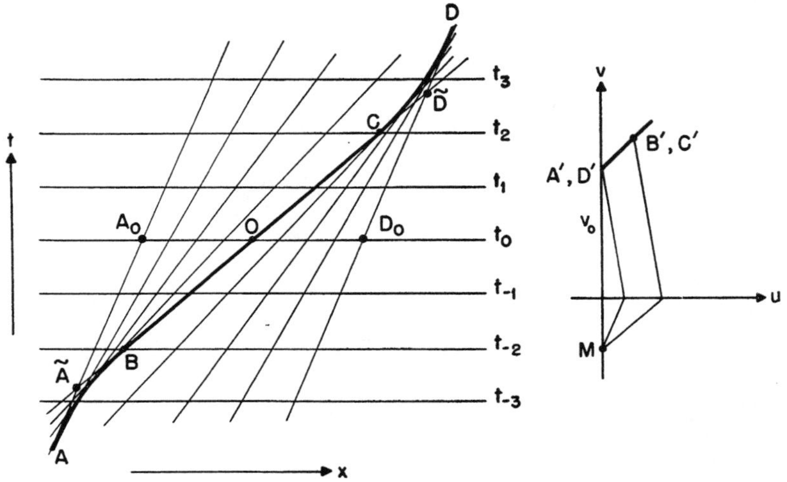

FIG. 68. Limited simple-wave disturbance.

We suppose further that the tangents at A and D are parallel and have slope a_0, the sound velocity at $u = 0$, measured from the t-axis. Then for a forward wave the image of that part of the x,t-plane covered by the tangents to E is the segment $A'B'$ of the 45° line, where A' ($= D'$) has the coordinates $u = 0$, $v = v_0$, and the endpoint B' ($= C'$) with coordinates u_1, v_1, say, may be found by reversing the usual construction: draw the parallel to BC through M in the speedgraph plane and then the line with slope -5 meeting the 45° line at B'. The flow pattern in the x,t-plane is uniquely determined in the region mapping into $A'B'$, i.e., in the region between the parallel tangents at A and D, with the exception of the two regions $A\tilde{A}B$ and $D\tilde{D}C$ in which tangents to E intersect each other. Since the tangents at A and D are characteristics, the solution in the adjoining region may be analytically quite different provided merely that the combined solution is continuous along these lines. Thus we may assume that $u = 0$, $v = v_0$ outside these parallels. The map in the speedgraph of these regions is then the point A'.

In the time interval determined by the two horizontals through B and C, say $t = t_{-2}$ and $t = t_2$, the motion is determined for all x. The particle lines can be constructed easily by means of the speedgraph or can be computed according to the formulas developed in Sec. 1. We wish however to study how the u-values, considered as functions of x, change in time (see Fig. 69).

At the time $t = t_0$, the midpoint of the interval mentioned above, the velocity is zero for all points to the left of A_0, it increases as we cross the straight characteristics from zero at A_0 to the maximum value u_1 at O

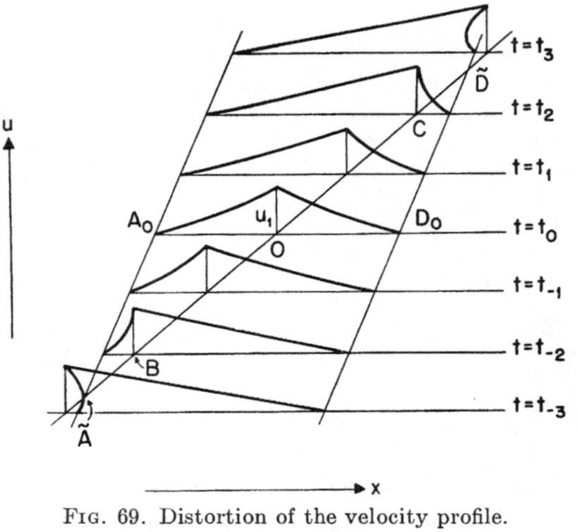

Fig. 69. Distortion of the velocity profile.

(which maps into B'), and then drops to zero again at D_0. If the arcs AB and CD are symmetric with respect to the point O, the curve of u versus x will be symmetric with respect to the line x = constant through O (see the middle diagram of Fig. 69). Now the u-values are constant along a straight characteristic. Thus, if we examine the u-values along $t = t_1$, the same u-values are found, but further to the right, corresponding to greater values of x. It is clear, however, that the rate of movement to the right is largest for the maximum value u_1 and least for the end values. The analogous phenomenon, only with shifting to the left, takes place if we pass from $t = t_0$ to $t = t_{-1}$ (see Fig. 69).

This flow pattern obviously represents the behavior of a limited disturbance within a fluid otherwise at rest and it shows clearly that the expression "wave" is appropriate. The two endpoints of the disturbance progress at the rate a_0 while the culmination point moves at a higher speed, (see $D_0\tilde{D}$ and $\tilde{A}\tilde{D}$ in Fig. 69). Thus, the front part of the wave is steepened, while the rear part is flattened out. This result may be compared with the results of Sec. 4.2 for a *small* perturbation where we found that each part of the initial disturbance progresses, unchanged in shape, at the sound speed a_0. There, however, in omitting the terms of higher order in the basic equations, we made the problem in the x,t-plane a linear one, so that the characteristics were the same for all solutions and, in fact, the parallels of slope $\pm a_0$. The exact solution, here considered, shows that the characteristics are divergent to the rear of the flow and convergent to the front; this brings about a *distortion* of the curves of u versus x. Further, $v - u$ is constant for the whole flow, so that the distortion of the curves for v, and therefore for p and ρ, is similar.

If we extend the time value t to the lower or upper limit of the time interval (t_{-2}, t_2) in which the flow pattern is determined for all x, we find that $\partial u/\partial x$ becomes infinite on one side of the crest of the wave.[34] The solution breaks down completely for values of t outside the time interval (t_{-2}, t_2). For instance, at $t = t_{-3}$ and $t = t_3$ there is more than one value of u determined by some values of x.*

Another example of a simple wave solution is shown in Fig. 70. In a tube extending to infinity in both directions the fluid is initially at rest; the value of a is the same everywhere, a_0; and v is constant, say $v = v_0$. Suppose that at $t = 0$ a piston starts moving toward the right from the position $x = 0$ according to the law $x = x_1(t)$, represented by the curve C.[35] The tangent to C at the origin is vertical, so that the velocity of the piston is

* The breakdown is not due to the discontinuity in slope of the u,x-curve, and clearly occurs in every case of a limited simple wave disturbance. In fact, it is easily seen that, for increasing t, $\partial u/\partial x$ first becomes infinite at a point corresponding to that of maximum negative slope on the initial u,x-curve.

continuous at $t = 0$. The curve C is the particle line for all particles adjacent to the piston on either side. The initial conditions $u = 0$, $v = v_0$ uniquely determine the state of rest $u = 0$, $v = v_0$ at all points within any characteristic quadrangle determined by a segment of the x-axis to the right or left of the origin. The characteristics for this state of rest are the straight lines whose slopes, measured from the t-axis, are $\pm a_0$. Thus the two straight lines OA and OB with slopes $\mp a_0$ limit the outside regions in which the velocity is identically zero. The flow pattern between OA and OB is determined as follows. Each point P on the x_1-curve has a velocity u, represented by the slope of the tangent at P. Then P is mapped in the speedgraph into one of the points P' or P'' with abscissa u and lying on the $+45°$ or $-45°$ line through O' with coordinates $0, v_0$. Since OA is a $(u - a)$-characteristic and OB a $(u + a)$-characteristic, the area between OA and C is covered by a backward wave and maps into the $-45°$ line, while the area between OB and C is covered by a forward wave, mapped onto the $+45°$ line. Using P' for the right and P'' for the left part of the flow, one can construct the two straight characteristics at P, using the lines with slopes ± 5 in the speedgraph. Along each of these lines u and v are constant, so that one can proceed to the construction of the particle lines.

The pressure values (determined by the v-values) are different on the two sides of the piston. The difference in v-values is given by the distance

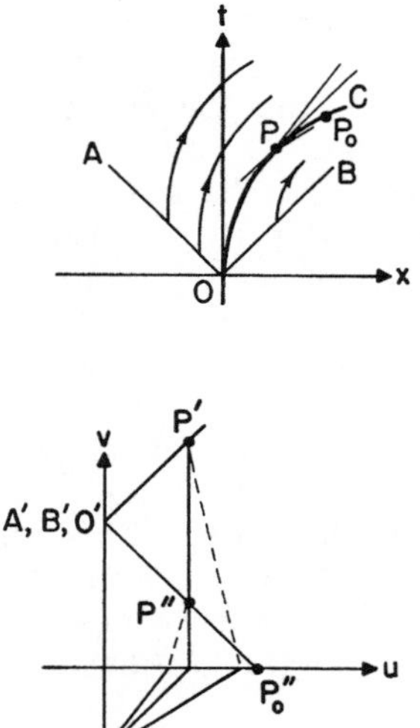

FIG. 70. Construction of the two centered simple waves in the case of a disturbance caused by a piston moving in a gas at rest along a path C.

$P'P''$. To bring about this flow, it is necessary to apply to the piston an external force toward the right, gradually increasing it to give the appropriate difference $P'P''$.

The flow to the right is seen to be a compression wave. This type of solution of a flow problem breaks down after a finite time interval. We shall discuss this in Art. 14.

The rarefaction wave to the left continues indefinitely. When the piston reaches the speed $u = v_0$ (point P_0'' in the speedgraph), the particles adjacent to the piston have attained maximum speed; any further increase in the velocity of the piston leaves a cavity, in which $\rho = 0$, between these particles and the piston, while the velocities of all other particles tend asymptotically to the value $u = v_0$.

4. Combination of simple waves

If the values of u and ρ are constant in any finite region of the x,t-plane, the flow in any adjacent neighborhood can only be a simple wave. This is seen immediately from the fact that a region of constant u,ρ is mapped into a single point in the speedgraph plane. Then, any adjacent neighborhood cannot map into an area in the speedgraph plane and must correspond to an element of arc passing through that point.* (Of course, the notion of simple wave includes the limiting case of uniform flow.)

A simple wave can form the transition between any uniform state: $u = u_1$, $\rho = \rho_1$, and another such state: $u = u_2$, $\rho = \rho_2$, provided that either $u + v$ or $u - v$ has the same value in both states. (Since v cannot be negative, $u_2 - u_1$ must be less than v_1 or greater than $-v_1$, respectively). By combining a forward and a backward wave, and inserting a uniform motion in between, any final state can be reached. An example is shown in Fig. 71. Here $u = 0$ at the beginning and at the end of the motion, but pressure and density have increased corresponding to the change in v from A' to C'. The flow in the x,t-plane is not unique, however, since both envelopes E_1 and E_2 of the straight characteristics can be chosen arbitrarily. The only restriction is that the critical regions, bounded for each wave by the two extreme characteristics and the envelope, be outside the flow region. For the first simple wave the two extreme straight characteristics in question are the ones which are parallel to the line segments MA_1' and MB_1' in the speedgraph; and for the second, the ones parallel to MB_2' and MC_2'. In our example v_0 and v_2, that is, the ordinates of A' and C', have the ratio 1:1.2; with $\kappa = 1.4$, this corresponds to a compression ratio of $(1.2)^5$:1 $\sim$ 5:2. This ratio appears clearly in the figure as the concentration of the vertical particle lines at the end of the flow compared to that at the begin-

* A more formal proof follows lines similar to those for the plane case (Sec. 18.1).

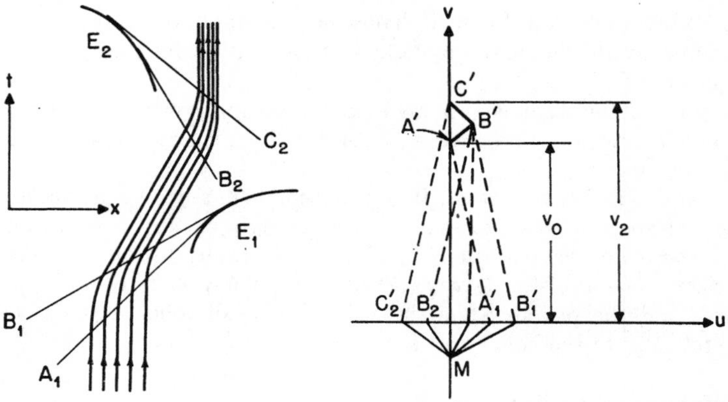

FIG. 71. The use of a pair of simple waves to effect a given compression without resultant motion.

ning of the flow. To induce the flow indicated in Fig. 71, two pistons must be used, having, respectively, the motions of the first and last particle lines.

In order to choose, from among all possible flows in the x,t-plane, one which achieves a given compression as quickly as possible, we use centered waves with the centers located as shown in Fig. 72. (The subscripts 0, 1, and 2, when attached to u, v, or a, will signify the values of that quantity in the initial, intermediate, and final uniform flow regions, respectively.) Here the left piston is first accelerated up to the velocity $u_1 = (v_2 - v_0)/2$ and then moved at constant speed, with v-value $v_1 = (v_2 + v_0)/2$, until the desired compression is attained; the right piston remains at rest until the time $t = t_1 = l_0/a_0$, when the first wave reaches it, and it must then start moving with speed u_1 and be decelerated down to $u_2 = 0$. The compression ratio is l_0/l_1. The time interval t_2 (see Fig. 72) can be computed by applying Eq. (8) to the second wave, and the time required for

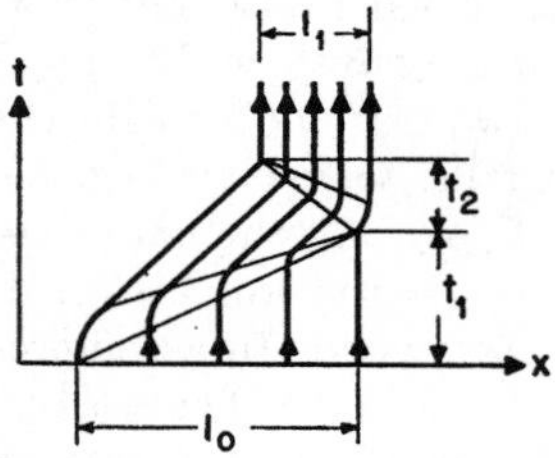

FIG. 72. Effecting a given compression in the shortest time.

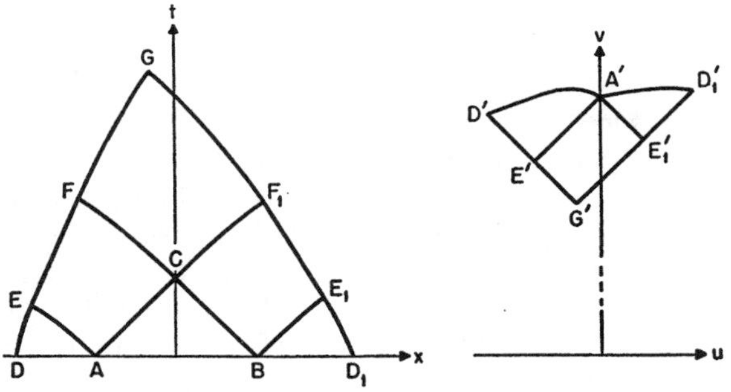

FIG. 73. Penetration of two simple waves. $AEFC$ and BE_1F_1C are the simple wave regions, and $CFGF_1$ the region of penetration.

the whole process turns out to be[36]

$$t = t_1 + t_2 = \frac{l_0}{a_0}\left[1 + \left(\frac{2a_0}{a_0 + a_2}\right)^{h^2}\right]. \tag{16}$$

With a combination of a forward and a backward wave any compression ratio can be realized. It will be seen later (Sec. 14.6) that no solution of similar type exists if the right-hand piston is kept entirely at rest.

As another example we study the penetration of two simple waves. It is assumed (see Fig. 73) that at $t = 0$ the interval AB contains gas at rest: $u = 0$, $v = v_0$. This uniform state then obtains throughout the triangle ABC, where AC and BC have slopes $\pm a_0$ measured from the t-axis. In the speedgraph, the entire triangle ABC maps onto one point A'. To the left, along AD, and to the right, along BD_1, some disturbance is imposed. The given values of u and v along these segments map them in the speedgraph onto two curves $A'D'$ and $A'D_1'$. (B' coincides with A', as does C'.) By means of the methods described in Sec. 12.5 and Sec. 12.7, the flow pattern corresponding to the areas $A'D'E'$ and $A'D_1'E_1'$ can be developed. In the areas adjacent to AC and BC the flows must be simple waves mapping onto the characteristics $A'E'$ and $A'E_1'$, respectively. The areas in which these simple waves occur are bounded by straight characteristics EF and E_1F_1, which are mapped into the points E' and E_1', respectively, and by the cross-characteristics CF and CF_1, which must map into $A'E'$ and $A'E_1'$. At C the two simple waves meet and the problem to be solved is the subsequent flow in the penetration region $CFGF_1$, which maps into the rectangle $A'E'G'E_1'$ in the speedgraph.[37]

The data for this problem include: the curves CF and CF_1 in the physical plane and the compatible u,v-values along these curves, with $v - u = v_0$ along the first and $v + u = v_0$ along the second. From these data we know the functions $\alpha(v)$ and $\beta(v)$ of Eq. (12.49), namely,

(17) $$x - ut = \alpha(v) \text{ along } CF, \qquad x - ut = \beta(v) \text{ along } CF_1 .$$

The formulas developed in Art. 12 then supply an explicit solution of the problem,* giving, for example, the coordinates of the point G, etc.

If the two simple waves to the left and to the right are symmetrical centered waves, the computation can be carried out easily. By shifting the x-axis if necessary, we may place the centers on the x-axis and use these points as A and B. It was shown in Sec. 2, Eq. (11″), that along a cross-characteristic for a centered wave the value of $x - ut$ is proportional to the $-(3 - \kappa)/2(\kappa - 1)$ power of v, e.g., for $\kappa = 1.4$, it is proportional to v^{-2}. In deriving this equation it was assumed that the center of the wave was at $x = 0$. Here, if we choose the t-axis through C, whose ordinate is t_0, the centers are at $x = \pm a_0 t_0 = \pm v_0 t_0/5$. Then the value of x in Eq. (11″) is to be corrected by this amount, so that for the left-hand wave

(18) $$x - ut = \alpha(v) = \frac{v_0 t_0}{5}\left(\frac{v_0^2}{v^2} - 1\right);$$
$$A(v) = \int_{v_0}^{v} \alpha(v)\, dv = -\frac{v_0 t_0}{5v}(v - v_0)^2.$$

It is not necessary to examine $\beta(v)$, etc., because of the symmetry of the whole problem, which means that the functions f and g of Sec. 12.6 are identical. For convenience we choose the constant C in Eq. (12.54) to be $v_0^2 t_0/20$. Then since $\xi_1 = \eta_1 = v_0$, Eq. (12.54) gives

(19) $$f'(\xi) = \frac{2}{5} v_0 t_0 v(v - v_0)\left(v - \frac{v_0}{2}\right) = \frac{v_0 t_0}{20}\xi(\xi^2 - v_0^2)$$
$$g'(\eta) = \frac{2}{5} v_0 t_0 v(v - v_0)\left(v - \frac{v_0}{2}\right) = \frac{v_0 t_0}{20}\eta(\eta^2 - v_0^2).$$

By differentiation and integration we derive from (19)

(20) $$f'' = \frac{v_0 t_0}{20}(3\xi^2 - v_0^2), \qquad f = \frac{v_0 t_0}{80}(\xi^2 - v_0^2)^2$$

and analogous expressions for g'' and g. If these are introduced in (12.47) we find, after some algebraic rearrangement,[38]

(21) $$x - ut = \frac{v_0 t_0 u}{10 v^3}(u^2 - 3v^2 - v_0^2)$$
$$t = \frac{3 v_0 t_0}{8 v^5}\left[v^4 + \frac{2 v_0^2}{3} v^2 + v_0^4 - 2u^2(v^2 + v_0^2) + u^4\right].$$

* Since $f(\xi)$ in (12.47) is determined by $\alpha(v)$ alone, see (12.54′), it represents the influence of the wave on the left, and, similarly, $g(\eta)$ that of the wave on the right.

These two equations supply the values of x and t corresponding to each point (u,v) of the rectangle $A'E'G'E_1'$. In particular, corresponding to G' we find the x,t-values for the "end" of the penetration.

The results expressed in (20) and (21) can be checked by computing the values of $x - ut$ and t along the characteristics CF and CF_1 as functions of v, setting $u = \pm(v - v_0)$, and comparing them with the values along cross-characteristics given by (11′) and (11″), corrected for the fact that the centers are not at the origin. Moreover, $x - ut$ satisfies the differential equation (12.34) with $n = -2$, and t satisfies the same equation with $n = -3$.

Article 14

Theory of Shock Phenomena

1. Nonexistence of solutions. Effect of viscosity

Assume that at $t = 0$ the interval AB of the x-axis is occupied by a fluid mass in a state of rest and at uniform pressure: $u = 0$, $v = v_0 =$ constant. If we draw through A and B (see Fig. 74a) the two straight lines of slope $\pm a_0$ (where a_0 is the sound velocity corresponding to v_0 and the slopes are measured from the t-axis) and the vertical particle lines representing the state $u = 0$, $v = v_0$ for all points (x,t) within the triangle ABC, this solution certainly satisfies the general differential equations (11.2), (11.3), and (11.4), regardless of whether viscosity and heat conduction are admitted, and also satisfies the initial conditions at $t = 0$. In the theory, developed in Art. 12, of an elastic, inviscid and nonconducting fluid, the solution is uniquely determined within the characteristic quadrangle, of which ABC is the upper half, by the conditions along AB. This means that the differential equations of an inviscid and nonconducting fluid have no solution in ABC consistent with the given initial conditions other than the solution $u = 0$, $v = v_0$ everywhere in ABC. [Such a fluid is necessarily elastic by virtue of the entropy being initially the same for all particles (see Sec. 2.3).]

On the other hand, it is undoubtedly possible to subject an actual fluid mass between A and B to some additional conditions. For example, the fluid may be contained in a cylindrical tube with a solid piston initially at rest at the point A. Suppose we start moving the piston to the right according to an arbitrary law $x = f(t)$ subject only to the restriction that $df/dt = 0$ at $t = 0$. This law may be represented by the curve AD in Fig. 74a, and this curve is necessarily the particle line for the fluid particle initially at A. Nothing can prevent us from moving the piston so that the point D falls within the triangle ABC.[39] Thus, a contradiction is apparent. The velocity at D must be that determined by the slope of the tangent to the curve AD at D and not the value $u = 0$ determined by the solution given above.

Note that this contradiction would not occur if the fluid were considered to be incompressible, so that the whole fluid mass moved with the piston, behaving as a rigid body. In fact, the triangle ABC, in which the solution given above is determined, would not exist, since for an incompressible fluid the sound velocity is infinite, causing C to fall on the x-axis.

The curve AD can cross over into the area ABC only if the velocity of the piston exceeds, at some time, the value a_0 determining the slope of AC. Thus one might suppose that the inconsistency is restricted to the occurrence of supersonic velocities. One can, however, superimpose on the flow of Fig. 74a a uniform motion with velocity u_0 toward the left, i.e., u_0 negative. This is shown in Fig. 74b. Here the characteristic triangle is no longer isosceles, the sides AC and BC having slopes $u_0 + a_0$ and $u_0 - a_0$, respectively. This time the initial velocity of the piston must be u_0 rather than 0. The curve AD can intersect AC if at some time the velocity of the piston exceeds $u_0 + a_0$, and if we take $u_0 > -a_0$ then both $|u_0|$ and $|u_0 + a_0|$ are subsonic.

We see then that *the differential equations for the theory of an inviscid and nonconducting fluid admit no solution corresponding to certain boundary conditions which can be enforced by simple physical arrangements.*

In considering the physical example mentioned above, we cannot give up the condition of at least simply adiabatic flow, nor can we restrict the

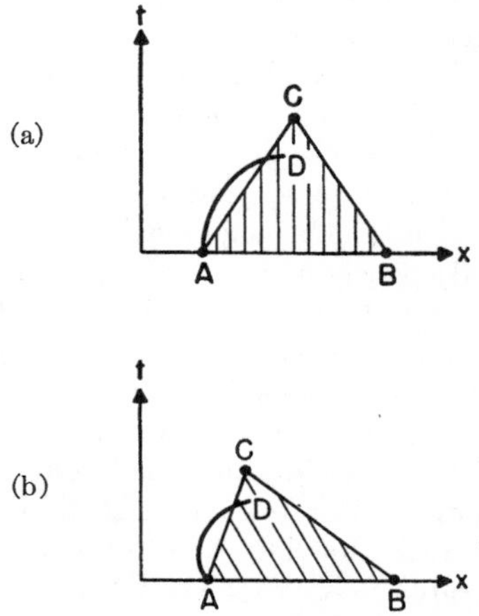

FIG. 74. Enforceable boundary conditions for which no inviscid solution exists.

freedom of choosing the boundary conditions; therefore the only way out of the impasse is to take viscosity and/or heat conduction into account.

One could object, for example, that for dry air under normal conditions the viscosity coefficient is very small, and that the influence of viscosity may be negligible. This objection, however, is invalidated by the results of Art. 11. There it was seen that a particular type of flow, much different from that occurring in an inviscid fluid, is possible in the viscous case, the difference becoming more pronounced, the smaller the value of the viscosity coefficient. We refer to the flow pattern with rapid change from a velocity value u_1 to a smaller value u_2, accompanied by similar changes in p and ρ. As we pass to the limit $\mu = 0$ in (11.26), we see that the extent L_0 of the transition region also tends to zero. In other words, the theory of a viscous compressible fluid yields, in the limit of vanishing viscosity, flow patterns with abrupt changes in the state variables, no indication of which is given in the theory based throughout on the assumption $\mu = 0$.[40] Since the forces caused by viscosity are proportional to the product of μ by velocity derivatives, here $\partial u/\partial x$, it is understandable that even a very small value of μ, when combined with an exceedingly large value of $\partial u/\partial x$, may produce a considerable effect upon the flow pattern.* All this suggests the following conception of the flow patterns possible in a fluid of low viscosity.

One may expect that in a fluid of low viscosity the ideal theory, assuming $\mu = 0$, will supply a reasonable approximation to the actual flow in general, but that the regions in which such approximations are valid are separated from each other by steady or moving thin layers within which occur rapid changes in the state variables, as suggested by the theory for a viscous fluid (partly) developed in Art. 11. Numerous observations of air in motion confirm this conception in all respects.[41] In particular, it is observed that flow patterns of this combined type occur in cases, such as the example mentioned at the beginning of this section, for which there exists no consistent solution based only on the ideal fluid theory.

This phenomenon of an almost abrupt change in the values of the state variables is known as *shock*.

A shock theory for one-dimensional flow in a form accessible to mathematical treatment is therefore based on the following principles. *It is assumed that the original differential equations for the motion of an ideal fluid: the equation of continuity, Newton's equation, and the specifying condition, are valid at all points of the x,t-plane with the exception of certain "shock lines"; across these lines the state variables are discontinuous, the sudden changes being governed by rules derived from the theory of viscous and/or heat-conducting*

* A somewhat analogous situation presents itself in the theory of an incompressible fluid with regard to the boundary layer solution of the Navier-Stokes equations.

fluids. In more general problems shock surfaces in x,y,z,t-space take the place of shock lines.

2. The shock conditions for a perfect gas

In Secs. 11.3 and 11.4 it was seen that in the steady flow of a viscous fluid a rapid change from the state u_1,p_1,ρ_1 to another state u_2,p_2,ρ_2 is possible only if these six values satisfy the three relations (11.9′), (11.22), and (11.23):

$$\rho_1 u_1 = \rho_2 u_2 \,, \tag{1a}$$

$$p_1 + m u_1 = p_2 + m u_2 \,, \tag{1b}$$

$$\frac{u_1^2}{2} + \frac{\gamma}{\gamma - 1}\frac{p_1}{\rho_1} = \frac{u_2^2}{2} + \frac{\gamma}{\gamma - 1}\frac{p_2}{\rho_2}, \tag{1c}$$ [42]

where m is an abbreviation for $\rho_1 u_1$ or $\rho_2 u_2$. If this same flow is viewed from a coordinate system moving in the negative x-direction at constant speed c and if u_1, u_2 now represent velocities with respect to the moving coordinate system, Eqs. (1) read

$$\rho_1(u_1 - c) = \rho_2(u_2 - c) = m, \tag{2a}$$

$$p_1 + m(u_1 - c) = p_2 + m(u_2 - c), \tag{2b}$$

$$\frac{(u_1 - c)^2}{2} + \frac{\gamma}{\gamma - 1}\frac{p_1}{\rho_1} = \frac{(u_2 - c)^2}{2} + \frac{\gamma}{\gamma - 1}\frac{p_2}{\rho_2}. \tag{2c}$$

The same situation arises, viewed from a fixed coordinate system, if the transition progresses at constant speed c toward the right. Thus, Eqs. (2) hold also for a particular kind of nonsteady motion. We shall now show that Eqs. (2) are the limiting transition conditions even in the most general case of nonsteady flow; we start from the general equations of Sec. 11.1.

Consider a small segment of the x-axis, say from x_1 to x_2, which progresses at speed c, where c may be positive or negative. Then the material derivative, which is $d/dt = u(\partial/\partial x) + \partial/\partial t$ in the fixed x,t-coordinate system, may be written

$$\frac{d}{dt} = (u - c)\frac{\partial}{\partial x} + \frac{\partial'}{\partial t}, \tag{3}$$

where $\partial'/\partial t$ means the derivative with respect to time at a point fixed in the moving segment, and x now refers to coordinates on the moving segment, while u and c are velocities relative to the original coordinate system.*

* This corresponds to the change of coordinates considered in Sec. 24.5.

Then, since c is independent of x, the continuity equation (11.2) can be written

$$\frac{\partial}{\partial x}[\rho(u - c)] + \frac{\partial' \rho}{\partial t} = 0. \tag{4}$$

Likewise, Newton's equation (11.3) takes the form

$$\rho(u - c)\frac{\partial u}{\partial x} + \rho\frac{\partial' u}{\partial t} + \frac{\partial}{\partial x}(p - \sigma_x') = 0, \tag{5}$$

and if we add Eq. (4) multiplied by u, this becomes

$$\frac{\partial}{\partial x}[\rho u(u - c) + p - \sigma_x'] + \frac{\partial'(\rho u)}{\partial t} = 0. \tag{5'}$$

Finally, the specifying condition (11.4′), which states that the flow is simply (but not strictly) adiabatic, is now

$$\begin{aligned} &\rho(u - c)\frac{\partial}{\partial x}\left(\frac{u^2}{2} + \frac{gRT}{\gamma - 1}\right) + \rho\frac{\partial'}{\partial t}\left(\frac{u^2}{2} + \frac{gRT}{\gamma - 1}\right) \\ &\qquad + \frac{\partial}{\partial x}\left[u(p - \sigma_x') - k\frac{\partial T}{\partial x}\right] = 0. \end{aligned} \tag{6}$$

Adding Eq. (4), this time multiplied by $u^2/2 + gRT/(\gamma - 1)$, we obtain

$$\begin{aligned} &\frac{\partial}{\partial x}\left[\rho(u - c)\left(\frac{u^2}{2} + \frac{gRT}{\gamma - 1}\right) + u(p - \sigma_x') - k\frac{\partial T}{\partial x}\right] \\ &\qquad + \frac{\partial'}{\partial t}\left[\rho\left(\frac{u^2}{2} + \frac{gRT}{\gamma - 1}\right)\right] = 0. \end{aligned} \tag{6'}$$

Now T may be replaced by its value from the equation of state (11.4); then Eqs. (4), (5′), and (6′) are three equations determining u, p, and ρ as functions of x and t.

Each equation is of the form

$$\frac{\partial A}{\partial x} + \frac{\partial' B}{\partial t} = 0$$

and when integrated over the interval x_1 to x_2 supplies a relation of the form

$$A(x_2) - A(x_1) + \int_{x_1}^{x_2} \frac{\partial' B}{\partial t}\, dx = 0. \tag{7}$$

We now consider solutions for which the time derivatives $\partial'/\partial t$ of the state variables remain bounded as μ_0 and k tend to zero. Then Eq. (7) holds also for the limit flow, with the integral tending to zero as x_2 approaches x_1.

Thus, if 1 and 2 refer to adjacent points on either side of the shock line, the difference $A_2 - A_1$ must vanish. In order to have x_1 and x_2 approach each other from opposite sides of the shock, with (7) remaining valid throughout, it is, of course, necessary that c be exactly the *velocity of the shock front* (i.e., the slope of the shock line). Introducing successively for A the three expressions from (4), (5′), and (6′), we obtain three conditions:

$$(8)\qquad \begin{gathered} [\rho(u - c)]_1^2 = 0, \qquad [\rho u(u - c) + p - \sigma_x']_1^2 = 0, \\ \left[\rho(u - c)\left(\frac{u^2}{2} + \frac{1}{\gamma - 1}\frac{p}{\rho}\right) + u(p - \sigma_x') - k\frac{\partial T}{\partial x}\right]_1^2 = 0. \end{gathered}$$

The first of these is exactly (2a). Since we assume that the fluid behaves like an ideal fluid on either side of the shock, the viscous stress σ_x' and the heat flux $k(\partial T/\partial x)$ must vanish at 1 and 2. Then subtracting c times the first relation (8) from the second, we have (2b). The third relation gives

$$\left[\rho(u - c)\left(\frac{u^2}{2} + \frac{1}{\gamma - 1}\frac{p}{\rho}\right) + up\right]_1^2 = 0.$$

Here we must subtract c times the second equation and simplify by means of the first to obtain Eq. (2c).

Thus the three equations (2a), (2b), and (2c) represent necessary conditions relating the initial and final values of an abrupt transition. One restriction must still be added to the conditions. It was seen that the flow of Art. 11, which in the limit supplies a special case, at least, of such a transition (namely one with $\partial'/\partial t = 0$ for all variables), is not reversible: it always goes from lower to higher values of Θ [see Eq. (11.13) and Fig. 54] and therefore of T or p/ρ. Thus, the program indicated at the end of the preceding section may be formulated more precisely as follows.

We consider flow patterns in the x,t-plane which fulfill the differential equations of ideal fluid theory everywhere except on certain curves (of unknown shape), *while along these "shock lines" occur discontinuities, in u, p, and ρ, which satisfy the three conditions* (2) *and the inequality*

$$(9)\qquad \frac{p_2}{\rho_2} \geqq \frac{p_1}{\rho_1},$$

where for any particle, state 1 *precedes state* 2.[43]

Flow patterns of this type are often called "discontinuous solutions of the ideal fluid equations". It should be remembered, however, that the discontinuity conditions (2) cannot be derived without taking viscosity into account. In fact, if it is assumed that the flow is inviscid even in the transition zone, then the value of p/ρ^γ cannot change for any particle. But Eqs. (2) are compatible with

$$\frac{p_2}{\rho_2^\gamma} \neq \frac{p_1}{\rho_1^\gamma},$$

as will be seen below in Sec. 3.

The first to study shock problems of compressible fluid flow was the mathematician B. Riemann (1860). He did not think in terms of viscosity, and, on the basis of observation, he took for granted the possibility of discontinuities. His shock conditions included (2a) and (2b), but he used $p_2/\rho_2^\gamma = p_1/\rho_1^\gamma$ rather than (2c). Although this procedure is not justified, the numerical results for ordinary conditions do not differ considerably from those obtained by the proper method (see Sec. 3). The correct shock conditions were first given by W. Rankine (1870) and then, independently, by H. Hugoniot (1889).[44]

Moreover, the essential point is not the derivation of necessary conditions to be fulfilled at a surface of discontinuity. The only justification for admitting solutions of the type considered here (regions of continuity separated by shock lines) is supplied by the existence of viscous flow solutions exhibiting transition regions whose width tends to zero simultaneously with the viscosity coefficient μ. Flow patterns including shock lines are not "discontinuous solutions of the ideal fluid equations" (see also Sec. 15.2), but rather asymptotic solutions of the viscous fluid equations for the limit case $\mu \to 0$.

3. Some properties of shocks

The shock conditions consist of the three equations (2a), (2b), (2c), and the inequality (9).[45] Before we work with these equations, some limiting cases will be mentioned. If $u_1 = u_2 = c$, then $m = 0$, while (2b) gives $p_1 = p_2$. In this case the third condition (2c) is fulfilled for an arbitrary value of $\rho_1 = \rho_2$. This possibility is not usually included in the concept of shock, since no particle crosses the line of discontinuity. Another limiting case is $u_1 = u_2 \neq c$. Then, as before, from (2b) it follows that $p_1 = p_2$, while the third condition leads to $\rho_1 = \rho_2$. No actual discontinuity occurs, and this case will be referred to as *zero shock*. The same conclusion follows if we know only that $p_1 = p_2$, or that $\rho_1 = \rho_2$, provided that the particles actually cross the shock line.

To bring the shock conditions into a more suitable form, we first introduce the velocities relative to the shock front, which moves at velocity c:

$$u_1' = u_1 - c, \qquad u_2' = u_2 - c.$$

Then Eqs. (2a) and (2b) become

(10a) $$\rho_1 u_1' = \rho_2 u_2' = m,$$

(10b) $$p_1 - p_2 = m(u_2' - u_1').$$

Here u_1', u_2', and m are assumed to be different from zero. Equation (2c) may be written as

(2c′) $$u_1'^2 - u_2'^2 = \frac{2\gamma}{\gamma - 1}\left(\frac{p_2}{\rho_2} - \frac{p_1}{\rho_1}\right).$$

The factor $u_1' - u_2'$ may be replaced from (10b) by $(p_2 - p_1)/m$. Furthermore, we may write u'/m for $1/\rho$, by (10a). After multiplying through by m, we obtain

$$(p_2 - p_1)(u_1' + u_2') = \frac{2\gamma}{\gamma - 1}(p_2 u_2' - p_1 u_1')$$

or, with the usual abbreviation $h^2 = (\gamma + 1)/(\gamma - 1)$,

(10c) $$p_2 u_1' - p_1 u_2' = h^2(p_2 u_2' - p_1 u_1').$$

In addition, we have the inequality (9). From (2c′) we therefore have $u_1'^2 - u_2'^2 > 0$, or

(11) $$| u_1' | > | u_2' | .$$

By (10a), the quantities u_1' and u_2' have the same sign. Thus if $u_1 > c$, we have $u_1' > 0$, $u_2' > 0$, and (11) becomes $u_1 - c > u_2 - c$ or $u_1 > u_2$; if $u_1 < c$, then we have $u_1' < 0$, $u_2' < 0$, and (11) becomes $c - u_1 > c - u_2$ or $u_1 < u_2$. In (10a), the inequality (11) gives $\rho_1 < \rho_2$; and in (10b), since m has the same sign as u_1' and u_2', it gives $p_1 < p_2$. Finally, the temperature T is proportional to p/ρ, so that (9) gives directly the inequality $T_1 < T_2$.

Thus, since we have assumed that state 1 precedes state 2 in time, i.e., that a particle enters the moving shock in state 1, we have the following result: *A physically possible shock* (that is, a rapid transition governed by viscous fluid theory) *is always a "compression shock"; pressure, density, and temperature increase, while the absolute value of the relative velocity decreases.* The two possibilities $u_1 < c$ and $u_1 > c$ are illustrated in Fig. 77. It was shown in Sec. 3.4 that in the strictly adiabatic flow of a viscous fluid the entropy of a particle cannot decrease. We shall see later that, in an actual compression shock, the entropy does in fact increase.

Another interesting fact can be derived if we consider the *relative Mach numbers* M_1', M_2' corresponding to the relative velocities u_1', u_2' before and after the shock. As usual, we define the sound velocity a and the Mach number M in strictly adiabatic inviscid flow by

$$(12) \qquad a^2 = \gamma \frac{p}{\rho}, \qquad M'^2 = \frac{u'^2}{a^2} = \frac{\rho u'^2}{\gamma p} = \frac{mu'}{\gamma p}.$$

Then, if u' in (10b) and (10c) is replaced by $\gamma p M'^2/m$, these equations become

$$\gamma p_1 M_1'^2 - \gamma p_2 M_2'^2 = p_2 - p_1,$$

$$p_1(h^2 p_1 + p_2)M_1'^2 - p_2(h^2 p_2 + p_1)M_2'^2 = 0.$$

Solving these for $M_1'^2$ and M'^2, we obtain

$$(13) \qquad M_1'^2 = \frac{\gamma + 1}{2\gamma}\frac{p_2}{p_1} + \frac{\gamma - 1}{2\gamma}, \qquad M_2'^2 = \frac{\gamma + 1}{2\gamma}\frac{p_1}{p_2} + \frac{\gamma - 1}{2\gamma}.$$

Since $p_2/p_1 \geqq 1$, it is seen that $M_1'^2$ cannot be less than 1 and $M_2'^2$ cannot be greater than 1; they can equal 1 only in the case of zero shock. Moreover, $M_2'^2$ cannot be less than $(\gamma - 1)/2\gamma$, the value corresponding to $p_1/p_2 = 0$ (infinite compression). Thus

$$(14) \qquad \frac{\gamma - 1}{2\gamma} \leqq M_2'^2 \leqq 1 \leqq M_1'^2 \leqq \infty.$$

If Eq. (10c) is solved for p_2, we find

$$(15) \qquad p_2 = p_1 \frac{u_2' - h^2 u_1'}{u_1' - h^2 u_2'} = p_1 \frac{h^2 \rho_2 - \rho_1}{h^2 \rho_1 - \rho_2},$$

using (10a). Thus, as p_1/p_2 decreases from 1 to 0, the ratio ρ_2/ρ_1 increases from 1 to h^2. Thus we have learned: *In a physically possible shock, the velocity relative to the shock front is supersonic before and subsonic after the shock. The density ratio* ρ_2/ρ_1 *cannot exceed* h^2 ($=6$ in air) *and the square of the relative Mach number after the shock cannot be less than* $(\gamma - 1)/2\gamma$ ($=\frac{1}{7}$), *the extreme values corresponding to an infinite pressure ratio* $p_2/p_1 = \infty$. Note that the actual velocities u_1, u_2, and c may all be subsonic.

Next we use Eq. (15) to show that the entropy increases during an actual shock transition. Since the entropy is essentially the logarithm of p/ρ^γ, it is sufficient to show that

$$(16) \qquad \frac{p_2}{\rho_2^\gamma} > \frac{p_1}{\rho_1^\gamma} \qquad \text{or} \qquad \frac{p_2}{p_1} - \left(\frac{\rho_2}{\rho_1}\right)^\gamma > 0.$$

The inequality is certainly true in the case $p_2/p_1 = \infty$, where $\rho_2/\rho_1 = h^2$.

We consider therefore the case $\rho_2/\rho_1 < h^2$. If we take the expression for p_2/p_1 from (15), we find

$$\frac{p_2}{p_1} - \left(\frac{\rho_2}{\rho_1}\right)^\gamma = \frac{h^2\rho_2 - \rho_1}{h^2\rho_1 - \rho_2} - \left(\frac{\rho_2}{\rho_1}\right)^\gamma = \frac{h^2(\rho_2/\rho_1) - 1}{h^2 - (\rho_2/\rho_1)} - \left(\frac{\rho_2}{\rho_1}\right)^\gamma.$$

Since $\rho_2/\rho_1 < h^2$, the denominator of the first term on the right cannot be negative. If suffices therefore to consider only the numerator of the combined fraction, i.e., to show that the function

$$z(\xi) = h^2\xi - 1 - (h^2 - \xi)\xi^\gamma, \qquad \text{where } \xi = \rho_2/\rho_1, \tag{17}$$

has positive values for $1 < \xi$ where $\xi < h^2$. By differentiation,

$$z'(\xi) = h^2 - \gamma h^2\xi^{\gamma-1} + (\gamma + 1)\xi^\gamma = h^2[1 - \gamma\xi^{\gamma-1} + (\gamma - 1)\xi^\gamma], \tag{17'}$$

$$z''(\xi) = -\gamma(\gamma - 1)h^2\xi^{\gamma-2} + \gamma(\gamma + 1)\xi^{\gamma-1} = \gamma(\gamma + 1)(\xi - 1)\xi^{\gamma-2}. \tag{17''}$$

From these it appears that z, z', and z'' vanish at $\xi = 1$ (corresponding to no increase of density, i.e., zero shock) and that z'' is positive for $\xi > 1$, from which it follows that $z(\xi)$ is positive for $\xi > 1$. Thus we have shown that the entropy is greater after the shock than before; the amount of the increase depends on the values of u', p, and ρ before and after the shock transition.*

The theory of the one-dimensional flow of a perfect fluid, as developed in Arts. 12 and 13, was based on the assumption that a relation of the form p/ρ^γ = constant held throughout the fluid. We now learn that when a shock occurs, the value of p/ρ^γ changes for each particle, and the amount of change will not, in general, be the same for all particles. In other words, *the results derived in the two preceding articles need not hold in the region of the x,t-plane beyond a shock front*, except in the case that all particles undergo the same change in entropy. In Art. 15 some consequences of this situation will be discussed.

It has already been indicated that the actual change in the value of p/ρ^γ is in most cases not considerable. To compute this difference, we may again use the expression for p_2/p_1 from (11), and we find

$$\frac{p_2}{\rho_2^\gamma} - \frac{p_1}{\rho_1^\gamma} = \frac{p_1}{\rho_1^\gamma}\left[\frac{p_2}{p_1}\left(\frac{\rho_1}{\rho_2}\right)^\gamma - 1\right] = \frac{p_1}{\rho_1^\gamma}\left[\frac{h^2\xi - 1}{h^2 - \xi}\,\xi^{-\gamma} - 1\right]. \tag{18}$$

Now $h^2 = 6$, so that for $\xi = 3$ (half the maximum value of ξ), for example, the bracket has the value 0.217. If the bracket is developed in a power series in $\xi - 1$, we find

$$\frac{\gamma(\gamma^2 - 1)}{12}(\xi - 1)^3\left[1 - \frac{3}{2}(\xi - 1) + \cdots\right]. \tag{19}$$

* See Fig. 148 in which $\lambda = \exp[(S_1 - S_2)/gR]$.

With $\xi = 1.2$, corresponding to $p_2/p_1 = 1.29$, using the first two terms gives 0.00063 as compared with the exact value 0.00068 found from (18). It is seen that if $\xi - 1$ is not too large, i.e., for not too strong a shock, the assumption that the value of p/ρ^γ does not change is not too bad an approximation.[46]

4. The algebra of the shock conditions

The shock conditions (2a), (2b), and (2c) are three algebraic equations in the seven variables u_1, p_1, ρ_1, u_2, p_2, ρ_2, and c. They determine a four-dimensional variety (hypersurface) in a space of seven dimensions. A complete study of this algebraic variety will not be attempted here. Equations (10) present a comparatively simpler case, in which two sets of three variables u_1', p_1, ρ_1 and u_2', p_2, ρ_2 are related by three equations; these may be considered as determining a point-to-point transformation of a three-dimensional space onto itself.

First of all, we note that the transformation is an involution, i.e., the equations do not change if the subscripts 1 and 2 are interchanged. We may solve for one set, say u_2', p_2, ρ_2, using $m = \rho_1 u_1'$, and obtain, in addition to the trivial solution $u_2' = u_1'$, $p_2 = p_1$, $\rho_2 = \rho_1$, the relations

$$\begin{aligned} u_2' &= \frac{1}{\gamma + 1}\left[2\gamma \frac{p_1}{m} + (\gamma - 1)u_1'\right] \\ p_2 &= \frac{1}{\gamma + 1}[-(\gamma - 1)p_1 + 2mu_1'] \\ \rho_2 &= \frac{(\gamma + 1)m^2}{2\gamma p_1 + (\gamma - 1)mu_1'}. \end{aligned} \tag{20}$$

The first of the following equalities is obtained by multiplying the first equation of (20) by $h^2 u_1'$, and the second follows by an interchange of subscripts:

$$h^2 u_1' u_2' = u_1'^2 + \frac{2\gamma}{\gamma - 1}\frac{p_1}{\rho_1} = u_2'^2 + \frac{2\gamma}{\gamma - 1}\frac{p_2}{\rho_2}. \tag{21}$$

In general, the square of a velocity divided by $2g$ is called a velocity head, and in the adiabatic case $\gamma p/(\gamma - 1)\rho g$ is the pressure head (see Sec. 2.5). If we disregard the influence of gravity, the sum of velocity head and pressure head is the "total head" H which occurs in the Bernoulli equation.* Using H' (to indicate that this refers to the relative velocities), we can write (21) in the form

$$H_1' = H_2' = h^2 \frac{u_1' u_2'}{2g}. \tag{22}$$

* The total head is constant for a particle only when the motion is steady,[47] see Sec. 2.5.

The equality of H_1' and H_2', which is actually (2c), lends itself to the (misleading) interpretation that the shock transition behaves like the steady flow of an inviscid fluid.

A dimensionless formulation of the conditions (10) is obtained by introducing the Mach numbers M_1' and M_2' as defined in (12) and the ratios

$$(23) \qquad \frac{p_2}{p_1} = \eta, \qquad \frac{u_2'}{u_1'} = \frac{\rho_1}{\rho_2} = \frac{1}{\xi} = \zeta.$$

Then (10c) becomes

$$(24a) \qquad \eta - \zeta = h^2(\eta\zeta - 1),$$

and (10b) may be written as

$$(24b) \qquad \frac{\eta - 1}{1 - \zeta} = \gamma M_1'$$

or as

$$(24c) \qquad \frac{\frac{1}{\eta} - 1}{1 - \frac{1}{\zeta}} = \gamma M_2'^2.$$

Equation (24a) is the equation of an equilateral hyperbola in the ζ,η-plane, having the asymptotes $\zeta = 1/h^2$ and $\eta = -1/h^2$ and intersecting the ζ-axis at $\zeta = h^2$ (Fig. 75).[48] The part of the hyperbola for which ζ or η is negative has no physical significance. Under our assumption that the subscript 1 refers to the values before the shock, we have $\zeta < 1$, $\eta > 1$; a particular shock determines a point $P_1(\zeta,\eta)$ to the left of A on the hyperbola, where $A(1,1)$ corresponds to zero shock. Equation (24b) is the equation of a straight line through A with slope $-\gamma M_1'^2$ and passing through P_1. If P_1 is given, then the slope is determined; if the slope is given, then P_1 is determined as the intersection of the straight line with the hyperbola.

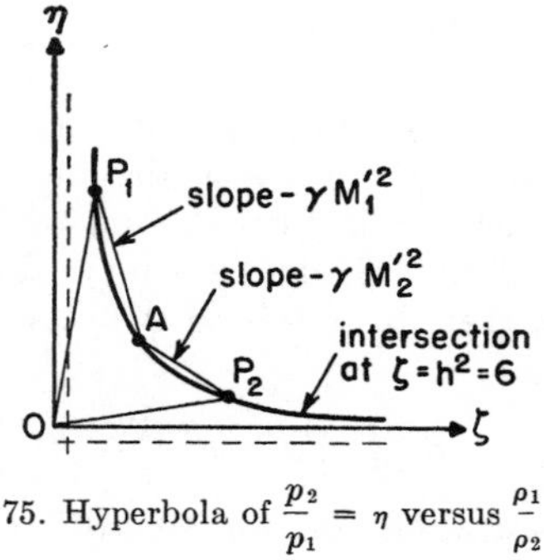

FIG. 75. Hyperbola of $\frac{p_2}{p_1} = \eta$ versus $\frac{\rho_1}{\rho_2} = \zeta$.

The point P_2 with coordinates $1/\zeta$, $1/\eta$ corresponding to a given shock P_1 also lies on the hyperbola (24a), but to the right of A. It may easily be located on the graph by using the fact that OP_2 makes the same angle with the ζ-axis as OP_1 does with the η-axis. From Eq. (24c) we see that the slope of AP_2 is $-\gamma M_2'^2$.

Solving Eqs. (24a) and (24b) for η and ζ, which amounts to finding the point at which the line through A with slope $-\gamma M_1'^2$ intersects the hyperbola, we find, of course, $\eta = \zeta = 1$ and as the second point of intersection

$$\eta = \frac{p_2}{p_1} = \frac{1}{\gamma + 1}\,[2\gamma M_1'^2 - (\gamma - 1)] = \frac{2\gamma}{\gamma + 1}\,M_1'^2 - \frac{1}{h^2},$$

(25)

$$\frac{1}{\xi} = \zeta = \frac{u_2'}{u_1'} = \frac{\rho_1}{\rho_2} = \frac{1}{\gamma + 1}\left[\frac{2}{M_1'^2} + (\gamma - 1)\right] = \frac{2}{\gamma + 1}\,\frac{1}{M_1'^2} + \frac{1}{h^2}.$$

Analogous equations, with the subscripts 1 and 2 interchanged and with ξ, η, and ζ replaced by their reciprocals, are obtained from (24a) and (24c). The first equation (25) and its analogue are identical with (13). By means of Eqs. (24) and (25) each of the quantities η, ζ, ξ $(= 1/\zeta)$, $M_1'^2$, and $M_2'^2$ can be expressed in terms of any one of the others as a linear or a linear fractional function. All five variables have the value 1 in the case of zero shock, where no change occurs in the values of u', p, and ρ. The deviation from unity of any of the quantities above can be taken as a measure of the *strength of the shock.*

If $1/\xi$ is used in place of ζ in the first equation (24), we find

(26) $$h^2(\eta - \xi) = \xi\eta - 1,$$

which is known as the *Hugoniot equation.*[49] In Fig. 76 is shown that part

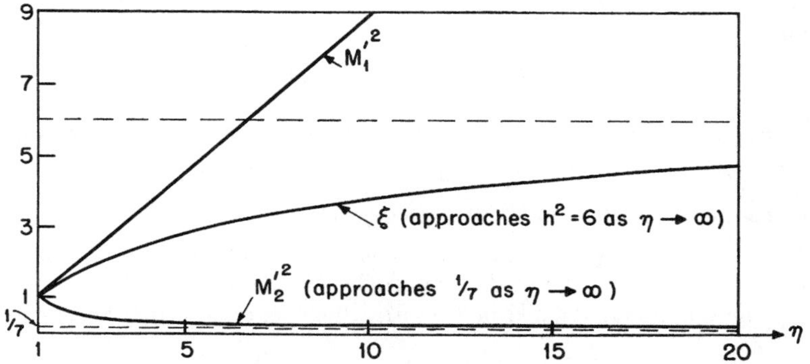

FIG. 76. Graphs of $M_1'^2$, $M_2'^2$, and ξ versus η.

of the equilateral hyperbola (26) for which $\xi > 1$, $\eta > 1$, together with the two curves giving $M_1'^2$ and $M_2'^2$ as functions of η, according to the equations (13).

If the first equation (25) is multiplied by its analogue, we obtain a symmetrical relation between the relative Mach numbers, namely,

$$1 = \left(\frac{2\gamma}{\gamma + 1} M_1'^2 - \frac{1}{h^2}\right)\left(\frac{2\gamma}{\gamma + 1} M_2'^2 - \frac{1}{h^2}\right),$$

or

$$2\gamma M_1'^2 M_2'^2 = 2 + (\gamma - 1)(M_1'^2 + M_2'^2), \tag{27}$$

which is the same as the relation (11.30′) between the actual Mach numbers in the steady case.

Practical problems are often such that some data concerning the actual velocities u_1, u_2 are known in advance. In these cases we must use relations which do not involve the relative velocities u_1', u_2'. One such relation is the Hugoniot equation (26), which can be written in the form

$$\frac{p_2 - p_1}{\rho_2 - \rho_1} = \gamma \frac{p_2 + p_1}{\rho_2 + \rho_1}. \tag{26′}$$

Of course, this follows directly from (15) without first introducing the dimensionless variables. If an expression for m not involving c is obtained from (10a) and substituted in (10b), one finds

$$p_2 - p_1 = \frac{\rho_1 \rho_2}{\rho_2 - \rho_1} (u_2 - u_1)^2. \tag{28}$$

Eliminating ρ_2 or ρ_1 from Eqs. (26′) and (28) leads to*

$$\begin{aligned}(p_2 - p_1)^2 &= \frac{\rho_1}{2} (u_2 - u_1)^2[(\gamma + 1)p_2 + (\gamma - 1)p_1] \\ &= \frac{\rho_2}{2} (u_2 - u_1)^2[(\gamma + 1)p_1 + (\gamma - 1)p_2].\end{aligned} \tag{29}$$

Another important relation of this type, linking the change in temperature (i.e., in p/ρ) to $(u_2 - u_1)^2$, will be derived in the next section.

5. Representation of a shock in the speedgraph plane

In Sec. 12.3 we saw that in one-dimensional flow the state of a moving particle at any moment is determined by two quantities: the velocity u and the quantity v defined in (12.19), which represents the density (and

* The curve of p_2 versus u_2, for fixed state 1, is sometimes used in place of the "shock curve" given in the next section.

pressure) and likewise has the dimensions of a velocity. It is assumed, of course, that a (p,ρ)-relation holds for the particle. In the speedgraph plane (u,v-plane) the continuous motion of a particle is represented by a curve. In the case of a simple wave this curve is one and the same $\pm 45°$ line for all particles.

When a particle passes through a shock front, its representative point (u,v) in the speedgraph plane undergoes a sudden jump. We shall now study the conditions governing this discontinuous transition. We shall restrict ourselves to the case when the sound velocity a is related to p and ρ by the equation $a^2 = \gamma p/\rho$, where γ is constant throughout. This is admissible if before and after the shock the motion is supposed strictly adiabatic and the fluid an inviscid perfect gas. The relation between v and p/ρ is

$$v^2 = \frac{4a^2}{(\gamma - 1)^2} = \frac{4\gamma}{(\gamma - 1)^2} \frac{p}{\rho} \qquad \left(= 35 \frac{p}{\rho}\right), \tag{30}$$

see (12.19′), and the inequality (9) is expressed by

$$v_2 \geqq v_1 . \tag{30′}$$

The relation between the pairs of coordinates u_1,v_1 and u_2,v_2, corresponding to the states before and after a shock, is determined by the three shock conditions (2a), (2b), and (2c). Now (2c) is a direct relation between the u- and p/ρ-values, i.e., between the u- and v-values. If (2b) is divided by m and m replaced by $\rho_1(u_1 - c)$ or $\rho_2(u_2 - c)$ from (2a), a second such relation follows. We see therefore that there are two equations in the five variables u_1, v_1, u_2, v_2, and c. In other words, there is a simple infinity of pairs u_2,v_2 corresponding to a given pair u_1,v_1, or: *The end points of a shock transition from a fixed initial point* (u_1,v_1) *lie on a curve in the speedgraph plane.* We shall call this the *shock curve.*

In order to find the curve explicitly it is necessary to eliminate c, and this is conveniently done if we start from the shock conditions in the form of Eqs. (21), which read, when expressed in terms of u and v,

$$\begin{aligned} &\frac{\gamma + 1}{\gamma - 1} (u_1 - c)(u_2 - c) \\ &\qquad = (u_1 - c)^2 + \frac{\gamma - 1}{2} v_1^2 = (u_2 - c)^2 + \frac{\gamma - 1}{2} v_2^2. \end{aligned} \tag{31}$$

The second equality is linear in c; if this is solved for c and the result substituted into the first equality, the condition on u_2,v_2 may be transformed into

$$(v_2^2 - v_1^2)^2 - 2(u_2 - u_1)^2(v_1^2 + v_2^2) - \frac{4\gamma}{(\gamma - 1)^2} (u_2 - u_1)^4 = 0. \tag{32}$$

This is a second-order relation between the two variables v_2^2/v_1^2 and $(u_2 - u_1)^2/v_1^2$; these variables can be given parametrically by

$$(32') \qquad \frac{v_2^2}{v_1^2} = \frac{\eta(\eta + h^2)}{h^2\eta + 1}, \qquad \frac{(u_2 - u_1)^2}{v_1^2} = \frac{\gamma - 1}{2\gamma}\frac{(\eta - 1)^2}{h^2\eta + 1},$$

where the parameter η is the pressure ratio introduced in (23) [see the derivation of (33) and (34) below]. In graphing the relation (32) we do not consider, of course, negative values of v. We also disregard values of v_2 less than v_1, since $v_2 \geqq v_1$ by (30′). The result is the *shock diagram*, Fig. 77, giving a graphical representation of all possible shock transitions. Each curve has a 90° corner at the point (u_1,v_1) and consists of all points (u_2,v_2) which are possible end points of a shock transition starting from the state represented by the corner point. The right-hand branch, $u_2 > u_1$, applies if $c > u_1$, and the left, $u_2 < u_1$, if $c < u_1$ (see p. 202). Each curve is symmetric with respect to the vertical line through the corner and all curves are asymptotic to the two lines through $(u_1,0)$ whose slopes, measured from the u-axis, are $\pm\sqrt{2\gamma/(\gamma - 1)} = \pm\sqrt{1 + h^2}$. The dimensionless parameters of the shock, $p_2/p_1 = \eta$, $\rho_2/\rho_1 = \xi$, M_1', and M_2' have constant

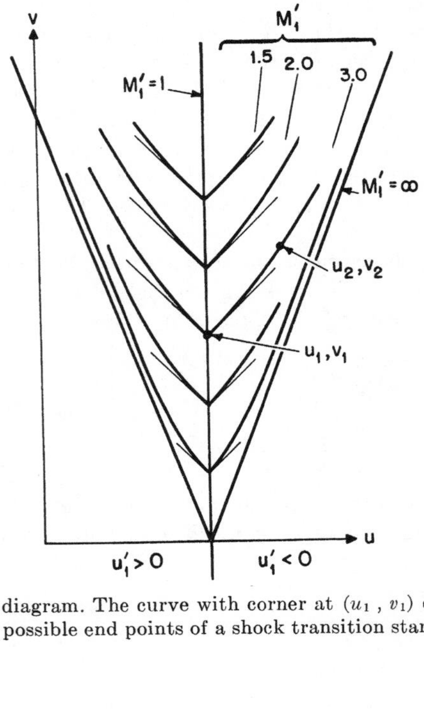

FIG. 77. Shock diagram. The curve with corner at (u_1, v_1) consists of all points (u_2, v_2) which are possible end points of a shock transition starting from (u_1, v_1).

values on any ray through $(u_1,0)$. In fact, if the slope of such a line is $\lambda = v_2/(u_2 - u_1)$, then

$$(33)\qquad \frac{1}{\lambda^2} = \frac{(u_2' - u_1')^2}{v_2^2} = (\xi - 1)^2 \frac{u_2'^2}{v_2^2} = \left(\frac{\gamma - 1}{2}\right)^2 (\xi - 1)^2 M_2'^2,$$

where (30) and $M_2' = u_2'/a_2$ have been used to obtain the last equality. Using (24) and (25) we then find

$$(34)\qquad \begin{aligned} \lambda^2 &= \frac{2}{\gamma - 1}\frac{h^2\xi - 1}{(\xi - 1)^2} = \frac{2\gamma}{\gamma - 1}\frac{\eta(\eta + h^2)}{(\eta - 1)^2} \\ &= \frac{[2\gamma M_1'^2 - (\gamma - 1)][(\gamma - 1)M_1'^2 + 2]}{(\gamma - 1)^2(M_1'^2 - 1)^2} = \frac{h^4 M_2'^2}{(1 - M_2'^2)^2}. \end{aligned}$$

The values of M_1' for various rays through $(u_1, 0)$ are indicated in the shock diagram, Fig. 77.

It is worth noting that the curves in the shock diagram have a second-order contact with the $\pm 45°$ lines through the corner point.[50] Now these lines are the images of particle lines in corresponding simple waves. Thus we see that, *for a weak shock* (all ratios close to 1), *the state variables of a particle undergo approximately the same kind of changes as they do in a simple wave.*

All possible shock transitions are obtained from the shock diagram of Fig. 77 by displacing it horizontally.

6. Example of a shock phenomenon. The Riemann problem[51]

We consider a straight tube of length $AB = l$, closed at both ends, and assume that at $t = 0$ the fluid in the tube has constant pressure p_0, and density ρ_0, and a uniform velocity u_0 directed from A to B. If we take as the origin the position of A at $t = 0$, the boundary conditions, namely,

$$(35a)\qquad u = u_0, \qquad v = v_0 \qquad \text{at } t = 0, \qquad 0 < x < l,$$

are realized if before time $t = 0$ the tube with the enclosed fluid mass has been moving (with velocity u_0) as one rigid body, and then the tube is suddenly stopped at $t = 0$. The stopping introduces the boundary conditions

$$(35b)\qquad u = 0 \qquad \text{at } x = 0 \quad \text{and} \quad x = l \qquad \text{for all } t.$$

Now the conditions (35a) determine a unique and continuous solution at each point within the characteristic triangle ABC, where AC and BC have slopes $u_0 + a_0$ and $u_0 - a_0$, respectively (the slopes being measured from the t-axis with a_0, the sound velocity, corresponding to v_0 or to p_0,

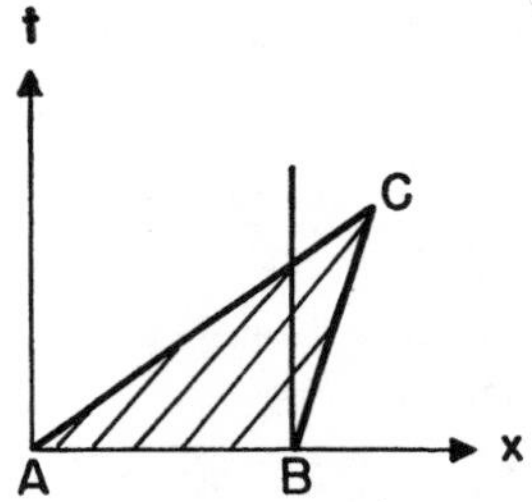

FIG. 78. Riemann's problem.

ρ_0). This solution is represented in Fig. 78 by the set of parallel particle lines of slope u_0. The figure is drawn under the assumption that u_0 is supersonic, so that $u_0 \pm a_0$ are both positive. In this case it is immediately evident that this continuous solution is inconsistent with the boundary conditions at $x = l$. Thus a flow pattern which includes a shock line must be sought.

In the speedgraph, Fig. 79, the point P_0 with coordinates u_0, v_0 represents the whole region of the physical plane in which u and v remain constant. So long as the laws of continuous motion hold, an adjacent region can only be mapped into a curve passing through P_0, and we saw in Sec. 13.1 that such a curve must necessarily be a characteristic, and the corresponding flow a simple wave. The $+45°$ line P_0P_1 ending at a point with abscissa $u_1 = 0$ and ordinate $v_1 < v_0$ supplies in fact an adequate solution for the region adjoining AC on the left. Figure 79 shows, in the x,t-plane,

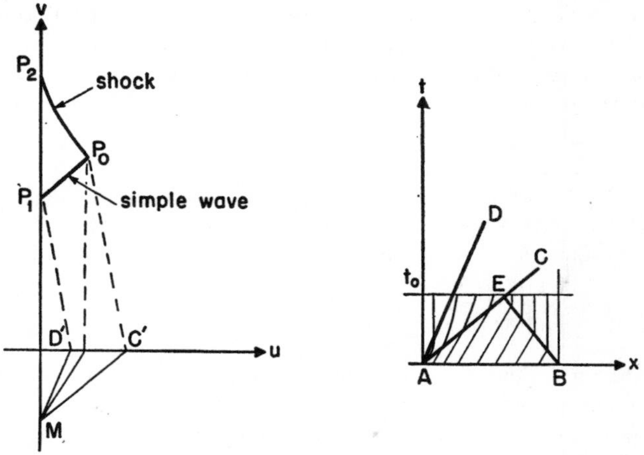

FIG. 79. Solution of the Riemann problem up to the time of interaction of shock and simple wave.

a centered simple wave (rarefaction wave) with center A between the characteristics AC (parallel to MC') and AD (parallel to MD'). (For the construction see Sec. 12.5.) All particles arrive at the line AD with zero velocity. Here the curved particle lines turn into straight lines parallel to the t-axis. In mechanical terms: a steadily increasing region of fluid at rest establishes itself at the left end of the tube.

No similar, continuous solution exists for the right side of the tube. Here the transition from velocity u_0 to zero velocity takes place by means of a shock. To find the end point P_2 of the shock transition we have to draw the shock curve with corner at P_0 up to the point where it meets the v-axis. The shock front BE in the physical plane is, by definition, the line of slope c measured from the t-axis. Since the shock diagram gives the value of $M_1' = |u_0 - c|/a_0$ for each point, and u_0 and a_0 are known, we can compute c and then plot the shock line BE along which the particle velocities jump from u_0 to 0. The solution up to the time t_0, the ordinate of E, is shown in Fig. 79. Here E is the point of intersection of the shock line and the lowest straight characteristic of the simple wave centered at A. The solution consists of four parts: fluid at rest to the left of AD and to the right of BE, uniform flow with $u = u_0$, $v = v_0$ within the triangle AEB, and finally the centered simple wave between AE and AD.

Analytically, the four regions are determined as follows. First, the simple wave extends from the ray of slope $u_0 + a_0 = u_0 + (\gamma - 1)v_0/2$, along which the velocity is u_0, to the ray of slope

$$a_1 = (\gamma - 1)v_1/2 = (\gamma - 1)(v_0 - u_0)/2,$$

along which the velocity is zero.* That $v_1 = v_0 - u_0$ may be seen by computing the ordinate of P_1 in Fig. 79. Accordingly, the equation of the particle lines [see Eq. (13.8)] is

$$t\left[\frac{x}{t} + (v_0 - u_0)\right]^{h^2} = \text{constant}. \tag{36}$$

Second, the two unknown parameters of the shock transition, the slope c of BE and the ordinate v_2 of P_2, can be found from the two equations (31), where u_0 and 0 are the values of u_1 and u_2, respectively. Choosing the negative value of c we find from these equations

$$\begin{aligned} c &= u_0\left[\frac{3-\gamma}{4} - \sqrt{\left(\frac{\gamma+1}{4}\right)^2 + \frac{1}{M_0^2}}\right] \\ v_2^2 &= v_0^2 + u_0^2\left[1 + \frac{4}{\gamma - 1}\sqrt{\left(\frac{\gamma+1}{4}\right)^2 + \frac{1}{M_0^2}}\right], \end{aligned} \tag{37}$$

* We assume $u_0 < v_0$. Otherwise, a cavity forms at the left end of the tube (see also end of Sec. 13.3.)

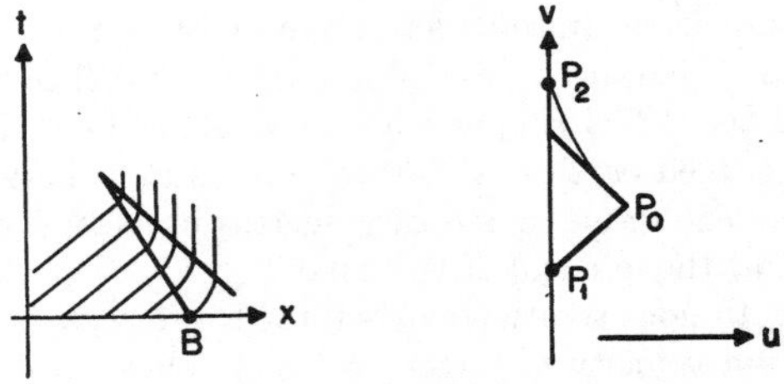

FIG. 80. The only possible continuous change to zero velocity.

where $M_0 = u_0/a_0$. Finally, the ordinate of the point E in Fig. 79, that is, the time t_0 up to which the solution has been constructed, is the ordinate of the point of intersection of $AC\colon x = (u_0 + a_0)t$ and $BE\colon x = l + ct$, and is $t_0 = l/(u_0 + a_0 - c)$.

All this holds whether M_0 is greater or smaller than 1, since it is also true in the subsonic case, when the point C of Fig. 78 falls between the vertical lines through A and B, that no continuous solution exists which satisfies the condition $u = 0$ at $x = l$ for all t. In fact, if there were a continuous solution to the right of BC, it would have to be a simple wave since it would be adjacent to the region of constant u,v within the triangle ABC. The map of this simple wave in the speedgraph plane, Fig. 80, must lie on the $-45°$ line through P_0 in the direction of decreasing u, tangent at P_0 to the shock curve P_0P_2 of the preceding figure. This corresponds to case IV of Fig. 67, that of a backward compression wave. But we have seen that in this case the straight characteristics in the x,t-plane converge to the left, so that the particle lines have the general shape shown in the first Fig. 80. That is, the only possible continuous change precludes from the start a vertical particle line through B.

Article 15

Further Shock Problems

1. Behavior of a shock at the end of a tube or a wall (shock reflection)[52]

In this section we discuss another example, on the basis of the theory developed in the preceding article. Supplements and modifications of the theory will be introduced below.

Assume that the tube in which the flow takes place is closed at its right end, and that a shock front moves toward this end with velocity c. This shock is represented in the x,t-plane (Fig. 81) by the straight line AB

whose slope is c measured from the t-axis. If we suppose the fluid in the tube to be at rest before the shock reaches it, the particle lines below AB are vertical straight lines which continue above AB in some other direction (see the example of a shock considered in Sec. 14.6, where, however, the shock was moving from right to left). If B represents the point at which the shock reaches the right end of the tube, where the velocity must be 0 for all value of t, then the particle line through B must remain vertical. A solution satisfying this condition is obtained by assuming another shock line BC with slope c' ($c' < 0$) along which the inclined particle lines again change direction, but now in the opposite sense, so as to become vertical again. This phenomenon, where a shock front moves forward and then back, is known as *shock reflection*. The expression "reflection" suggests the symme-

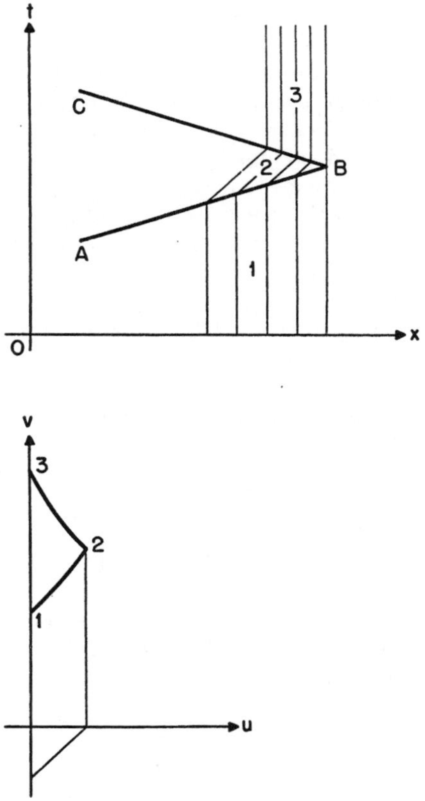

FIG. 81. Reflection of a shock front at a fixed wall.

try condition, $c' = -c$. We shall see, however, that this relation is not fulfilled even approximately, except for very weak shocks.

This same flow can be considered in a two- or three-dimensional space, rather than in a tube, all particles with the same x having the same velocity, etc. Then the "end" of the tube is represented by a wall normal to the x-direction.

In the speedgraph (Fig. 81) the fluid state prior to the first shock is represented by a point 1 on the v-axis. The point 2, corresponding to the state between the two shocks, must lie somewhere on the shock curve with its corner at 1, and in fact on the right half of this curve, since $u_1' < 0$. If the vertical axis of the shock diagram is then shifted so as to pass through 2, there is one curve with its vertex at 2, and on this must lie the point 3 representing the fluid state after the second shock. Since we know that the fluid is at rest after the second shock, the point 3 must also lie on the v-axis, so that 3 is uniquely determined once 2 is known. Thus, given the state 1, there exists a simply infinite number of possible transitions 1–2–3, each corresponding to a different increase of v or p/ρ (temperature). As the Mach numbers M_1 and M_3 are zero, we can characterize each possible transition by the value of M_2, the Mach number of this intermediate state. Now M_2 has the value $2/(\gamma - 1)$ times the slope u_2/v_2 (measured from the v-axis) of the radius vector from the origin to the point 2. Hence M_2^2 may range between 0 and $2/\gamma(\gamma - 1)$, the maximum value corresponding to the case when the point 2 is at infinity on the shock curve, with the radius vector becoming the common asymptote of the shock curves (see Sec. 14.5). Our task is now to compute the ratio c'/c and the pressure and density ratios p_3/p_1 and ρ_3/ρ_1 as functions of M_2.

To this end we apply Eq. (14.21); for the first shock we have $u_1' = -c$, $u_2' = u_2 - c$. Dividing all members by u_2^2 and introducing $M_2^2 = \rho_2 u_2^2/\gamma p_2$, we find

$$
(1)\qquad \begin{aligned} -h^2 \frac{c}{u_2}\left(1 - \frac{c}{u_2}\right) &= \frac{c^2}{u_2^2} + \frac{2\gamma}{\gamma - 1}\frac{p_1}{\rho_1 u_2^2} \\ &= \left(1 - \frac{c}{u_2}\right)^2 + \frac{2}{\gamma - 1}\frac{1}{M_2^2}. \end{aligned}
$$

For the second shock we must first replace the subscripts 1 and 2 in Eq. (14.21) by 2 and 3 and then use $u_2' = u_2 - c'$, $u_3' = -c'$. If we divide again by u_2^2, the equations read

$$
(2)\qquad \begin{aligned} -h^2 \frac{c'}{u_2}\left(1 - \frac{c'}{u_2}\right) &= \left(1 - \frac{c'}{u_2}\right)^2 + \frac{2}{\gamma - 1}\frac{1}{M_2^2} \\ &= \frac{c'^2}{u_2^2} + \frac{2\gamma}{\gamma - 1}\frac{p_3}{\rho_3 u_2^2}. \end{aligned}
$$

A comparison of the first and third members of (1) with the first and second members of (2) shows that c/u_2 and c'/u_2 satisfy the same quadratic equation. Remembering that c is positive and c' negative and that $h^2 = (\gamma + 1)/(\gamma - 1)$, we find

$$(3) \qquad \frac{c}{u_2} = \frac{3-\gamma}{4} + S, \quad \frac{c'}{u_2} = \frac{3-\gamma}{4} - S, \quad S = \sqrt{\left(\frac{\gamma+1}{4}\right)^2 + \frac{1}{M_2^2}},$$

so that*

$$(3') \qquad \frac{c'}{c} = \frac{3 - \gamma - 4S}{3 - \gamma + 4S}.$$

As M_2 goes to zero, S becomes infinite, and the ratio c'/c has the limit -1. At the upper limit, $M_2^2 = 2/\gamma(\gamma - 1)$, the value of S is $(3\gamma - 1)/4$, so that $c'/c = -2/h^2$, which is $-\frac{1}{3}$ in the case $\gamma = \frac{7}{5}$. As is to be expected, the formula for c'/u_2 agrees with the first result of (14.37), found in the simpler case studied in Sec. 14.6.

Using the ratios c/u_2 and c'/u_2, in terms of M_2, we can derive the density increase from the first shock condition, Eq. (14.10a), applied to the two shocks:

$$(4) \qquad -\rho_1 c = \rho_2(u_2 - c); \qquad \rho_2(u_2 - c') = -\rho_3 c'.$$

If the values from (3) are introduced into the last expression, the result can be reduced to

$$(4') \qquad \frac{\rho_3}{\rho_1} = \frac{4 + M_2^2(\gamma + 1 + 4S)}{4 + M_2^2(\gamma + 1 - 4S)}.$$

This quotient has the value 1 for $M_2 = 0$ and the value $\gamma(\gamma + 1)/(\gamma - 1)^2$ ($=21$ for $\gamma = \frac{7}{5}$) at the upper limit of M_2.

To compute the pressure increase we can solve in Eq. (1) for $p_1/\rho_1 u_2^2$ in terms of c/u_2 and in (2) for $p_3/\rho_3 u_2^2$ in terms of c'/u_2. In combination with (4) this leads to

$$(5) \qquad \frac{p_3}{p_1} = \frac{c' - u_2}{c - u_2} \cdot \frac{2c' - (\gamma + 1)u_2}{2c - (\gamma + 1)u_2},$$

which, by the use of Eqs. (3), may be reduced to

$$(5') \qquad \frac{p_3}{p_1} = \frac{4 + \gamma M_2^2(\gamma + 1 + 4S)}{4 + \gamma M_2^2(\gamma + 1 - 4S)}.$$

* From (3) also follows $(c - u_2)(u_2 - c') = a_2^2$. Hence the Mach numbers of the state 2 relative to the incident and reflected shocks are reciprocals.

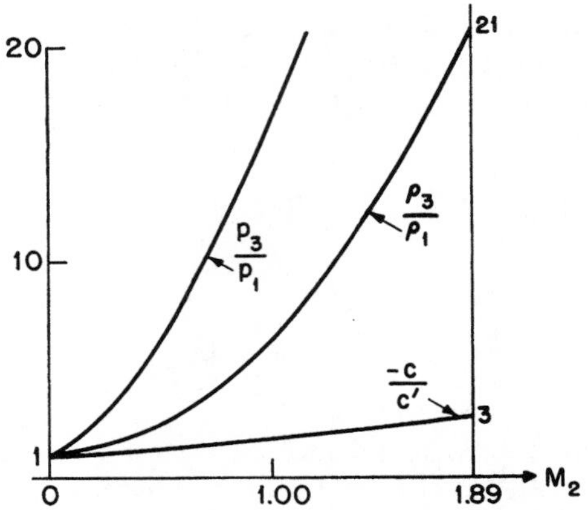

FIG. 82. Pressure, density, and shock speed ratios, produced by reflection of a shock front, as functions of the Mach number M_2 of the intervening flow.

This ratio increases from 1 to ∞ as M_2 increases from 0 to its maximum value.

In Fig. 82 the quantities $-c/c'$, ρ_3/ρ_1, and p_3/p_1 are shown as functions of M_2 in the case $\gamma = 1.4$.

In a similar way the pressure ratio p_3/p_2, across the reflected shock alone, may be determined. We find

$$\frac{p_3}{p_2} = 1 + \frac{\gamma M_2^{\,2}}{4}(\gamma + 1 + 4S),$$

which increases from 1 for very weak incident shocks (M_2 small) to $(3\gamma - 1)/(\gamma - 1) = 8$ for very strong ones (M_2 near its limiting value). Alternatively this formula may be combined with (5′) to give the ratio of the pressure increments across the two shocks,

$$\frac{p_3 - p_2}{p_2 - p_1} = \frac{4S + (\gamma + 1)}{4S - (\gamma + 1)}.$$

This ratio approaches 1 for very weak incident shocks—corresponding to the acoustic case—but increases to $2\gamma/(\gamma - 1) = 7$ for very strong shocks. Similar results are found for the density.

For very small M_2 the Taylor developments of (3′), (4′), and (5′) through terms of first order give

$$(6) \qquad -\frac{c}{c'} \sim 1 + \frac{3 - \gamma}{2} M_2\,, \qquad \frac{\rho_3}{\rho_1} \sim 1 + 2M_2\,, \qquad \frac{p_3}{p_1} \sim 1 + 2\gamma M_2\,.$$

The principal results are that *the reflected shock can have a velocity of propagation as small as one-third that of the incoming shock and that very considerable increases of pressure and density may be caused by the reflection.*

2. Discontinuous solutions of the equations for an ideal fluid

It was stressed in Sec. 14.2 that flow patterns which include shock fronts cannot rightly be considered "discontinuous solutions of the differential equations for an ideal fluid". These equations are, of course, satisfied by the u,p,ρ-distributions in the spaces between shock lines; but across a shock line the conditions of inviscid flow are violated. For example, in the adiabatic case inviscid flow requires that each particle keep its entropy value, but the value is changed as a particle crosses a shock line. This contradiction cannot be eliminated by any sophistry about the "nature of discontinuous transitions". The correct theoretical basis for admitting flow patterns which include shocks is the fact that these flows are asymptotic solutions, for $\mu \rightarrow 0$, of the equations for viscous (and/or heat-conducting) fluids.

Nevertheless, as in other branches of continuum mechanics (elasticity theory, incompressible fluid theory), there also exist true *discontinuous solutions* of the partial differential equations for an ideal fluid. The occurrence of such solutions was discussed previously (Arts. 9 and 10): across the so-called characteristic lines certain first-order derivatives of the dependent variables may change abruptly,* without any of the differential equations being violated. Essentially, this depends upon the fact that these equations do not lead to any conclusions about these derivatives across the characteristic. Note that *shock lines are not characteristics.*

We saw in Sec. 12.2 that in the one-dimensional flow of an ideal fluid there occur characteristics which have the slope $u + a$ or $u - a$, measured from the t-axis. Across such a characteristic line the variables u, p, and ρ remain continuous, but their derivatives normal to the line may change abruptly; the integration theory developed in Arts. 12 and 13 depended essentially on the use of these characteristics. These characteristics, when considered in the general theory of characteristics given in Art. 9, correspond to the vanishing of the second factor of Eq. (9.26). It was mentioned, however, at the end of Art. 9, that in all cases of ideal fluid motion another type of characteristic exists, corresponding to the vanishing of the first factor in (9.26). In the one-dimensional case this factor is $(u\lambda_1 + \lambda_4)$, indicating that the normal to the characteristic has the slope

* A discontinuity is usually called *of zero order* if the variables themselves undergo a sudden change, *of nth order* if the discontinuity appears first in the nth derivatives. Since, however, in fluid mechanics one may take as dependent variables either the velocity components, etc., or such functions as particle function and potential, this classification is not always unambiguous.

$\lambda_1/\lambda_4 = -1/u$ and the characteristic itself the slope u. Thus, all *particle lines in the x,t-plane are characteristics* in this sense.[53] We shall now discuss the possible discontinuities which may occur across the particle lines; they may appear in incompressible, as well as in compressible, flow.

The first to consider flow problems of this kind, in the case of incompressible fluids, was H. von Helmholtz.[54] In a famous paper of 1868 he studies particularly the free jet problem: a steady two-dimensional flow is bounded by two streamlines, along each of which the pressure has a constant value. In the absence of gravity or other external forces this flow is not affected by the presence of fluid at rest, bordering these streamlines, for clearly in this case both the continuity equation and Newton's equation are still satisfied for each fluid element. Then the velocity has a jump, while p remains continuous. A more general case is that of a surface of discontinuity separating two three-dimensional regions; the normal velocity vanishes on either side of the surface, and the pressure, but not the tangential velocity component, remains continuous across the surface. This type of discontinuity appears in the theory of wings of finite span.[55]

In our one-dimensional flow of an ideal perfect gas the following situation can occur. Consider a fluid mass within a tube, which is initially (at $t = 0$) situated between the points A and B, and whose particles move according to certain particle lines in the x,t-plane (Fig. 83). This flow satisfies the differential equations with some constant $C_1 = p/\rho^\gamma$. The pressure distribution is determined throughout, and in particular along the particle line BB' which starts at B. This motion is compatible with, and uninfluenced by, the presence of a weightless piston which is initially at B and which moves according to the space-time curve BB', subjected to a pressure from the right which at each moment just equals the computed p. Now we can imagine another fluid mass, initially between B and C, which moves in such a way that BB' is again the particle line of the particle initially at B and the pressure along BB' equal to that computed before, while in this section

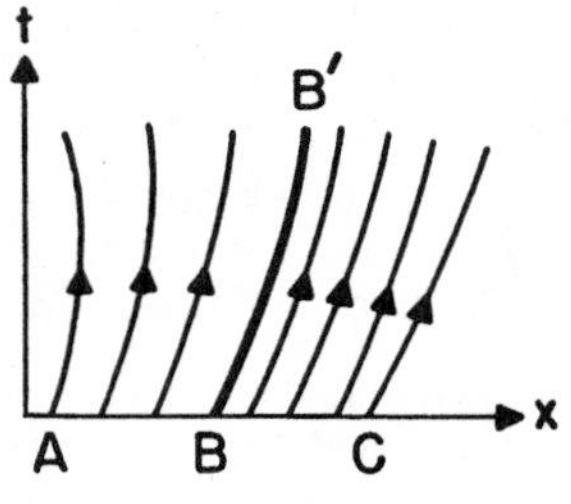

FIG. 83. Contact discontinuity BB′.

of the fluid p/ρ^γ has a different constant value C_2.* This moving mass would supply the required pressure on the piston from the right, while vice versa the mass initially between A and B would supply the pressure from the left necessary to maintain the motion of the second mass. Nothing would change if the piston were omitted. We would then have a flow pattern for the fluid mass initially between A and C, with a continuous set of particle lines and continuous p-distribution, but with a jump in the ρ-values along BB'. If ρ_1 is the density on the left side of BB' and ρ_2 the density on the right, we have

$$\frac{\rho_1}{\rho_2} = \left(\frac{C_2}{C_1}\right)^{1/\gamma} = \text{constant} \neq 1$$

for all points of BB'.

The simplest example of this kind is the case of all particles between A and C moving with constant velocity under the same pressure, as one rigid column, while the density has different constant values for the two masses to the right and to the left of B. It is evident that the equation of continuity and Newton's equation (steady flow, no external forces) are satisfied.

One may observe that the temperature, which is proportional to p/ρ, is not the same on both sides of BB'. But since we do not admit heat conduction, this has no influence on the motion. Thus we have seen that *in the strictly adiabatic flow of a perfect, inviscid fluid, flow patterns are possible in which the density jumps in a constant ratio along certain particle lines, while velocity and pressure remain continuous.* This phenomenon is nowadays called a *contact discontinuity*.

In the speedgraph plane the line BB' is mapped into two distinct curves the ordinates of which, for the same abscissa, have the constant ratio

$$\frac{v_2}{v_1} = \sqrt{\frac{\rho_1}{\rho_2}} = \left(\frac{C_2}{C_1}\right)^{1/2\gamma}.$$

3. Example of a contact discontinuity: collision of two shocks[56]

A simple example of a contact discontinuity occurs in the case we shall consider next, the head-on collision of two shock fronts. Assume that fluid along the interval AB of the x-axis (Fig. 84) is initially in a uniform state of rest: $u = 0$, $p = p_0$, $\rho = \rho_0$ at $t = 0$. Consider two shock fronts moving in opposite directions, one along AC causing the state $0, p_0, \rho_0$ to change into a state u_1, p_1, ρ_1, and one along BC changing the initial state to u_2, p_2, ρ_2.[57] At the time t_0, represented by the ordinate of C, the particles

* Analytically, the required flow on the right is determined by the prescribed values of u and v on BB', and can be constructed as in Sec. 10.3.

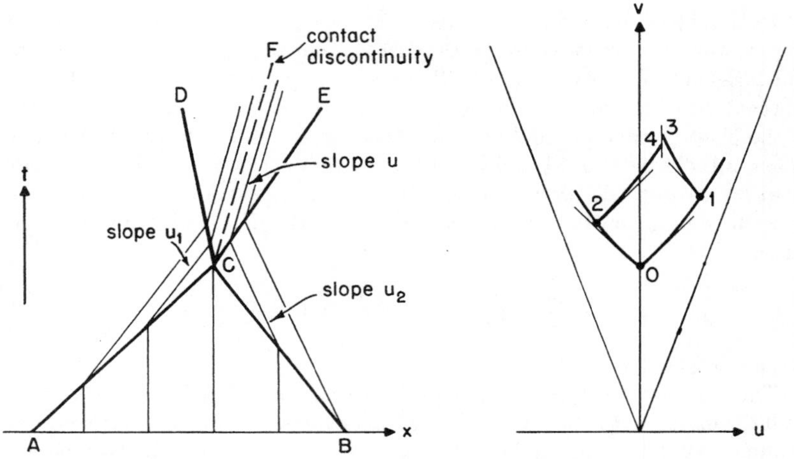

FIG. 84. Collision of two shocks AC, BC, production of a contact discontinuity.

to the left of C have velocity u_1, pressure p_1, and density ρ_1, while those to the right have u_2, p_2, ρ_2. In the speedgraph plane these two states are represented by two points 1 and 2 lying on the two branches of the shock curve with its corner, 0, at $(0, v_0)$, where $v_0^2 = 4\gamma p_0/\rho_0(\gamma - 1)^2$. What happens for $t > t_0$?

We can find a flow pattern for $t > t_0$ which satisfies all conditions if we assume that two new shock fronts (reflected shocks) form at C and move along appropriate lines CD and CE. The two states of particles after passing through the second shock fronts will be represented by two points 3 and 4, with 3 lying on the shock curve starting at 1, and 4 on the shock curve starting at 2. The two points 3 and 4 must fulfill two conditions: they must have the same abscissa u in order that the final particle velocity in the wedge between CD and CE be one and the same for all particles, and second, the pressure value p must be equal for all particles. Now it is possible, in general, to satisfy two conditions by choosing the slopes c_3 and c_4 of the shock lines CD and CE appropriately. The values of the density after the reflected shocks, however, will not necessarily be the same for particles to the left of C as for those to the right. Thus the final particle line through C (broken line CF in Fig. 84) will, in general, be a discontinuity line of the type described in the preceding section.

It is easy, in principle, to find the numerical solution, that is, the values of u, p, and ρ after the second shocks, when the two states u_1,p_1,ρ_1 and u_2,p_2,ρ_2 are given. We simply apply to the transitions across CD and CE the equation (14.29) which gives the relation between the pressure and velocity changes across a shock. Using u and p to denote the final values of velocity and pressure, we have

$$(7)\qquad \begin{aligned} (p - p_1)^2 &= \frac{\rho_1}{2}(u - u_1)^2[(\gamma + 1)p + (\gamma - 1)p_1], \\ (p - p_2)^2 &= \frac{\rho_2}{2}(u - u_2)^2[(\gamma + 1)p + (\gamma - 1)p_2]. \end{aligned}$$

Eliminating p between these two equations, we obtain one equation determining u; one root of this equation is the common abscissa of 3 and 4 and therefore determines the slope of the final particle lines between CD and CE. Using the equation (14.32) of the shock curve, applied to the arcs $\widehat{24}$ and $\widehat{13}$, we can then find the ordinates, corresponding to the abscissa u, of the points 3 and 4, and from these we know the values of p/ρ_3 and p/ρ_4. On the other hand, the common value of p is determined by (7), now that u is known, so that finally the two densities ρ_3 and ρ_4, prevailing to the left and to the right of the discontinuity line CF, can also be computed.

The occurrence of such contact discontinuities after a shock collision has been observed, although in most cases the effect is rather weak unless the original shocks are of great, but very different, strengths.

To carry out the computation, we may suppose that the pressure ratios η_1, η_2 of the initial shocks are given. Then the density ratios ξ_1, ξ_2 are determined from the Hugoniot equation (14.26). Finally, the second of Eqs. (14.32′) yields, with $u_0 = 0$,

$$(8)\qquad \frac{u_1^2}{v_0^2} = \frac{\gamma - 1}{2\gamma}\frac{(\eta_1 - 1)^2}{h^2\eta_1 + 1}, \qquad \frac{u_2^2}{v_0^2} = \frac{\gamma - 1}{2\gamma}\frac{(\eta_2 - 1)^2}{h^2\eta_2 + 1}.$$

Thus u_1/v_0 and u_2/v_0 are also expressed in terms of the compression ratios. If we write x and y for the unknowns u/v_0 and p/p_0, the equations (7) become

$$(7')\qquad \begin{aligned} (y - \eta_1)^2 &= \frac{2\gamma}{(\gamma - 1)^2}\xi_1\left(\frac{u_1}{v_0} - x\right)^2[(\gamma + 1)y + (\gamma - 1)\eta_1], \\ (y - \eta_2)^2 &= \frac{2\gamma}{(\gamma - 1)^2}\xi_2\left(x - \frac{u_2}{v_0}\right)^2[(\gamma + 1)y + (\gamma - 1)\eta_2]. \end{aligned}$$

Once y has been found, the density ratios follow from the Hugoniot equation:

$$(9)\qquad \frac{\rho_3}{\rho_0} = \xi_1\frac{h^2y + \eta_1}{y + h^2\eta_1}, \qquad \frac{\rho_4}{\rho_0} = \xi_2\frac{h^2y + \eta_2}{y + h^2\eta_2}.$$

Taking, for example, $\eta_1 = 8$, $\eta_2 = 4$, and $u_1 > 0$, $u_2 < 0$, with $\gamma = \frac{7}{5}$, we find $\xi_1 = \frac{7}{2}$, $\xi_2 = \frac{5}{2}$, $u_1/v_0 = 1/\sqrt{7}$, $u_2/v_0 = -3/5\sqrt{7}$. The solution of (7′) gives

$$x = 0.1405, \qquad y = 21.87.$$[58]

When these values are substituted in (9), we obtain for the density ratios $\rho_3/\rho_0 = 6.97$, $\rho_4/\rho_0 = 7.37$, so that $\rho_4/\rho_3 = 1.057$. These values were used in the example in Fig. 84. The reader might compute for himself, from the given data η_1, η_2, the four shock velocities c_1, c_2, c_3, c_4, all as multiples of v_0. It is seen from the figure, and from the value of ρ_4/ρ_3, that the discontinuity is insignificant, even though we have started with high pressure ratios η_1, η_2.

4. Numerical method of integration

In all the examples we have discussed, solutions could be given consisting of regions of uniform flow and simple waves, separated, respectively, by straight shock lines or straight characteristics. It is clear that cases with general boundary conditions cannot be treated in this way. For example, in the Riemann problem considered in Sec. 14.6 the solution was obtained only up to the moment when the shock wave to the right meets the simple wave to the left [point $E(x_0, t_0)$ in Fig. 79]. The flow after this, for $t > t_0$, is determined by the conditions $u = 0$ at $x = 0$ and $x = l$, together with the conditions within the fluid for $t = t_0$, namely,

$$\text{for } x < x_0: \qquad u = f(x), \qquad p = g(x), \qquad \rho = h(x);$$
$$\text{for } x > x_0: \qquad u = 0, \qquad p = p_1, \qquad \rho = \rho_1.$$

In these conditions x_0, t_0 are the (known) coordinates of E, and p_1, ρ_1 the (known) values of p and ρ after the shock. The functions $f(x)$, $g(x)$, and $h(x)$ are constant to the left of AD, but then continue with the nonconstant values of u, p, and ρ determined by the equations for the simple centered wave. It can be anticipated that, for $t > t_0$, a curved shock line will develop, starting at E, but neither its shape nor the flow pattern on either side of it can be expressed in terms of elementary functions. In such cases one has to resort to numerical methods of integration, that is, one substitutes finite-difference equations for the differential equations of the flow and then solves the resulting algebraic problem by the methods of practical analysis. We shall sketch this procedure here for the case of one-dimensional nonsteady flow of a viscous fluid, neglecting gravity and heat conduction. Including heat conduction, however, would not essentially change the setup.

If we write σ rather than σ_x', Eqs. (2) to (6) of Art. 11, with $k = 0$, give

$$\text{(10)} \qquad \begin{gathered} \frac{d\rho}{dt} + \rho \frac{\partial u}{\partial x} = 0, \qquad \rho \frac{du}{dt} + \frac{\partial (p - \sigma)}{\partial x} = 0, \\ \rho \frac{d}{dt}\left[\frac{u^2}{2} + \frac{1}{\gamma - 1}\frac{p}{\rho}\right] + \frac{\partial}{\partial x}[u(p - \sigma)] = 0, \qquad \sigma = \mu_0 \frac{\partial u}{\partial x}. \end{gathered}$$

The first is the equation of continuity, the second is Newton's equation for a viscous fluid, the third expresses the fact that the motion is (strictly) adiabatic, and the last represents the usual (Navier-Stokes) assumption as to the form of the viscous stresses. The boundary conditions will be stated later.

The third equation in (10) may be simplified by subtracting from it u times the second equation, giving

$$\frac{1}{\gamma - 1}\left(\frac{dp}{dt} - \frac{p}{\rho}\frac{d\rho}{dt}\right) + (p - \sigma)\frac{\partial u}{\partial x} = 0,$$

and then replacing $d\rho/dt$ by its value from the first equation, giving

$$\frac{dp}{dt} = \frac{\partial u}{\partial x}[(\gamma - 1)\sigma - \gamma p]. \tag{11}$$

The first equation in (10) is automatically satisfied if we introduce the particle function ψ [see (12.11)]. This function is constant along particle lines: $d\psi/dt = 0$. Then the second equation (10) becomes

$$\frac{\partial \psi}{\partial x}\frac{du}{dt} + \frac{\partial(p - \sigma)}{\partial x} = 0. \tag{12}$$

The system (10) is thus replaced by (11) and (12), together with $\sigma = \mu_0 \partial u/\partial x$.

To carry out numerical integration, we let $u_\nu(t)$, $p_\nu(t)$, and $x_\nu(t)$ denote the values of u, p, and x as functions of t on particle lines corresponding to equally spaced values of ψ: $\psi_\nu(t) - \psi_{\nu-1}(t) = m =$ constant. Then all derivatives with respect to x are replaced by difference quotients, e.g., $\partial p/\partial x$ becomes $(p_\nu - p_{\nu-1})/(x_\nu - x_{\nu-1})$, so that our three equations yield for each value of ν the equations

$$\begin{aligned} m\frac{du_\nu}{dt} &= (p_{\nu-1} - \sigma_{\nu-1}) - (p_\nu - \sigma_\nu), \\ \frac{dp_\nu}{dt} &= \frac{u_{\nu+1} - u_\nu}{x_{\nu+1} - x_\nu}[(\gamma - 1)\sigma_\nu - \gamma p_\nu], \\ \sigma_\nu &= \mu_0 \frac{u_{\nu+1} - u_\nu}{x_{\nu+1} - x_\nu}. \end{aligned} \tag{13}$$

These equations can be interpreted as follows: in a tube of finite length and unit cross-sectional area, the continuously distributed fluid mass is considered as broken up into a finite number, say n, of mass points, each of mass m. The abscissa of the νth mass point at time t is $x_\nu(t)$, and its velocity is

$$u_\nu(t) = \frac{dx_\nu(t)}{dt}. \tag{14}$$

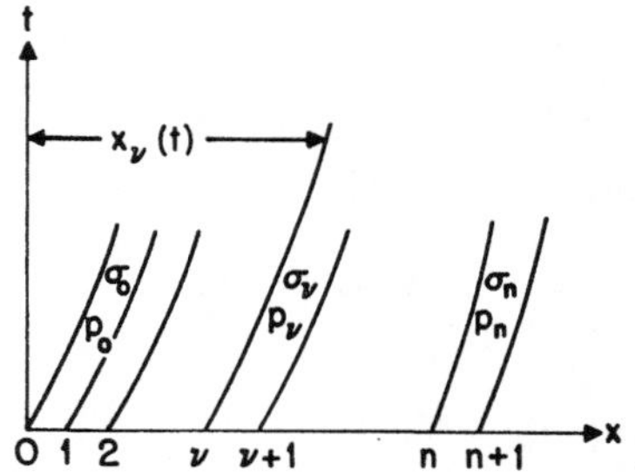

FIG. 85. Discrete particle model for one-dimensional nonsteady viscous flow.

The pressure and the viscous stress are considered constant in the interval between the νth and $(\nu + 1)$st particle and denoted by p_ν and σ_ν. Then the first equation (13) is Newton's Second Law for the νth mass point; the second equation (13) gives the rate of change of p_ν as determined by the condition of adiabatic flow, while the third expresses σ_ν in terms of the variables x_ν, u_ν. In the x,t-plane the motion is represented by n distinct particle lines (Fig. 85), each defined by one of the functions $x_\nu(t)$, $\nu = 1, 2, \cdots, n$. The figure also includes two curves marked 0 and $n + 1$ on the x-axis; these lines $x_0(t)$ and $x_{n+1}(t)$ are supposed given as boundary conditions.

If the value of σ_ν from the last equation (13) is substituted into the other two equations, and u_ν is expressed in terms of x_ν by (14), we have, for each ν $(\nu = 1, 2, \cdots, n)$, a system of two simultaneous ordinary differential equations in x_ν and p_ν, the first equation being of second order in x_ν, the second of first order in p_ν. The independent variable is t, and the equations include a constant parameter m. The right-hand member of the second equation for $\nu = n$ includes x_{n+1} and u_{n+1}, which are given by the boundary condition specifying $x_{n+1}(t)$. The first equation for $\nu = 1$ involves σ_0, which is determined from the boundary condition specifying $x_0(t)$: $\sigma_0 = \mu_0(u_1 - u_0)/(x_1 - x_0)$. It also contains p_0, and to make the system of equations complete, we add a $(2n + 1)$st equation determining p_0:

$$\frac{dp_0}{dt} = \frac{u_1 - u_0}{x_1 - x_0}[(\gamma - 1)\sigma_0 - \gamma p_0]. \tag{13'}$$

Thus we have $(2n + 1)$ differential equations for the $(2n + 1)$ unknowns $x_1, x_2, \cdots, x_n$; $p_0, p_1, \cdots, p_n$ as functions of t.

The integration of the system (13) requires, in addition to the boundary conditions $x_0(t)$ and $x_{n+1}(t)$, a knowledge of the $3n$ initial values $x_\nu(0)$, $u_\nu(0)$, and $p_\nu(0)$. Now the initial conditions of the original problem are given as three functions $u_0(x)$, $p_0(x)$, and $\rho_0(x)$. This last function serves to give the value of m and the initial abscissas $x_\nu(0)$. For example, one may plot the integral curve $\int_0^x \rho_0(x)dx = \psi(x)$ from $x = x_0(0) = 0$ to $x = x_{n+1}(0) = l$, say, and take for $x_\nu(0)$ the abscissa of the point on the curve with ordinate $\nu\psi(l)/(n+1)$. The mass in each of the $n+1$ intervals from $x = x_{i-1}(0)$ to $x = x_i(0)$, $i = 1, 2, \cdots, n+1$, is then $\psi(l)/(n+1)$ and this is taken for the mass m of each of the particles. Finally, we have $u_\nu(0) = u_0[x_\nu(0)]$, and $p_\nu(0) = p_0[x_\nu(0)]$ for $\nu = 1, 2, \cdots, n$.

The computation procedure is now as follows. For $t = 0$, the right-hand members of all equations (13) and (13′) are known, and from them follow the initial values of dp_0/dt, du_ν/dt, and dp_ν/dt, while $dx_\nu/dt = u_\nu(0)$. Taking a small time interval Δt, and assuming that the rates of change remain constant during this interval, we can compute the values of p_0, x_ν, p_ν, and u_ν at the time $t = \Delta t$ for all $\nu = 1, 2, \cdots, n$. From the values now found, we can go on to compute the values for $t = 2\,\Delta t$, etc. The technique of numerical integration provides certain rules on how to improve the accuracy of the procedure by using successive approximations. Other rules to be observed concern the order of magnitude of Δt relative to the size of the segments $(x_{\nu+1} - x_\nu)$, etc. These details, however, cannot be discussed here.[59]

5. Some remarks on the application of the preceding method

When the step-by-step integration method just described is applied to cases where the boundary conditions do not lead to the occurrence of shock phenomena, no serious difficulty arises. If an empirical value is used for μ_0, as indicated in Sec. 11.5, one will find, in general, that the influence of viscosity is practically negligible. Therefore one may omit the σ-terms in (13) and work with the simpler system

$$
\begin{aligned}
m\,\frac{du_\nu}{dt} &= p_{\nu-1} - p_\nu\,, && (\nu = 1, 2, \cdots, n)\,,\\
\frac{dp_\nu}{dt} &= -\,\frac{u_{\nu+1} - u_\nu}{x_{\nu+1} - x_\nu}\,\gamma p_\nu\,, && (\nu = 0, 1, \cdots, n).
\end{aligned}
\tag{15}
$$

In particular, when the flow is initially isentropic, this furnishes a numerical approximation to the solution for which an analytic form was also given in Sec. 12.4. The approximation method must be used, however, whenever the boundary conditions do not lead to explicit expressions for the functions f and g appearing in the general integral (12.42). Straightforward computation based on Eqs. (15) will then solve the problem—if a solution exists.

We saw, however, in Sec. 14.1, that a solution of the differential equations of an ideal fluid does not exist for certain boundary conditions. In such cases, pursuing the computation procedure based on (15) would lead to inconsistencies, or to nonconvergence for increasing n and diminishing Δt.[60] For these cases one could try to apply the principle developed in Sec. 14.1, that is, assume that the whole flow pattern consists of regions in which ideal fluid theory applies, separated by shock lines across which occur discontinuities governed by the shock conditions. For instance, in the Riemann problem mentioned at the beginning of the preceding section, one has to suppose that for $t > t_0$ the flow pattern includes a curved shock line starting at E in an unknown direction (denoted by its slope c, measured from the t-axis). It can then be seen that the Eqs. (15), combined with the shock conditions, allow us to compute c and the values u and p for the time $t_0 + \Delta t$, etc.[61] This procedure, however, is cumbersome and, although some attempts to find solutions in this way were made, it seems to have been abandoned.*

It must be noted that the Eqs. (15), or the corresponding differential equations

$$(16) \qquad \rho \frac{du}{dt} = - \frac{\partial p}{\partial x}, \qquad \frac{dp}{dt} = - \gamma p \frac{\partial u}{\partial x},$$

are not based on the assumption that p/ρ^γ is an over-all constant (which would not be admissible in the case of curved shock lines). In fact, (15) was derived from (10), with $\sigma = 0$, and the third equation (10) then states only that p/ρ^γ remains constant along each particle line (see also the following section).

On the other hand, using the unabridged system (13) which includes viscous terms, or even equations including heat-conduction terms, has proved successful. As was mentioned in Sec. 14.1, the concept of regions of ideal fluid flow separated by lines, across which the shock conditions hold, is only an expedient for representing, to a certain degree of approximation, the solutions of the equations (10) for viscous flow with small μ_0. If it were possible to find exact integrals of (10) satisfying given boundary conditions, for the empirical values of μ_0, such integrals would supply more realistic flow patterns which would include narrow regions within which the variables u, p, and ρ change rapidly. Thus one would expect an approximate integral of (10) derived by means of (13) and (14) to yield a good approximation to the flow pattern. In an attempt to find an approximate integral of (10), however, the following difficulty arises.

When a differential quotient, such as $\partial u/\partial x$, is replaced by a difference quotient $(u_{\nu+1} - u_\nu)/(x_{\nu+1} - x_\nu)$, a reasonable approximation is obtained only if the intervals $(x_{\nu+1} - x_\nu)$ are small compared to any interval in which u experiences a sizable change (see Fig. 86). This means that at least several x_ν-values must fall within the transition region. But we saw in Sec. 11.3 that under average conditions the thickness of a transition layer is of the order of magnitude of 0.1 mm or less. Now it is hardly possible to operate

* Probably this method could be resurrected now that the high-speed computing machines can do the calculations involved.

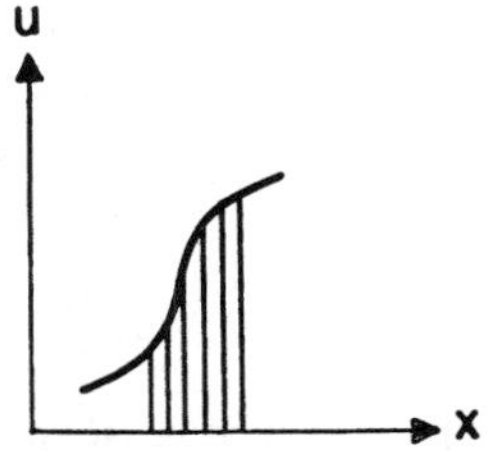

FIG. 86. Approximation of differential coefficient by difference quotient.

with intervals measured in hundredths of millimeters, even with the best computing machines available today.

One way out of this difficulty that has been found convenient is to increase the value of μ_0 appearing in Eqs. (13).[62] If, for example, μ_0 is taken one hundred times larger than the experimental value, one still finds that the greater part of the flow is hardly affected by the viscosity terms. On the other hand, there appear transition regions with a thickness of the order of magnitude of 10 mm, which are well represented by a computation based on intervals of the order of 1 mm. It is important to notice that, for small values of μ_0, the relation between the initial and final values of the transition are effectively the shock conditions, and these are independent of μ_0.

6. The inviscid flow behind a curved shock line

In the simple case of a straight shock line, a flow that is uniform before the shock is still isentropic after the shock. We shall now consider the more general case where, in a one-dimensional flow, a discontinuity point moves at a nonconstant velocity c. This corresponds, in the x,t-plane, to a curved shock line. It is assumed that the motion prior to the shock is uniform, but it is no longer isentropic after the shock, since the magnitude of the sudden change of entropy at the shock transition depends, for each particle, on the instantaneous velocity c of the shock point at the moment when the particle reaches the transition line.

The theory for inviscid flow developed in Arts. 12 and 13 was based on the assumption that the specifying equation for the flow was an over-all relation between p and ρ. This is no longer true of the flow after the shock, since different particles have different values of entropy. The condition of strictly adiabatic flow for the region behind the shock leads, as was seen in Sec. 1.5, only to the condition

(17) $$\frac{dS}{dt} = 0,$$

where S, the entropy or a given function of it, is a known function of p and ρ. To study the inviscid flow behind the shock—or any inviscid flow for which p is a given function of ρ only for each particle line—we must go back to Eq. (11.2) and the first component of Newton's equation, namely,

$$\frac{\partial}{\partial x}(\rho u) + \frac{\partial \rho}{\partial t} = 0, \qquad \rho \frac{du}{dt} = -\frac{\partial p}{\partial x}, \tag{18}$$

and proceed as in Art. 12, but this time with the less restrictive condition (17) in place of the over-all relation.

Exactly as in Sec. 12.2, the first equation can be satisfied by introducing the particle function $\psi(x,t)$:

$$\rho = \frac{\partial \psi}{\partial x}, \qquad \rho u = -\frac{\partial \psi}{\partial t}. \tag{19}$$

When $\psi(x,t)$ is introduced into the second equation (18), the left-hand member becomes, as in Sec. 12.2,

$$-\left(\frac{\partial^2 \psi}{\partial t^2} + 2u \frac{\partial^2 \psi}{\partial x\, \partial t} + u^2 \frac{\partial^2 \psi}{\partial x^2}\right),$$

but this time the expression for the right-hand side is not so simple.

The specifying equation (17) expresses the fact that $S(p,\rho)$ remains constant along each particle line. Thus $S(p,\rho)$ is a function of ψ determined by the boundary conditions (here the conditions along the shock line). From $S(p,\rho) = F(\psi)$ we derive

$$\frac{\partial S}{\partial p}\frac{\partial p}{\partial x} + \frac{\partial S}{\partial \rho}\frac{\partial \rho}{\partial x} = F' \frac{\partial \psi}{\partial x} = F'\rho. \tag{20}$$

Here F' is the derivative of F and is therefore also a known function. If S were an over-all constant, we would have $F' = 0$. It is customary to introduce the notion of sound velocity, a, even in this case, where $dp/d\rho$ is not defined (its value depending on dx/dt). An acceptable definition (see Secs. 5.2 and 9.6) is

$$a^2 = -\frac{\partial S/\partial \rho}{\partial S/\partial p} = \frac{dp/dt}{d\rho/dt}. \tag{21}$$

The second equality is an immediate consequence of (17). Thus a^2 is a known function of p and ρ as soon as $S(p,\rho)$ is known, and can be expressed in terms of ψ and ρ if $F(\psi)$ is also known. Solving for $\partial p/\partial x$ in (20) and using a^2 for the quotient of the derivatives of S, we have

$$\frac{\partial p}{\partial x} = a^2 \frac{\partial \rho}{\partial x} + \frac{\rho F'}{\partial S/\partial p} = a^2 \frac{\partial^2 \psi}{\partial x^2} + \frac{\rho F'}{\partial S/\partial p}. \tag{22}$$

Thus the second equation (18) becomes

$$\frac{\partial^2\psi}{\partial t^2} + 2u\,\frac{\partial^2\psi}{\partial x\,\partial t} + (u^2 - a^2)\,\frac{\partial^2\psi}{\partial x^2} = \frac{\rho F'}{\partial S/\partial p}\,. \tag{23}$$

The left-hand member of (23) is identical with that of Eq. (12.11′), but there the right-hand side was zero.

In (23) the right-hand side includes ρ, which equals $\partial\psi/\partial x$; and F', which is a known function of ψ once $F(\psi)$ is determined; and finally $\partial S/\partial p$, a given function of p and ρ which, by virtue of $S(p,\rho) = F(\psi)$, can be expressed as a function of ψ and ρ. Since u is given by $(-\partial\psi/\partial t):(\partial\psi/\partial x)$, we see that the coefficients in (23), as well as the right-hand side, depend only on ψ and its first-order derivatives. Thus: *Eq. (23) is a planar* nonhomogeneous differential equation of second order in ψ,* differing from (12.11′) only by the term on the right.

For a perfect gas the entropy is proportional to the logarithm of p/ρ^γ. Choosing S to be a simple function of the entropy, rather than the entropy itself, we may write

$$S(p,\rho) = \frac{p}{\rho^\gamma}, \qquad \frac{\partial S}{\partial p} = \frac{1}{\rho^\gamma}, \qquad \frac{\partial S}{\partial \rho} = -\frac{\gamma p}{\rho^{\gamma+1}}, \qquad a^2 = \gamma\,\frac{p}{\rho},$$

and in this case the right-hand member of (23) is $\rho^{\gamma+1}\,F'(\psi)$.

The principal result which we can derive from (23) is the following. Since the characteristics of the differential equation depend only on the second-order terms, they are the same in the present case as in that considered in Arts. 12 and 13. Thus, *the characteristic lines in the x,t-plane are the lines with the slopes $u + a$ and $u - a$, where a is the same function of p,ρ as before.* (See also comments at the end of Sec. 24.2.) Further conclusions analogous to those of Arts. 12 and 13, however, cannot be drawn. The interchange of dependent and independent variables, the use of the speedgraph, etc., are no longer of avail, since Eq. (23) is nonhomogeneous.

It was shown in Sec. 14.3 that the actual change in entropy across a shock is in most cases very small. Thus, if the flow is uniform before the shock, if the shock is not too strong and if, at the same time, the variation of slope along the shock line is slight, the derivative F' will be small. In these circumstances one may, as a rule, consider the flow after the shock to be isentropic.[63]

7. A second approach[64]

There is an alternative approach to this problem of nonisentropic flow. If u times the first equation in (18) is added to the second, the latter is re-

* That is, linear in the derivatives of highest order, see Sec. 9.4.

placed by

$$\frac{\partial}{\partial t}(\rho u) + \frac{\partial}{\partial x}(p + \rho u^2) = 0.$$

This equation allows us to introduce a new function $\bar{\xi}(x,t)$ such that

$$d\bar{\xi} = \rho u\, dx - (p + \rho u^2)dt, \tag{24}$$

just as the first equation in (18) permits the introduction of $\psi(x,t)$ such that

$$d\psi = \rho\, dx - \rho u\, dt. \tag{25}$$

Substituting from this into Eq. (24) we have

$$d\bar{\xi} = u\, d\psi - p\, dt. \tag{26}$$

We have seen that in strictly adiabatic flow the entropy $S(p,\rho)$ is a function of ψ alone, $F(\psi)$, determined by the boundary conditions. Prescription of F therefore provides an algebraic relation between p,ρ,ψ throughout the flow. If any two of these three variables are selected as new independent variables in place of x and t, then the third may be considered a known function of these two for any given problem. Moreover, Eq. (26) can be rewritten as

$$d\xi = u\, d\psi + t\, dp, \tag{27}$$

where $\xi = \bar{\xi} + pt$. We are thus led to select ψ and p as the two new independent variables in place of x and t, and to replace $\bar{\xi}$ by ξ. The functions $u(\psi,p)$ and $t(\psi,p)$ are then given by

$$u = \frac{\partial \xi}{\partial \psi}, \qquad t = \frac{\partial \xi}{\partial p}, \tag{28}$$

and Eq. (25) yields for $x(\psi,p)$:

$$\begin{aligned} \frac{\partial x}{\partial \psi} - u\frac{\partial t}{\partial \psi} &= \frac{1}{\rho}, \\ \frac{\partial x}{\partial p} - u\frac{\partial t}{\partial p} &= 0. \end{aligned} \tag{29}$$

Substitution from (28) in (29) gives

$$\begin{aligned} \frac{\partial x}{\partial \psi} &= \frac{\partial \xi}{\partial \psi}\frac{\partial^2 \xi}{\partial \psi \partial p} + \frac{1}{\rho}, \\ \frac{\partial x}{\partial p} &= \frac{\partial \xi}{\partial \psi}\frac{\partial^2 \xi}{\partial p^2}, \end{aligned} \tag{30}$$

and if $x = x(\psi, p)$ is eliminated between these two equations, the following Monge-Ampère equation is obtained for $\xi = \xi(\psi,p)$:

$$\text{(31)} \qquad \frac{\partial^2 \xi}{\partial \psi^2} \frac{\partial^2 \xi}{\partial p^2} - \left(\frac{\partial^2 \xi}{\partial \psi \partial p} \right)^2 = \frac{\partial}{\partial p} \left(\frac{1}{\rho} \right).$$

Here, since ρ is a known function of ψ and p, the right-hand side is a known function of the same variables. Once a suitable solution $\xi(\psi,p)$ of this equation is determined, the remaining variables u, t, and x are given as functions of ψ and p by Eqs. (28) and, according to (30), by

$$x = \int \left[\left(\frac{\partial \xi}{\partial \psi} \frac{\partial^2 \xi}{\partial \psi \partial p} + \frac{1}{\rho} \right) d\psi + \frac{\partial \xi}{\partial \psi} \frac{\partial^2 \xi}{\partial p^2} \, dp \right],$$

respectively.

Unlike Eq. (23) the Monge-Ampère equation (31) is not planar. However, it has a very simple form in that its right-hand member is a function of the independent variables alone. This member may be rewritten as $-1/\rho^2$ times $\partial\rho/\partial p$, and since this derivative is to be taken with ψ (or S) fixed, we may write

$$\frac{\partial \rho}{\partial p} = - \frac{\partial S/\partial p}{\partial S/\partial \rho} = \frac{1}{a^2},$$

according to Eq. (21). Hence

$$\text{(32)} \qquad \frac{\partial}{\partial p} \left(\frac{1}{\rho} \right) = -\lambda^2 < 0, \qquad \lambda = \frac{1}{a\rho}.$$

Conversely, if λ is given, corresponding functions $S(p,\rho)$ and $F(\psi)$ can be obtained by integrating this last equation and expressing the result in the form $S(p,\rho) = F(\psi)$.

For a perfect gas we may take, as before, $S = p/\rho^\gamma$, so that $1/\rho = p^{-1/\gamma}[F(\psi)]^{1/\gamma}$ and

$$\text{(33)} \qquad \lambda = p^{-\nu}\delta(\psi), \qquad \nu = \frac{\gamma + 1}{2\gamma}, \qquad \delta(\psi) = \frac{1}{\sqrt{\gamma}} [F(\psi)]^{1/2\gamma}.$$

In the case of isentropic motion, ρ is a function of p alone whether or not the gas is perfect. Then the right-hand side of Eq. (31), and hence λ, depends on p alone. For a perfect gas in isentropic motion both are proportional to powers of p by (33) since $\delta(\psi)$ is then an over-all constant.

8. Nonisentropic simple waves. Linearization

The Monge-Ampère equation (31) possesses two families of characteristics, defined, as before, to be lines in the plane of the independent variables across which the analytic character of a solution may change. Along a characteristic either

$$\text{(34)} \qquad \begin{aligned} \lambda \, d\psi - dt &= 0, \\ \lambda \, dp + du &= 0, \end{aligned} \qquad \text{or} \qquad \begin{aligned} \lambda \, d\psi + dt &= 0, \\ \lambda \, dp - du &= 0, \end{aligned}$$

depending on the family to which the characteristic belongs. The top equations in (34) reduce, by virtue of Eq. (25), to

$$dx = (u \pm a)\,dt,$$

respectively, in agreement with the result concerning the characteristics of Eq. (23). For isentropic motion we have seen that λ is a function of p alone, in which case the bottom equations in (34) may be integrated to give

$$v \pm u = \text{constant}, \qquad v = \int \frac{dp}{a\rho} = \int \frac{a\,d\rho}{\rho},$$

which is in agreement with Eq. (12.22).

This suggests investigating other conditions under which one or another of the pairs of equations (34) possesses an integral. More explicitly, we wish to find those functions $\lambda(\psi,p)$ for which there exists a function $W(\psi,p,\xi,u,t)$ such that $W =$ constant holds by virtue of, say, the first pair of equations (34) and Eq. (27). This question can be answered quite generally, but we restrict ourselves here to those functions λ which correspond to the case of a perfect gas, see (33). We find that $\delta(\psi)$ must either be constant, corresponding to the isentropic case discussed above, or else have the form

$$\delta(\psi) = k(\psi - \psi_0)^{\nu-2}, \tag{35}$$

where k and ψ_0 are arbitrary constants. The corresponding integral is

$$W_1 \equiv u\psi + tp - \xi - \frac{k}{(\nu - 1)}\left(\frac{\psi}{p}\right)^{\nu-1} = \text{constant}, \tag{36}$$

where ψ_0 has been suppressed. This is easily checked, for

$$dW_1 = (u\,d\psi + t\,dp - d\xi) + \left(\psi\,du + \frac{k\psi^{\nu-1}}{p^{\nu}}\,dp\right) + \left(p\,dt - \frac{k\psi^{\nu-2}}{p^{\nu-1}}\,d\psi\right) = 0$$

along the characteristic, since there the first bracket vanishes by Eq. (27) and the second and third by the first pair of equations in (34). For this $\delta(\psi)$, the corresponding distribution of entropy from particle to particle is given by

$$F(\psi) = A\psi^{1-3\gamma}, \qquad A = (\gamma k^2)^{\gamma}. \tag{37}$$

For any isentropic motion, $v + u$ remains constant on each $(u + a)$-characteristic. Flows for which the constant does not change from characteristic to characteristic are called simple waves. Correspondingly, we may equate W_1, given by Eq. (36), to an over-all constant and thereby generate a class of solutions of Eq. (31) which may be called *nonisentropic simple waves*. It should be emphasized that for a perfect gas these solutions are defined only for the entropy distribution given by (37). We may easily verify that this procedure does generate solutions of Eq. (31). For with u and t defined by (28), so that (27) holds, we have on equating dW_1 to zero:

$$\psi\, du + k\,\frac{\psi^{\nu-1}}{p^{\nu}}\, dp + p\, dt - \frac{k\psi^{\nu-2}}{p^{\nu-1}}\, d\psi = 0.$$

Using (28) once more and remembering that from (35): $\lambda = k\psi^{\nu-2}/p^{\nu}$, we obtain

$$\left[\psi\,\frac{\partial^2\xi}{\partial\psi^2} + p\left(\frac{\partial^2\xi}{\partial\psi\partial p} - \lambda\right)\right] d\psi + \left[\psi\left(\frac{\partial^2\xi}{\partial\psi\partial p} + \lambda\right) + p\,\frac{\partial^2\xi}{\partial p^2}\right] dp = 0,$$

which holds identically in ψ and p. Equating the coefficients of $d\psi$ and dp to zero and eliminating ψ and p themselves from the resulting equations, we find

$$\frac{\partial^2\xi}{\partial\psi^2}\,\frac{\partial^2\xi}{\partial p^2} - \left[\left(\frac{\partial^2\xi}{\partial\psi\partial p}\right)^2 - \lambda^2\right] = 0,$$

in agreement with Eqs. (31) and (32). Thus we have shown that any solution of the first-order equation W_1 = constant—with u and t as in Eqs. (28)—is also a solution of the original second-order equation (31). In the classical theory of the Monge-Ampère equation such first-order equations are called intermediate integrals.

A similar discussion can be made for the second pair of equations in (34). Again $\delta(\psi)$ is either a constant or given by Eq. (35). Now W_1 is replaced by

$$W_2 = u\psi + tp - \xi + \frac{k}{(\nu - 1)}\left(\frac{\psi}{p}\right)^{\nu-1},$$

and equating this to a constant throughout the flow leads to anisentropic simple waves of a second kind, in the same way that equating $v - u$ to an over-all constant does in the isentropic case.

We may introduce new variables α and β by means of the equations

$$\alpha = W_1, \qquad \beta = W_2,$$

and for flows other than simple waves these may be taken as new independent variables. The two families of curves α = constant, β = constant, in the ψ,p-plane are the characteristics, so that the variables α and β play a similar role to that of ξ and η in Art. 12. In these new variables it may be shown that $1/\psi$, $1/p$, u, and t all satisfy linear second-order equations, of the form

$$\frac{\partial^2 z_n}{\partial\alpha\,\partial\beta} + \frac{n}{\alpha - \beta}\left(\frac{\partial z_n}{\partial\alpha} - \frac{\partial z_n}{\partial\beta}\right) = 0.$$

For $\gamma = \frac{7}{5}$ we have $n = 3$ for $z_n = 1/\psi$, $n = -4$ for $z_n = 1/p$, $n = 4$ for $z_n = u$, and $n = -3$ for $z_n = t$. For general γ we have $n = N, -(N + 1), (N + 1), -N$, respectively, where $N = (\gamma + 1)/2(\gamma - 1)$. The analogy with Eq. (12.34) becomes evident if that equation is rewritten with the characteristic variables $\xi = v + u$ and $\eta = v - u$ as new independent variables [see Eq. (12.43)]. Thus, *for the entropy distribution* (37) *the equations governing the motion may, by appropriate change of independent variables, be replaced by a linear equation in* $1/\psi$, $1/p$, u, *or* t *of the type considered in* Sec. 12.4.

Alternatively the same conclusions concerning the characteristics, Eqs. (34), can be reached by considering Eqs. (17) and (18) as a planar system of three homogeneous first-order equations for u, p, and ρ. For this purpose, Eq. (17) is written out as

$$\frac{dp}{dt} - a^2 \frac{d\rho}{dt} = 0,$$

according to Eq. (9.25). The general theory of characteristics, as developed in Art. 9, will then yield the characteristic conditions (34).

CHAPTER IV

PLANE STEADY POTENTIAL FLOW

Article 16

Basic Relations

1. Direct approach

We deal in this article with the plane, steady, irrotational flow of an elastic, inviscid fluid, neglecting the influence of gravity. The basic hypothesis for steady plane flow is that all quantities are independent of z and t, and that q_z, the z-component of the velocity vector $\mathbf{q}$, vanishes. The components of $\mathbf{q}$ are then functions of the coordinates x and y. The relations valid for this type of motion are contained in, or can be derived from, various discussions in Chapters I and II. For the convenience of the reader, however, we start here by setting up the main equations of our problem without making use of previous results, except, of course, for the very first principles.

The fact that a motion in the x,y-plane is steady gives the equation of continuity (1.II) the form

$$\frac{\partial(\rho q_x)}{\partial x} + \frac{\partial(\rho q_y)}{\partial y} = 0. \tag{1}$$

The fact that it is irrotational is expressed by

$$\frac{\partial q_y}{\partial x} - \frac{\partial q_x}{\partial y} = 0. \tag{2}$$

Both equations can be transformed in such a way that they become independent of the choice of the coordinate axes. Denote by $\partial/\partial s$ differentiation in the direction of the velocity vector $\mathbf{q}$, and by $\partial/\partial n$ differentiation in the direction which makes an angle of $+90°$ with the first. If we identify, at a point P, the x- and y-directions in (1) and (2) with the s- and n-directions, respectively, (1) and (2) yield

$$q\frac{\partial \rho}{\partial s} + \rho\frac{\partial q}{\partial s} + \rho\frac{\partial q_n}{\partial n} = 0, \qquad \frac{\partial q_n}{\partial s} - \frac{\partial q}{\partial n} = 0,$$

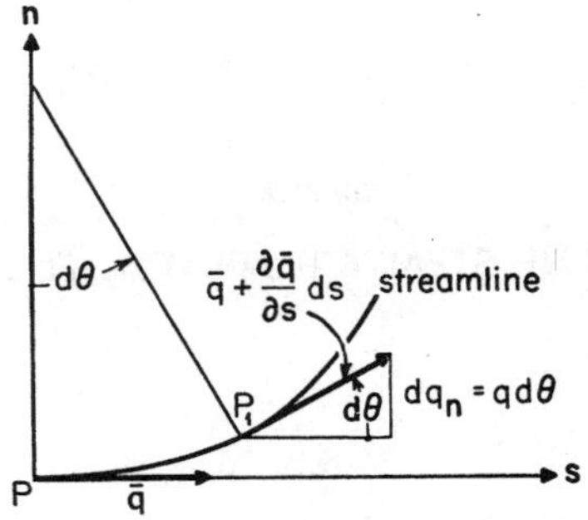

FIG. 87. Streamline and normal.

since $q_x = q$, $q_y = 0$ at P. Although $q_y = 0$, the derivatives of $q_y = q_n$ are not zero. If θ denotes the angle between $\mathbf{q}$ and any fixed direction, it is seen from Fig. 87 that the increment of q_y is $q\,d\theta$; thus,

$$\frac{\partial q_y}{\partial x} = \frac{\partial q_n}{\partial s} = q\frac{\partial \theta}{\partial s} = \frac{q}{R}, \qquad \frac{\partial q_y}{\partial y} = \frac{\partial q_n}{\partial n} = q\frac{\partial \theta}{\partial n}, \tag{3}$$

where $1/R = \partial\theta/\partial s$ denotes the curvature, and R the radius of curvature of the streamline. The equations of continuity and irrotationality, (1) and (2), respectively, then take the forms

$$\begin{aligned} \frac{\partial(\rho q)}{\partial s} + \rho q\frac{\partial \theta}{\partial n} = \rho\frac{\partial q}{\partial s} + q\frac{\partial \rho}{\partial s} + \rho q\frac{\partial \theta}{\partial n} &= 0, \\ \frac{\partial q}{\partial n} - q\frac{\partial \theta}{\partial s} &= 0, \end{aligned} \tag{4}$$

or, equivalently,

$$\frac{\partial(\log \rho)}{\partial s} + \frac{\partial(\log q)}{\partial s} + \frac{\partial \theta}{\partial n} = 0, \qquad \frac{\partial(\log q)}{\partial n} - \frac{\partial \theta}{\partial s} = 0. \tag{4'}$$

The variable ρ still appears in (4), but may be eliminated as follows. The fluid, assumed elastic, satisfies a (p,ρ)-relation. Thus there is a derivative $dp/d\rho$, which we have called a^2, where a is the sound velocity, and

$$\frac{\partial \rho}{\partial s} = \frac{d\rho}{dp}\frac{\partial p}{\partial s} = \frac{1}{a^2}\frac{\partial p}{\partial s}. \tag{5}$$

The s-component of the Newton equation—which in the absence of gravity and viscosity is equivalent to the differential form (2.21′) of the Bernoulli equation—is

$$\frac{\partial p}{\partial s} = -\rho\frac{dq}{dt} = -\rho q\frac{\partial q}{\partial s}. \tag{6}$$

Then, if we introduce the Mach number $M = q/a$, the Eqs. (4) read

$$\frac{\partial q}{\partial s} = \frac{q}{M^2 - 1}\frac{\partial \theta}{\partial n}, \qquad \frac{\partial q}{\partial n} = q\frac{\partial \theta}{\partial s}. \tag{7}$$

Here the first equation is independent of the assumption of irrotationality.

However, M involves a as well as q. To eliminate a we use the Bernoulli equation in the integrated form [see Eq. (8.3)]:

$$\frac{q^2}{2} + \int \frac{dp}{\rho} = \text{constant}, \tag{8}$$

where, in the irrotational flow of an elastic fluid, the constant is the same for the entire flow, as shown in Sec. 6.5.

The explicit computation will be carried out only for a polytropic fluid: $p = C\rho^\kappa$, where κ is a constant >1. Then

$$a^2 = \frac{\kappa p}{\rho}, \qquad \int \frac{dp}{\rho} = \frac{\kappa}{\kappa - 1}\frac{p}{\rho} + \text{constant} = \frac{a^2}{\kappa - 1} + \text{constant},$$

and the Bernoulli equation is

$$\frac{q^2}{2} + \frac{a^2}{\kappa - 1} = \frac{a_s^2}{\kappa - 1},$$

where a_s denotes the value of the sound velocity at a stagnation point. Thus,

$$a^2 = a_s^2 - \frac{\kappa - 1}{2}q^2, \tag{9}$$

and

$$M^2 = \frac{q^2}{a^2} = \frac{q^2}{a_s^2 - \frac{\kappa - 1}{2}q^2}, \qquad M^2 - 1 = \frac{\frac{\kappa + 1}{2}q^2 - a_s^2}{a_s^2 - \frac{\kappa - 1}{2}q^2}. \tag{10}$$

The equations (7), with $M^2 - 1$ expressed by (10), and a scale factor a_s chosen, are two differential equations of first order for the unknowns q and θ. *The set of Eqs.* (7) *and* (10) *includes all the information on which the analysis of the flow pattern under discussion will be based.** If, in addition to the velocity distribution, the values of pressure, density, and absolute temperature are required, these follow from the Bernoulli equation in connection with the (p,ρ)-relation and the equation of state. (See Art. 8, and in particular Sec. 8.4, where in Table I, numerical values of the variables are given for the polytropic case with $\kappa = \gamma = 1.4$.)

* In the general non-polytropic case Eq. (8) takes the place of (10).

Reviewing our earlier results we note that the fact that ρ and p can be expressed in terms of q justifies the use of q as a dependent variable. Then the other natural variable to determine $\mathbf{q}$ is θ, i.e., the angle that $\mathbf{q}$ makes with some fixed direction; moreover, θ determines the streamline pattern, which is fixed for steady flow. It is then likewise natural to introduce the rate of change along a streamline, $\partial/\partial s$, and the rate of change in the direction normal to the s-direction.*

From (9) it is seen that a decreases steadily as q increases from zero, and that a vanishes for $q = q_m$ where

$$q_m = \sqrt{\frac{2}{\kappa - 1}}\, a_s . \tag{11}$$

This value of q cannot be exceeded. Introducing q_m in (9) gives the Bernoulli equation in the form

$$a^2 = \frac{\kappa - 1}{2}(q_m^2 - q^2). \tag{9'}$$

On the other hand, from (9), q equals a if both have the value

$$q_t = a_t = \sqrt{\frac{2}{\kappa + 1}}\, a_s . \tag{11'}$$

This value, defined by the condition $q = a$, has been called the transition velocity, also sonic or transonic speed. The polytropic relation in the form

$$\frac{p}{p_1} = \left(\frac{\rho}{\rho_1}\right)^{\kappa} = \left(\frac{a}{a_1}\right)^{2\kappa/(\kappa-1)},$$

where p_1, ρ_1, a_1 are some corresponding values of p, ρ, a, shows that p, ρ, and (since $\kappa > 1$) also p/ρ vanish for $q = q_m$.

From the polytropic relation and (9′) the relation between pressure and velocity [see Eq. (8.28)] takes the form

$$\frac{p}{p_s} = \left[1 - \frac{q^2}{q_m^2}\right]^{\kappa/(\kappa-1)}, \tag{12}$$

where p_s is the value of p at a stagnation point.† The relation between density and velocity corresponding to (12) is

$$\frac{\rho}{\rho_s} = \left[1 - \frac{q^2}{q_m^2}\right]^{1/(\kappa-1)}. \tag{12'}$$

In the following we shall let $\rho_s = 1$, unless otherwise specified.

* If the equations are considered in a region rather than at a single point, we are actually using curvilinear coordinates, namely, the system of streamlines and their orthogonal trajectories.

† In Sec. 8.1 the relation between p and q was discussed for a general (p,ρ)-relation, while in Sec. 8.3 a polytropic relation was assumed.

Dimensionless variables can be introduced in various ways, dividing q either by q_m, or by a_s, or by $a_t = q_t$. With the dimensionless quantity $v = q/a_s$ Eqs. (7) and (10) become:

$$(M^2 - 1)\frac{\partial v}{\partial s} = v\frac{\partial \theta}{\partial n}, \qquad \frac{\partial v}{\partial n} = v\frac{\partial \theta}{\partial s},$$

(13)

$$M^2 - 1 = \frac{(\kappa + 1)v^2 - 2}{2-(\kappa-1)v^2},$$

which include no parameter except κ.

2. Equations for the potential and stream functions

The differential equation for the potential of an irrotational inviscid flow has been derived in its most general form in Sec. 7.2 and later considered for steady plane motion in Sec. 7.5. Writing now φ instead of Φ we quote Eq. (7.31):

$$\text{(14)} \qquad \frac{\partial^2\varphi}{\partial x^2}\left(1 - \frac{q_x^2}{a^2}\right) - 2\frac{\partial^2\varphi}{\partial x\partial y}\frac{q_xq_y}{a^2} + \frac{\partial^2\varphi}{\partial y^2}\left(1 - \frac{q_y^2}{a^2}\right) = 0.$$

Here

$$\text{(15)} \quad \mathbf{q} = \operatorname{grad}\varphi, \qquad q_x = \frac{\partial\varphi}{\partial x}, \qquad q_y = \frac{\partial\varphi}{\partial y}, \qquad q^2 = \left(\frac{\partial\varphi}{\partial x}\right)^2 + \left(\frac{\partial\varphi}{\partial y}\right)^2,$$

and a^2 is to be determined from (9). It follows from (15) that

$$\text{(15')} \qquad \frac{\partial\varphi}{\partial s} = q, \qquad \frac{\partial\varphi}{\partial n} = 0.$$

To derive (14) directly and at the same time obtain its form for the natural coordinate system introduced in Sec. 1, we obtain from the first of Eqs. (7) and (15′),

$$\text{(16)} \qquad \Delta\varphi = \operatorname{div}\mathbf{q} = \frac{\partial q}{\partial s} + q\frac{\partial\theta}{\partial n} = M^2\frac{\partial q}{\partial s} = M^2\frac{\partial^2\varphi}{\partial s^2},$$

and since $\Delta\varphi$ is the sum of second derivatives in any two directions orthogonal to each other, Eq. (16) gives

$$\text{(16')} \qquad \frac{\partial^2\varphi}{\partial n^2} = (M^2 - 1)\frac{\partial^2\varphi}{\partial s^2}.$$

If, on the other hand, $q^2\partial q/\partial s$ is replaced by $q\,\partial(q^2/2)/\partial s$ in (16), q^2 is taken from (15), and $q\,\partial/\partial s$ is replaced by $q_x\,\partial/\partial x + q_y\,\partial/\partial y$, we find immediately

$$\Delta\varphi = \frac{1}{a^2}\left(q_x^2\frac{\partial^2\varphi}{\partial x^2} + 2q_xq_y\frac{\partial^2\varphi}{\partial x\partial y} + q_y^2\frac{\partial^2\varphi}{\partial y^2}\right),$$

which is identical with (14).

In the case of two-dimensional steady motion another approach is possible. Instead of satisfying Eq. (2) by introducing the function φ, we may satisfy Eq. (1) by setting

$$(17) \qquad \rho q_x = \frac{\partial \psi}{\partial y}, \qquad \rho q_y = -\frac{\partial \psi}{\partial x}.*$$

Hence,

$$(18) \qquad \frac{\partial \varphi}{\partial x} = \frac{1}{\rho}\frac{\partial \psi}{\partial y}, \qquad \frac{\partial \varphi}{\partial y} = -\frac{1}{\rho}\frac{\partial \psi}{\partial x}.$$

In the natural coordinate system we have

$$(18') \qquad \frac{\partial \varphi}{\partial s} = \frac{1}{\rho}\frac{\partial \psi}{\partial n}, \qquad \frac{\partial \varphi}{\partial n} = -\frac{1}{\rho}\frac{\partial \psi}{\partial s}.$$

The function ψ, the *stream function*, has the property that *its value is constant along a streamline*; in fact, using (15′) and (18′) we obtain

$$(19) \qquad \frac{\partial \psi}{\partial s} = 0, \qquad \frac{\partial \psi}{\partial n} = \rho q.$$

The last equation states that $d\psi$ is the mass flux moving through the cross section of a stream tube of unit height between the lines ψ and $\psi + d\psi$.

Equations (18) and (18′) show that the lines φ = constant, the *equipotential lines*, or *potential lines*, are orthogonal to the lines ψ = constant, the streamlines.

Next it is seen that ψ satisfies an equation of second order of exactly the same form as (14). To show this, we compute $\partial^2\psi/\partial s^2$, and obtain

$$\frac{\partial^2 \psi}{\partial s^2} = -\frac{\partial}{\partial s}(\rho q_n) = -\rho\frac{\partial q_n}{\partial s},$$

according to the definition of ψ, Eq. (17). The n-component of the Newton equation then yields

$$(20) \qquad \frac{\partial^2 \psi}{\partial s^2} = \frac{1}{q}\frac{\partial p}{\partial n} = \frac{a^2}{q}\frac{\partial \rho}{\partial n},$$

where, as before, $a^2 = dp/d\rho$. Using the second expression for $\partial^2\psi/\partial s^2$ in these last two equations, and the condition of irrotationality (7) in the form $\partial q_n/\partial s = \partial q/\partial n$, we have

$$(20') \qquad (M^2 - 1)\frac{\partial^2 \psi}{\partial s^2} = q\frac{\partial \rho}{\partial n} + \rho\frac{\partial q}{\partial n} = \frac{\partial(\rho q)}{\partial n} = \frac{\partial^2 \psi}{\partial n^2}.$$

* If φ and ψ are to be of equal dimensions we have to set $\rho q_x = \rho_0\, \partial\psi/\partial y$, $\rho q_y = -\rho_0\, \partial\psi/\partial x$, where ρ_0 is some standard density. We set $\rho_0 = \rho_s = 1$, in accordance with the remark at the end of Sec. 1.

Thus ψ satisfies the same equation in natural coordinates as does ϕ [see Eq. (16′)].

If now $\partial(\rho q)/\partial n$ in Eq. (20′) is replaced by $(1/\rho q)\partial(\frac{1}{2}\rho^2 q^2)/\partial n$, also $\rho^2 q^2$ by $(\partial\psi/\partial x)^2 + (\partial\psi/\partial y)^2$, and $q\,\partial/\partial n$ by $q_x\,\partial/\partial y - q_y\,\partial/\partial x$, we obtain

$$q^2 \frac{\partial^2\psi}{\partial n^2} = q_y^2 \frac{\partial^2\psi}{\partial x^2} - 2q_xq_y \frac{\partial^2\psi}{\partial x\partial y} + q_x^2 \frac{\partial^2\psi}{\partial y^2}.$$

Moreover, $\Delta\psi = \partial^2\psi/\partial s^2 + \partial^2\psi/\partial n^2 = \partial^2\psi/\partial x^2 + \partial^2\psi/\partial y^2$. Hence, Eq. (20′) becomes

$$\text{(21)}\qquad \frac{\partial^2\psi}{\partial x^2}\left(1 - \frac{q_x^2}{a^2}\right) - 2\frac{\partial^2\psi}{\partial x\partial y}\frac{q_xq_y}{a^2} + \frac{\partial^2\psi}{\partial y^2}\left(1 - \frac{q_y^2}{a^2}\right) = 0.$$

Although this is apparently the same equation as (14), it cannot be used in the same way, for a^2 is expressible in terms of the derivatives of φ by (9) and (15), while the relation between a^2 and ψ is less direct.*

A few comments may be added. While the existence of a potential depends on the assumption of irrotationality, the stream function is based on the always-valid continuity equation in cases where this equation is essentially two-dimensional. This is so in the problem of the present chapter and in steady three-dimensional flow with axial symmetry (see Sec. 7.7). If the axis of symmetry has the x-direction and y is the radial direction, the equation of continuity has the form

$$\frac{\partial}{\partial x}(y\rho q_x) + \frac{\partial}{\partial y}(y\rho q_y) = 0,$$

and a stream function may be defined by

$$\rho y q_x = \frac{\partial\psi}{\partial y}, \qquad \rho y q_y = -\frac{\partial\psi}{\partial x}.^1$$

We shall not consider this problem, although many results and procedures are very similar to corresponding ones for the plane (x,y)-problem (see Sec. 9.4). Stream functions cannot be defined in the corresponding unsteady problems, in contrast to the situation in the theory of incompressible fluids.

Various analogies exist between the (x,t)-problem of Chapter III and the (x,y)-problem considered in the present chapter. In both cases the equations of the problem form a system of two homogeneous, planar, nonlinear partial differential equations of first order, whose coefficients contain only the dependent variables ρ, u and q_x, q_y, respectively; such systems are sometimes called "reducible". Thus by an interchange of variables a linear

* The present derivation was chosen in view of a generalization of Eq. (21) (see Sec. 24.2) which will cover cases where irrotationality is not assumed.

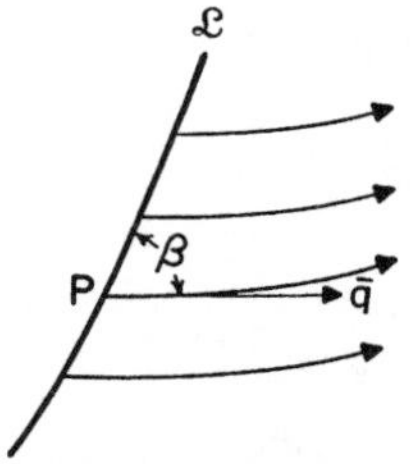

FIG. 88. Determination of flow from data along $\mathfrak{L}$.

system may be obtained in the speedgraph or hodograph planes. The potential and particle functions of Sec. 12.2 correspond, of course, to the present potential and stream functions, and the apparently identical Eqs. (12.11′) and (12.12′) to the Eqs. (14) and (21) of the present section. The (x,t)-problem, which is always hyperbolic (so that there are always real characteristics), corresponds mathematically, as we shall see later in more detail, to the supersonic (x,y)-problem. Accordingly, we shall have to expect difficulties not existing in the (x,t)-problem.[2]

3. Subsonic and supersonic flow. Characteristics[3]

The general theory of characteristics developed in Art. 9, when applied to the potential equation (14), led to the following results. The equation is elliptic (no real characteristics) in a region where the velocity q is smaller than the local sound velocity a $(M < 1)$; when $q > a$ $(M > 1)$, the equation is hyperbolic and the lines crossing the streamlines at the angles $\pm \alpha$ (where $\sin \alpha = 1/M$) are characteristics (Mach lines); the transition occurs for $q = a$ $(M = 1)$. A short derivation of these results, independent of the previous arguments and based on Eqs. (7), follows.

Assume that a velocity distribution satisfying Eqs. (7) is known on one side of a line $\mathfrak{L}$ (Fig. 88). One may ask, to what extent is the flow on the other side of $\mathfrak{L}$ determined? Obviously the derivatives of q and θ in the direction of $\mathfrak{L}$ are known. If $\mathfrak{L}$ crosses the streamline at the point P at the angle β, these given quantities are (if $\partial/\partial l$ denotes the directional derivative in the $\mathfrak{L}$-direction)

$$(22) \qquad \frac{\partial q}{\partial l} = \frac{\partial q}{\partial s} \cos \beta + \frac{\partial q}{\partial n} \sin \beta, \qquad \frac{\partial \theta}{\partial l} = \frac{\partial \theta}{\partial s} \cos \beta + \frac{\partial \theta}{\partial n} \sin \beta.$$

When, by use of (7), derivatives of θ are replaced by derivatives of q, Eqs. (22) become

$$(22') \qquad \begin{aligned} \frac{\partial q}{\partial s} \cos \beta + \frac{\partial q}{\partial n} \sin \beta &= \frac{\partial q}{\partial l}, \\ \frac{\partial q}{\partial s} (M^2 - 1) \sin \beta + \frac{\partial q}{\partial n} \cos \beta &= q \frac{\partial \theta}{\partial l}, \end{aligned}$$

two linear equations for $\partial q/\partial s$ and $\partial q/\partial n$; they have a unique solution, except when the determinant of their coefficients vanishes, that is, except when

$$\begin{vmatrix} \cos\beta & \sin\beta \\ (M^2-1)\sin\beta & \cos\beta \end{vmatrix} = 1 - M^2\sin^2\beta = 0.$$

The same condition is found if, in (22), one eliminates the derivatives of q by means of (7) and then tries to solve the two equations for $\partial\theta/\partial s$ and $\partial\theta/\partial n$.

The present result follows also from Eq. (10.2′) if we identify in turn x, y, u, v, and ϕ with s, n, θ, q, and β.

The conclusion is that whereas, in general, the values of q and θ given along a curve $\mathfrak{L}$ determine the first-order derivatives of both q and θ at each point of $\mathfrak{L}$, no such inference is possible if, in the case $M > 1$, $\mathfrak{L}$ crosses the streamlines at an angle whose sine is $\pm 1/M$. Along such a line segment any two solutions of the partial differential equations (7) with the same values of q and θ along that line can be patched together, since the derivatives of q and θ in an "exterior" direction, i.e., in a direction different from that of $\mathfrak{L}$, are not determined by the data along $\mathfrak{L}$ and the differential equations, and may therefore be different on the two sides. For example, we shall see (Art. 18) that a uniform flow on parallel straight streamlines can be continued as a flow along curved streamlines if and only if the transition takes place along a line that intersects the streamlines at the Mach angle $\alpha = \text{arc}\sin 1/M$ (see Fig. 44). (Here the transition line $\mathfrak{L}$ is a straight line, since in the uniform flow region q, a, and consequently α are constant along the line.) The line which we obtain by turning $\mathbf{q}$ at each point in the positive (negative) sense through the angle α is called the plus characteristic C^+, or plus Mach line (minus characteristic C^-, or minus Mach line).

Denoting by ϕ^+ (ϕ^-) the angle which a C^+ (C^-) makes with a fixed direction, say with the x-axis, and measuring θ likewise from the x-axis, we arrive at the characteristic conditions or direction conditions

$$\phi^+ = \theta + \alpha, \qquad \phi^- = \theta - \alpha$$

$$\sin\alpha = \frac{1}{M}, \quad \text{or} \quad \cot\alpha = \sqrt{M^2 - 1}. \tag{23}$$

Here α depends only on q or, by Eq. (10), on M. The formulas show clearly how the characteristic directions depend on the solution q, θ under consideration.[4]

Since $\sin\alpha$ is smaller than 1, except at a sonic point, and is nonnegative, the positive direction on each characteristic may be defined as the one that makes an acute angle with the velocity vector. It is then obvious that prescribing the two positive characteristic directions at a point, except where $M = 1$, is equivalent to prescribing $\mathbf{q}$. In other words, the velocity distribution in a region is uniquely determined by the two sets of directed characteristics.

We know from the general theory that, in contrast to noncharacteristic curves, along a characteristic the values of q and θ cannot be chosen arbitrarily. In fact, from Eqs. (22′) we find

$$\frac{\partial q}{\partial l} \cos \alpha - q \frac{\partial \theta}{\partial l} \sin \alpha = \frac{\partial q}{\partial s} [\cos \beta \cos \alpha - (M^2 - 1) \sin \beta \sin \alpha]$$
$$+ \frac{\partial q}{\partial n} [\sin \beta \cos \alpha - \cos \beta \sin \alpha].$$

Since $M^2 - 1 = \cot^2 \alpha$, both brackets vanish if $\beta = \alpha$, that is, if $\partial/\partial l$ refers to differentiation along a plus characteristic. In the same way we see that $\cos \alpha \, \partial q/\partial l + q \sin \alpha \, \partial \theta/\partial l = 0$ if $\partial/\partial l$ refers to a minus characteristic. Hence one has, with an obvious notation, the following relations between derivatives of q and θ along a Mach line:

$$(24) \qquad \frac{\partial q}{\partial l^+} = q \tan \alpha \frac{\partial \theta}{\partial l^+}, \qquad \frac{\partial q}{\partial l^-} = - q \tan \alpha \frac{\partial \theta}{\partial l^-},$$

or [see also Eqs. (9.17)]

$$(24') \quad \frac{dq}{d\theta} = q \tan \alpha, \quad \text{along a } C^+; \qquad \frac{dq}{d\theta} = - q \tan \alpha, \quad \text{along a } C^-.$$

In rectangular coordinates, letting $q_x = u$ and $q_y = v$, we obtain from Eq. (9.15) or by direct computation

$$(24'') \qquad \left(\frac{dv}{du}\right)^{\pm} = - \cot (\theta \mp \alpha), \quad \text{along a } C^{\pm}.$$

Equations (24) are the compatibility relations, introduced in a general form in Sec. 9.3 and discussed in detail for $k = n = 2$ in Sec. 10.2, which hold between the derivatives of the dependent variables along a C^+ and along a C^-, respectively. They show that along a characteristic, θ is a function of q, and uniquely determined by specifying q and θ at a single point. We see that here in contrast to the linear case a characteristic may be any geometrically given curve, with compatible values of the dependent variables prescribed along it such as to make it "characteristic". Although it follows from the general theory of Arts. 9 and 10, as well as from the derivation at the beginning of this section, we state explicitly: For a steady plane potential flow of an elastic fluid, the streamlines are not exceptional or characteristic; the only characteristics are the Mach lines.

4. Basic boundary-value problems

The role played by the characteristics in the solution of boundary-value problems (see Secs. 10.2 and 10.3) can now be explained in our present case as follows. Along an arc AB *values of* q *and* θ *are given* (Cauchy problem) in such a way that the curve AB has at no point the direction of either of the characteristics; this means that at no point is the angle between the curve and the x-axis equal to $\theta + \alpha(q)$ or $\theta - \alpha(q)$, where $\alpha(q)$ is a known function of q whose form depends on the (p,ρ)-relation.* Consider the neighboring points 1 and 2 on AB (Fig. 89) and draw the characteristics through them, or rather, straight lines in the characteristic directions. (Knowing q and θ, we know their directions $\theta \pm \alpha$). They have an intersection, 3, on the upper side of AB, and an intersection, 4, on the lower

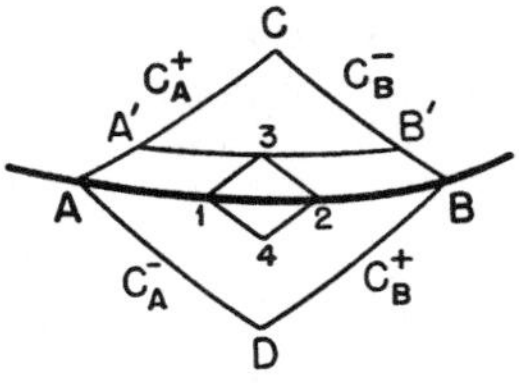

FIG. 89. Cauchy problem.

side. Considering all distances 12, 13, and 23 as infinitesimal and neglecting terms of higher order, we conclude from (24) that, with an obvious notation,

$$
\begin{aligned}
q_3 - q_1 &= +(\theta_3 - \theta_1)q_1 \tan \alpha_1 \\
q_3 - q_2 &= -(\theta_3 - \theta_2)q_2 \tan \alpha_2 .
\end{aligned} \tag{25}
$$

For the point 4 the opposite signs hold in Eqs. (25). With respect to the unknown q_3 and θ_3, the determinant of these equations is

$$(q_1 \tan \alpha_1 + q_2 \tan \alpha_2),$$

which is different from zero since $\tan \alpha$ cannot change sign; thus both q_3 and θ_3 can be computed. In this way, starting from a sequence of neighboring points along AB, one can derive from the given values of q and θ on AB the values along a second row of points, $A'B'$. Continuing in the same manner, q- and θ-values are found for all lattice points within a curvilinear triangle ABC, where AC and BC are characteristics, one belonging to the

* This section is, mathematically, a straightforward application of the considerations of Sec. 10.3 to our flow problem. In order to avoid repeated interruption of the presentation we refer the reader to Art. 10 for more careful formulations, for comments and Notes.

plus set and one to the minus set—provided the procedure does not break down earlier, which can happen if the direction of a cross line such as $A'B'$, $A''B''$, $\cdots$ approaches somewhere a characteristic direction (see Sec. 10.3). However, due to the noncharacteristic nature of AB, and the continuity of all functions involved, this can happen only at a finite distance from AB. All these conclusions apply, of course, also to the triangle ABD on the lower side of AB.

As a second case, consider *data given along two intersecting lines AB and AC, one of them a characteristic.* Assume that AB is a minus characteristic and that the noncharacteristic arc AC is in one of the angular spaces between AB and the plus characteristic through A. Values of q and θ along AB must be given in such a way that at each point its angle with the x-axis is $\phi^- = \theta - \alpha$, and such that the second equation (24) holds. It follows that if the geometric shape of AB is given, we may prescribe only the value of either q or θ at *one* point of AB; then q and θ are determined along AB. We further suppose that either θ or q (or some component of $\mathbf{q}$) is given along AC.

From the data along AB the initial elements of the plus characteristics at all points of AB can be derived (Fig. 90), and we assume that they are plotted in the direction toward AC. If the point 1 is adjacent to A, the characteristic element through 1 will cut the line AC in some point 3. From a given value at 3 and from $q_3 - q_1 = (\theta_3 - \theta_1)q_1 \tan \alpha_1$, both the quantities θ_3 and q_3 can be derived. This enables us to find the beginning 34 of the minus characteristic through 3, and the compatibility relations applied to the segments 34 and 24 give the values q_4, θ_4. In this way, step by step, the whole quadrangle $ABDC$, where CD and BD are characteristics, can be filled by a net of points at which $\mathbf{q}$ is known.

A slight modification of the procedure has to take place if AC is likewise a characteristic, here a plus characteristic. If we know that of two intersecting arcs, given geometrically, one is a C^+ and the other a C^-, then neither θ nor q can be arbitrarily prescribed anywhere along these two curves. Their values follow from the two direction conditions and the two compatibility conditions, with both q_A and θ_A determined by the two direction conditions at A. The stepwise construction of the net inside the char-

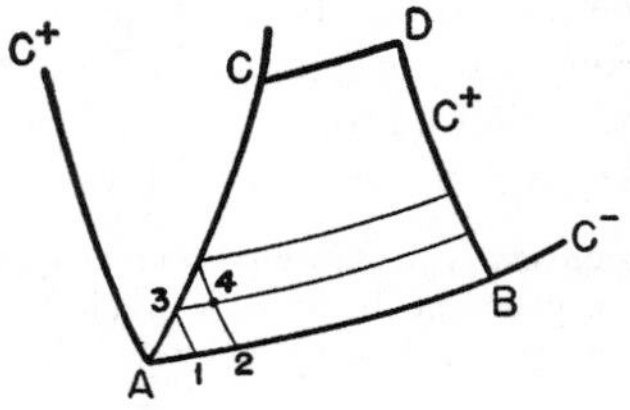

FIG. 90. Data along two intersecting lines, one of them a characteristic.

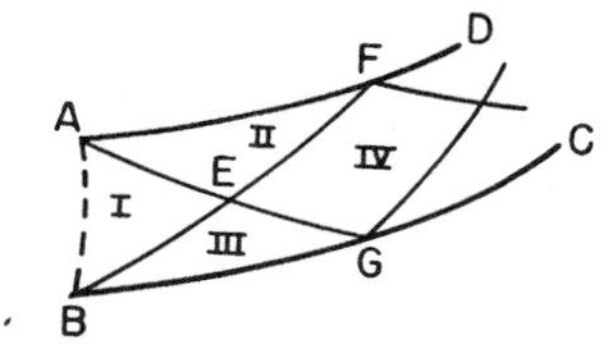

FIG. 91. Plane duct.

acteristic quadrangle $ABDC$ is analogous to the preceding one and to the procedure in Art. 10.

As an example, consider the duct $ABCD$ of Fig. 91. The entrance velocities along AB determine, independently of the shape of the walls, a supersonic flow in triangle I. Since the walls are streamlines, we know θ along them. The values of q and θ found for AE with the known θ-values along AD determine $\mathbf{q}$ in AEF, region II, where AE and EF are characteristics. In the same way, knowing θ along BC leads to the velocity distribution in BEG, region III. From the values of q and θ along EF and EG (two characteristics) follow those in region IV, and so on. All this holds, of course, only in the case of a purely supersonic motion. In a subsonic flow it is not possible to attribute the flow pattern in any partial region to the influence of specific data.

5. Hodograph[5]

In Sec. 8.2 the notion of hodograph has been introduced as follows. In a steady two-dimensional flow a point-to-point correspondence was made between a point $P(x,y)$ in the x,y-plane, at which the velocity is $\mathbf{q}$, and a point P' in a q_x,q_y-plane with rectangular coordinates q_x, q_y or polar coordinates q,θ. This mapping of the points P onto the points P' is known as the *hodograph transformation*. In general, a streamline in the physical plane, or plane of flow, is mapped onto a line in the q,θ-plane, the hodograph plane, which we again call *a streamline*. Our first task is to find the differential equations which the stream function and the potential function have to satisfy as functions of q_x and q_y, or of q and θ, rather than of x and y or other coordinates in the flow plane. The passage to the hodograph plane, that is, the introduction of the previously dependent variables q and θ as independent variables, is equivalent to the interchange of the pairs x,y and q,θ as studied in Sec. 10.6.

Let f be a differentiable function of $\mathbf{q}$, and consider its value at the point P of the physical plane and at a neighboring point P_1. Let $\partial/\partial l$ designate differentiation in the direction PP_1; then

$$\frac{\partial f}{\partial l} = \frac{\partial f}{\partial q}\frac{\partial q}{\partial l} + \frac{\partial f}{\partial \theta}\frac{\partial \theta}{\partial l}. \tag{26}$$

We apply this formula four times, taking φ and ψ for f and the directions ds and dn (Fig. 87) for dl. Then recalling (15′) and (19),

$$\frac{\partial\varphi}{\partial s} = q, \qquad \frac{\partial\varphi}{\partial n} = 0, \qquad \frac{\partial\psi}{\partial s} = 0, \qquad \frac{\partial\psi}{\partial n} = \rho q, \tag{27}$$

we find the following two pairs of linear equations for $\partial q/\partial s$, $\partial\theta/\partial s$ and $\partial q/\partial n$, $\partial\theta/\partial n$, as the respective unknowns:

$$\begin{cases} q = \dfrac{\partial\varphi}{\partial q}\dfrac{\partial q}{\partial s} + \dfrac{\partial\varphi}{\partial\theta}\dfrac{\partial\theta}{\partial s}, \\[2ex] 0 = \dfrac{\partial\psi}{\partial q}\dfrac{\partial q}{\partial s} + \dfrac{\partial\psi}{\partial\theta}\dfrac{\partial\theta}{\partial s}, \end{cases} \qquad \begin{cases} 0 = \dfrac{\partial\varphi}{\partial q}\dfrac{\partial q}{\partial n} + \dfrac{\partial\varphi}{\partial\theta}\dfrac{\partial\theta}{\partial n}, \\[2ex] \rho q = \dfrac{\partial\psi}{\partial q}\dfrac{\partial q}{\partial n} + \dfrac{\partial\psi}{\partial\theta}\dfrac{\partial\theta}{\partial n}. \end{cases} \tag{28}$$

Each of these two pairs has the determinant

$$D = \frac{\partial\varphi}{\partial q}\frac{\partial\psi}{\partial\theta} - \frac{\partial\varphi}{\partial\theta}\frac{\partial\psi}{\partial q} \equiv \frac{\partial(\varphi,\psi)}{\partial(q,\theta)}, \tag{29}$$

the Jacobian of φ,ψ with respect to q,θ. Supposing that D is different from zero, we obtain

$$\begin{aligned} \frac{\partial q}{\partial s} &= \frac{1}{D}\, q\, \frac{\partial\psi}{\partial\theta}, & \frac{\partial\theta}{\partial s} &= -\frac{1}{D}\, q\, \frac{\partial\psi}{\partial q}, \\ \frac{\partial q}{\partial n} &= -\frac{1}{D}\, \rho q\, \frac{\partial\varphi}{\partial\theta}, & \frac{\partial\theta}{\partial n} &= \frac{1}{D}\, \rho q\, \frac{\partial\varphi}{\partial q}. \end{aligned} \tag{30}$$

If these values are introduced in the two basic equations (7), the factor $1/D$ drops out, and we obtain the fundamental equations in the hodograph plane

$$\frac{\partial\varphi}{\partial q} = \frac{M^2 - 1}{\rho q}\frac{\partial\psi}{\partial\theta}, \qquad \frac{\partial\varphi}{\partial\theta} = \frac{q}{\rho}\frac{\partial\psi}{\partial q}. \tag{31}$$

When (31) is compared with equations (18) and (18′), which follow immediately from the definitions of potential and stream function, its superiority is obvious. Since ρ and M are given functions of q, and q is an independent variable in (31), *these equations are linear.*

By another differentiation we can easily eliminate either φ or ψ and obtain a single second-order equation for the other:

$$\frac{\partial}{\partial q}\left(\frac{q}{\rho}\frac{\partial\psi}{\partial q}\right) = \frac{M^2 - 1}{\rho q}\frac{\partial^2\psi}{\partial\theta^2}, \tag{32}$$

$$\frac{\partial}{\partial q}\left(\frac{\rho q}{M^2 - 1}\frac{\partial\varphi}{\partial q}\right) = \frac{\rho}{q}\frac{\partial^2\varphi}{\partial\theta^2}. \tag{33}$$

Carrying out one differentiation in (32) we find

$$\frac{\partial^2\psi}{\partial q^2} - \frac{M^2 - 1}{q^2}\frac{\partial^2\psi}{\partial\theta^2} + \frac{\rho}{q}\frac{d}{dq}\left(\frac{q}{\rho}\right)\frac{\partial\psi}{\partial q} = 0.$$

Now using (from the Bernoulli equation)

$$\frac{d}{dq}\left(\frac{1}{\rho}\right) = -\frac{1}{\rho^2}\frac{d\rho}{dp}\frac{dp}{dq} = \frac{1}{\rho}\frac{q}{a^2},$$

we obtain *Chaplygin's equation*

$$\frac{\partial^2\psi}{\partial q^2} + \frac{1 - M^2}{q^2}\frac{\partial^2\psi}{\partial\theta^2} + \frac{1}{q}(1 + M^2)\frac{\partial\psi}{\partial q} = 0. \tag{32'}$$

This equation is true for an arbitrary elastic fluid. If, for a polytropic gas, M^2 is expressed in terms of q by means of (10), the result is

$$\begin{aligned} q^2\left(1 - \frac{\kappa - 1}{2}\frac{q^2}{a_s^2}\right)\frac{\partial^2\psi}{\partial q^2} &+ \left(1 - \frac{\kappa + 1}{2}\frac{q^2}{a_s^2}\right)\frac{\partial^2\psi}{\partial\theta^2} \\ &+ q\left(1 - \frac{\kappa - 3}{2}\frac{q^2}{a_s^2}\right)\frac{\partial\psi}{\partial q} = 0. \end{aligned} \tag{32''}$$

Here the first and third parentheses can be written as $(1 - q^2/q_m^2)$ and $(1 - q^2/q_t^2)$, respectively.

In a similar way we find

$$\frac{\partial^2\varphi}{\partial q^2} + \frac{1 - M^2}{q^2}\frac{\partial^2\varphi}{\partial\theta^2} + \frac{1 - M^2}{\rho q}\frac{d}{dq}\left(\frac{\rho q}{1 - M^2}\right)\frac{\partial\varphi}{\partial q} = 0. \tag{33'}$$

If we carry out the differentiation, a term dM/dq or da/dq will eventually remain. In the case of a polytropic gas we obtain

$$q^2(1 - M^2)\frac{\partial^2\varphi}{\partial q^2} + (1 - M^2)^2\frac{\partial^2\varphi}{\partial\theta^2} + q(1 + \kappa M^4)\frac{\partial\varphi}{\partial q} = 0, \tag{33''}$$

where M^2 may be replaced by means of (10). Equations (32) for ψ are simpler than (33) for φ.

For an incompressible fluid, $\rho = \rho_s = 1$, $a \to \infty$, $M \to 0$, the equations (31) reduce to the Cauchy-Riemann equations in the variables $\log q$ and $-\theta$, and both (32′) and (33′) reduce to Laplace's equation in these variables. In this case a *complex potential* $w(z) = \varphi + i\psi$, where $z = x + iy$ is introduced. If w is differentiated with respect to z the *complex velocity* ζ

$$\zeta = w'(z) = \frac{\partial\varphi}{\partial x} + i\frac{\partial\psi}{\partial x} = q_x - iq_y = qe^{-i\theta}$$

is obtained, and $w(z)$ is an analytic function of ζ, although in general not given by only *one* series in the whole field of flow.

To the equations (31) through (33) we may apply the methods of integration valid in a linear problem, especially the method of combining particular solutions. But it would not be correct to say that by transformation to the hodograph plane our original problem has been "linearized". (This can be done only by taking recourse to approximations.) We have, however, split off one portion of the total problem that can be treated by methods of linear analysis.

As seen in Sec. 1 (for a polytropic fluid) and previously in Sec. 8.2, the hodograph mapping of the flow lies inside a circle of radius q_m, the supersonic part falling in the annular region between this circle and the sonic circle of radius q_t, the subsonic part being inside the sonic circle. The ratio of the two radii is, for a polytropic fluid,

$$\frac{q_m}{q_t} = \sqrt{\frac{\kappa + 1}{\kappa - 1}} = h; \tag{34}$$

$h^2 = 6$, $h = 2.45$ for $\kappa = 1.4$, and $h = 2.437$ for $\kappa = 1.405$. All this is in agreement with the discussion in Sec. 8.3.

6. Characteristics in the hodograph plane[6]

Since the transformed equations (31) are linear, there exist *fixed* characteristics, Γ, in the hodograph plane. By the formulas given in Art. 10, equations of these characteristics can be written down for the linear system (31) or read from (32′) or (33′) by formulas found in Art. 9:

$$\frac{d\theta}{dq} = \pm \frac{\sqrt{M^2 - 1}}{q} = \pm \frac{1}{q \tan \alpha}. \tag{35}$$

Equations (35) also follow from the compatibility relations (24) [see Eq. (10.5) with $a = b = 0$ and the coefficients a_i and b_i ($i = 1, \cdots, 4$) depending only on u and v]. Computing (35) in these two ways, by means of the linear equations (31) and from the equations (24), gives, incidentally, a proof of the fact (studied in full in Sec. 10.7) that the Γ-characteristics of the linear equations in the hodograph are the images of the C-characteristics in the x,y-plane.

For an independent and direct derivation of this fundamental relation we consider the mapping of the characteristics in the physical plane, the Mach lines, onto the hodograph plane. We ask: if the points P, P_1 of the physical plane are mapped onto P', P_1' of the hodograph, what is the direction of $P'P_1'$ when PP_1 makes the angle $\pm\alpha$ with the streamline through P? The changes of q and θ from P to P_1 are

$$dq = \frac{\partial q}{\partial s} ds + \frac{\partial q}{\partial n} dn = \left(\frac{\partial q}{\partial s} \pm \tan \alpha \frac{\partial q}{\partial n}\right) ds, \qquad d\theta = \left(\frac{\partial \theta}{\partial s} \pm \tan \alpha \frac{\partial \theta}{\partial n}\right) ds.$$

Using Eqs. (7) and $M^2 - 1 = \cot^2 \alpha$, we have

$$(36)\qquad dq = \left(\frac{\partial q}{\partial s} \pm q \tan \alpha \frac{\partial \theta}{\partial s}\right) ds, \qquad d\theta = \left(\frac{\partial \theta}{\partial s} \pm \frac{1}{q} \cot \alpha \frac{\partial q}{\partial s}\right) ds,$$

from which Eq. (35) again follows, or

$$(37)\qquad q \frac{d\theta}{dq} = \pm \cot \alpha .$$

The quotient $q\, d\theta/dq$ is the tangent of the angle between the tangent to Γ at P' and the radius vector $O'P'$ (Fig. 92). Thus (37) states: The characteristics Γ^+, Γ^- in the hodograph plane form the angles $\pm\alpha'$, where $\alpha' = 90° - \alpha$, with the radius-vector direction (the direction of $\mathbf{q}$). In other terms, the C^+ in the physical plane through P is perpendicular to the Γ^- in the hodograph through P', and similarly for C^- and Γ^+ [see also Eq. (24″)].

The essential property expressed in (35) or (37) is not, however, this metric relation between the directions of the C- and Γ-characteristics but rather the fact that the slopes of the Γ-characteristics at P' are determined by the coordinates q and θ of P'. Therefore the hodograph characteristics can be computed and drawn once and for all, without further use of the Eqs. (7) or (18). They depend only on the (p,ρ)-relation.

In polytropic flow, $\cot \alpha = \sqrt{M^2 - 1}$ is given by the square root of the second expression of (10), and (37) becomes

$$(38)\qquad d\theta = \pm \left[\frac{(\kappa + 1)\, q^2 - 2a_s^2}{2a_s^2 - (\kappa - 1)\, q^2}\right]^{\frac{1}{2}} \frac{dq}{q} = \pm \left(\frac{h^2 q^2 - q_m^2}{q_m^2 - q^2}\right)^{\frac{1}{2}} \frac{dq}{q}.$$

By integration we obtain

$$(39)\qquad \theta = \pm(\alpha + h\sigma + \text{constant}),$$

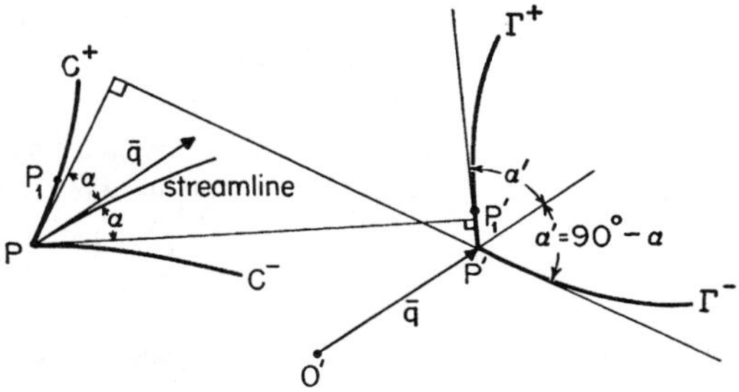

FIG. 92. Mach lines and hodograph characteristics. Physical Plane. Hodograph Plane.

where σ is defined by

$$(39') \qquad \cot \sigma = h \tan \alpha = \frac{h}{(M^2 - 1)^{\frac{1}{2}}} = h \left(\frac{q_m^2 - q^2}{h^2 q^2 - q_m^2} \right)^{\frac{1}{2}},$$

as can be verified by differentiation, or, in one formula,

$$(39'') \quad \pm \theta = \text{arc tan} \frac{1}{\sqrt{M^2 - 1}} + h \text{ arc tan} \frac{\sqrt{M^2 - 1}}{h} + \text{constant}.$$

The expression to the right can be written in various other equivalent forms.* As q increases from $q_t = q_m/h$ to q_m, M goes from 1 to ∞, σ from 0° to 90°, α from 90° to 0°, α' (the angle of the characteristic with the radius vector) from 0° to 90°, and θ increases (upper sign) along a plus

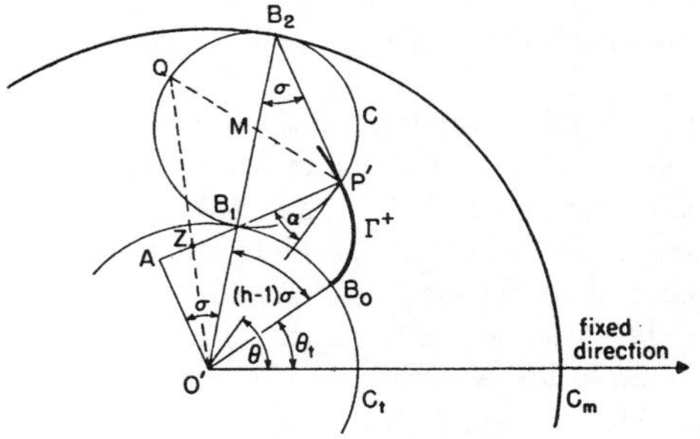

FIG. 93. The hodograph characteristics as epicycloids.

characteristic or decreases (lower sign) along a minus characteristic by $(h - 1)$ times 90°, which is 130.45° for $\kappa = 1.4$, and 129.32° for $\kappa = 1.405$.

The geometrical nature of these characteristics is explained in Fig. 93. Let P' be an arbitrary point in the annular region corresponding to supersonic flow between the sonic circle C_t and the maximum circle C_m. Let C with center M and radius $(q_m - q_t)/2$ be that circle through P' tangent to both C_t and C_m, for which the inclination of $O'M$ is greater than θ, i.e., we consider a Γ^+-characteristic. Let B_1 and B_2 be the points of contact of C with C_t and C_m, respectively. B_2P' and $O'A$ are perpendicular to the straight line $P'B_1A$. Then $B_2B_1/O'B_1 = h - 1$; hence $AP'/AB_1 = h$ and

$$q^2 - (O'A)^2 = h^2[q_t^2 - (O'A)^2], \qquad (O'A)^2 = \frac{\kappa - 1}{2}(q_m^2 - q^2) = a^2.$$

Hence $O'P'A$ is the Mach angle α, and AP' is normal to the Γ^+-characteris-

* This equation, which gives the relation between θ and M or between θ and q along a characteristic line, is of course identical with the compatibility relation (24), which here could be integrated.

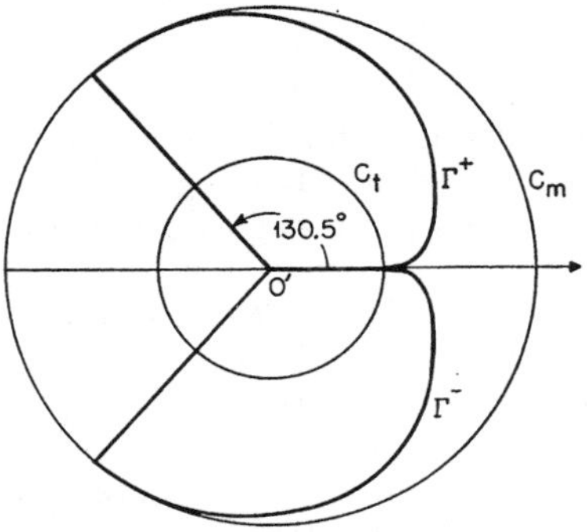

FIG. 94. Two epicycloids with horizontal initial direction.

tic through P'. Also, $\cot AO'B_1 = h \tan O'P'A$, and therefore $AO'B_1$ is the angle σ defined in (39'). From (39), $\theta = \alpha + h\sigma - 90° + \theta_t$, where θ_t denotes the angle of the initial direction $O'B_0$ to the fixed direction. The arc B_1P' equals $2\sigma(q_m - q_t)/2 = (h - 1)q_t\sigma$, while the arc B_0B_1 equals $q_t[(\theta - \theta_t) + (90° - \alpha) - \sigma] = q_t(h\sigma - \sigma)$ and the two arcs are equal.

The characteristic Γ^+ is thus generated by a point P' fixed on the circle C as it rolls without slipping on the exterior of the circle C_t starting at B_0. Such curves are called *epicycloids*. Reversing the sense of the rolling, we generate the Γ^--characteristics. Both sets of characteristics are *congruent epicycloids*, and any curve of each set can be obtained from a fixed one by rotation around O'.

Incidentally, if h is an irrational number, indefinite continuation of the rolling would finally generate epicycloids arbitrarily close to all epicycloids (of both sets). It is also useful to know that the center of curvature corresponding to the point P' is the point Z where the straight line from O' to Q, the second end point of the diameter $P'M$, crosses the normal $P'A$.

In Fig. 94 are shown the two characteristics with horizontal initial direction, $\theta_t = 0$. They form an apparent cusp at the sonic circle where $\alpha = 90°$, i.e. where the angle α' between each characteristic and the horizontal radius vector is zero, and are tangent to the maximum circle where $\alpha' = 90°$, $\alpha = 0°$, $|\theta| = 130.5°$, for $\kappa = 1.4$.

A further helpful relation is the following. If we designate (Fig. 93) by u and v the components $AP' = q \cos \alpha$ and $O'A = q \sin \alpha$ of the velocity vector, normal and parallel to the tangent at P', then it is seen from the figure that

$$O'A = (O'B_1) \cos \sigma = q_t \cos \sigma = v, \qquad AP' = (O'B_2) \sin \sigma = q_m \sin \sigma = u,$$

and we obtain

$$\frac{u^2}{q_m^2} + \frac{v^2}{q_t^2} = 1. \tag{40}$$

In this ellipse E, with semiaxes q_t, q_m, the radius vector $\mathbf{q} = \overline{O'P'}$ makes the angle α with the u-axis (Fig. 95a). By using E the directions of the Mach lines corresponding to a point P' in the hodograph can easily be found (see Fig. 95b).

We have seen that the right side of (39) depends on q only and can therefore be expressed in terms of any of the variables q, M, p, or ρ (except, of course, for a scale factor). The relations between these quantities were developed in Secs. 8.3 and 8.4 and partly rederived in the present article, Sec. 1. Tables for different independent variables, as well as diagrams, are available (see Note 27 to Chapter II). A table serving as an illustration rather than for technical purposes was given at the end of Sec. 8.4. There we tabulated, versus M, the magnitudes p/p_s, ρ/ρ_s, T/T_s, q/q_m, and $\rho_t q_t/\rho q$. In the following small table we show in addition some values of α, σ, and

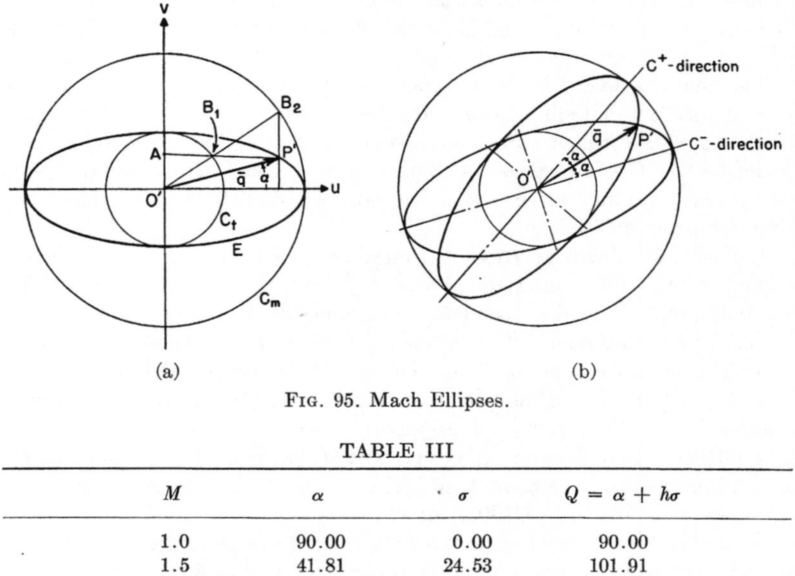

FIG. 95. Mach Ellipses.

TABLE III

M	α	σ	$Q = \alpha + h\sigma$
1.0	90.00	0.00	90.00
1.5	41.81	24.53	101.91
2.0	30.00	35.26	116.38
2.5	23.58	43.08	129.12
3.0	19.47	49.10	139.75
3.5	16.60	53.85	148.53
4.0	14.48	57.67	155.78
∞	0.00	90.00	220.45

$Q = \alpha + h\sigma$ (measured in degrees) computed from the second formula of (23) and from (39′) in terms of M, for $\kappa = 1.4$.

It should be said that the geometrical fact that these hodograph characteristics happen to be known curves, namely, epicycloids, is without much bearing on the problem. The main point is that these fixed characteristics are known and can be plotted and tabulated. Incidentally, we recall that another, actually more interesting, geometrical property of the Γ-lines has been proved in Sec. 8.2, which in our present terminology can be stated thus: The fixed characteristics in the hodograph plane, here epicycloids, are the projections of the asymptotic lines of the pressure hill (see Fig. 38).

7. The nets of characteristics in the physical and hodograph planes

In order to facilitate computations, it is convenient to adopt a coordinate system which will serve for the fixed characteristics in the hodograph plane, as well as for the Mach lines. This can be done in various ways. We have seen that the differential equation (35) of the Γ-curves can be explicitly integrated yielding (39). Guided by Eq. (35), we now introduce the function Q of q, defined by

$$dQ = \frac{dq}{q \tan \alpha} = \frac{dq}{q} \sqrt{M^2 - 1}, \tag{41}$$

or (see Table III) with restriction to a polytropic fluid, by

$$Q = \int_{q_t}^{q} \frac{dq}{q \tan \alpha} + 90° = \alpha + h\sigma. \tag{41′}$$

Then the compatibility relations (24) and (24′) along the C^+, C^- on the one hand, and the equations of the fixed characteristics Γ^+, Γ^- on the other, may be written as follows:

$$\begin{aligned} Q(q) - \theta &= \text{constant}, \qquad \text{along } C^+ \text{ and on } \Gamma^+, \\ Q(q) + \theta &= \text{constant}, \qquad \text{along } C^- \text{ and on } \Gamma^-. \end{aligned} \tag{42}$$

Now denote the Mach lines C^-, C^+ as ξ-lines and η-lines, respectively. If, accordingly, we introduce coordinates ξ,η by

$$Q(q) - \theta = 2\xi, \qquad Q(q) + \theta = 2\eta, \tag{43}$$

then from (42) the Mach lines C^- (and the Γ^--characteristics) are the lines η = constant (the ξ-lines), and the Mach lines C^+ (and the Γ^+-characteristics) are the lines ξ = constant (the η-lines); for each individual line $\xi = \xi_0$ (or $\eta = \eta_0$) we have $2\xi_0 = 90° - \theta_t$ (or $2\eta_0 = 90° + \theta_t$), with θ_t the value of θ, for this line, at the sonic circle where the Γ^+-direction, the Γ^--direction, and the flow direction coincide.[7]

From (43) follow the important relations

$$\theta = \eta - \xi, \qquad Q = \eta + \xi, \tag{44}$$

which, incidentally, show that the sum $(\eta + \xi)$ is constant along the concentric circles q = constant in the hodograph, while the difference $(\eta - \xi)$ remains constant along the radial lines. In the physical plane $\theta = \eta - \xi =$ constant and $Q = \eta + \xi =$ constant designate the *lines of constant inclination* θ, and *the lines of constant speed* q, respectively.

As an immediate consequence of (42), we obtain the following property of the Mach lines (see Fig. 96). Consider two fixed Mach lines of the same family, say the two ξ-lines, $\eta = \eta_0$, $\eta = \eta_1$; let points on these two lines having the same ξ be called "corresponding points", and let ϵ denote the angle between the flow directions at corresponding points. Then from (44)

$$\Delta\theta = \epsilon = \theta(\xi, \eta_1) - \theta(\xi, \eta_0) = (\eta_1 - \xi) - (\eta_0 - \xi) = \eta_1 - \eta_0 . \tag{45}$$

Thus this angle depends only upon the two fixed C^--lines, and not on the variable ξ. Also, interchanging the role of the two families, we obtain

$$\Delta'\theta = \epsilon' = \theta(\xi_1, \eta) - \theta(\xi_0, \eta) = \xi_0 - \xi_1 . \tag{45'}$$

Similarly for the angle $Q(q)$, which is a function of q, M, p, or ρ:

$$\begin{aligned} \Delta Q &= Q(\xi, \eta_1) - Q(\xi, \eta_0) = \eta_1 - \eta_0 = -\epsilon, \\ \Delta' Q &= Q(\xi_1, \eta) - Q(\xi_0, \eta) = \xi_1 - \xi_0 = -\epsilon'. \end{aligned} \tag{45''}$$

Thus, it is seen that *the angle between the flow directions at corresponding points of two fixed characteristics remains the same all along these characteristics*; and an analogous constancy relation holds for the differences in Q.[8]

In Sec. 4 we indicated how the net of Mach lines is determined in a certain domain by specific types of boundary conditions. To construct this net we used the compatibility relations (24) in the form of finite difference

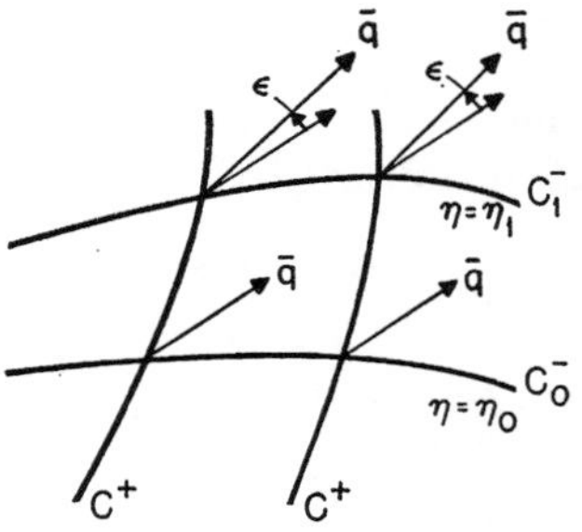

FIG. 96. Constancy of $\Delta\theta$ along characteristics.

equations (25) and the direction conditions (23). This most direct procedure, essentially due to J. Massau, becomes a practical approximation method if the compatibility relations are used in the form (43) together with (23) and with a tabulation of θ (or Q) as a function of q along a Mach line (see end of preceding section). A first approximation thus found can be improved by iteration. In the following we describe a few other useful procedures, both numerical and graphical, for the determination of the Mach lines, if the net of Γ-characteristics is considered known.

To fix ideas, consider the characteristic boundary-value problem where we know that two arcs OA, OB in the x,y-plane are parts of a ξ-line and an η-line, respectively. The case where the whole or part of one of these curves is a straight line will be considered in Art. 18 where we shall study so-called simple waves; here they are assumed to be curved. Consider a subdivision $\xi_0, \xi_1, \cdots, \xi_m, \cdots$ on OA and $\eta_0, \eta_1, \cdots, \eta_n, \cdots$ on OB and a corresponding lattice where a general nodal point has the coordinates (ξ_m, η_n) or briefly (m,n). Then from Eqs. (44), $\theta_{mn} = \eta_n - \xi_m$, we have

$$\theta_{mn} - \theta_{m0} - \theta_{0n} + \theta_{00} = 0. \tag{46}$$

Similarly

$$Q_{mn} - Q_{m0} - Q_{0n} + Q_{00} = 0. \tag{46'}$$

Hence knowing θ and Q at the subdivision points on OA, OB, we can determine them at all nodal points.

We still need, however, the coordinates x_{mn}, y_{mn} in the physical plane corresponding to (ξ_m, η_n) to which $\mathbf{q}_{mn}$ belongs. To find these coordinates, i.e., our original independent variables, we use a step-by-step procedure. The simplest is the recurrence procedure

$$\begin{aligned} y_{mn} - y_{m-1,n} &= \tan \frac{\overset{+}{\phi}_{m-1,n} + \overset{+}{\phi}_{mn}}{2} (x_{mn} - x_{m-1,n}), \\ y_{mn} - y_{m,n-1} &= \tan \frac{\overset{-}{\phi}_{m,n-1} + \overset{-}{\phi}_{mn}}{2} (x_{mn} - x_{m,n-1}), \end{aligned} \tag{47}$$

where the $x_{m,0}$, $y_{m,0}$, $x_{0,n}$, $y_{0,n}$ are the known coordinates of the subdivision points along OA and OB, respectively; the $\overset{+}{\phi}_{mn}$, $\overset{-}{\phi}_{mn}$ are known at all nodal points since θ and q are known there. Thus (47) are two simultaneous linear equations which determine x_{mn}, y_{mn} once $x_{m,n-1}$, $y_{m,n-1}$ and $x_{m-1,n}$, $y_{m-1,n}$ have been found. According to the quasi-linear character of our problem the original unknowns θ, q are found first through the linear compatibility relations, or in other words through our knowledge of the net in the hodograph, while for the independent variables x and y we need a stepwise procedure based on the direction conditions. The indicated procedure, which is, of course, essentially a rearrangement of that in Sec. 4, can likewise be made more accurate by means of iterations.

The best known *graphical procedure* for constructing the Mach net is based on the property that the C^+, C^- through P are perpendicular to the Γ^-, Γ^+ through P' in the hodograph. Consider a C^- in the physical plane with points P_1, P_2, P_3, $\cdots$ on it and known velocity vectors $\mathbf{q}_1$, $\mathbf{q}_2$, $\mathbf{q}_3$, $\cdots$. If we plot these in the hodograph, their endpoints P_1', P_2', P_3', $\cdots$ lie on the corresponding Γ^-. Then the tangent at P_i' to this Γ^- is perpendicular to the characteristic C_i^+ through P_i (Fig. 97). This is used systematically in a procedure due to A. Busemann (see Fig. 98). Denote lattice points in the hodograph by P_{mn}'. Then, starting from the given boundary values, the Mach net of points P_{mn} is constructed successively according to the rule

$$P_{mn}P_{m,n+1} \perp P_{mn}'P_{m-1,n}', \qquad P_{mn}P_{m+1,n} \perp P_{mn}'P_{m,n-1}'.$$

In this way the Γ^--curves correspond to the C^-, and the Γ^+ to the C^+; dotted line segments in the hodograph which have Γ^--direction are perpendicular to dotted segments in the flow plane which have C^+-direction,

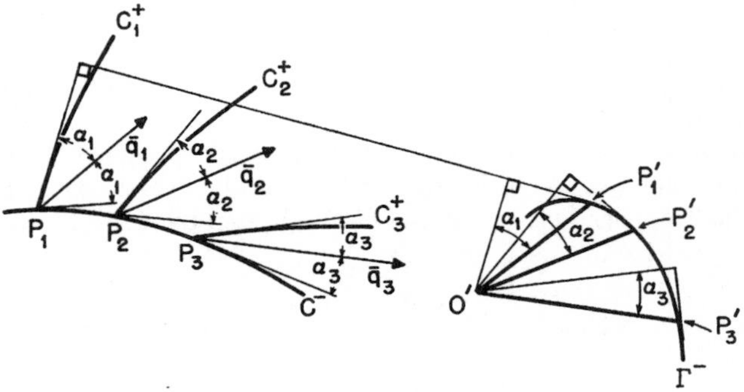

FIG. 97. Graphical construction of Mach lines.

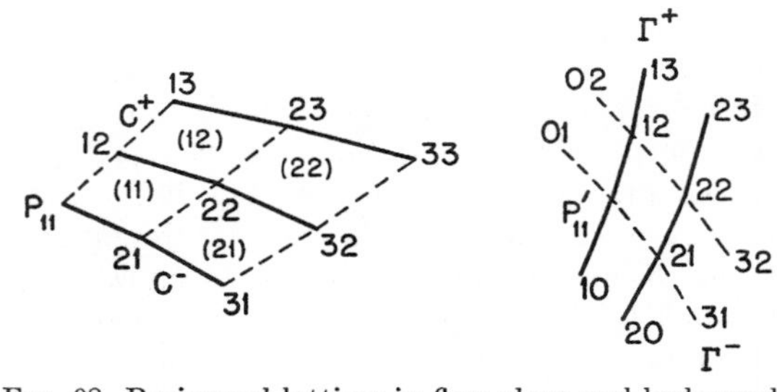

FIG. 98. Reciprocal lattices in flow plane and hodograph.

and similarly for Γ^+ and C^-. If in the physical plane each mesh is denoted by the two numbers equal to the smallest pair of subscripts of the four corners P_{ik}, it is seen that *the mesh* (i,k) *corresponds to the point* P'_{ik} in the sense that the four sides around the mesh (i,k) are respectively normal to the four sides through P'_{ik} (reciprocal lattices).

Article 17

Further Discussion of the Hodograph Method

1. Differential equations for the Legendre transforms

In the preceding article we started with the two nonlinear partial differential equations of first order, (16.1) and (16.2), which express continuity and irrotationality. Transforming these to natural or intrinsic coordinates, we obtained the basic nonlinear equations (16.7) for q and θ, which together with Eq. (16.10) defined the problem completely. In Sec. 16.2 the potential φ and stream function ψ were introduced by means of (16.15), (16.17), and (16.18). The equations (16.18) with φ and ψ as dependent, x and y as independent variables, or their natural form (16.18′) again constitute a *pair of nonlinear partial differential equations of first order*. These, although simple in appearance, are quite complicated since ρ has to be expressed in terms of q, and q in terms of φ. Next we derived a *nonlinear second-order* equation for φ alone and one for ψ alone, Eqs. (16.14) and (16.21), and the corresponding intrinsic forms (16.16′) and (16.20′). In Sec. 16.5 the hodograph transformation was introduced and it led to various *linear* equations. Using now q and θ (the previous dependent variables) as independent variables, we obtained the basic *linear equations of first order* (16.31), and also the *equations of second order* (16.32) or (16.32′), (16.33) or (16.33′) for ψ alone or φ alone, respectively.

In this and the following section we shall give still other useful equations of the problem. It is left to the reader to consider and determine various analogies with the equations of Chapter III.

One pair of nonlinear equations in the physical plane for the velocity components $q_x = u$, $q_y = v$ consists of Eqs. (16.2) and (16.14):

$$(1) \qquad (a^2 - u^2)\frac{\partial u}{\partial x} - uv\left(\frac{\partial u}{\partial y} + \frac{\partial v}{\partial x}\right) + (a^2 - v^2)\frac{\partial v}{\partial y} = 0,$$

$$(2) \qquad \frac{\partial v}{\partial x} - \frac{\partial u}{\partial y} = 0.$$

These equations can be linearized by the direct interchange of variables introduced in Sec. 10.6 and used in Chapter III. If the Jacobian

$$j = \frac{\partial u}{\partial x}\frac{\partial v}{\partial y} - \frac{\partial u}{\partial y}\frac{\partial v}{\partial x} = \frac{\partial(u,v)}{\partial(x,y)}$$

is different from zero we obtain the *linear* system

$$\text{(3)} \qquad \begin{aligned} (a^2 - u^2)\frac{\partial y}{\partial v} + uv\left(\frac{\partial x}{\partial v} + \frac{\partial y}{\partial u}\right) + (a^2 - v^2)\frac{\partial x}{\partial u} &= 0, \\ \frac{\partial x}{\partial v} - \frac{\partial y}{\partial u} &= 0. \end{aligned}$$

To satisfy the second of these equations we introduce a function $\Phi(u,v)$ such that

$$\text{(4)} \qquad \frac{\partial \Phi}{\partial u} = x, \qquad \frac{\partial \Phi}{\partial v} = y,$$

then the first Eq. (3) yields

$$\text{(5)} \qquad (a^2 - u^2)\frac{\partial^2 \Phi}{\partial v^2} + 2uv\frac{\partial^2 \Phi}{\partial u\,\partial v} + (a^2 - v^2)\frac{\partial^2 \Phi}{\partial u^2} = 0.$$

The Φ introduced in this way is the *Legendre transform*[9] *of the potential* φ. From Eqs. (4) and (16.15) we have

$$\text{(4')} \qquad d\Phi = x\,du + y\,dv, \qquad d\varphi = u\,dx + v\,dy,$$

and by integration we obtain

$$\begin{aligned} \Phi = \int(x\,du + y\,dv) &= \int d(xu + yv) - \int(u\,dx + v\,dy) \\ &= xu + yv - \varphi. \end{aligned}$$

Hence, between φ considered as a function of x and y and its transform Φ considered as a function of u and v, the relations

$$\text{(6)} \qquad \begin{aligned} \Phi &= xu + yv - \varphi = x\frac{\partial \varphi}{\partial x} + y\frac{\partial \varphi}{\partial y} - \varphi \\ \varphi &= ux + vy - \Phi = u\frac{\partial \Phi}{\partial u} + v\frac{\partial \Phi}{\partial v} - \Phi \end{aligned}$$

hold.

Using q, θ as independent variables instead of u, v and introducing the abbreviations

$$\text{(7)} \qquad x\cos\theta + y\sin\theta = X, \qquad y\cos\theta - x\sin\theta = Y$$

for the components of the radius vector $\mathbf{r}$ in the direction of the flow and normal to it, we obtain from (6): $\Phi = q\,(x \cos\theta + y \sin\theta) - \varphi = qX - \varphi$, so that

$$\frac{\partial \Phi}{\partial q} = X, \qquad \frac{\partial \Phi}{\partial \theta} = qY, \tag{4''}$$

and

$$\varphi = q\frac{\partial \Phi}{\partial q} - \Phi, \qquad \Phi = X\frac{\partial \varphi}{\partial s} - \varphi. \tag{6'}$$

The following relations, derived from Eqs. (16.17) and (4), serve to determine the stream function ψ, if $\Phi(u, v)$ is known:

$$\frac{\partial \psi}{\partial u} = \rho\left(-v\frac{\partial^2 \Phi}{\partial u^2} + u\frac{\partial^2 \Phi}{\partial u\,\partial v}\right), \qquad \frac{\partial \psi}{\partial v} = \rho\left(-v\frac{\partial^2 \Phi}{\partial u\,\partial v} + u\frac{\partial^2 \Phi}{\partial v^2}\right), \tag{8}$$

or

$$\frac{\partial \psi}{\partial q} = \rho\left(\frac{\partial^2 \Phi}{\partial q\,\partial \theta} - \frac{1}{q}\frac{\partial \Phi}{\partial \theta}\right), \qquad \frac{\partial \psi}{\partial \theta} = \rho\left(q\frac{\partial \Phi}{\partial q} + \frac{\partial^2 \Phi}{\partial \theta^2}\right). \tag{8'}$$

Likewise, a *Legendre transform* Ψ *of the stream function* ψ may be defined. If we interchange the variables x, y for the variables ρu, ρv, the continuity equation becomes

$$\frac{\partial x}{\partial(\rho u)} + \frac{\partial y}{\partial(\rho v)} = 0.$$

This equation may be satisfied by introducing a function Ψ of ρu and ρv such that

$$\frac{\partial \Psi}{\partial(\rho u)} = y, \qquad \frac{\partial \Psi}{\partial(\rho v)} = -x; \qquad \text{also,} \qquad \frac{\partial \Psi}{\partial(\rho q)} = Y. \tag{9}$$

By integration of (9), the following relations, analogous to (6) and (6′), obtain:

$$\begin{aligned} \Psi &= \rho u y - \rho v x - \psi = x\frac{\partial \psi}{\partial x} + y\frac{\partial \psi}{\partial y} - \psi = Y\frac{\partial \psi}{\partial n} - \psi, \\ \psi &= \rho u y - \rho v x - \Psi = \rho u\frac{\partial \Psi}{\partial(\rho u)} + \rho v\frac{\partial \Psi}{\partial(\rho v)} - \Psi = \rho q\frac{\partial \Psi}{\partial(\rho q)} - \Psi. \end{aligned} \tag{10}$$

We now derive *a pair of first-order equations for* Φ *and* Ψ *analogous to Eqs.* (16.31). Using the last of Eqs. (9) and the relation $d\rho/dq = -\rho q/a^2$, we find immediately

$$\frac{\partial\Psi}{\partial q} = \frac{\partial\Psi}{\partial(\rho q)}\frac{d(\rho q)}{dq} = Y\left[q\frac{d\rho}{dq} + \rho\right] = Y\rho\left(1 - \frac{q^2}{a^2}\right),$$

$$\frac{\partial\Psi}{\partial\theta} = \frac{\partial\Psi}{\partial(\rho u)}\frac{\partial(\rho u)}{\partial\theta} + \frac{\partial\Psi}{\partial(\rho v)}\frac{\partial(\rho v)}{\partial\theta} = -\rho q X.$$

Hence,

$$(9') \qquad \frac{\partial\Psi}{\partial q} = \rho Y(1 - M^2), \qquad \frac{\partial\Psi}{\partial\theta} = -\rho q X,$$

and substituting from (4″) we obtain the desired equations:

$$(11) \qquad \frac{\partial\Psi}{\partial q} = \frac{\rho}{q}(1 - M^2)\frac{\partial\Phi}{\partial\theta}, \qquad \frac{\partial\Psi}{\partial\theta} = -\rho q\frac{\partial\Phi}{\partial q}.$$

Elimination of Φ or Ψ in the usual way, yields the linear second-order equations:

$$(12) \qquad \frac{q^2}{1 - M^2}\frac{\partial^2\Phi}{\partial q^2} + \frac{\partial^2\Phi}{\partial\theta^2} + q\frac{\partial\Phi}{\partial q} = 0,$$

$$(13) \qquad \frac{q^2}{1 - M^2}\frac{\partial^2\Psi}{\partial q^2} + \frac{\partial^2\Psi}{\partial\theta^2} + \rho q\frac{d}{dq}\left[\frac{q}{\rho(1 - M^2)}\right]\frac{\partial\Psi}{\partial q} = 0.$$

Comparing the second-order equations for φ, ψ, Φ, and Ψ, we note that those for ψ and Φ, namely, Eqs. (16.32′) and (12), are simpler than the other two.[10]

2. Other linear differential equations

(a) *Equations for X and Z = qY.* If $\Phi(q,\theta)$ is a solution of (12), then, since all coefficients in (12) depend only on q, any derivative $\partial\Phi/\partial\theta$, $\partial^2\Phi/\partial\theta^2$, $\cdots$ will likewise satisfy (12). In particular, since $\partial\Phi/\partial\theta = qY$, the function

$$(14) \qquad Z = qY = q(y\cos\theta - x\sin\theta)$$

satisfies the same equation as Φ:

$$(15) \qquad \frac{q^2}{1 - M^2}\frac{\partial^2 Z}{\partial q^2} + \frac{\partial^2 Z}{\partial\theta^2} + q\frac{\partial Z}{\partial q} = 0.$$

From Eqs. (4″) and (15) we obtain for X and Z the linear system

$$(16) \qquad \frac{\partial Z}{\partial q} = \frac{\partial X}{\partial\theta}, \qquad \frac{\partial Z}{\partial\theta} = \frac{q^2}{M^2 - 1}\frac{\partial X}{\partial q} - qX.$$

Elimination of X leads back to (15), and elimination of Z to an equation for X alone. The equation (15) is as simple as Eqs. (12) or (16.32′). Equations (16) compare with Eqs. (11) and (16.31).

(b) *Equations with independent variables Q, θ or ξ, η.* If we consider the various linear equations of second order (16.32′), (16.33′), (12), (13), and (15), and for the moment denote by F any of the dependent variables, we see that, with respect to the second-order terms, all these equations are of the form

$$\frac{\partial^2 F}{\partial q^2} - \frac{M^2 - 1}{q^2}\frac{\partial^2 F}{\partial \theta^2} \cdots = 0.$$

This is to be expected since these terms determine the fixed characteristics in the q,θ-plane (real and distinct only where $M > 1$), which are the same no matter which function of $\mathbf{q}$ we use to describe the flow. From each of them we find again Eq. (16.35). The same result follows, of course, from each of the pairs of equations (11), (16), or (16.31), by means of the technique of Art. 10.

The various differential equations become simpler, if instead of q, the variable Q, defined in Eq. (16.41) for supersonic flow, is used; more generally, in the subsonic and supersonic cases, respectively, we define the new variables λ and Q by

$$\text{(17)} \qquad \frac{dQ}{dq} = \frac{\sqrt{M^2 - 1}}{q}, \qquad \frac{d\lambda}{dq} = \frac{\sqrt{1 - M^2}}{q}.$$

The second-order terms in the respective equations of second order are then simply

$$\frac{\partial^2 F}{\partial Q^2} - \frac{\partial^2 F}{\partial \theta^2} \qquad \text{and} \qquad \frac{\partial^2 F}{\partial \lambda^2} + \frac{\partial^2 F}{\partial \theta^2},$$

while the first-order equations (16.31) take the more symmetric form:

$$\text{(18)} \quad \frac{\partial \varphi}{\partial Q} = \frac{\sqrt{M^2 - 1}}{\rho}\frac{\partial \psi}{\partial \theta}, \quad \frac{\partial \varphi}{\partial \theta} = \frac{\sqrt{M^2 - 1}}{\rho}\frac{\partial \psi}{\partial Q}, \qquad M > 1,$$

$$\text{(19)} \quad \frac{\partial \varphi}{\partial \lambda} = -\frac{\sqrt{1 - M^2}}{\rho}\frac{\partial \psi}{\partial \theta}, \quad \frac{\partial \varphi}{\partial \theta} = \frac{\sqrt{1 - M^2}}{\rho}\frac{\partial \psi}{\partial \lambda}, \qquad M < 1.$$

From Eqs. (18) and (19) follow equations of second order for φ and for ψ, as, e.g., Eq. (21.6) for ψ derived from (19).

In the supersonic case we may also use the characteristic variables ξ,η given by Eqs. (16.43), (16.44). The equations (18) then take the form

$$\text{(20)} \qquad \frac{\partial \varphi}{\partial \xi} = -\frac{\sqrt{M^2 - 1}}{\rho}\frac{\partial \psi}{\partial \xi}, \qquad \frac{\partial \varphi}{\partial \eta} = \frac{\sqrt{M^2 - 1}}{\rho}\frac{\partial \psi}{\partial \eta}.$$

These equations have the nature of compatibility relations. The variables Q, λ, ξ, η will be much used in Arts. 19 and 21. In the second-order equations derived from Eq. (20) the second-order terms reduce to the mixed derivative with respect to ξ and η, and if we deal similarly with the Legendre transforms we obtain, for example,

$$\frac{\partial^2\Phi}{\partial\xi\,\partial\eta} = \frac{q + q''}{q'}\left(\frac{\partial\Phi}{\partial\xi} + \frac{\partial\Phi}{\partial\eta}\right), \tag{21}$$

where $q' = dq/dQ = q \tan \alpha$ and q, q', q'' are to be expressed in terms of ξ, η.[11] From a solution $\Phi(\xi,\eta)$ of (21) we obtain with the notation (7)

$$X = \frac{1}{2q'}\left(\frac{\partial\Phi}{\partial\eta} + \frac{\partial\Phi}{\partial\xi}\right), \qquad Y = \frac{1}{2q}\left(\frac{\partial\Phi}{\partial\eta} - \frac{\partial\Phi}{\partial\xi}\right), \tag{21'}$$

from which x and y follow in terms of ξ, η. An equation of the same simple form as (21) holds for the stream function ψ.*

For some purposes it is advantageous (see Art. 20) to use, instead of the variables q, θ, the variables q^2, θ. This will be considered when needed.

(c) *Equations with independent variables σ and θ.* We mention one more important transformation of the basic equations due also to Chaplygin. Let us introduce a new variable

$$\sigma = \int_q^{q_t} \frac{\rho\, dq}{q}, \qquad \frac{d\sigma}{dq} = -\frac{\rho}{q}. \tag{22}$$

It is immediately seen that Eqs. (16.31) become

$$\frac{\partial\varphi}{\partial\sigma} = K\frac{\partial\psi}{\partial\theta}, \qquad \frac{\partial\varphi}{\partial\theta} = -\frac{\partial\psi}{\partial\sigma}, \tag{23}$$

and the second-order equation (16.32) takes the form

$$\frac{\partial^2\psi}{\partial\sigma^2} + K\frac{\partial^2\psi}{\partial\theta^2} = 0, \tag{24}$$

where

$$K = \frac{1 - M^2}{\rho^2} \tag{24'}$$

is a complicated function of σ. (See Fig. 99.) It is seen from (22) that σ decreases with increasing q and that for $q \to 0$, $\sigma \to \infty$ as $-\log q$; also $\sigma = 0$ for $q = q_t$ and σ is negative for supersonic q-values, positive for subsonic q-values. The function K, depending on σ, tends towards unity as $\sigma \to +\infty$, $q \to 0$; it equals zero for $\sigma = 0$, $q = q_t$, $M = 1$; and tends to $-\infty$ as $q \to q_m$ and as σ tends to its negative minimum value. (The second derivative of σ as function of q, that is, the first derivative of $-\rho/q$ equals $(\rho/q)(1 + M^2)$, and is therefore always positive.)

* Other linear equations with ξ, η as independent variables are Eqs. (19.7).

This transformation is used in two different contexts. On the one hand, an expansion of $\sqrt{K} = \sqrt{1 - M^2}/\rho$ in powers of M shows that, for a polytropic fluid with $\kappa = 1.4$, this expression differs from unity only by terms of the order of M^4, i.e., $\sqrt{K} = 1 - 0.3\, M^4 \cdots$.* This suggests *the approximation* $K = 1$, invented by Chaplygin and later elaborated in the v. Kármán-Tsien method (see Secs. 5 and 6). On the other hand, the simple form of (24) with $K \gtreqless 0$ for subsonic, sonic, supersonic flow is a convenient starting point for the study of *transonic flow*, the flow in the neighbor-

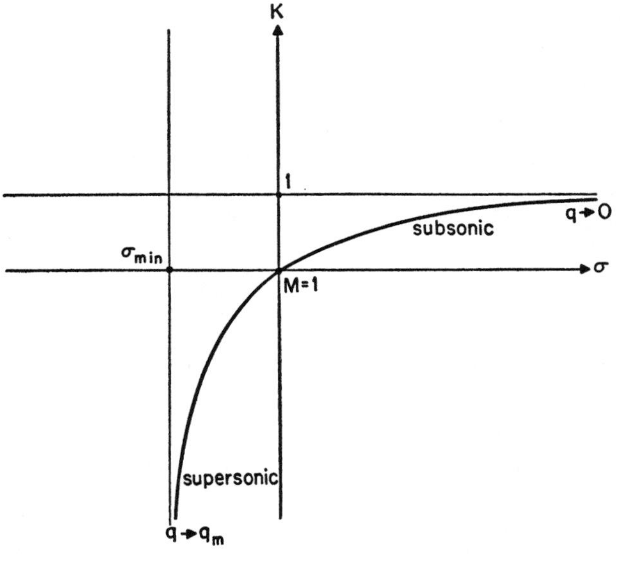

FIG. 99. $K = (1 - M^2)/\rho^2$ as function of $\sigma = \int_q^{q_t} \rho\, dq/q$.

hood of $M = 1$ (see Art. 25).

3. Transition from the hodograph to the physical plane

In Sec. 16.5 we remarked that it would not be correct to say that by the hodograph transformation the original nonlinear problem has been linearized. This can be done only by having recourse to approximations. We have merely split off one portion of the total problem which can be treated by methods of linear analysis. Once a solution in the hodograph plane has been found we still have to transfer it back to the physical plane.

* We obtain $\sqrt{K} = (1/\rho_s)(1 - 0.3\, M^4 \cdots)$, but, as before, $\rho_s = 1$.

Suppose that we know a solution $\psi(q,\theta)$ of Eq. (16.32′) in some region in the hodograph plane.[12] Then the functions $\partial\psi/\partial q$, $\partial\psi/\partial\theta$ can be computed, and from (16.31) the functions $\partial\varphi/\partial q$, $\partial\varphi/\partial\theta$. Next, we have

$$q_x\,dx + q_y\,dy = d\varphi, \qquad -\rho q_y\,dx + \rho q_x\,dy = d\psi$$

and from these, if $\rho q^2 \neq 0$, dx and dy follow:

$$(25)\quad dx = \frac{1}{q}\left[\cos\theta\,d\varphi - \frac{1}{\rho}\sin\theta\,d\psi\right], \qquad dy = \frac{1}{q}\left[\sin\theta\,d\varphi + \frac{1}{\rho}\cos\theta\,d\psi\right],$$

and

$$(25')\quad \begin{aligned} \frac{\partial x}{\partial q} &= \frac{\cos\theta}{q}\frac{\partial\varphi}{\partial q} - \frac{1}{\rho}\frac{\sin\theta}{q}\frac{\partial\psi}{\partial q}, & \frac{\partial x}{\partial\theta} &= \frac{\cos\theta}{q}\frac{\partial\varphi}{\partial\theta} - \frac{1}{\rho}\frac{\sin\theta}{q}\frac{\partial\psi}{\partial\theta},\\ \frac{\partial y}{\partial q} &= \frac{\sin\theta}{q}\frac{\partial\varphi}{\partial q} + \frac{1}{\rho}\frac{\cos\theta}{q}\frac{\partial\psi}{\partial q}, & \frac{\partial y}{\partial\theta} &= \frac{\sin\theta}{q}\frac{\partial\varphi}{\partial\theta} + \frac{1}{\rho}\frac{\cos\theta}{q}\frac{\partial\psi}{\partial\theta}; \end{aligned}$$

the coordinates x,y in the physical plane may then be found as functions of q,θ by quadratures. If, in a formal simplification, $z = x + iy$ is used we obtain

$$(25'')\quad dz = \frac{e^{i\theta}}{q}\left(d\varphi + i\frac{d\psi}{\rho}\right),$$

and if the derivatives of φ are expressed in terms of those of ψ by (16.31), the result may be given in condensed form by

$$z = \int \frac{e^{i\theta}}{\rho}\left[\left(\frac{\partial\psi}{\partial q} + \frac{i}{q}\frac{\partial\psi}{\partial\theta}\right)d\theta + \left(\frac{M^2-1}{q^2}\frac{\partial\psi}{\partial\theta} + \frac{i}{q}\frac{\partial\psi}{\partial q}\right)dq\right].$$

The streamlines in the x,y-plane are the images of the curves $\psi(q,\theta) =$ constant $= k$ in the hodograph. The computation of the streamlines may be arranged so that the necessary integration is simplified, since along a streamline Eqs. (25) take the form

$$(26)\quad dx = \frac{\cos\theta}{q}\,d\varphi, \qquad dy = \frac{\sin\theta}{q}\,d\varphi, \qquad d\psi = 0.$$

Also $dq = -[(\partial\psi/\partial\theta)/(\partial\psi/\partial q)]\,d\theta$ along a streamline, and therefore with $D = \partial(\varphi,\psi)/\partial(q,\theta)$,

$$d\varphi = -D\,d\theta \Big/ \frac{\partial\psi}{\partial q} = A(q,\theta)\,d\theta = B(\theta,k)\,d\theta,$$

since $q = q(\theta,k)$ along the streamline $\psi(q,\theta) = k$. If $B(\theta,k) \neq 0$, the functions $x(\theta,k)$, $y(\theta,k)$, which define the streamlines, are then obtained by integrating

$$dx = \frac{\cos\theta}{q(\theta,k)} B(\theta,k)\, d\theta, \qquad dy = \frac{\sin\theta}{q(\theta,k)} B(\theta,k)\, d\theta. \tag{26'}$$

The whole computation involves considerable work, and is based on the assumption that $\psi(q,\theta) = k$ can be solved for q.

If we know a solution $\Phi(u,v)$ or $\Phi(q,\theta)$ of Eqs. (5) or (12), the determination of $x = x(q,\theta)$, $y = y(q,\theta)$ by Eqs. (4) in the first case, and by Eqs. (4″) and (7) in the second case, is easier. However, in order to find the stream function $\psi(q,\theta)$, a quadrature is needed. We may, e.g., find $\partial\Psi/\partial q$, $\partial\Psi/\partial\theta$ from Eqs. (11), then Ψ by a quadrature, and ψ from the second Eq. (10); or we may use Eqs. (8) or (8′) to find derivatives ot ψ, and then determine ψ itself by a quadrature.

A situation where Eqs. (15) and (16) are useful arises, e.g., in the case of a Cauchy problem. If along a curve $\mathcal{K}$ (noncharacteristic) we know q,θ, then we know the image $\mathcal{K}'$ of $\mathcal{K}$ in the hodograph and x,y along $\mathcal{K}'$; therefore we know X, Y, and $qY = Z$, i.e., we have for the pair of linear equations (16) the Cauchy data Z, X along $\mathcal{K}'$ and this determines a solution within a characteristic quadrangle. Then, from X and Z in terms of q and θ, we know $x(q,\theta)$ and $y(q,\theta)$. A similar approach applies to the characteristic boundary-value problem.

In this connection let us recall that the transition from the hodograph to the physical plane was already considered in Sec. 16.4 and 16.7, when we attempted in various ways to derive the Mach lines from the Γ-lines, corresponding to certain given initial data. Since the transition there, however, was based on the use of characteristics, the considerations were limited to the supersonic case. Also—in contrast to the present approach—the methods of those sections were numerical or graphical methods.

Let us now comment on the final step, the determination of q,θ in terms of x,y. We return to our starting point where a particular solution $\psi(q,\theta)$ of Eq. (16.32′) was known in some region R of the q,θ-plane, where $q \leqq q_m$. Our objective is to find, if possible, the velocity $\mathbf{q}$ over the x,y-region corresponding to R. We saw that, if $\rho q^2 \neq 0$, the four derivatives $\partial x/\partial q$, $\cdots$, $\partial y/\partial\theta$ are determined by (25′), in terms of $\partial\varphi/\partial q$, $\cdots$, $\partial\psi/\partial\theta$, and integration gives $x = x(q,\theta)$, $y = y(q,\theta)$.

Now, these last equations must be inverted. The flow in the physical plane, in the region corresponding to R, is determined if q and θ are given as single-valued functions of x,y, i.e. if

$$I = \frac{\partial(x,y)}{\partial(q,\theta)} = \frac{\partial(x,y)}{\partial(\varphi,\psi)}\frac{\partial(\varphi,\psi)}{\partial(q,\theta)} = \frac{1}{\rho q^2} D \neq 0,$$

where, from (16.31),

$$(27) \qquad D = \frac{\partial(\varphi,\psi)}{\partial(q,\theta)} = -\frac{1}{\rho q}\left[q^2\left(\frac{\partial\psi}{\partial q}\right)^2 + (1 - M^2)\left(\frac{\partial\psi}{\partial\theta}\right)^2\right].$$

Since ρ and M are given functions of q, the vanishing of D depends only on the given solution $\psi(q,\theta)$ in R. These circumstances will be discussed in detail in Art. 19. One result, however, is obvious, namely, that in subsonic flow, I can vanish only if all four derivatives $\partial\psi/\partial q, \cdots, \partial\varphi/\partial\theta$ vanish; then, also $\partial x/\partial q, \cdots, \partial y/\partial\theta$ vanish. It can be seen that this can happen only at isolated points. (For a more complete discussion see Sec. 19.2.)

On the other hand, if $M > 1$ the determinant D, considered as a function of q and θ, can vanish as the difference of two positive terms. The locus $D(q,\theta) = 0$ defines in general a curve in the hodograph whose image in the flow plane we call a limit line. We found such lines in certain instances in Chapters II and III. It was seen that in the neighborhood of such a limiting line there are two flows which meet there; by that we mean that both flows have there the same q and θ. These circumstances, and also the situation for $M = 1$, will become clear when we study more examples and the general theory. At any rate, as long as there is no singularity in the mapping of the flow plane onto the hodograph and vice versa there is equivalence between results in these planes, but whenever the mapping is singular, we need a study of the mapping in the neighborhood of the singularity.

In actual problems one is faced with additional difficulties. So far we made the assumption that a particular solution, say $\psi(q,\theta)$, of the hodograph equation had been found, and we found a difficulty in the possible occurrence of limit lines. Important practical problems, however, such as flow in a duct or around a profile, *are boundary-value problems*, and here new circumstances arise. Only a few indications are in order at this time.

First, in such problems, the boundary conditions are given in the physical plane, and in general it is not possible to derive from them boundary conditions in the hodograph, which would determine the corresponding linear hodograph problem. We may think of flow around a profile, with the typical boundary condition that the contour in the physical plane forms a streamline i.e. that $\psi = 0$ along the given contour. Since we cannot derive from these data the velocity distribution $\mathbf{q}$ along the given contour, we do not know the image of the contour in the hodograph and cannot set up the corresponding linear hodograph problem. Moreover, and this is perhaps an even more fundamental difficulty, in many cases we are not certain whether we are in possession of correct boundary conditions even in the physical plane, correct in the sense that a uniquely determined solution of the problem exists which satisfies the differential equations and the boundary conditions, and depends, in an appropriate sense, continuously upon the boundary data. Much of this difficulty follows from the nonlinearity: in fact the elliptic or hyperbolic character of our nonlinear equations *depends upon the solution* under consideration, which, in turn, should be singled out by means of the boundary conditions; and it is well-known that in the two cases of elliptic and hyperbolic problems quite different sets of boundary conditions must be given in order to determine a solution. (See Art. 25).

Consider, e.g., a duct with supersonic flow at the entrance section (see end of Sec. 16.4). In this case the flow will be uniquely determined as long as it remains supersonic. If, however, the flow does not remain supersonic, then it is no longer determined by the conditions at the entrance. Or consider the flow around a profile: if we know or assume that the problem is entirely subsonic, the analogy to incompressible problems leads to the formulation of boundary conditions in the physical plane, for which definite mathematical results have recently been reached (see Sec. 25.2). If, however, we cannot exclude the possibility of a mixed problem, we do not know whether the assumed boundary conditions, suggested by physical intuition and mathematical analogy, determine a solution (i.e., whether we are not prescribing either too little or too much), quite apart from the task of obtaining the solution.

Given this state of affairs we shall to a great extent have recourse to *indirect methods*. Starting with examples of exact solutions, one may then characterize a posteriori the physical situation to which the flow defined by this exact solution actually corresponds. These examples are interesting as such; in addition, they illustrate certain general theoretical facts that may be typical. This holds for the simple examples which will be considered in the following section as well as for the problems which we shall study in Art. 20 after having added to our knowledge of the basic theory. Actually, even the more powerful methods to be presented in Art. 21 constitute only an indirect approach, although there the object is to solve boundary value problems.

4. Radial flow, vortex flow, and spiral flow obtained as exact solutions in the hodograph

We shall now consider as first examples of exact hodograph solutions, some of the flows which have already been obtained in the physical plane in Art. 7.[13] This will lead to new insight into certain properties of these flows.

(a) *Compressible vortex flow.* This is a particular case of the axially symmetric flow of Sec 7.5, with radial velocity $q_r = 0$. Here we take as a starting point equation (12) and the particular solution

$$\Phi = -C\theta \qquad (C > 0). \tag{28}$$

From Eqs. (7) and (4″) we find

$$\frac{\partial \Phi}{\partial q} = 0 = x \cos \theta + y \sin \theta, \qquad \frac{\partial \Phi}{\partial \theta} = -C = q(y \cos \theta - x \sin \theta),$$

and consequently

$$x = \frac{C \sin \theta}{q} = \frac{Cq_y}{q^2}, \qquad y = -\frac{C \cos \theta}{q} = -\frac{Cq_x}{q^2}. \tag{29}$$

It is easy to invert these equations and to obtain

$$q_x = -\frac{Cy}{r^2}, \qquad q_y = \frac{Cx}{r^2}, \qquad q = \frac{C}{r}, \tag{30}$$

from which follows

$$\frac{dy}{dx} = \tan \theta = \frac{q_y}{q_x} = -\frac{x}{y}, \qquad \text{or} \qquad x^2 + y^2 = \frac{C^2}{q^2}, \tag{31}$$

as the equation of the streamlines, which are here concentric circles with constant value of the speed q along each circle. To the maximum value $q = q_m$ corresponds, by (31), a minimum value $r_{\min} = r_1$ of the radius r, and $C = q_m r_1$. The circulation has the same value Γ for all streamlines:

$$\Gamma = \oint \mathbf{q} \cdot d\mathbf{l} = \oint qr\, d\theta = 2\pi rq = 2\pi C, \tag{32}$$

or

$$\frac{\Gamma}{2\pi} = rq, \qquad q = \frac{\Gamma}{2\pi r},$$

and $r_1 = \Gamma / 2\pi q_m$. Also

$$r = \frac{r_1 q_m}{q} = r_1 \sqrt{1 + \frac{2}{(\kappa - 1)M^2}} = \frac{r_1}{\sqrt{1 - \rho^{\kappa - 1}}}. \tag{30'}$$

We see that on the circle $r = r_1$, where $q = q_m$, density and pressure are zero. As r increases indefinitely, q decreases towards $q = 0$, and ρ and p increase from zero toward their stagnation values (see Fig. 100a,b).[14] On the circle $r = hr_1$ the velocity is sonic, $M = 1$; hence there is supersonic flow in the ring between $r = r_1$ and $r = hr_1$ and subsonic flow outside. From Eq. (6′) we find $\varphi = -\Phi = C\theta$, which confirms that the equipotential lines are the orthogonal trajectories of the streamlines.

(*b*) *Radial motion. Source and sink.* Next, we wish to consider radial motion, i.e., a flow that goes radially from or toward a center (see Sec. 7.3). The solution of (16.32′)

$$\psi = k\theta, \tag{33}$$

where k is an arbitrary constant, corresponds to straight hodograph streamlines through O'. By means of (16.31) and (25′) we find the coordinate functions

$$x = \frac{k}{\rho q} \cos \theta, \qquad y = \frac{k}{\rho q} \sin \theta. \tag{34}$$

and see that the physical streamlines are indeed straight lines through the

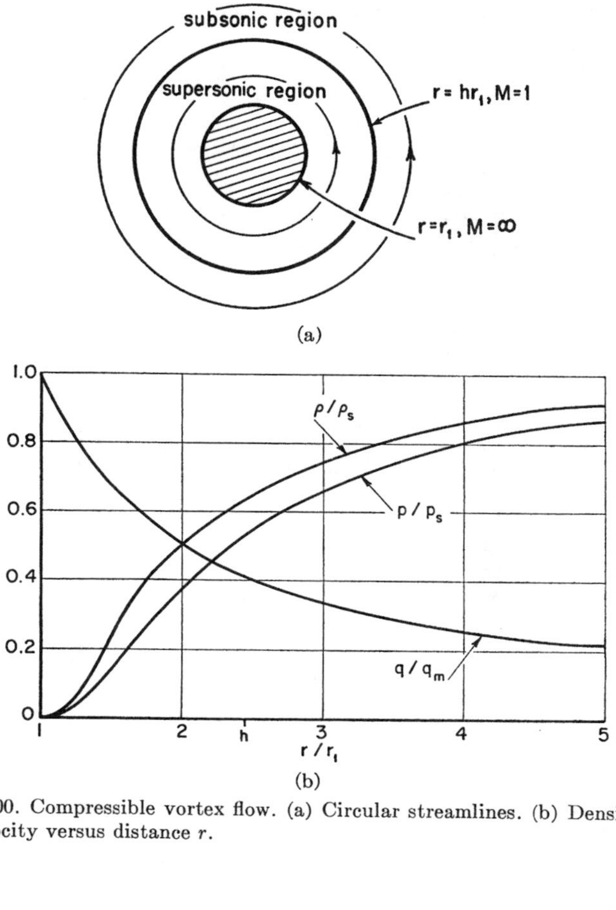

FIG. 100. Compressible vortex flow. (a) Circular streamlines. (b) Density, pressure, velocity versus distance r.

origin (see Fig. 101). From (34)

$$r = \pm\frac{k}{\rho q} = \frac{m}{2\pi\rho q}, \tag{34'}$$

where $\pm\, m$ is the strength per unit length of the source, and the $\pm$ signs apply according as to whether $k \gtrless 0$. The circles: r = constant, are lines of constant q; they are perpendicular to the radial streamlines; hence they are the equipotential lines which are here concentric circles. Also, substituting in (34′) for ρ or q, respectively, we obtain, with $\rho_s = 1$,

$$r = \pm\frac{k}{q}\left(1 - \frac{q^2}{q_m^{\,2}}\right)^{-1/(\kappa-1)} = \pm\,\frac{k}{\rho q_m}\,(1 - \rho^{\kappa-1})^{-\frac{1}{2}}. \tag{34''}$$

The relation (34″) between q and r has been discussed in Sec. 7.3. (There

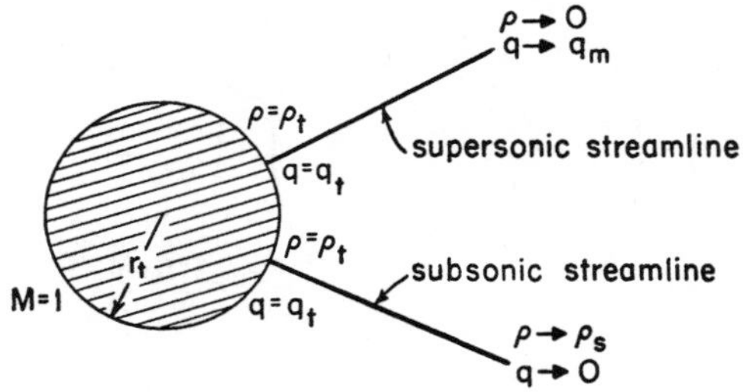

(a)

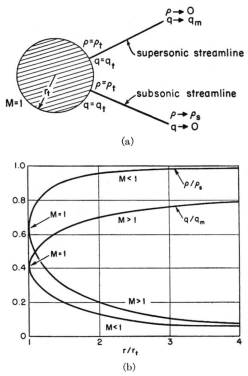

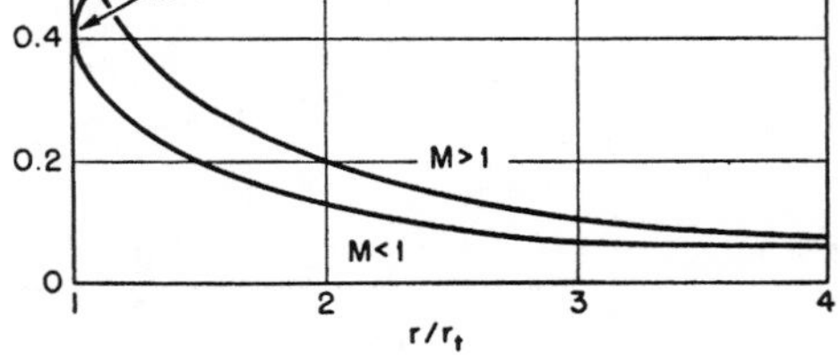

(b)

FIG. 101. Radial flow. (a) Two radial flows and limit line. (b) Density and velocity versus distance in plane radial flow.

$q_r = \pm q$ denoted the velocity in the direction of increasing r.) The result was: The curve $q = q(r)$ has two branches, with no real point for r less than a certain value $r_{\min} = r_t$, where $M = 1$. The upper branch of the lower curve in Fig. 101b represents the velocity q/q_m in a purely supersonic flow (of the source or sink type), with velocities ranging from sonic speed q_t at $r = r_t$ to maximum velocity q_m at $r = \infty$. The lower branch represents the velocity in a subsonic flow (of source or sink type), with velocities going from $q = 0$ at infinity to sonic speed q_t at $r = r_t$; there the two flows meet. As q increases from $q = 0$ to $q = q_t$, the density decreases from its stagnation value $\rho = \rho_s = 1$ to its sonic value $\rho = \rho_t$, and as q goes from $q = q_t$ to $q = q_m$, the density decreases from ρ_t to zero. The same conclusions are reached by direct consideration of the flow intensity ρq in (34′) [see Eq. (8.5) and Fig. 40]. Introducing sonic values, (34′) can also be written

$$\frac{r}{r_t} = \frac{\rho_t q_t}{\rho q}.$$

At the sonic circle, $r = r_t$, where the two flows meet, *the Jacobian* $\partial(\varphi, \psi)/\partial(q, \theta)$ *vanishes* since, from (33) and (27),

$$D = \frac{\partial(\varphi,\psi)}{\partial(q,\theta)} = \frac{M^2 - 1}{\rho q} k^2.$$

Also the *acceleration* b becomes *infinite* there. In fact,

$$b = dq/dt = q(\partial q/\partial s) = \pm q(dq/dr);$$

using (34′) and the differentiation formula (8.5) for ρq, we obtain

$$b = \frac{\rho q^3}{k(M^2 - 1)}, \tag{35}$$

which becomes infinite for $M = 1$. Across the limit circle $r = r_t$ the flows cannot be continued and in the vicinity of this line they must be regarded as physically impossible. Our solution offers the simplest example of the fact that an extremely simple single-valued solution in the hodograph $(\theta_1 \leqq \theta \leqq \theta_2)$ may lead to different physical flows which meet along a limiting line.

(c) *Superposition. Spiral flow.* Since the equations for φ, ψ, Φ, Ψ in the hodograph are linear, a linear combination of two solutions of each of these equations is again a solution. Next, x and y follow from Φ or Ψ by the relations (4) or (9), and from φ, ψ by (25′). We conclude that if $\psi_1(q,\theta)$ and $\psi_2(q,\theta)$ are two solutions of the stream-function equation, or $\Phi_1(q,\theta)$ and $\Phi_2(q,\theta)$ are two solutions of the equation for Φ then $\psi = c_1\psi_1 + c_2\psi_2$ or $\Phi = c_1\Phi_1 + c_2\Phi_2$ are likewise solutions of the respective equations. The new coordinates which correspond, e.g., to ψ are given by

$$
(36) \qquad \begin{aligned} x(q,\theta) &= c_1x_1(q,\theta) + c_2x_2(q,\theta), \\ y(q,\theta) &= c_1y_1(q,\theta) + c_2y_2(q,\theta), \end{aligned}
$$

where x_1,y_1 and x_2,y_2 are the coordinates corresponding to ψ_1 and ψ_2, respectively. Applying this principle to the two preceding solutions discussed in (a) and (b) we obtain (with $C > 0$, $k > 0$) by means of (29) and (34), the new solution

$$
(37) \qquad \begin{aligned} x &= \frac{C}{q}\sin\theta + \frac{k}{\rho q}\cos\theta = \frac{1}{q^2}\left(Cq_y + \frac{k}{\rho}q_x\right), \\ y &= -\frac{C}{q}\cos\theta + \frac{k}{\rho q}\sin\theta = \frac{1}{q^2}\left(-Cq_x + \frac{k}{\rho}q_y\right), \end{aligned}
$$

$$
(37') \qquad r^2 = \frac{C^2}{q^2} + \frac{k^2}{(\rho q)^2} \qquad \text{or} \qquad r = \frac{1}{q}\sqrt{C^2 + \frac{k^2}{\rho^2}}.
$$

This last equation is the same as Eq. (7.40) and the discussion of the latter given in Sec. 7.6 and illustrated in Fig. 31 applies. To see that the flows are the same we write Eqs. (37) in the form:

$$
x = r\cos(\theta - \bar{\theta}), \qquad y = r\sin(\theta - \bar{\theta}),
$$

where r is given in (37′) and

$$
\cos\bar{\theta} = \frac{k}{\rho q r}, \qquad \sin\bar{\theta} = \frac{C}{qr},
$$

so that $\bar{\theta}$ is the angle that the streamlines make with the radius vector. Then with the usual meaning of q_r and q_θ, we find

$$
(37'') \qquad \begin{aligned} \sin\bar{\theta} &= \frac{q_\theta}{q} = \frac{C}{qr} \qquad \text{or} \qquad rq_\theta = C, \\ \cos\bar{\theta} &= \frac{q_r}{q} = \frac{k}{\rho q r} \qquad \text{or} \qquad rq_r\rho = k, \end{aligned}
$$

i.e., the equations (7.36) and (7.39). Hence the flows are identical, and we merely add a few remarks.

The main result obtained in Sec. 7.6 was that there exist for each pair of constants C, k two different flows, both extending over the same region from $r = r_l$ to $r = \infty$, with the same supersonic velocity q_l at r_l, one entirely supersonic, the other of the "mixed" type. To compute q_l we differentiate (37′) and obtain

$$
(38) \qquad \frac{C^2}{q^3} + \frac{k^2}{(\rho q)^3}(1 - M^2)\rho = 0, \qquad \text{or} \qquad M_l^2 = 1 + \frac{C^2}{k^2}\rho_l^2 > 1;
$$

from this M_l the value q_l follows. Next, we compute the Jacobian $D = \partial(\varphi,\psi)/\partial(q,\theta)$. For the stream function ψ_1 of the vortex flow we find from $\phi_1 = C\theta$, by Eqs. (16.31): $\partial\psi_1/\partial\theta = 0$, $\partial\psi_1/\partial q = C\rho/q$. Therefore, using (33), we obtain for the present flow

$$(39) \qquad \frac{\partial\psi}{\partial\theta} = k, \qquad \frac{\partial\psi}{\partial q} = C\frac{\rho}{q}, \qquad \psi = C\int^q \frac{\rho}{q}\,dq + k\theta,$$

and from (27)

$$D = -\frac{k^2}{\rho q}\left(\frac{C^2\rho^2}{k^2} + 1 - M^2\right),$$

and this is zero for the value $M = M_l$ of (38). We also see that the value of the acceleration

$$(40) \qquad \frac{dq}{dt} = q\frac{\partial q}{\partial s} = q^2\frac{\partial q}{\partial\phi} = q^2\frac{\partial\psi}{\partial\theta}\frac{1}{D} = q^2\frac{k}{D}$$

tends to infinity at the limit circle $r = r_l$.

Also from (37″) and (38), $M_l \cos\bar{\theta}_l = 1$ follows. Hence, from $\cos\bar{\theta}_l = 1/M_l = \sin\alpha_l$, we conclude that the Mach lines of one of the two families are tangent to the limit circle (see Sec. 19.3), and that the streamlines of either flow meet the limit circle under the angle α_l. Since the component of the velocity normal to a characteristic equals a, it follows that the radial velocity q_r is sonic at the limit circle.

All streamlines of the supersonic flow are congruent and so are those of

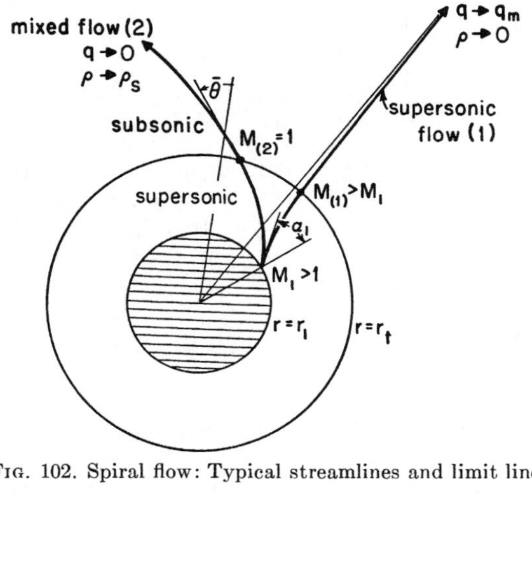

FIG. 102. Spiral flow: Typical streamlines and limit line.

the mixed flow. Figure 102 shows one streamline of each flow. These are to be considered as two streamlines, each for a different flow, not as one streamline with a cusp; the two flows should not be thought of as continuations of each other. They appear simultaneously in our example, where a single stream function $\psi(q,\theta)$ yields two different flows which meet at $r = r_l$ with the same $\mathbf{q} = \mathbf{q}_l$. Let us add, however, the following remark. The reader should not be left with the impression that the appearance of a limiting line is an essentially mathematical phenomenon, introduced by the use of the hodograph representation. He should rather remember that these same flows exhibiting the same singularities were obtained in Art. 7 by means of considerations exclusively in the physical plane.

5. The Chaplygin-Kármán-Tsien approximation

(a) *Definitions.* In this book, approximations have not been dealt with; we mean by an approximation a modification of the basic differential equations with the purpose of simplifying the mathematics of a problem without changing too much its physical meaning. In this and the next section we shall take up a method which in relation to the basic equations of steady potential flow must be considered an approximation in the above sense. It is however, likewise possible to interpret the procedure as the exact theory of a gas with a particular specifying equation. This specifying equation will turn out to be the linearized (p,ρ)-relation $p = A - B/\rho$, $B > 0$ mentioned already in Eqs. (1.5c) and (2.17d). The method, while it is of an elementary character, is interesting from various points of view; one is that it can be considered as a simplified model of the general plane problem posed in Art. 16.

We start with the equations (19)

$$\frac{\partial\varphi}{\partial\lambda} = -\frac{(1 - M^2)^{\frac{1}{2}}}{\rho}\frac{\partial\psi}{\partial\theta}, \qquad \frac{\partial\varphi}{\partial\theta} = \frac{(1 - M^2)^{\frac{1}{2}}}{\rho}\frac{\partial\psi}{\partial\lambda}, \qquad M < 1,$$

where $1/\rho$ stands for ρ_s/ρ with $\rho_s = 1$. Expansion for polytropic gas of

$$\frac{\rho_s}{\rho}(1 - M^2)^{\frac{1}{2}} = \left(1 + \frac{\kappa - 1}{2}M^2\right)^{1/(\kappa-1)}(1 - M^2)^{\frac{1}{2}}$$

in powers of M^2 yields, with $\kappa = \gamma = 1.4$

$$\frac{\rho_s}{\rho}(1 - M^2)^{\frac{1}{2}} = \left(1 + \frac{M^2}{2} + (2 - \kappa)\frac{M^4}{8} + \cdots\right)\left(1 - \frac{M^2}{2} - \frac{M^4}{8}\cdots\right)$$

$$= 1 - \frac{\kappa + 1}{8}M^4 + \cdots = 1 - 0.3M^4 + \cdots.$$

It is thus seen that $(\rho_s/\rho)\,(1 - M^2)^{\frac{1}{2}} = (1/\rho)(1 - M^2)^{\frac{1}{2}}$ differs from unity only by terms of the order of M^4. If—pending further discussion—*we approximate* $(1/\rho)(1 - M^2)^{\frac{1}{2}}$ *by unity,* Eqs. (19) simplify to

$$(41) \qquad \frac{\partial\varphi}{\partial\lambda} = -\frac{\partial\psi}{\partial\theta}, \qquad \frac{\partial\varphi}{\partial\theta} = \frac{\partial\psi}{\partial\lambda}, \qquad M < 1.$$

Next consider this approximation in relation to the variables $K = (1 - M^2)/\rho^2$ and $\sigma = \int_q^{q_t} \rho\, dq/q$ [see (22) and (24′)] introduced and used by Chaplygin. Clearly K then becomes equal to unity and Eqs. (23) reduce to

$$(41') \qquad \frac{\partial\varphi}{\partial\sigma} = \frac{\partial\psi}{\partial\theta}, \qquad \frac{\partial\varphi}{\partial\theta} = -\frac{\partial\psi}{\partial\sigma},$$

which are the same equations as (41) except that $-\sigma$ takes the place of λ. Indeed the definitions (17) and (22) of λ and σ, namely,

$$(42) \qquad \frac{d\lambda}{dq} = \frac{\sqrt{1 - M^2}}{q}, \qquad \frac{d\sigma}{dq} = -\frac{\rho}{q},$$

show that $d\lambda = -d\sigma$ if $\sqrt{1 - M^2}$ is put equal to ρ.[15]

To justify the replacement of $\sqrt{1 - M^2}/\rho$ by unity we may also, following Chaplygin, write K in terms of $\tau = q^2/q_m{}^2$:*

$$K = \frac{1 - M^2}{\rho^2} = \frac{1 - h^2\tau}{(1 - \tau)^{h^2}}, \qquad \text{where } h^2 = \frac{\kappa + 1}{\kappa - 1}.$$

Then,

$$\frac{dK}{d\tau} = -\frac{h^2\tau}{(1 - \tau)^{h^2+1}},$$

and this shows that $K(\tau)$, which decreases with increasing τ, varies so slowly for small τ-values that it is practically constant and may be put equal to unity.[16]

Finally, consider the expression of ρ in terms of M^2, namely,

$$\rho = \left(1 + \frac{\kappa - 1}{2} M^2\right)^{-1/(\kappa-1)}.$$

If the right side is to equal $(1 - M^2)^{\frac{1}{2}}$, κ must equal -1. We review: *The approximation* $\sqrt{1 - M^2} = \rho$, *which we shall briefly call the Chaplygin approximation, is obtained by taking for* κ, *the exponent in the polytropic relation, the value* $\kappa = -1$. *In this case,* $d\lambda = -d\sigma$ *where* λ *and* σ *are given by Eqs.* (42), *and the basic equations* (16.31) *become Cauchy-Riemann equa-*

* This variable will be widely used later, particularly in Arts. 20 and 21. Formulas for M and ρ in terms of τ are given in Eq. (20.3′).

tions (41) *or* (41′) *in the Cartesian coordinates* $(\lambda, -\theta)$, *or* (σ, θ). *Consequently* $\varphi + i\psi$ *is an analytic function of* $\lambda - i\theta$ (*or* $\sigma + i\theta$), *and the Laplace equation holds for both* φ *and* ψ.[17]

The assumption $\kappa = -1$ can be interpreted to mean that we consider the pressure-density relation $p = -B/\rho + A$, $B > 0$ (note that a constant can be added to any polytropic relation) as an approximation to the usual relation $p = C\rho^{\kappa}$, $\kappa = 1.4$. The choice of A and B is still arbitrary, and it is in the choice of these constants that von Kármán and Tsien deviate from Chaplygin. Chaplygin chose A and B so as to make the line $p = -B/\rho + A$ (Fig. 103) tangent to the usual isentrope of a gas at a stagnation point,

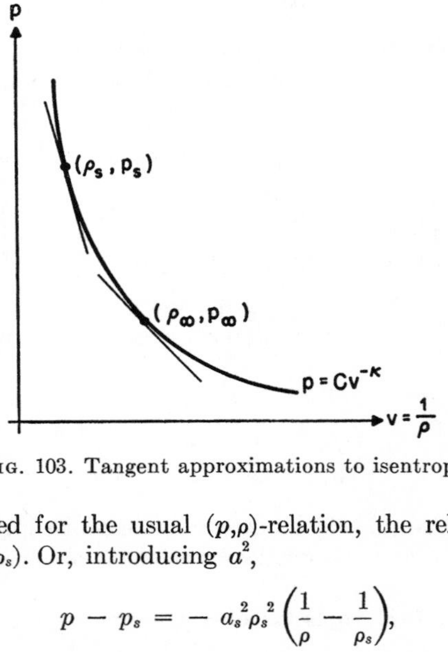

FIG. 103. Tangent approximations to isentrope.

i.e., he substituted for the usual (p,ρ)-relation, the relation $p - p_s = -\kappa\rho p_s(1/\rho - 1/\rho_s)$. Or, introducing a^2,

$$(43) \qquad p - p_s = -a_s^2\rho_s^2\left(\frac{1}{\rho} - \frac{1}{\rho_s}\right),$$

and hence, for the constants A, B,

$$(43') \qquad A = p_s + \rho_s a_s^2, \qquad B = a_s^2\rho_s^2.$$

Von Kármán and Tsien (Fig. 103) take for the point of tangency one with undisturbed stream conditions*, so that, using the subscript ∞,

$$(43'') \qquad A = p_\infty + \rho_\infty a_\infty^2, \qquad B = a_\infty^2\rho_\infty^2.$$

For either choice of A and B such a gas has some simple but unusual

* We may have in mind the flow past a profile with uniform stream conditions at infinity.

properties which are easily verified. We find from Bernoulli's equation*

$$(44) \qquad a^2 - q^2 = a_s^2, \qquad M = q(a_s^2 + q^2)^{-\frac{1}{2}}.$$

The relation between ρ and q is now

$$(44') \quad \frac{\rho}{\rho_s}\sqrt{1 + \frac{q^2}{a_s^2}} = 1, \qquad \text{or} \qquad \frac{\rho}{\rho_s} = \sqrt{1 - M^2} = \frac{a_s}{\sqrt{a_s^2 + q^2}}.^{18}$$

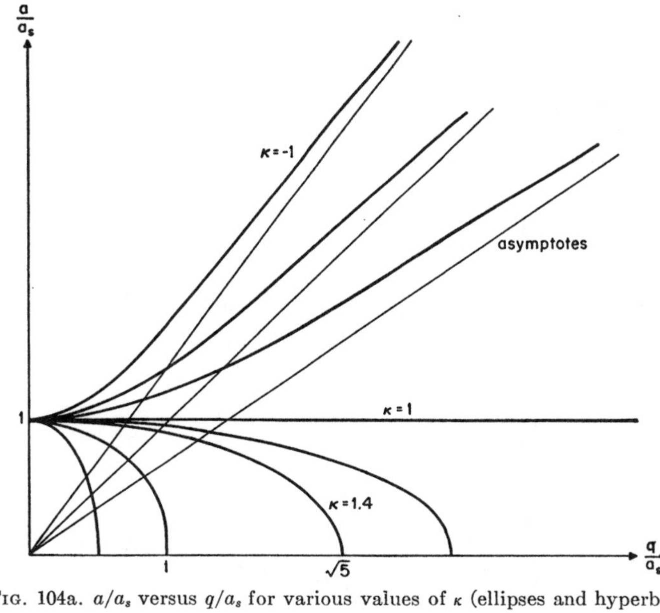

FIG. 104a. a/a_s versus q/a_s for various values of κ (ellipses and hyperbolas).

We note that there is now no restriction as to the value of q: in fact, $q \to \infty$ as $M \to 1$, as seen from the second Eq. (44), and in this case, as seen from (44′), $\rho \to 0$. Hence, as $M \to 1$, $\rho \to 0$, $q \to \infty$, $a \to \infty$. It is thus seen that in the case of the Chaplygin approximation, the sound velocity a/a_s increases with increasing rather than decreasing q, in contrast to the case $\kappa > 1$. Figure 104a shows a/a_s versus q/a_s for various values of κ, in particular for $\kappa = 1.4$ and $\kappa = -1$, whereas Fig. 104b shows M versus q/a_s for various values of κ. For $\kappa = -1$ the curve has $M = 1$ as an asymptote. It is thus seen that an initially subsonic flow will remain subsonic if $\kappa = -1$; hence transonic flow cannot be studied by this method.

* We have in Kármán-Tsien's case $a_s^2 = a_\infty^2 - q_\infty^2$.

From the first equation (42) we have, using (44),

$$\frac{d\lambda}{dq} = \frac{a_s}{q\sqrt{a_s^2 + q^2}}, \qquad \lambda = \log \frac{q}{a_s + \sqrt{a_s^2 + q^2}}. \tag{45}$$

If the integration constant in the expression for λ is chosen as in (45), it is seen that $\lambda \to -\infty$, as $q \to 0$, and $\lambda \to 0$ as $q \to \infty$, $M \to 1$; thus λ increases monotonically. From the second formula (45) we see that for q-values small compared to a_s :

$$\lambda = \log \frac{q}{2a_s} + O\left(\frac{q^2}{a_s^2}\right), \qquad \text{hence} \qquad e^{\lambda} \sim \frac{q}{2a_s}. \tag{45'}$$

(b) *Relation to an incompressible flow problem.*[19] Eqs. (41) are Cauchy-Riemann equations in the Cartesian coordinates λ, $-\theta$. It will however be more convenient to introduce the new independent variable,

$$v = 2a_s e^{\lambda} = \frac{2q}{1 + \left(1 + \frac{q^2}{a_s^2}\right)^{\frac{1}{2}}}. \tag{46}$$

The choice of the multiplicative constant $2a_s$ is such that for $a_s \to \infty$, or q

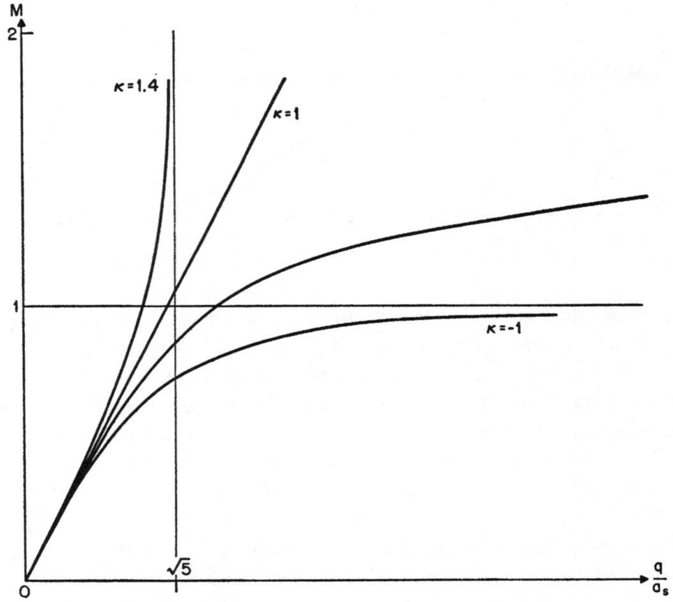

FIG. 104b. M versus q/a_s for various values of κ.

small compared to a_s, the $v \sim q$ [as in (45′)]. The equations (41) then take the form:

$$(47) \qquad v\frac{\partial\varphi}{\partial v} = -\frac{\partial\psi}{\partial\theta}, \qquad v\frac{\partial\psi}{\partial v} = \frac{\partial\varphi}{\partial\theta}, \qquad M < 1.$$

These are Cauchy-Riemann equations in polar coordinates $v, -\theta$, i.e., exactly the form which Eqs. (16.31) take on as $M \to 0$. Consequently $\varphi + i\psi$ is an analytic function of $ve^{-i\theta}$ and the real and imaginary parts of any analytic function of $ve^{-i\theta}$ will be solutions of (47). These equations may be interpreted as the equations of an incompressible fluid with complex velocity $ve^{-i\theta} = \zeta$ where the speed v varies between 0 and $2a_s$. We state: If we apply the Chaplygin approximation to the basic equations (16.31) and introduce in (41) instead of λ, a new variable v by (46), these equations take the form of incompressible flow equations in the polar coordinates v, $-\theta$.

If q and ρ are expressed in terms of v we obtain, by (46) and (44′),

$$(46') \qquad q = \frac{4a_s^2 v}{4a_s^2 - v^2}, \qquad \rho = \frac{4a_s^2 - v^2}{4a_s^2 + v^2}.$$

We see that as v increases monotonically from 0 to $2a_s$, q goes from 0 to ∞, and ρ from 1 to 0. The transformations (46), (46′) are thus interpreted as relations between an incompressible flow and a compressible Chaplygin flow.

6. Continuation

(a) *Flow past a profile.* Let us try to construct the compressible Chaplygin flow past a given profile P_0. We assume that this flow does not involve circulation. This restriction will be removed later in the section. We assume as known the *complex potential* $w(z)$ of the incompressible flow around P_0 (where $z = x + iy$) and introduce the complex velocity $dw/dz = \zeta(z) = ve^{-i\theta}$. We use this last equation to express z in terms of ζ and obtain $w(z) = w_0(\zeta) = \varphi_0 + i\psi_0$ where φ_0 and ψ_0 depend on v and θ, and $\psi_0 = 0$ along P_0. We then define a stream function $\psi(q,\theta)$ of a compressible Chaplygin flow by setting

$$(48) \qquad \psi(q,\theta) = \psi_0(v,\theta),$$

where (46) holds between v and q, and similarly for $\varphi(q,\theta)$. These φ and ψ thus defined are solutions of (16.31) under the Chaplygin approximation. We shall however see that these $\psi(q,\theta)$, $\varphi(q,\theta)$ do not in general provide a solution of the given boundary-value problem of flow past P_0. Denote by X, Y ($Z = X + iY$) the coordinates in the physical plane of the compressible flow in order to distinguish them from the x, y in the plane of incompressible flow. Then from $\psi(q,\theta)$, $\varphi(q,\theta)$ follow, in the usual way, $X(q,\theta)$, $Y(q,\theta)$ (see Sec. 3). Some curve P in the Z-plane will correspond to the given profile P_0 in the z-plane, along which $\psi_0 = 0$. We shall show that this contour in the Z-plane differs in shape from the given contour P_0, but reduces to P_0 as $a_s \to \infty$, $M \to 0$, $v \to q$. Thus, even in the present simplified situation, we do not solve an exact boundary-value problem. It is, however, possible here (in contrast to the polytropic case which we shall study in following articles) to indicate a very simple formula for the deviation between the two contours, the so-called *shape correction*.

We obtain from Eq. (25″)

$$dZ = dX + i\,dY = \frac{1}{q} e^{i\theta} \left(d\varphi + \frac{i}{\rho}\, d\psi \right),$$

and substitute for ρ and q from (46′) in terms of v. Then, denoting by $\bar{w}$, $\bar{\zeta}$, etc., the conjugate complex functions to w, ζ, we have

$$\begin{aligned} dZ &= \frac{e^{i\theta}}{q}\left(d\varphi + \frac{i}{\rho}\, d\psi\right) = \frac{e^{i\theta}}{4a_s^2 v}(4a_s^2 - v^2)\,d\varphi + i\,\frac{e^{i\theta}}{4a_s^2 v}(4a_s^2 + v^2)\,d\psi \\ &= \frac{e^{i\theta}}{v}(d\varphi + i\,d\psi) - \frac{ve^{i\theta}}{4a_s^2}(d\varphi - i\,d\psi) \\ &= \frac{1}{v} e^{i\theta}\,dw - \frac{1}{4a_s^2} ve^{i\theta}\,d\bar{w}. \end{aligned}$$

Therefore,

$$dZ = \frac{dw}{\zeta} - \frac{1}{4a_s^2}\,\bar{\zeta}\,d\bar{w}. \tag{49}$$

Replacing ζ by dw/dz and $\bar{\zeta}$ by $d\bar{w}/d\bar{z}$ we obtain:

$$dZ = dz - \frac{1}{4a_s^2}\left(\frac{d\bar{w}}{d\bar{z}}\right)^2 d\bar{z}. \tag{49′}$$

Then

$$Z = z - \frac{1}{4a_s^2} \int \overline{\left(\frac{dw}{dz}\right)^2 dz}\,. \tag{49″}$$

We thus obtain for each $w(z)$ the Z for the compressible flow which corresponds to a z in the incompressible flow; the respective speeds v and q are linked by Eq. (46).*

We note that from Eq. (49) $\varphi + i\psi$ is a nonanalytic function of $X + iY$.

* This simple formula compares, in the exact theory, with such involved results as (67) and (72) in [24] pp. 242, 243.

(b) *Circulation.* For flow around a profile a difficulty arises if the incompressible flow around the profile involves circulation, since the function to the right in (49″) will be single-valued only if

$$\oint \overline{\left(\frac{dw}{dz}\right)^2 dz} = 0,$$

for any path around P_0, and this is true only in the absence of circulation. In other words, when the flow has circulation the profile shapes in the Z-plane furnished by the above theory are not closed. Hence, our method as explained here is valid only if in the incompressible flow circulation is absent. Several authors have generalized the procedure to take care of this difficulty. We indicate here a method due to C. C. Lin.[20]

Denote by $w(z)$ the complex potential of the incompressible flow with circulation, and introduce instead of ζ, more generally,

$$\xi(z) = k(z)ve^{-i\theta} \tag{50}$$

$$\frac{dw}{dz} = \xi. \tag{50'}$$

Here $k(z)$ is an analytic function of z, regular and without zeros in the exterior of the given profile P_0 (including the point $z = \infty$), and such that

$$\oint k(z)dz - \frac{1}{4a_s^2}\oint \overline{\xi^2 \frac{dz}{k(z)}} = 0 \tag{51}$$

for any closed contour enclosing the given profile P_0. In addition

$$|\tfrac{1}{2}\xi(z)| < |k(z)| < \infty \qquad \text{on } P_0. \tag{51'}$$

It may then be proved in the same way as (49″) was obtained that the Z of the compressible flow is given by

$$Z = \int k(z)\, dz - \frac{1}{4a_s^2}\int \overline{\xi^2 \frac{dz}{k(z)}}, \tag{50''}$$

to be compared to (49″), and with an analogous interpretation.

Thus to a complex potential which represents an incompressible flow with circulation, a related compressible Chaplygin flow with circulation is given by (50), (50′), and (50″), *and we are still free in the choice of* $k(z)$. In these equations z can be considered a parameter; and after it is eliminated we obtain relations between X, Y, q, θ, the compressible flow coordinates and the velocity.

The essential point in the above method is that $ve^{-i\theta}$ is equated to a function of z which is not directly the complex velocity $\zeta(z)$ as in Kármán-Tsien's scheme, but equal to $\xi(z)/k(z)$ where $k(z)$ is still widely arbitrary.

For $k(z) = 1$ the condition (51) of the closure of P in the Z-plane is not satisfied, if circulation is present. On the other hand, $k(z)$ should not depart too much from unity if the profiles P and P_0 are not to differ too much from each other.

Finally, discussing the direct problem, Lin relates the choice of $k(z)$ to the mapping of P_0 onto a circle: with $w(z)$ now the complex potential for incompressible flow (with same circulation) past the circle, the function $k(z)$, which must satisfy the above requirements, should "not differ too much" from $k_0(z)$, the derivative of the mapping function. This idea is then related to a well-known method of v. Mises[21] who has shown (in incompressible flow) how to transform a circle into an airfoil of very general shape.

(c) *Additional remarks.* The influence of compressibility upon the pressure distribution may be easily estimated. We refer the reader to the literature.[22]

In order to demonstrate the change of contour by means of an example, Tsien has computed the compressible flow about an approximately elliptic profile by starting with an ellipse in the z-plane.[23] The deviation of the new contour from the ellipse tends to zero* as $a_s \to \infty$, or $M_\infty \to 0$; it is however shown that for not too small M_∞-values, and if the given ellipse is nearly a circle, the deviation is quite appreciable.

We have presented here the Chaplygin approximation not only because it is widely and successfully used, but also because it can be considered from various aspects: on the one hand, it constitutes an approximation which greatly simplifies the usual equations; on the other hand, it is an easily understood *exact* theory. This simplified theory does not however avoid the basic non-linearity of the problem and therefore may play, in certain respects, the role of a mathematical model. In this sense, it is of interest that here most steps can actually be carried out explicitly and without too great difficulty, in contrast to the general situation where more difficult and delicate methods are required (see Art. 21). Finally, an exact existence proof for subsonic flow past a body has been given in this case, in a comparatively simple way.[24]

* This limit behaviour holds not only for an ellipse in the z-plane but for fairly general profiles, as seen by (49″). It holds true even for analogous constructions in the polytropic case, as we shall see in Art. 21.

Article 18

Simple Waves

1. Definition and basic properties

As in the preceding article we shall consider here plane potential flow (see Sec. 16.1). *Simple wave* or *simple wave solution* (see also Art. 13 regarding the (x,t)-problem) is the name of an important type of solution of the basic equations (16.7).[25] Simple waves can be derived and introduced in several essentially equivalent ways. Here we take the following starting point. We ask for a flow in which the lines q = constant and θ = constant coincide; we call such lines $\mathfrak{L}$-lines. Each $\mathfrak{L}$-line is thus mapped, by definition, onto one point of the hodograph. We assume that these points do not all coincide, i.e., that the solution does not merely represent a region of constant state, but that the whole set of $\mathfrak{L}$-lines, or the whole region R in the x,y-plane covered by $\mathfrak{L}$-lines, is mapped onto one line Λ of the hodograph. This fact, namely, that a two-dimensional flow region has a one-dimensional hodograph, was the starting point in Sec. 13.1 (where the speedgraph served the same purpose as does the hodograph in the present problem).

The existence of the line Λ in the hodograph plane implies the existence of a relation between q and θ, and, as a consequence, the vanishing throughout R of the Jacobian

$$(1) \qquad i = \frac{\partial(q,\theta)}{\partial(x,y)}.$$

Now using Eqs. (16.7) and $\partial(x,y)/\partial(s,n) = 1$, we obtain

$$(1') \qquad i = \frac{\partial q}{\partial s}\frac{\partial \theta}{\partial n} - \frac{\partial q}{\partial n}\frac{\partial \theta}{\partial s} = \frac{1}{q}\left[\left(\frac{\partial q}{\partial s}\right)^2 (M^2 - 1) - \left(\frac{\partial q}{\partial n}\right)^2\right].$$

The right side of (1′) cannot vanish in R, for $M < 1$, unless $\partial q/\partial s = \partial q/\partial n = 0$ and, by (16.7), either $q = 0$ or $\partial\theta/\partial n = \partial\theta/\partial s = 0$ which means constant flow. Therefore, the flow under consideration cannot be subsonic, and real characteristics exist in R.

Among the lines crossing the $\mathfrak{L}$-lines there must be at least one set of characteristics, and the image of each of these must lie on Λ. (In fact such a characteristic C intersects each $\mathfrak{L}$-line at a point. Each of these points must map onto Λ since each $\mathfrak{L}$-line is mapped as a whole onto a point of Λ; hence, each point of C is mapped onto a point of Λ.) It follows that Λ is a Γ^+ or a Γ^-, and that each Mach line of the second set—each $C^\pm$ if Λ is a $\Gamma^\mp$—is mapped onto a single point of Λ. Thus the $\mathfrak{L}$-lines form this second set of characteristics. Since on each of them both θ and q, and therefore α

as a function of q, are constant, it follows that $\theta \mp \alpha$, that is, the slope of each $\mathcal{L}$-line, is constant. Hence the $\mathcal{L}$-lines are straight.

Since the whole region R covered by the $\mathcal{L}$-lines is mapped onto one characteristic, Γ^+ or Γ^-, the equation of this characteristic, $Q \mp \theta =$ constant, [see (16.41) and (16.42)] is valid throughout R. On the other hand, it will now be proved that this supersonic pattern represents a flow, i.e., that q and θ satisfy Eqs. (16.7). Consider an arbitrary point in the x,y-plane; through it passes an $\mathcal{L}$-line, say a C^+, which makes the angle α with the stream direction. On account of the constancy of q and θ along this line, we have $\partial q/\partial l = 0$, $\partial\theta/\partial l = 0$ along it, or

$$\frac{\partial q}{\partial s}\cos\alpha + \frac{\partial q}{\partial n}\sin\alpha = 0, \qquad \frac{\partial\theta}{\partial s}\cos\alpha + \frac{\partial\theta}{\partial n}\sin\alpha = 0,$$

and from $Q(q) + \theta =$ constant everywhere in the wave region

$$Q'\frac{\partial q}{\partial s} + \frac{\partial\theta}{\partial s} = 0, \qquad \text{where} \qquad Q' = \frac{dQ}{dq} = \frac{\cot\alpha}{q}.$$

Then, from the first and third of these equations,

$$\frac{\partial q}{\partial n} = -\cot\alpha\,\frac{\partial q}{\partial s} = \cot\alpha\,\frac{1}{Q'}\frac{\partial\theta}{\partial s} = q\,\frac{\partial\theta}{\partial s},$$

and from the second and third equations,

$$\frac{\partial\theta}{\partial n} = -\cot\alpha\,\frac{\partial\theta}{\partial s} = \cot\alpha\,Q'\,\frac{\partial q}{\partial s} = \frac{\cot^2\alpha}{q}\frac{\partial q}{\partial s},$$

and these are the equations (16.7).

We now state the definition: *A plane, steady, irrotational flow in a region R is called a simple wave if one set of Mach lines consists of straight lines on each of which* $\mathbf{q}$ = *constant. The image of R in the hodograph is an arc of a Γ-characteristic. If it is a Γ^+ ("forward" wave), then $Q - \theta$ has a constant value throughout R; if it is a Γ^- ("backward" wave), the same will hold for $Q + \theta$.**

Since pressure, density, and absolute temperature are functions of q, the straight Mach lines are also the isobars, isotherms, etc. For a forward wave, the *straight characteristics* are the C^- and the others, the *cross-characteristics*, are the C^+. In a backward wave the straight Mach lines are the C^+ and the C^- are the cross-characteristics.[26]

The flow pattern introduced here forms the intermediate case between the general case, where a region of the physical plane is mapped onto an area of the hodograph, and the completely degenerate case, where $\mathbf{q}$ =

* The terms "forward" and "backward", which have here no particular physical meaning, are used in analogy to Art. 13.

constant in a region of the x,y-plane and this whole region is mapped onto a single point. We have seen that throughout the simple-wave region $i = \partial(q,\theta)/\partial(x,y) = 0$. Therefore, the exchange of the variables x,y and q,θ, essential for the hodograph method, is not possible here, and simple waves are indeed "lost solutions", which cannot be obtained as solutions of linear equations in the hodograph plane.

The following is a basic property of simple waves: *Adjacent to a region of constant state is either another region of constant state or a simple wave.* In other words, a region of constant state which maps onto a single hodograph point cannot be directly adjacent to a region of general flow which corresponds to an area in the hodograph. A simple wave must form the link between them. To prove this statement, consider a region R_1 of constant state and some region R_2 of nonconstant state adjacent to it. The line separating these two flow regions of different types must be an arc S of a characteristic, say a C^+, since, as was shown in Art. 9, it is only across such a curve that the derivatives of the flow variables can change discontinuously. Since S belongs to R_1, it carries constant values of $\mathbf{q}$, and therefore its image in the hodograph is a point P. Through each point of R_2, passes a characteristic C^-. Consider the subregion R of R_2 whose C^--characteristics intersect the segment S. The image of each C^- in R must lie on a characteristic Γ^-, passing through the point P. Since there is, however, only *one* Γ^- passing through a given point, the images of all these C^- lie on one and the same $\Gamma^- = \Gamma_0^-$. Hence the flow in R is a simple wave.

A simple wave can connect any uniform supersonic state $q = q_1$, $\theta = \theta_1$ with another uniform supersonic state $q = q_2$, $\theta = \theta_2$ provided either $Q + \theta$ or $Q - \theta$ has the same value in both states. By combining a forward wave and a backward wave, and inserting a uniform state between the two, a given final state q_2, θ_2 can be reached, in general, and in many cases in two ways.

An individual wave may be specified in several ways. We may, e.g., *designate a certain characteristic Γ_0^- as the image of the whole backward wave,* and in addition give in the x,y-plane *a family of straight lines to represent the straight C^+*. If these C^+ all have a point in common, that is, if their envelope degenerates to a point, we speak of a *centered wave.*[27] The velocity distribution for the wave, centered or not, follows easily from the above definitions. Call ϕ the angle which a straight Mach line, a C^+ in a backward wave, a C^- in a forward wave, makes with the positive x-direction. Then the two relations hold:

$$(2) \qquad Q(q) \mp \theta = \text{constant}, \qquad \theta \mp \alpha = \phi,$$

where, as in the remainder of this article, the upper (lower) sign holds for a forward (backward) wave. The constant is known since the Γ_0^+ (Γ_0^-) is assumed given and ϕ, the angle made by the straight C^- (C^+) with the x-axis, is known. Therefore, along each straight Mach line the velocity $\mathbf{q}$ is determined by (2).

We know that the C^- (C^+) through a point P is perpendicular to the Γ^+ (Γ^-) through the corresponding point P' in the hodograph. Hence a straight C^- in a forward wave is perpendicular to the tangent to the fixed Γ_0^+ at the point P' which corresponds to that whole line C^-. To sum up: denote by θ_t the angle between the q_x-axis and the "initial direction" of the Γ_0^+ in the hodograph (see Fig. 105), by ϕ_t^- the angle from the x-axis to the "initial" C^- (i.e., the C^- which corresponds to the sonic point of Γ_0^+), and by 2ξ the given constant which singles out the particular Γ_0^+, as in Eq. (2) or in Eq. (16.43). Then, under the assumption of a polytropic gas, we have the following relations.

In a *forward wave* (Fig. 105)

$$Q - \theta \equiv \alpha + h\sigma - \theta = 2\xi, \qquad \theta - \alpha = \phi^-, \tag{3}$$

and consequently, using the first line of Table III (Sec. 16.6), or Fig. 105,

$$\begin{aligned} 2\xi &= 90° - \theta_t = -\phi_t^-, \qquad & \phi^- - \phi_t^- &= h\sigma, \\ Q &= 90° + (\theta - \theta_t), \qquad & \phi^- &= h\sigma - 2\xi. \end{aligned} \tag{3'}$$

In a *backward wave* (Fig. 106)

$$Q + \theta \equiv \alpha + h\sigma + \theta = 2\eta, \qquad \theta + \alpha = \phi^+, \tag{4}$$

and consequently

$$\begin{aligned} 2\eta &= 90° + \theta_t = \phi_t^+, \qquad & \phi^+ - \phi_t^+ &= -h\sigma, \\ Q &= 90° - (\theta - \theta_t), \qquad & \phi^+ &= 2\eta - h\sigma. \end{aligned} \tag{4'}$$

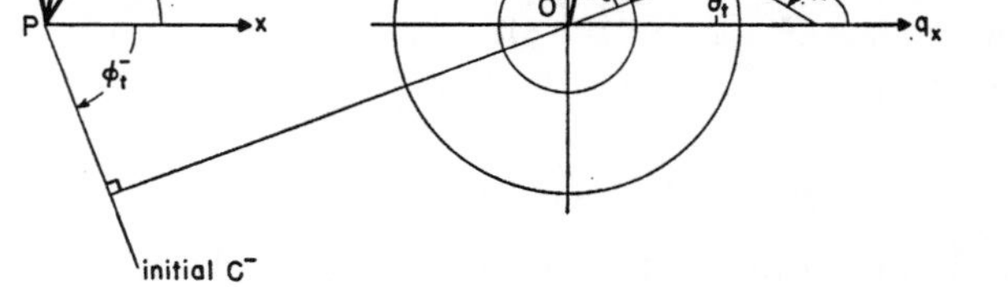

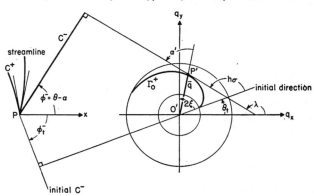

FIG. 105. Forward wave in physical plane and hodograph plane.

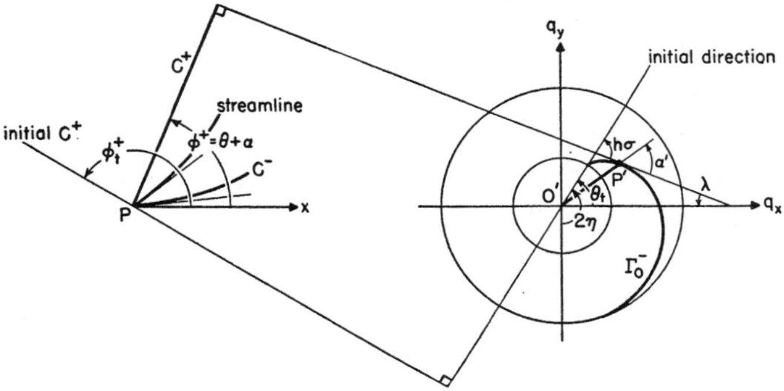

FIG. 106. Backward wave in physical plane and hodograph plane.

The last equation (4′) [and similarly (3′)] gives the relation between the angle of inclination ϕ^+ of any straight C^+ and the velocity q along it, since by Eq. (16.39′), σ depends only on q.* Another useful relation is

$$(5) \qquad \theta - \theta_t = \pm(h\sigma - \alpha'),$$

where $\alpha' = 90° - \alpha$ is the angle, introduced in Sec. 16.6, between the direction of a Γ-characteristic at a point P' and the radius vector $O'P'$.

2. Numerical data. Streamlines and cross Mach lines

We now take $\kappa = \gamma = 1.4$ in a polytropic (p,ρ)-relation. Consider the particular forward wave with $2\xi = 90°$ which corresponds to the Γ_0^+: $\theta = \alpha + h\sigma - 90°$. Remembering that $\sigma_t = 0°$ and $\alpha_t = 90°$, we see that $\theta_t = 0°$, and that $\phi^- = \theta - \alpha = h\sigma - 90°$.

Likewise, for the Γ_0^--wave with $2\eta = 90°$, or $\theta = 90° - \alpha - h\sigma$, we have $\theta_t = 0°$, $\phi^+ = \theta + \alpha = 90° - h\sigma$.

While the velocity varies from q_t to q_m and the Mach number M from 1 to ∞, the Mach angle α varies from 90° to 0° and σ from 0° to 90°. Taking the above waves, both with $\theta_t = 0°$, as representative, we see from the formulas that, for the forward wave, ϕ^- turns from $-90°$ to $(h - 1) \times 90° = 130.45°$, i.e., through 220.45°, and θ from 0° to 130.45°; for the backward wave, ϕ^+ turns from 90° to $-(h - 1) \times 90° = -130.45°$, i.e., through $-220.45°$, and θ from 0° to $-130.45°$. These facts are collected in Table IV.

* v. Mises, [26], uses $\lambda = 90° \pm \phi$ rather than ϕ, the angle λ being shown in our figures; in terms of λ the relation corresponding to the second equation in each of the sets (3′) and (4′) is then

$$h\sigma \pm \theta_t = \lambda,$$

with our previous sign convention.

TABLE IV
THE VARIATION OF SIGNIFICANT QUANTITIES IN FORWARD AND BACKWARD SIMPLE WAVES

	Forward Wave $\Gamma_0^+: \theta = \alpha + h\sigma - 90°, \phi^- = \theta - \alpha$		*Backward Wave* $\Gamma_0^-: \theta = -\alpha - h\sigma + 90°, \phi^+ = \theta + \alpha$
q	q_t	$\rightarrow$	q_m
M	1	$\rightarrow$	∞
α	90°	$\rightarrow$	0°
α'	0°	$\rightarrow$	90°
σ	0°	$\rightarrow$	90°
Q	90°	$\rightarrow$	220.45°
ϕ	$-90° \rightarrow 130.45°$		$+90° \rightarrow -130.45°$
θ	$0° \rightarrow 130.45°$		$0° \rightarrow -130.45°$

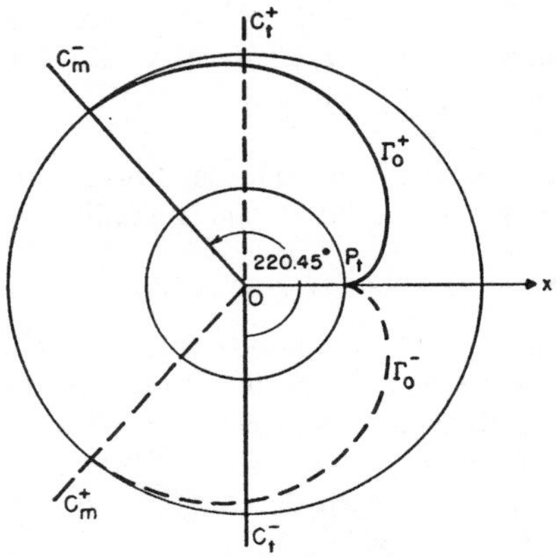

FIG. 107. Forward and backward simple waves with $\theta_t = 0$.

Figure 107 shows in one and the same figure the hodographs of the Γ_0^+ and the Γ_0^-, with $\theta_t = 0°$ for both, as well as the pairs of straight characteristics in the physical plane, C_t^+, C_m^+, and C_t^-, C_m^-. The backward wave Γ_0^- extends from the dashed line C_t^+ (which corresponds to the point P_t), turning clockwise by 220.45° towards the dashed line C_m^+; the complete forward wave begins at the solid line C_t^- (likewise corresponding to P_t) and turns in the counterclockwise sense by 220.45° towards C_m^-.

In Fig. 108 the deflection angle θ versus Mach number M, and likewise θ versus α, is plotted.

For *any* wave, with arbitrary θ_t, the first six lines of Table IV remain the same, while in the seventh line we must then write as entry $\phi - \phi_t \mp 90°$

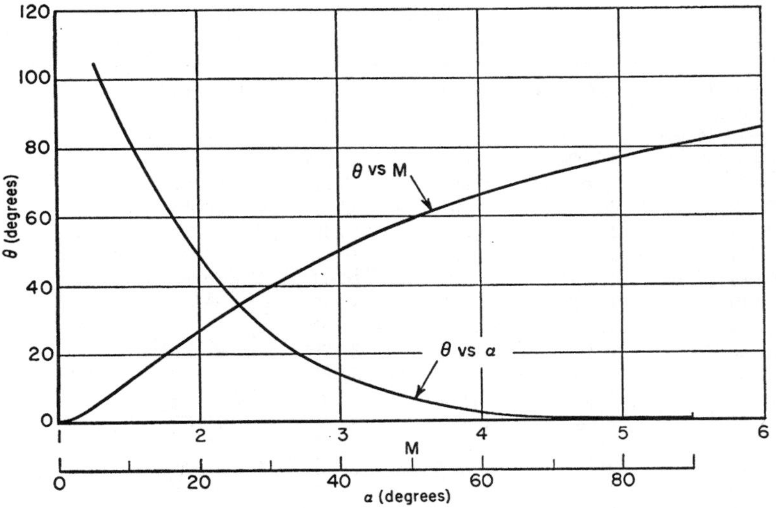

FIG. 108. Deflection angle θ versus Mach number M and versus Mach angle α.

instead of ϕ, and in the last line $\theta - \theta_t$ instead of θ; in (3′) and (4′) we still have $\phi_t^+ = \theta_t + 90° = 2\eta$ and $\phi_t^- = \theta_t - 90° = -2\xi$, respectively.

For supersonic flow and $\kappa = \gamma = 1.4$, Table V gives a tabulation of various quantities which characterize a simple wave: $\theta - \theta_t$, M, α, $\phi - \phi_t$, p/p_t and Q. The relations between the various angles are given by Eqs. (3) and (4). Of course, we could add to the table corresponding values of ρ/ρ_t, T/T_t, etc. In Table I (Sec. 8.4) we tabulated p/p_s, ρ/ρ_s and T/T_s and

$$p/p_t = (p/p_s)/0.5283, \quad (\rho/\rho_t) = (\rho/\rho_s)/0.6339, \quad T/T_t = (T/T_s)/0.8333.$$

We have seen in the preceding section that the velocity distribution over the set of straight Mach lines is determined if we know the epicycloid in the hodograph which is the image of the simple wave (that is, its "label" 2ξ or 2η) and the inclination ϕ of each straight Mach line. Consider, for example, $2\eta = 120°$, i.e., the backward wave $Q + \theta = 2\eta = 120°$. We want the velocity vector $\mathbf{q}$ along the straight Mach line of slope, say, tan 28°. Here $\phi_t = 120°$, $\theta_t = 120° - 90° = 30°$, and the range of ϕ-values in the physical plane is from 120° to $-100.45°$; hence $\phi = 28°$ is in this range. Corresponding to $|\phi - \phi_t| = 92°$ we find in the table the values $M = 2.132$, $\theta_t - \theta = 30°$ (for in a backward wave $\theta_t \geqq \theta$), and $\alpha = 27.97°$. Hence, since $\theta_t = 30°$, $\theta = 0°$, and for $M = 2.132$ we find, e.g., from Table I, $q/q_m = 0.6897$; thus the velocity vector along that Mach line is determined.

TABLE V

VARIOUS QUANTITIES WHICH CHARACTERIZE A SIMPLE WAVE (WHEN $\kappa = 1.4$)

$\lvert\theta - \theta_t\rvert$ (deg)	M	p/p_t	α (deg)	$\lvert\phi - \phi_t\rvert$ (deg)	Q (deg)
0	1.000	1.000	90	0	90
1	1.082	0.907	67.57	23.43	91
2	1.133	0.851	62.00	30.00	92
3	1.177	0.805	58.18	34.72	93
4	1.218	0.762	55.20	38.80	94
5	1.257	0.723	52.74	42.26	95
10	1.435	0.566	44.18	55.82	100
15	1.605	0.442	38.55	66.45	105
20	1.775	0.342	34.29	75.71	110
25	1.950	0.261	30.85	84.15	115
30	2.134	0.197	27.95	92.05	120
35	2.329	0.145	25.43	99.57	125
40	2.538	0.104	23.21	106.79	130
45	2.765	0.074	21.21	113.79	135
50	3.013	0.051	19.39	120.61	140
55	3.287	0.034	17.71	127.29	145
60	3.594	0.022	16.15	133.85	150
65	3.941	0.013	14.70	140.30	155
70	4.339	0.008	13.32	146.68	160
75	4.801	0.005	12.02	152.98	165
80	5.348	0.002	10.78	159.22	170
85	6.007	0.001	9.58	165.42	175
90	6.819	0.001	8.43	171.57	180
95	7.851	0.000	7.32	177.68	185
100	9.210	0.000	6.23	183.77	190
105	11.091	0.000	5.17	189.83	195
130.45	∞	0.000	0	220.45	220.45

Next let us consider *streamlines* and *cross-characteristics*. If we want to find the equation of either family we have to give the particular family of lines which form the straight characteristics. Let it be given in the form

$$y = \beta x + y_0(\beta). \tag{6}$$

So far, such information was not needed since the relation between the slope of a straight characteristic and the flow variables along it depends only on the label 2ξ or 2η.

Let (6) define C^+-lines $\beta = \tan(\theta + \alpha)$; since $Q(q) + \theta = 2\eta$, with η given, α and θ are known as functions of q, and hence of β. The differential equations of streamlines and cross-characteristics are

$$\frac{dy}{dx} = \tan\theta = k(\beta) \quad \text{and} \quad \frac{dy}{dx} = \tan(\theta - \alpha) = k_1(\beta),$$

respectively. By differentiation of (6)

$$dy = x\,d\beta + \beta\,dx + y_0'(\beta)d\beta$$

results, and substituting this in the equations of the streamlines and cross-characteristics, one obtains

$$(7) \qquad \frac{dx}{d\beta} = \frac{x + y_0'(\beta)}{k(\beta) - \beta} \qquad \text{and} \qquad \frac{dx}{d\beta} = \frac{x + y_0'(\beta)}{k_1(\beta) - \beta},$$

respectively.

Each of these is a linear differential equation of first order for $x = x(\beta)$, which upon integration provides, together with (6), a parametric representation of the respective family of curves. The constant of integration determines the particular streamline or cross-characreristic. In case of a wave centered at the origin, $y_0(\beta) = 0$ in (6).

It seems more practical, however, to give the family of straight Mach lines in a way better adapted to the problem, and to use a kind of generalized polar coordinates. It is easily seen that (in addition to the knowledge of 2η or 2ξ) the family of straight Mach lines is determined if merely *one* streamline or *one* cross-characteristic is known in the x,y-plane. In fact, knowing one streamline in the flow plane, we know θ at each point (x,y) of this line; then, in the case of a backward wave, $Q = 2\eta - \theta$ determines q and $\alpha(q)$. Therefore, at each point of the given streamline, the direction $\theta + \alpha$ of the C^+ through this point is known. The set can be determined analogously from the knowledge of one cross-characteristic, C^-; now $\theta - \alpha$ is known along this C^-, and $Q + \alpha = 2\eta - (\theta - \alpha)$ provides q. Then α, and finally $\theta + \alpha$ follow at all points of the C^-.

Consider now the first case where one streamline is given: $x = a(t)$, $y = b(t)$, $db/da = \tan\theta$, where t is a parameter (see Fig. 109). Consider

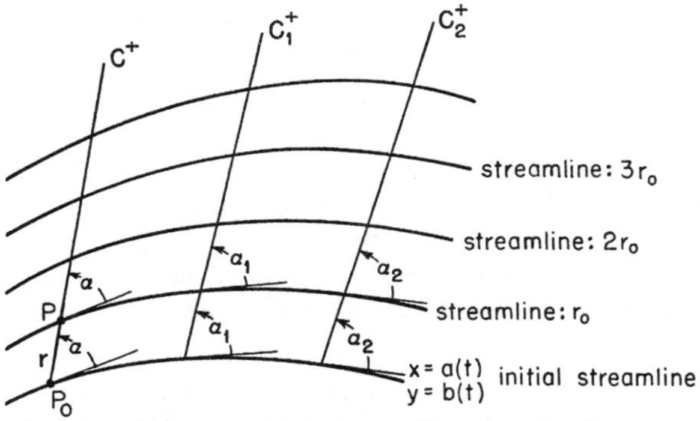

FIG. 109. A property of the streamlines in a simple wave.

again a backward wave. The straight characteristics are then the C^+. For a point $P(x,y)$ on the C^+ through P_0, $x = a + r\cos\phi^+$ and $y = b + r\sin\phi^+$, where $r = P_0P$. If we consider a, b, r, and $\phi^+ = \phi$ as functions of t, the equation of the streamline through P is

$$(db + r\cos\phi\, d\phi + dr\sin\phi)\cos\theta - (da - r\sin\phi\, d\phi + dr\cos\phi)\sin\theta = 0,$$

and, since $db\cos\theta - da\sin\theta = 0$, it follows, with $\phi - \theta = \alpha$, that $r\,d\phi\cos\alpha + dr\sin\alpha = 0$, or

$$\frac{r\,d\phi}{dr} = -\tan\alpha. \tag{8}$$

We note that this is the same equation which can be written immediately for a centered wave, using ordinary polar coordinates (r,ϕ). To integrate (8) we use Eq. (16.39′): $\tan\alpha = (1/h)\cot\sigma = (1/h)\cot[(2\eta - \phi)/h]$. Hence,

$$\frac{dr}{r} = h\tan\left(\frac{\phi - 2\eta}{h}\right)d\phi, \qquad r = r_0\left(\cos\frac{\phi - 2\eta}{h}\right)^{-h^2}. \tag{9}$$

It is seen that for $\phi = 2\eta = \phi_t$, $r = r_0$, while for the other extreme, $|\phi - \phi_t|$ tending to $h \times 90° \approx 220°$, the distance between the streamlines tends toward infinity.

In the same way, we find for the cross-characteristics the differential equation

$$\frac{r\,d\phi}{dr} = -\tan 2\alpha, \tag{10}$$

with the integral

$$r^2 = r_0^2\left(\sin\frac{\phi - 2\eta}{h}\right)^{-1}\left(\cos\frac{\phi - 2\eta}{h}\right)^{-h^2}. \tag{11}$$

Here r tends towards infinity when $\phi - \phi_t \to 0$ as well as when $\phi - \phi_t \to -220.45°$ (see Fig. 110a, b).*

From Eqs. (9) and (11) the following property of both the set of all streamlines and of all cross-characteristics follows: If the constant r_0 is changed to $2r_0$, $3r_0$, $\cdots$, the respective curves intersect any straight characteristic at equidistant points (see Fig. 109 and the end of Sec. 13.1).

In adapting the preceding discussion to a forward wave, we must change 2η to -2ξ, etc.

* We note that, even if we take an extremely small value for r_0, we can only sketch a small part of the complete streamline since r increases rapidly with $|\phi - \phi_t|$.

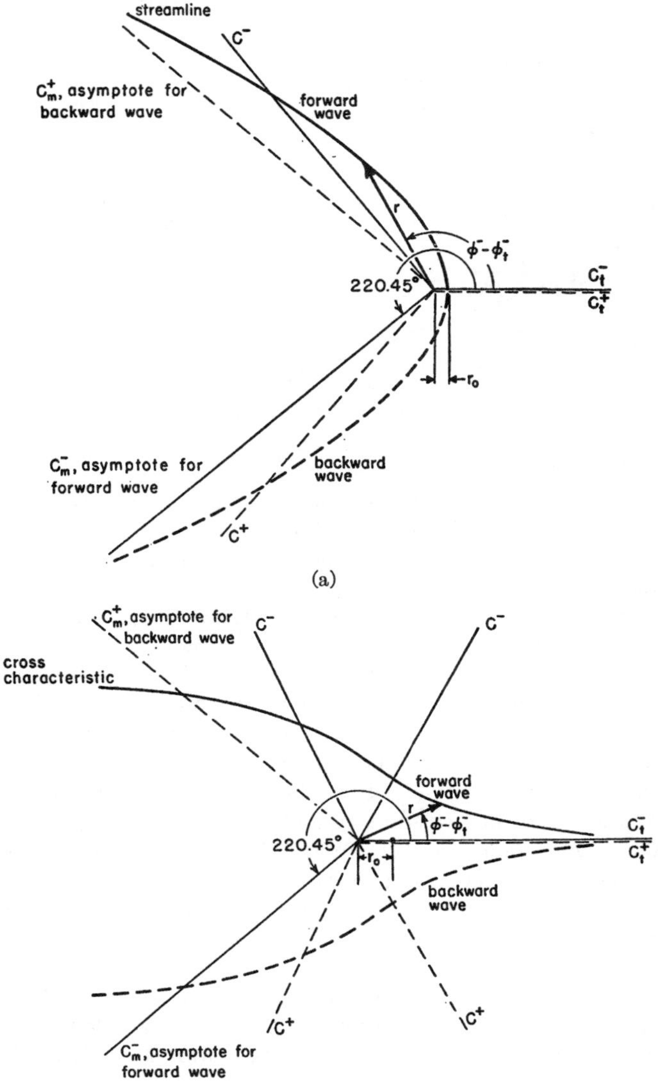

FIG. 110. (a) Streamline in a centered simple wave. (b) Cross-characteristic in a centered simple wave.

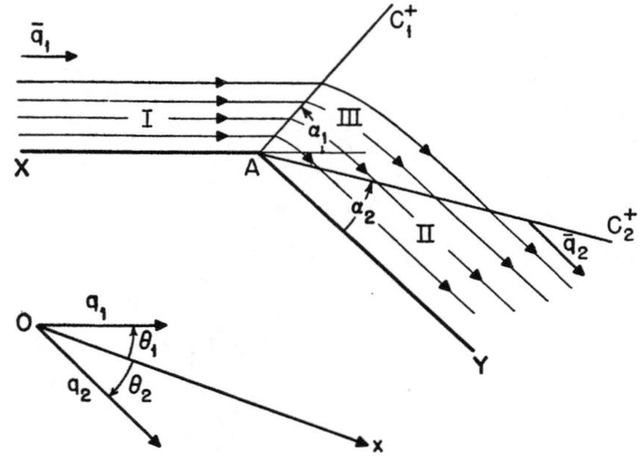

FIG. 111. Flow around a convex corner.

3. Examples of simple waves

(a) The most important example of a simple wave solution is the *flow around a convex corner* (see Fig. 111). We assume that the oncoming supersonic flow for which the straight wall XA is a streamline is *uniform* with given velocity $\mathbf{q}_1$, where θ_1 equals the angle of XA with the x-axis. The angle θ_2 of the velocity $\mathbf{q}_2$ is known along the streamline AY. [We note that these are not complete Cauchy data as discussed in Chapters II and III and in Sec. 16.4, since only θ is known along AY.] The flow is completely determined, as a uniform flow, to the left of the characteristic C_1^- through A, and this flow can change to a simple wave across a characteristic. The only condition beyond A is that AY be a streamline. The simplest type of solution compatible with these conditions is one in which the flow along AY is also uniform and the two states of uniform flow, I and II, are linked by a single centered simple wave III. With these data there exist in principle, as we shall see, *two* different simple wave solutions III, one a forward wave and one a backward wave, joining I and II. The numerical data, however, may be such that in a particular case only *one* solution, or perhaps none, exists.

This can easily be seen both arithmetically and geometrically. Since q_1 and θ_1 are known along XA, in particular at A, and a scale factor a_s or q_t or q_m is given, M_1 and $\alpha_1 = \text{arc sin } 1/M_1$ are known, and $Q(q_1)$ follows. To be specific we consider a backward wave. There is at A a $C^+{}_1$ which makes the angle α_1 with XA. Next, $Q(q_1) + \theta_1 = 2\eta$ determines η, and this holds throughout the wave.

On the other hand, the angle α_2, valid along AY, depends upon q_2 which we do not know; but if a backward wave III joins the two regions I and II, then

$$Q_1 + \theta_1 = Q_2 + \theta_2 ,$$

and, for the convex corner

$$\Delta\theta = \theta_1 - \theta_2 > 0,$$

where $\Delta\theta$ denotes the deflection angle. Since, according to Table IV, Q is always between 90° and 220.45°, we find the condition

$$Q_2 = Q_1 + \Delta\theta < 220.45°.$$

Similarly, in the case of a forward wave, $Q_1 - \theta_1 = Q_2 - \theta_2$ must hold, and, therefore

$$Q_2 = Q_1 - \Delta\theta > 90°.$$

If both inequalities are satisfied, we have two different solutions, which we shall characterize presently. It is now clear that the given data may also be such that only one, or neither of these solutions exists.

In geometrical terms, from a point P_1 with coordinates (q_1, θ_1) in the epicycloidal ring one can reach a point with given $\theta_2 < \theta_1$ (i.e., on a given radius) by moving along the Γ^- through P_1 until the radius in question is reached, provided this happens before being stopped at the maximum circle. Or one can move along the Γ^+ through P_1 until the radius in question is reached, provided this happens before meeting the sonic circle.

Numerical illustrations (for $\kappa = 1.4$, $h = \sqrt{6}$).[28] The above inequalities may be visualized by plotting in a $\Delta\theta, M$-plane the curves $Q = 90° + \Delta\theta$ and $Q = h \times 90° - \Delta\theta$ which intersect for $\Delta\theta \approx 65°$, $M \approx 3.94$. These curves together with the line $M = 1$ and the two asymptotes delimit four regions corresponding to the above-mentioned four possibilities (Fig. 112).

1. Assume, e.g., $\theta_1 = 0°$, and $\Delta\theta = \theta_1 - \theta_2 = 38°$, $M_1 = 1.5$. From trigonometric tables or from Table V, $\alpha_1 = 41.8°$, $\sigma_1 = 24.6°$, $Q_1 = h\sigma_1 + \alpha_1 = 2\eta = 101.9° = -38° + Q_2$; thus, $Q_2 = 139.9°$, which gives $\alpha_2 = 19.5°$, $M_2 = 3$. Hence everything is determined. The streamlines in the angular space between C_1^+ and C_2^+ are given explicitly by Eq. (9) with $2\eta = \phi_t = 101.9°$. Here q and M increase through the wave while p, ρ, T, and α decrease. Such a wave is called an *expansion wave* (see Fig. 113.1). For these numerical data, $Q_1 - \Delta\theta = 101.9° - 38° < 90°$; hence, *no second solution exists*. In fact, the point P_1 with coordinates $\Delta\theta = 38°$, $M = 1.5$ lies, in Fig. 112, in the region where only a backward wave exists.

2. Consider the same corner, $\theta_1 = 0°$, $\theta_2 = -38°$, but now let $M_1 = 3$, so that $\alpha_1 = 19.5°$. Then, $\sigma_1 = 49.1°$, $Q_1 = \alpha_1 + h\sigma_1 = 139.9°$ (equal to the

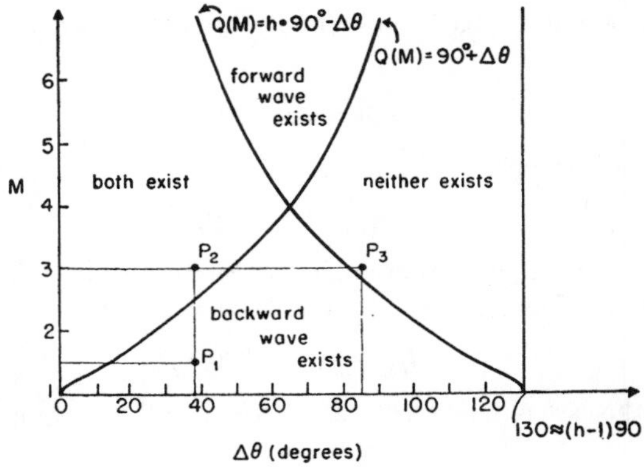

FIG. 112. Regions of two, one or no simple wave solutions.

previous Q_2). (a) $Q_1 - \Delta\theta = 139.9° - 38° = 101.9° = Q_2$; $M_2 = 1.5$, $\alpha_2 = 41.8°$. This forward wave is the reverse of the backward wave considered before. Here Q and M decrease through the wave, while pressure, density, and Mach angle increase through this *compression wave* (see Fig. 113.2a). (b) For these same data: $\theta_1 = 0°$, $M_1 = 3$, ($\alpha_1 = 19.5°$, $Q_1 = 139.9°$), $\Delta\theta = 38°$, a *second* solution, an expansion wave, also exists since $Q_1 + \Delta\theta = 139.9° + 38° = 177.9° = Q_2 < 220.5°$. To this Q_2 corresponds an M greater than 6, $M_2 = 6.4$, with an α_2 close to 9° (see Fig. 113.2b). In this case, therefore, we have *two different simple wave solutions* of the type under consideration (see in Fig. 112, the point P_2 with coordinates $\Delta\theta = 38°$, $M = 3$). More data are necessary to determine which of these two solutions will materialize.

3. Next, if $M_1 = 3$, $\Delta\theta = 85°$, then $Q_1 - \Delta\theta = 139.9° - 85° < 90°$, and for these initial data no compression-wave solution exists. Also $Q_1 + \Delta\theta = 139.9° + 85° = 224.9° > 220.5°$. Hence no expansion-wave solution exists either. In fact, the point P_3 in Fig. 112 with $\Delta\theta = 85°$, $M = 3$ is in the right-hand region where no solution exists. If we use, as on previous occasions, the term cavitation or vacuum for a state with $p = \rho = 0$, $q = q_m$, we can say that a zone of cavitation will exist between the C_2^+ which makes the angle $-80.6°(=139.9° - 220.5°)$ with the x-axis and the given second wall which makes the angle $-85°$ with the x-axis.

(b) *Flow around a convex bend.* The boundary is now a convex polygon or a continuous bend. We consider first the latter case. Neither presents an essentially new problem compared to case (a). Again, we assume the oncoming flow as uniform. At the point where the bend begins q_1, θ_1 are given, as well as a scale constant, a_s or a_t, of the oncoming uniform flow. The bend extends between two straight walls XA and BY (see Fig. 114), and the geometric shape of the whole wall $XABY$ is given; since this wall is a streamline, θ is given along it. Again $2\eta = Q(q_1) + \theta_1 = Q(q_2) + \theta_2$ determines q_2 at B for an expansion-wave solution, provided $Q(q_1) + \Delta\theta <$ 220.45°, where $\Delta\theta = \theta_1 - \theta_2 > 0°$; likewise if $Q(q_1) - \Delta\theta > 90°$, a compression wave can link the two constant states. The first Mach line, say C_A^+ in the case of the expansion wave, is determined by the angle $\alpha_1 = \alpha(q_1)$ which it makes with XA at A. The last, C_B^+, makes the angle $\alpha_2 = \alpha(q_2)$ with BY. At each intermediate point of the bend the angle θ is known and, always by means of $2\eta = \theta + Q(q)$, the corresponding q, M, and α follow.

If we prefer a geometric procedure, then in an epicycloidal diagram constructed as in Art. 16 we identify the radius of the inner circle with a_t (thus determining the scale); next, we look in the diagram for the point A' with the given polar coordinates q_1, θ_1. Through A' goes a Γ_0^- in the direction of decreasing θ toward the maximum circle, and a Γ_0^+ (not shown in figure) in the direction of increasing θ toward the sonic circle. Consider first the Γ_0^-. In order to find graphically the α for an arbitrary point P of the bend, i.e., the direction of C_P^+, we draw the tangent to the bend at

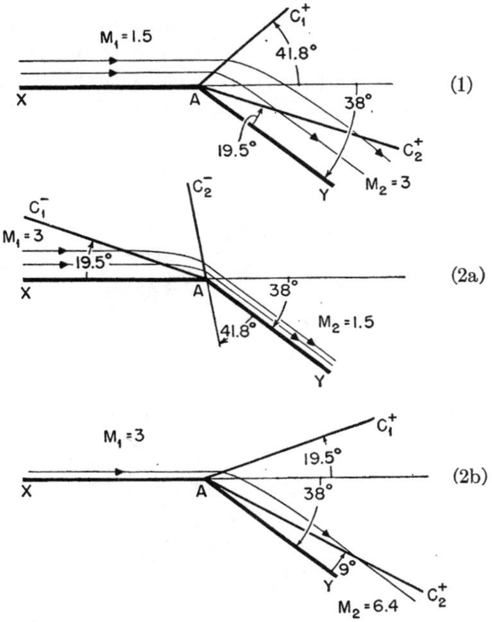

FIG. 113. Simple waves turning the same convex corner: (1) $M_1 = 1.5$, backward expansion wave, only solution. (2a) $M_1 = 3$, forward compression waves; (2b) $M = 3$, backward expansion wave.

P and a radius parallel to that tangent through O', the origin in the hodograph plane. This parallel intersects Γ_0^- at a point P' and q_P is equal to the distance $O'P'$. The tangent at P' to Γ_0^- is perpendicular to the direction of C_P^+ (since $C_P^+ \perp \Gamma_{P'}^-$).

In exactly the same way a compression solution may be constructed using the Γ_0^+ through A'. Of course, it can be decided beforehand by the abovementioned criteria whether two solutions exist or one or none. We note that in both cases the straight Mach lines diverge: as we proceed from A toward B, the angle θ decreases. In the first case, the C^+-lines point from the bend downstream, and the angle α which they make with the positive (downstream) direction of the bend decreases (see Fig. 115a); in the second case (Fig. 115b), the C^--lines make with the negative (up-stream) direction an increasing angle α, and hence, with the positive direction a decreasing angle.

The approach is the same for a convex polygon. If the oncoming uniform flow is completely given, several solutions may exist for which $ABCD$ (see Fig. 116) is a streamline and which are such that three different uniform states, K_1, K_2, K_3 are linked by centered waves, W_1 and W_2.

It is now easy to realize the connection and complete agreement between our present results and the uniqueness theorems for hyperbolic boundary-value problems (Arts. 10 and 16). Let us reason in terms of the continuous bend (Fig. 114), where the position is perhaps clearer than for the polygon or

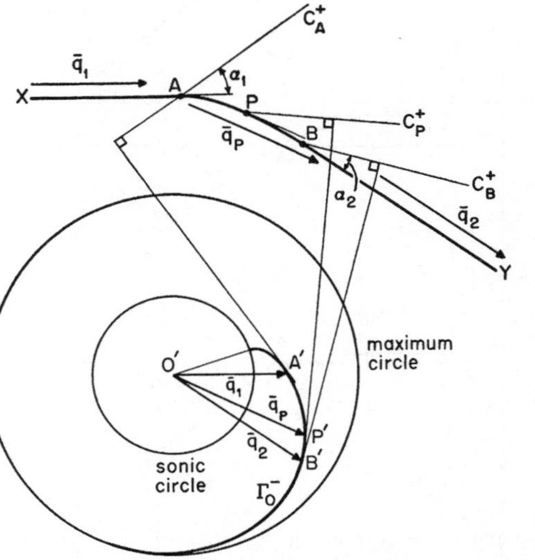

FIG. 114. Simple wave turning a bend.

corner. Consider, e.g., a case with two solutions. The situation is the following. As long as we know only the shape of the streamline $XABY$ and the q_1 on XA, we do not have Cauchy data; to have Cauchy data we must know $\mathbf{q}$ along the whole noncharacteristic curve $XABY$. This may be prescribed in many ways. The type of solution we are considering here, which consists of a simple wave joining two constant states, is just one such solution, though a particularly simple one. Thus, if we decide to look for a solution which is a simple wave along AB, in particular an expansion wave, we must give $\mathbf{q}$ along AB in a way compatible with an expansion wave. In other words, we have to prescribe $\mathbf{q}$ along AB so that we are led from the given q_1, θ_1 to the given θ_2 along a characteristic Γ_0^-. This is actually achieved by the above-explained construction, which provides $\mathbf{q}$ all along $XABY$. Thus, in a sense, the simple-wave construction serves to supplement the boundary

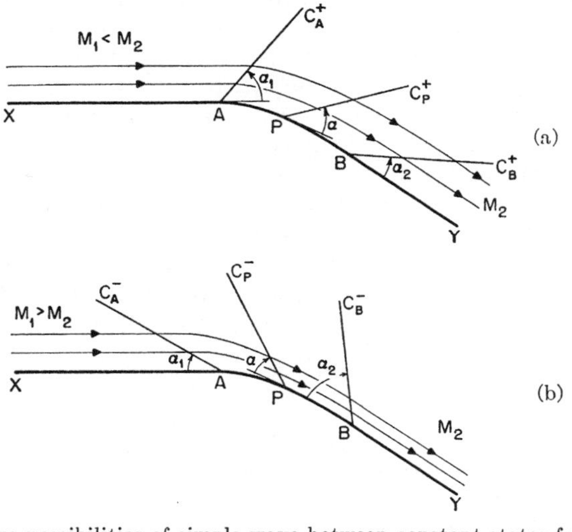

FIG. 115. Two possibilities of simple wave between constant states for the same bend: (a) Backward expansion wave; (b) Forward compression wave.

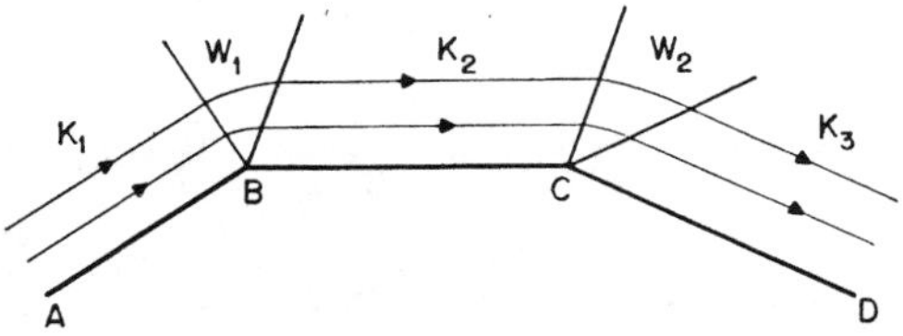

FIG. 116. Flow along a convex polygon.

data for the Cauchy problem; the situation is obscured by the fact that in a simple wave the distribution of $\mathbf{q}$ everywhere in the wave region follows in an immediate way from the distribution along the curve AB. On the other hand, construction of a compression wave along the bend involves other Cauchy data, namely, data compatible with a compression wave; corresponding to these different boundary data, there is then again one solution. If, as seen in a preceding example, a compression wave turns out to be impossible, this simply means that one cannot prescribe initial data compatible with that shape of bend and a compression wave, just as—to quote a trivial instance—one cannot "prescribe" a varying value of θ along a straight wall. If, finally, $\mathbf{q}_1$ and the shape of the bend are such that both an expansion and a compression wave are possible, it means (far from contradicting the uniqueness theorem) merely that of two quite different sets of Cauchy data along $XABY$, one furnishes an expansion wave, the other a compression wave. And, these lead to two different constant states.

(c) *Flow turning in a concave bend* (see Fig. 117). Here $\theta_1 - \theta_2 = \Delta\theta < 0$. We start with two numerical examples.

1. Let $M_1 = 1.22$; hence $\alpha_1 = 55.9°$, $Q_1 = 94°$. Let $\theta_1 = 0°$, $\theta_2 = 10°$, so that $\Delta\theta = \theta_1 - \theta_2 = -10°$ (Fig. 117.1). Consider a *forward* wave (straight C^--lines inclined upstream). Then from $Q_1 - \theta_1 = Q_2 - \theta_2$, $Q_2 = 94° + 10° = 104°$, from which follow $M_2 = 1.57$, $\alpha_2 = 39.6°$. This is an *expansion* wave which can turn the flow in the desired direction, since $Q_2 = Q_1 + \theta_2 - \theta_1 < 220.45°$. In this example, i.e., for this bend and $M_1 = 1.22$, a backward wave cannot be constructed since $Q_1 + \Delta\theta = 94° - 10° < 90°$.

2. Now take $M_1 = 1.57$, $\alpha_1 = 39.6°$, $Q_1 = 104°$ (Fig. 117.2a). Here

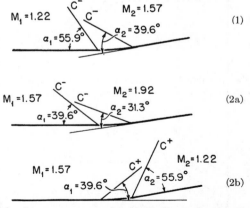

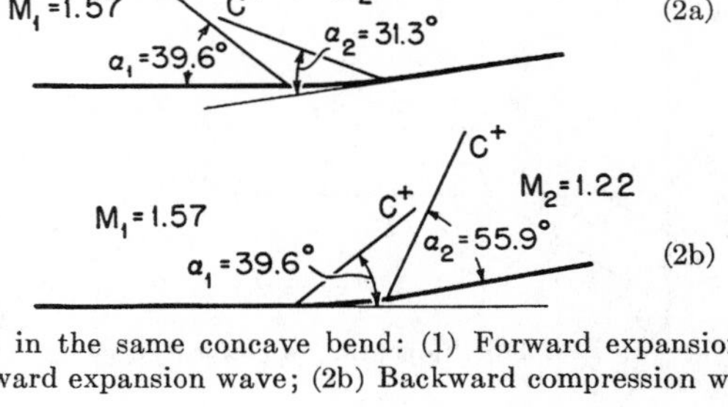

FIG. 117. Flows in the same concave bend: (1) Forward expansion wave, only solution; (2a) Forward expansion wave; (2b) Backward compression wave.

$Q_1 - \theta_1 = Q_2 - \theta_2$, with $\theta_1 = 0°$ and $\theta_2 = 10°$, gives $Q_2 = 114°$, $M_2 = 1.92$, $\alpha_2 = 31.3°$, a forward wave. In the case shown in Fig. 117.2b a backward wave can also be constructed, the reverse of the first forward wave, leading to $Q_2 = 94°$, $\alpha_2 = 55.9°$, $M_2 = 1.21$ through a compression wave.

It is seen that in both these cases the straight Mach lines converge. Figure 118 shows this schematically. For the compression wave and the C^+ (Fig. 118a), both θ and α increase, while in the expansion case (Fig. 118b) the angle $180° - \alpha$, which the downstream-inclined C^- makes with the positive direction of the bend likewise increases. Hence in both cases the straight Mach lines converge; they will, in general, have an envelope, and a simple wave solution exists only in a certain neighborhood of the bend, such that the envelope lies outside this neighborhood. The extent of such a neighborhood depends on the ratio of $\Delta\theta$ to the increment in the arc length, i.e., upon the curvature of the bend; in particular, *no* simple wave solution exists in the case of a sharp concave corner. This case can be dealt with by means of a solution involving a shock (see Art. 22). Figure 118c shows such an envelope or limit line, $\mathfrak{L}$, in the case of a smooth concave bend; the limit line is cusped. Through the point P passes *one* straight Mach line, and on it velocity, pressure, etc., are uniquely determined. Through the point Q, however, pass in general three straight Mach lines, each defining a different "flow"—a situation with no physical reality.

(d) *Successive simple waves.* Consider a combination of the cases considered in the two preceding subsections. Assume, e.g., a contour of the shape indicated in the first Fig. 119. Two regions of constant state XA and DY are joined by a contour which has first positive, then negative, then again positive curvature along AB, BC, and CD, respectively. Assume q_{I} known. Again we have incomplete Cauchy data: both q and θ are given along XA, but θ alone is known along $ABCDY$. We can link the regions of constant state I and II by three simple waves. The two possible solutions of this kind are shown in the hodograph: (a) 12, 213, 31: compression wave, expansion wave, compression wave; (b) 14, 415, 51: expansion, compression, expansion. The end velocity q_{II} is in both cases equal to q_{I}.

4. More elaborate examples involving simple waves

We consider here several problems of *two-dimensional ducts.*[29] At the end of Sec. 16.4 we saw that the supersonic flow in a duct is determined if the velocity $\mathbf{q}$ along an entrance section AB and the form of the two walls is given. Carrying out the required steps (see Fig. 91) for an arbitrary velocity distribution $\mathbf{q}$ along AB and arbitrary shapes of walls is difficult, and in general approximations will be needed. However, under certain circumstances a solution involving simple waves exists, and this simplifies the problem.

(a) *Uniform entrance velocity. General case.* Consider a duct whose walls are straight and parallel at the entrance, then begin to bend (see Fig. 120). At the entrance a uniform velocity $\mathbf{q}_0$ is given; hence $XAD\bar{A}\bar{X}$ will be a region of uniform velocity, where AD and $\bar{A}D$ are straight Mach lines

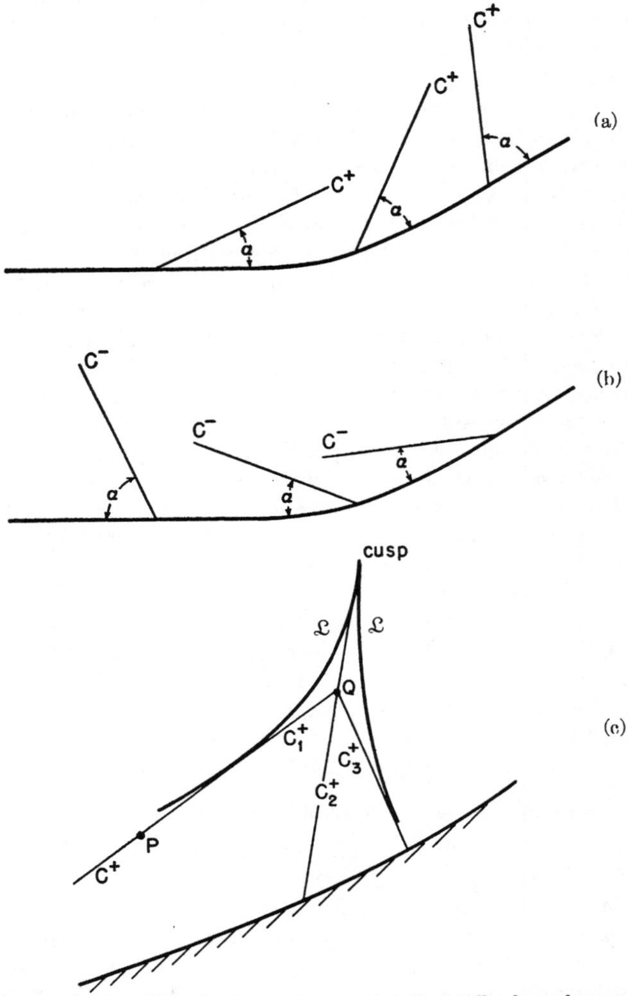

FIG. 118. Convergent Mach lines in concave bend: (a) Backward compression wave; (b) Forward expansion wave; (c) Envelope with cusp—three straight characteristics through Q, one through P.

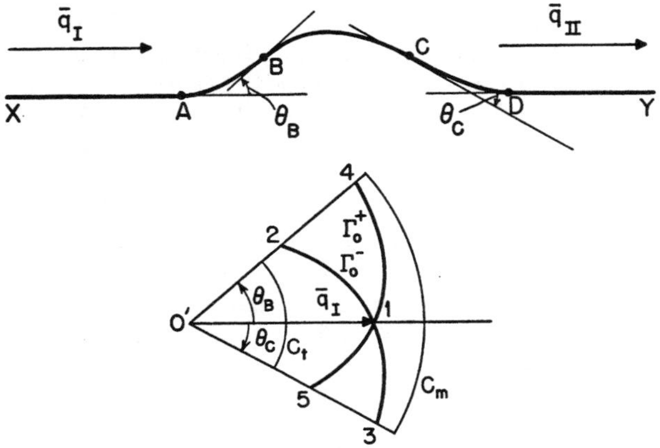

FIG. 119. Successive simple waves in physical plane, and hodograph plane.

whose slopes are given, since $\mathbf{q}_A = \mathbf{q}_{\bar{A}} = \mathbf{q}_0$. Adjacent to the straight characteristic AD is a simple forward wave with straight Mach lines C^-, and adjacent to $\bar{A}D$ a backward wave with straight Mach lines C^+. The cross-characteristics C_D^+ and C_D^- issuing from D are both curved; they intersect the walls at the points E and $\bar{E}$ respectively. Assume first that the walls are still curved at these points. Since we know $\mathbf{q}$ at D, we know the constant 2ξ of the forward wave ADE, and since θ is known along the streamline AE, the q-values follow there. Therefore we know $\mathbf{q}$ along each of the straight C^- in ADE, and this determines C_D^+ (see end of Sec. 16.4). A corresponding graphical procedure would be the following. We proceed from D in continuation of the direction $\bar{A}D$ to the intersection with a straight C^- close to D; at this point of intersection we again know q and θ, hence the direction of a short segment of the line, and we can go on. Similar considerations hold for the wave $\bar{A}D\bar{E}$.

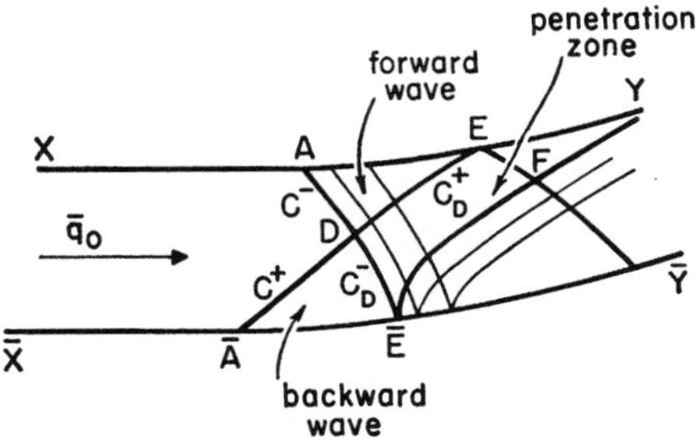

FIG. 120. Duct with uniform entrance velocity: general case.

Now if, as assumed, the walls at E and $\bar{E}$ are curved, there are in general no other simple waves, and we are again faced with the general problem. Knowledge of q (and compatible θ) along the characteristics ED, $\bar{E}D$ determines the flow in the characteristic quadrangle $ED\bar{E}F$, the "penetration zone". Next, the flow may be found in the triangular region limited by EF, by the characteristic which continues $\bar{E}F$, and by a piece of the upper wall along which θ is given, and similarly for the lower wall. This is the type of "mixed" problem discussed in Arts. 10 and 16 where values are given along two arcs, one of which is characteristic. As the next step we have to solve again a characteristic boundary-value problem, etc.

(b) *Short bend between straight walls.* The situation becomes simpler if, after a short bend starting at A, the wall again becomes rectilinear, so that E and $\bar{E}$ lie again on straight walls (see Fig. 121). Up to $AD\bar{A}$ there is uniform flow K_0 ; then, as before, two simple waves follow, the forward wave W_A and the backward wave $W_{\bar{A}}$. Again the cross-characteristics DC and $D\bar{C}$ can be found. So far it is the same as before. Now, however, since the wall beyond B is straight and the C^--characteristic BC is also straight,

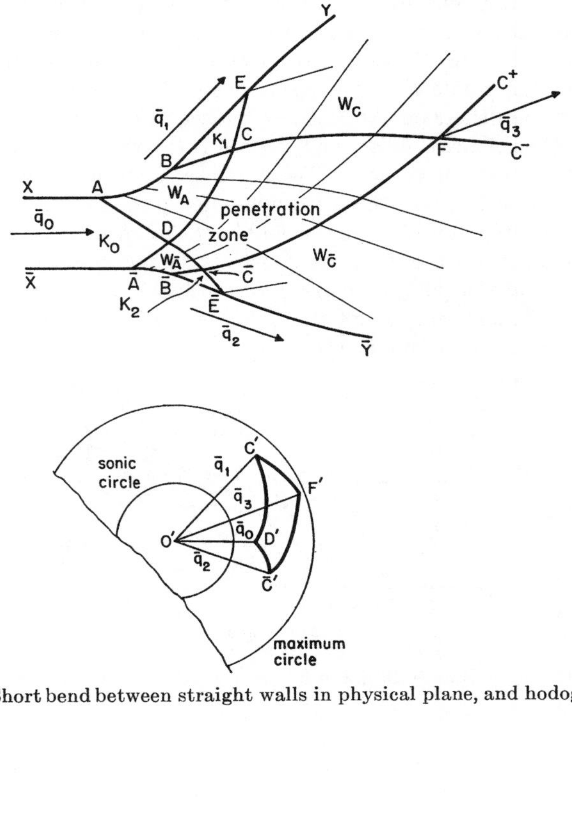

FIG. 121. Short bend between straight walls in physical plane, and hodograph plane.

a region of constant flow is determined in the triangle BCE, where CE is a straight C^+ of known direction. Everything is similar in $\bar{B}\bar{C}\bar{E}$. There follows, next, a penetration zone, $\bar{C}DCF$, where CF and $\bar{C}F$ are characteristics. In this zone there is general flow, determined by the characteristic boundary-value problem. Now, however, adjacent to the triangle BCE of constant state, there is again a simple wave, W_C, with straight C^+-lines, and similarly there is a forward wave $W_{\bar{C}}$ adjacent to $\bar{C}\bar{E}$. Since q and θ are known at C, we find the constant 2η of W_C by $Q_C + \theta_C = 2\eta$.

However the wave W_C with the cross-characteristic CF is not completely known, unless the penetration problem has been solved. Nevertheless the velocity $\mathbf{q}_3$ at F is known without the penetration problem being solved, as can be seen by considering the hodograph (Fig. 121). First the vector $\mathbf{q}_0 = \overline{O'D'}$ is located. From D' start the epicycloids $D'C'$ and $D'\bar{C}'$ limited by the radii $O'C'$ and $O'\bar{C}'$ which are parallel to BE and $\bar{B}\bar{E}$, respectively; these arcs $D'C'$ and $D'\bar{C}'$ are images of the waves W_A and $W_{\bar{A}}$, and the points C', $\bar{C}'$ represent the constant flows in BCE and $\bar{B}\bar{C}\bar{E}$. The images of the waves W_C and $W_{\bar{C}}$ are the epicycloidal arcs $C'F'$ and $\bar{C}'F'$, which intersect at F', and determine the vector $\overline{O'F'} = \mathbf{q}_3$.

In W_C the correspondence between θ and q is known but this is not enough. If, however, the shapes of the characteristics CF and $\bar{C}F$ are known (after the penetration problem has been solved), this determines the velocity distribution along CF, which is a cross-characteristic for W_C. Then, the velocity distribution everywhere in the "quadrilateral" W_C is known, in particular on the cross-characteristic issuing from E which limits W_C. At F there again begins a uniform flow, etc. Figure 122 shows schematically a flow in a duct whose walls are partly straight, where regions of constant flow, simple waves and general flows (shaded) alternate.

(c) *Particular cases.* Assume that in a situation such as that considered in (b) the C^- through A meets the opposite wall at a point where it is still straight, i.e., to the left of $\bar{A}$ (see Fig. 123). The characteristic $\bar{E}G$ is a cross-characteristic of the simple wave $A\bar{E}G = W_A$, and $W_{\bar{A}}$ drops out.

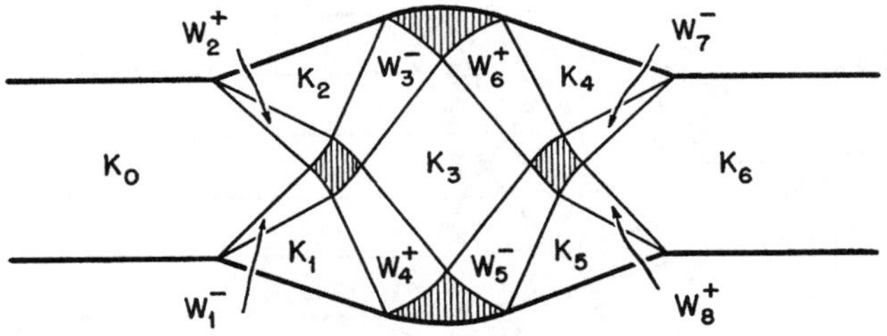

FIG. 122. Succession of constant flows, simple waves, and general flows in a symmetric duct.

Beyond $\bar{E}G$ the flow is no longer a simple wave and is determined as explained before. $\bar{E}G$ is sometimes called the *reflected Mach line of* $A\bar{E}$.

A further simplification arises if at the point $\bar{E}$ the wall has the shape of a streamline of the extended simple wave obtained from $A\bar{E}G$ by continuing the straight C^--characteristics across $\bar{E}G$; then this latter wave provides the continuation of the flow beyond $\bar{E}G$. This applies in the following construction.

(d) *Parallel exit out of a duct.* Consider a symmetric duct with otherwise general supersonic flow (neither uniform nor simple wave) in some region limited on the right by the characteristics ED, $\bar{E}D$ (see Fig. 124). We consider now the problem of *determining the shape* of the remaining part of the duct in such a way that the flow at the end is uniform and parallel to the horizontal axis of symmetry. A simple wave, or rather a symmetric pair of simple waves, may provide the transition. Beyond D, we want the characteristic line $\bar{E}D$ to continue as a straight C^+ of W_A, and similarly for ED and $W_{\bar{A}}$. This will be achieved if the wall EA is constructed so that it is a streamline of the simple wave W_A for which ED is a cross-characteris-

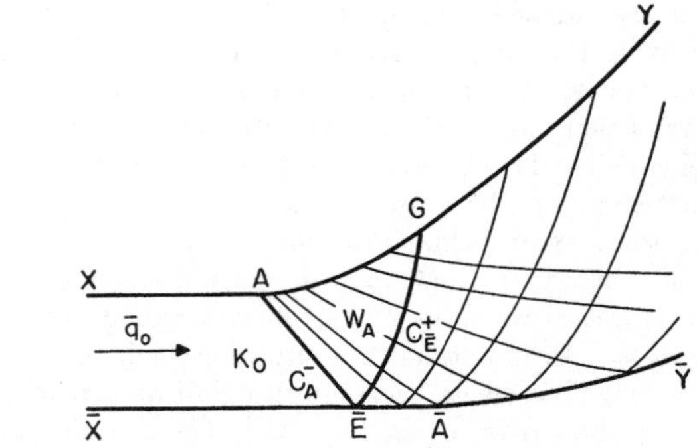

FIG. 123. $\bar{E}G$ reflected Mach line of $A\bar{E}$.

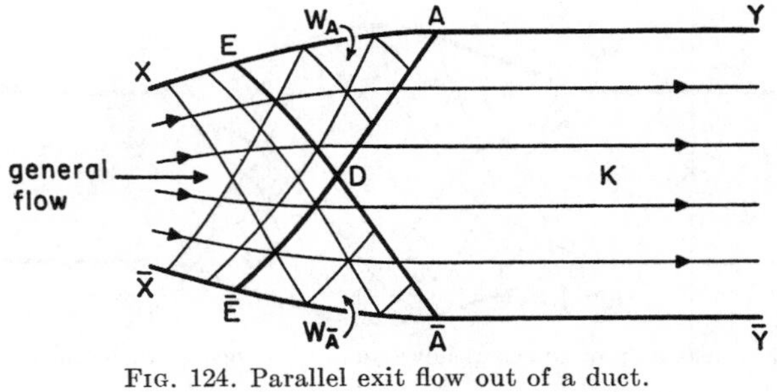

FIG. 124. Parallel exit flow out of a duct.

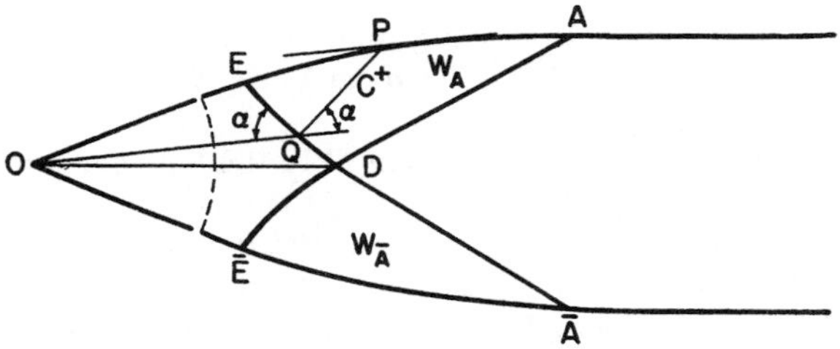

FIG. 125. Parallel exit flow out of a duct with radial entrance flow.

tic. Since q and θ are assumed to be known everywhere along ED, the constant $2\eta = \theta_E + Q_E$ follows, as well as the velocity distribution along all straight C^+ crossing ED in given directions; in particular, the streamline through E can be computed or constructed. By reason of symmetry, $\theta = 0$ at D; hence it is also zero at A and at $\bar{A}$; the walls beyond A and $\bar{A}$ can be continued horizontally, and the emerging flow will have constant horizontal velocity to the right.

A particular case may be mentioned. Consider a duct where the general flow (to the left of $ED\bar{E}$ in the preceding figure) is *radial* [see subsection (b) of Sec. 17.4]. This flow is explicitly known, including the shape of ED and the velocity values along it (see Fig. 125). The radii, such as OQ, are streamlines, and since the values of q and θ along the straight C^+ through Q remain constant, and EA is to be a streamline, it follows that the tangent to the wall at P is parallel to OQ.

Other problems which include combinations of simple waves and shocks will be discussed in Art. 23.

Article 19

Limit Lines and Branch Lines

1. Singularities of the hodograph transformation

Certain singularities found first in some of the simple examples of Art. 7 (radial flow, spiral flow) were reconsidered in Sec. 17.4 as singularities of the hodograph transformation. We found them connected with the vanishing of certain functional determinants at points or rather along lines. Along such a line in the physical plane, the *limiting line* or *limit line* (which in these examples happened to be a circle), (see also Fig. 118c) the acceleration became infinite, the streamlines showed apparent cusps, etc. In the present article we shall study limiting lines, and their counterpart branch lines, in some detail, and again from the point of view of singularities of the hodograph transformation.[30]

A physical solution assigns to a point P in the flow plane one and only one $\mathbf{q} = \mathbf{q}(\mathbf{r})$; in such a solution however, there is nothing to prevent the same $\mathbf{q}$ from appearing at several points P. If now instead of the flow $\mathbf{q} = \mathbf{q}(\mathbf{r})$, which we want to determine, we obtain from one or other of the inverted linear hodograph equations a hodograph solution $\mathbf{r} = \mathbf{r}(\mathbf{q})$, where $\mathbf{r}$ is a single-valued function of $\mathbf{q}$, such a solution may not be everywhere equivalent to the desired flow $\mathbf{q} = \mathbf{q}(\mathbf{r})$ since, roughly speaking, it may give too much as well as too little: (a) a solution $\mathbf{r} = \mathbf{r}(\mathbf{q})$ may well associate the same $\mathbf{r}$ with various $\mathbf{q}$-values; (b) a solution $\mathbf{q} = \mathbf{q}(\mathbf{r})$ in which the same $\mathbf{q}$ corresponds to different $\mathbf{r}$-values cannot appear as a single-valued function $\mathbf{r} = \mathbf{r}(\mathbf{q})$.

An illustration of the situation (a) is provided in the simple examples of radial flow and spiral flow (Sec. 17.4). We found that to each point in the flow field, except to points on the limit circle, there belonged two different velocities, which is physically impossible. We had to distinguish between *two different solutions* which meet at the limit line, i.e., which assume the same $\mathbf{q}$ at every point of this line. We saw in each case that the occurrence of a limit line was connected with the vanishing of the Jacobian $D = \partial(\varphi,\psi)/\partial(q,\theta)$ or of an equivalent determinant; on the other hand, in Sec. 17.3 we found $D \neq 0$ (D considered as a function of q and θ) as the condition for deriving a solution $\mathbf{q} = \mathbf{q}(\mathbf{r})$ from the hodograph solution $\mathbf{r} = \mathbf{r}(\mathbf{q})$.

If, however, $d = D^{-1}$ or an equivalent Jacobian is zero, then the transition from the physical plane to the hodograph plane is not possible. [More generally, this implies that from equations (10.1), with right sides zero and the coefficients dependent only on u and v, we cannot obtain equations (10.22).] We found this situation in the case of a simple wave, where the same $\mathbf{q}$ corresponds to infinitely many points $\mathbf{r}$, and where the Jacobian $\partial(q,\theta)/\partial(x,y)$ is zero in a two-dimensional region. At any rate, the examples point to two types of singularities, corresponding to the remarks (a) and (b), each type connected with the vanishing of certain (essentially equivalent) Jacobians.

A word about terminology may be added. If in the mapping of an X,Y-plane onto a U,V-plane one calls M-lines the lines of the X,Y-plane along which the Jacobian $\Delta = \partial(U,V)/\partial(X,Y) = 0$ and N-lines the lines of the X,Y-plane along which Δ becomes infinite, it is obvious, as a consequence of $[\partial(U,V)/\partial(X,Y)]\cdot[\partial(X,Y)/\partial(U,V)] = 1$, that the image of an M-line of the X,Y-plane is an N-line of the U,V-plane, and conversely. There are mathematically only two concepts. In our physical problem, however, the physical plane and the hodograph plane play very different roles, and four different names are in use. We shall call a line in the flow plane along which $i = \partial(q,\theta)/\partial(x,y)$ vanishes* a *branch line* of the flow plane, and a line in the flow plane along which $I = i^{-1} = \partial(x, y)/\partial(q, \theta) = 0$ we shall call a *limit line* of the flow plane. If terms are needed for the hodograph images of these lines, we call the hodograph image of a limit line a *critical curve* and that of a branch line an *edge*.[31]

As has been done in previous articles, we shall denote by capitals the Jacobians in which hodograph coordinates appear in the denominator, and by (corresponding) small letters those for which hodograph coordinates appear in the numerator. Hence, e.g.,

$$\frac{\partial(q_x,q_y)}{\partial(x,y)} = e, \qquad \frac{\partial(\varphi,\psi)}{\partial(q,\theta)} = D, \qquad \frac{\partial(x,y)}{\partial(\xi,\eta)} = J, \qquad \frac{\partial(x,y)}{\partial(q_x,q_y)} = E$$

and, e.g.,

$$I = \frac{\partial(x,y)}{\partial(q,\theta)} = J\,\frac{1}{2q\tan\alpha} = D\,\frac{1}{\rho q^2}.$$

Since $\partial(q,\theta)/(\partial(\xi,\eta) = 2q \tan \alpha$, the mapping of the supersonic region of the hodograph onto the ξ,η-plane is locally one to one, except when $\alpha = 90°$ or $\alpha = 0°$, i.e., $q = q_t$ or $q = q_m$. The determinants I, J, D and E are equivalent for supersonic flow and $\alpha \neq 0°, 90°$. Both branch and limit lines are essentially supersonic phenomena. A brief consideration of the subsonic cases will be given in the next section before starting the main discussion.

2. Some basic formulas. Subsonic cases

We first collect some simple formulas which will be needed. We have denoted the Mach lines in the flow plane by C^- and C^+ or by ξ-lines and η-lines, (Sec. 16.7). We use this last term for both the Mach lines in the x,y-plane and the rectangular coordinate lines in the ξ,η-plane. Denote by $\mathbf{r}$ the radius vector of a point P, and by $\hat{\boldsymbol{\xi}},\hat{\boldsymbol{\eta}}$ unit vectors which form angles $-\alpha$ and $+\alpha$, respectively, with the flow direction. Then $\partial\mathbf{r}/\partial\xi$ and $\partial\mathbf{r}/\partial\eta$ will have the ξ- and η-directions respectively, and we may define functions $h_1(\xi,\eta)$ and $h_2(\xi,\eta)$ by the equations

$$\text{(1)} \qquad \frac{\partial\mathbf{r}}{\partial\xi} = h_1\hat{\boldsymbol{\xi}}, \qquad \frac{\partial\mathbf{r}}{\partial\eta} = h_2\hat{\boldsymbol{\eta}}.$$

* This Jacobian is named here as a representative of several, and in general equivalent, Jacobians which we shall use according to the situation.

Hence, if ds_1 and ds_2 denote* line elements of ξ-lines and η-lines, respectively,

$$\frac{\partial \mathbf{r}}{\partial s_1} = \hat{\boldsymbol{\xi}}, \qquad \frac{\partial \mathbf{r}}{\partial s_2} = \hat{\boldsymbol{\eta}},$$

and

$$h_1 = \frac{ds_1}{d\xi}, \qquad h_2 = \frac{ds_2}{d\eta}; \tag{2}$$

with the notation $\phi^+ = \theta + \alpha, \quad \phi^- = \theta - \alpha$ we have

$$\begin{aligned} \frac{\partial x}{\partial \xi} &= h_1 \cos \phi^-, & \frac{\partial x}{\partial \eta} &= h_2 \cos \phi^+ \\ \frac{\partial y}{\partial \xi} &= h_1 \sin \phi^-, & \frac{\partial y}{\partial \eta} &= h_2 \sin \phi^+. \end{aligned} \tag{3}$$

Hence

$$J = \frac{\partial(x,y)}{\partial(\xi,\eta)} = h_1 h_2 \sin 2\alpha. \tag{4}$$

In the characterization of the singularities of the transformation it is preferable to use h_1 and h_2, and their reciprocals, rather than the Jacobian (4) since the vanishing of each of them has a distinct geometric meaning.[32] We note

$$\begin{aligned} I &= \frac{\partial(x,y)}{\partial(q,\theta)} = \frac{h_1 h_2}{q} \cos^2 \alpha, \\ D &= \frac{\partial(\varphi,\psi)}{\partial(q,\theta)} = \rho q h_1 h_2 \cos^2 \alpha. \end{aligned} \tag{4'}$$

Also, with letter subscripts denoting partial differentiations,

$$\begin{aligned} \psi_\xi &= -\rho q h_1 \sin \alpha, & \psi_\eta &= \rho q h_2 \sin \alpha \\ \varphi_\xi &= q h_1 \cos \alpha, & \varphi_\eta &= q h_2 \cos \alpha. \end{aligned} \tag{4''}$$

The h_1, h_2 are basic in the differential geometry of the net of characteristics in the physical plane.

Introducing, with a usual sign convention,† the radii of curvature R_1,

* In contrast to Arts. 9 and 16 the line elements of Mach lines are here denoted by ds_1 and ds_2.

† The sign of R_1 (of R_2) is chosen positive if the center of curvature of a ξ-line (η-line) lies in the direction of increasing η (increasing ξ).

R_2 of the ξ- and η-lines, respectively, and the corresponding curvatures κ_1, κ_2, as well as the curvature κ of a streamline, we find:

$$
\begin{aligned}
\kappa_1 &= \frac{1}{R_1} = \frac{\partial(\theta - \alpha)}{\partial s_1} = \frac{\partial(\theta - \alpha)}{h_1\,\partial \xi}, \\
\kappa_2 &= \frac{1}{R_2} = -\frac{\partial(\theta + \alpha)}{\partial s_2} = -\frac{\partial(\theta + \alpha)}{h_2\,\partial \eta}, \qquad (5)\\
\kappa &= \frac{\partial \theta}{\partial s} = \frac{1}{2\cos\alpha}\left(\frac{\partial\theta}{\partial s_1} + \frac{\partial\theta}{\partial s_2}\right) = \frac{1}{2\cos\alpha}\left(\frac{1}{h_2} - \frac{1}{h_1}\right).
\end{aligned}
$$

Denoting by a prime differentiation with respect to q, we find (always with $Q = \eta + \xi$, $\theta = \eta - \xi$) that

$$
h_1 = -R_1\left(1 + \frac{\alpha'}{Q'}\right), \qquad h_2 = -R_2\left(1 + \frac{\alpha'}{Q'}\right).^{*} \qquad (6)
$$

By equating in (3) the mixed derivatives of x and y, respectively, and then simplifying, we obtain a pair of linear equations for h_1, h_2:

$$
\begin{aligned}
\frac{\partial h_1}{\partial \eta}\sin 2\alpha + \left(\frac{\alpha'}{Q'} - 1\right)(h_2 + h_1\cos 2\alpha) &= 0, \\
\frac{\partial h_2}{\partial \xi}\sin 2\alpha + \left(\frac{\alpha'}{Q'} - 1\right)(h_1 + h_2\cos 2\alpha) &= 0,
\end{aligned} \qquad (7)
$$

with $Q' = 1/(q\tan\alpha)$.† While these equations do not seem to offer any particular advantages with respect to the general integration problem, as compared to other pairs of linear equations (see Secs. 17.1 and 17.2), they will help in our present discussion.[33]

We finish this preparatory section by considering the vanishing of representative Jacobians in the subsonic cases. Dealing first with the limit type singularities, we consider the Jacobian

$$
E = \frac{\partial(x,y)}{\partial(q_x,q_y)} = \frac{1}{q\rho^3}D = -\frac{1}{\rho q^4}\left[q^2\psi_q^{\,2} + (1 - M^2)\psi_\theta^{\,2}\right],
$$

* Note that $R_1 = 0$ does not follow from $h_1 = 0$ if, exceptionally, $1 + \alpha'/Q' = 0$; in the polytropic case this happens for the exceptional $M = 2/\sqrt{3 - \kappa}$).

† In polytropic flow:

$$
1 - \frac{\alpha'}{Q'} = \frac{\kappa + 1}{2\cos^2\alpha},
$$

which shows how the expression becomes infinite as $\alpha \to 90°$.

which vanishes only if

$$\frac{\psi_q}{q} = 0, \qquad \text{and} \qquad \frac{\psi_\theta}{q^2} = 0,$$

and this implies, according to (16.31),

$$\frac{\varphi_\theta}{q^2} = 0, \qquad \frac{\varphi_q}{q} = 0.$$

Hence, all four derivatives $\psi_q, \psi_\theta, \varphi_q, \varphi_\theta$ vanish. Then also $x_q, y_q, x_\theta, y_\theta$ vanish. Clearly, such a singularity is isolated. Incidentally it can then be shown that this singular point is a saddle point for both $\varphi(q,\theta)$ and $\psi(q,\theta)$ unless all second-order derivatives of φ and ψ vanish there. In fact, from (16.31) we conclude that

$$(8) \qquad \psi_{qq}\psi_{\theta\theta} = -\frac{\rho^2}{1 - M^2}\varphi_{q\theta}^2.$$

It is also seen from (16.31) that if $\varphi_{q\theta} = 0$, both ψ_{qq} and $\psi_{\theta\theta}$ vanish. Hence $\psi_{qq}\psi_{\theta\theta} - \psi_{q\theta}^2 < 0$ unless all three second-order derivatives of ψ vanish; a similar argument holds for $\varphi(q,\theta)$.

To study the branch type singularities for subsonic flow, we consider for $q \neq 0$ the Jacobian

$$e = \frac{\partial(q_x, q_y)}{\partial(x,y)} = -\left[\left(\frac{\partial q}{\partial n}\right)^2 + (1 - M^2)\left(\frac{\partial q}{\partial s}\right)^2\right].$$

It can vanish only if $\partial q/\partial n = 0$, $\partial q/\partial s = 0$, and it follows from Eqs. (16.7) that then, in addition $\partial\theta/\partial s = \partial\theta/\partial n = 0$. We may show again that the singular point is a saddle point for q and θ unless all second-order derivatives are zero. Hence, at any rate, these singularities occur only at isolated points.

3. Limit lines $\mathfrak{L}_1$ and $\mathfrak{L}_2$

Next, let us consider the supersonic region of a flow using the hodograph coordinates ξ,η. We shall assume in this and the next section that α is neither 90° nor 0°. In Sec. 5 the case $\alpha = 90°$ will be studied.

We consider for a given solution in the ξ,η-plane the locus $h_1(\xi,\eta) = 0$, and in particular the mapping onto the x,y-plane in the neighborhood of this locus. Let p be a point* with coordinates ξ_0, η_0, where $h_1(\xi_0, \eta_0) = 0$, $(\partial h_1/\partial\eta)_{\xi_0,\eta_0} \neq 0$. Then by the implicit function theorem there is a curve $\eta = g(\xi)$, with $\eta_0 = g(\xi_0)$ on which $h_1(\xi,\eta) = 0$. We call it a *critical curve* and denote it by l_1, and we call the curve $\mathfrak{L}_1$ in the flow plane correspond-

* We use here $p, m, q, \cdots$, for points in the ξ,η-plane rather than, as in previous articles, $P', M', Q\ldots$

ing to l_1 a *limit line*. A point on $\mathfrak{L}_1$, or rather a point P such that $h_1 = 0$ at its image p, will be termed a *limit point.*

We consider first points m on the critical curve where *both* $\partial h_1/\partial \xi$ *and* $\partial h_1/\partial \eta$ *are different from zero*. From the first of these conditions we conclude that l_1 does not have the ξ-direction at m, and from the second, using (7) and $\alpha \neq 0°, \neq 90°$, that $h_2 \neq 0$ at m. Then (1) shows that at M, the image of m, the limit line $\mathfrak{L}_1$ has the η-direction, i.e., $\mathfrak{L}_1$ is tangent to the characteristic C^+ at M. The same conclusion holds for any curve $\mathfrak{C}$ through M whose image c, say $\eta = f(\xi)$, does not have the ξ-direction at m. All such curves $\mathfrak{C}$ are tangent to $\mathfrak{L}_1$ at M.

For example, the images in the ξ,η-plane of lines of constant speed or of lines of constant direction do not have at m the *exceptional direction*, i.e., the ξ-direction. In fact, for the latter, $d\theta$ vanishes, and if $\alpha \neq 90°$, $d\eta = \frac{1}{2}(dQ + d\theta) \neq 0$; a similar conclusion holds for the former. Hence these lines are tangent to $\mathfrak{L}_1$ at M. A consequence is that the direction of the vector grad q at M is normal to $\mathfrak{L}_1$, and hence to the C^+ through M.

If, however, the curve c has the exceptional direction at m, i.e., $d\eta/d\xi = f'(\xi) = 0$, we can no longer conclude that the corresponding $\mathfrak{C}$ has the η-direction at M. We prove that in this case $\mathfrak{C}$ *has a cusp at* M. In fact, along $\mathfrak{C}$

$$\frac{dx}{d\xi} = x_\xi + f'(\xi)x_\eta, \qquad \frac{dy}{d\xi} = y_\xi + f'(\xi)y_\eta$$

so that since both x_ξ and y_ξ vanish with h_1 and $f'(\xi) = 0$, both $dx/d\xi$ and $dy/d\xi$ are zero at M. On the other hand,

$$\frac{d^2x}{d\xi^2} = x_{\xi\xi} + f''x_\eta, \qquad \frac{d^2y}{d\xi^2} = y_{\xi\xi} + f''y_\eta$$

and not both right sides vanish since, using (3),

$$x_{\xi\xi}y_\eta - y_{\xi\xi}x_\eta = h_2 \frac{\partial h_1}{\partial \xi} \sin 2\alpha \neq 0.$$

Hence the singular point of $\mathfrak{C}$ at M is a cusp. Since the ξ-line at m has the exceptional direction, the C^- at M has a cusp there, its tangent making the angle 2α with the direction of $\mathfrak{L}_1$. We know that streamlines and equipotential lines bisect the angles of the C^--and C^+-directions. Since they are therefore not tangent to $\mathfrak{L}_1$ at M, we infer that their images in the ξ,η-plane must have the exceptional ξ-direction at m and that therefore at M they must have cusps. This may also be seen directly. From $h_1 = 0$ and (4″), $\psi_\xi = 0$ and $\varphi_\xi = 0$ follow; hence at m, for a streamline, $d\psi = \psi_\eta d\eta = 0$, and as $\psi_\eta \neq 0$ (since $h_2 \neq 0$), $d\eta = 0$; hence at m the streamline has the ξ-direction. The same conclusion holds for potential lines.

Next consider the acceleration at M. Using again $Q = \eta + \xi$, $Q' = (q \tan \alpha)^{-1}$ we find

$$(9) \qquad b = \frac{dq}{dt} = q\frac{\partial q}{\partial s} = \frac{q}{2 \cos \alpha}\left(\frac{\partial q}{\partial s_1} + \frac{\partial q}{\partial s_2}\right) = \frac{q^2 \sin \alpha}{2 \cos^2 \alpha}\left(\frac{1}{h_1} + \frac{1}{h_2}\right),$$

which is thus seen to be *infinite at a limit line*, as found before in particular examples. The same holds true for the velocity gradient.

All our results, of course, have their counterparts for points of an $\mathfrak{L}_2$: $h_2 = 0$, at which $\partial h_2/\partial \xi \neq 0$, $\partial h_2/\partial \eta \neq 0$. We review: *Consider in the characteristic plane the locus* $h_1(\xi,\eta) = 0$, *i. e., the critical curve* l_1, *and points of* l_1 *at which* $\partial h_1/\partial \xi \neq 0$, $\partial h_1/\partial \eta \neq 0$. *Its image,* $\mathfrak{L}_1$, *the limit line, is the envelope of plus Mach lines* C^+, *of curves of constant speed and of curves of*

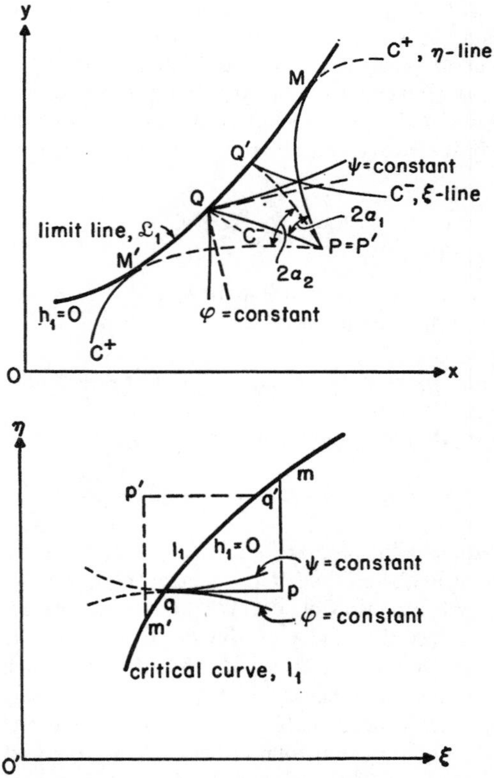

FIG. 126. Limit line and critical curve.

constant direction; it is the locus of cusps of the C^- Mach lines, of streamlines and of equipotential lines, all of which have in the ξ,η-plane the ξ-direction at their intersection with l_1. For the critical curve, $h_2 = 0$, the roles of C^+ and C^-, of η and ξ, are reversed.

We have seen in examples that two physical solutions meet at a limit line. We now consider the correspondence between the x,y- and ξ,η-planes near such a line (see Fig. 126). Consider a point P in the physical plane. It is the image of two points p and p'. In fact through P pass two characteristics of the nonexceptional kind, two C^+-lines, which touch the $\mathfrak{L}_1$-line at M and M', respectively, with corresponding points m and m' on l_1. On the η-line through m lies p, on that through m' lies p'; the two points p, p' are on opposite sides of the critical curve l_1 since η varies in the opposite sense along PM and along PM'. Likewise two ξ-lines (C^--characteristics), PQ and PQ', pass through P making angles $2\alpha_1$ and $2\alpha_2$ with the C^+-characteristics PM and PM', respectively. The point q corresponding to the point Q lies on the ξ-line through p and on l_1, while the point q' lies on the ξ-line through p' and on l_1.

We obtain a one-to-one correspondence in the usual way by considering separately the two sheets in the physical plane, which are each the image of *one* side of the critical curve $h_1 = 0$. On each sheet in the physical plane there is *one* flow.[34]

4. Special points of the limit line

(a) *Cusps of the limit line.* Our discussion is not yet finished. So far we have assumed that at the point of the curve $h_1(\xi,\eta) = 0$ considered, *both* $\partial h_1/\partial\xi$ and $\partial h_1/\partial\eta$ are different from zero. Now we drop this assumption. If $h_1(\xi,\eta) = 0$ and $\partial h_1/\partial\eta = 0$, it follows from (7), since $(\alpha'/Q') - 1 \neq 0$, that $h_2 = 0$; hence $h_1 = h_2 = 0$, and from (3) that all four derivatives $\partial x/\partial\xi, \cdots$ vanish. Such a point will be called a *double limit point.* The point of intersection of two critical curves $h_1 = 0$, $h_2 = 0$ is such a point; the corresponding point in the flow plane is the point of intersection of two limit lines $\mathfrak{L}_1$ and $\mathfrak{L}_2$ [see subsection (b) below and example in Sec. 20.5.] We lay aside this case for the moment.

Now, let us consider a point m of $h_1 = 0$ where $\partial h_1/\partial\eta \neq 0$, $\partial h_1/\partial\xi = 0$, $\partial^2 h_1/\partial\xi^2 \neq 0$; *the critical curve has the ξ-direction.* If the relation $h_1(\xi,\eta) = 0$ is written in the form $\eta = g(\xi)$, then $g' = d\eta/d\xi = (-\partial h_1/\partial\xi)/(\partial h_1/\partial\eta) = 0$ at m, while $g''(\xi) = (-\partial^2 h_1/\partial\xi^2)/(\partial h_1/\partial\eta) \neq 0$. As before, at the point M, the image of m, $dx/d\xi = x_\xi + x_\eta g'(\xi) = 0$ from $h_1 = 0$ and $g' = 0$, and likewise $dy/d\xi = 0$, while

$$\frac{d^2x}{d\xi^2} = x_{\xi\xi} + g''x_\eta, \qquad \frac{d^2y}{d\xi^2} = y_{\xi\xi} + g''y_\eta.$$

Here $x_{\xi\xi} = 0$, $y_{\xi\xi} = 0$, $g'' \neq 0$ and x_η, y_η cannot both vanish since $h_2 \neq 0$. Hence, not both derivatives of second order are zero, and $\mathfrak{L}_1$ *is seen to have a cusp at M*. We further conclude (since $h_2 \neq 0$) that again the C^+ at M touches $\mathfrak{L}_1$ (see Fig. 127), i.e., has the cusp tangent there, but does not itself have a cusp, and that the same holds for any curve through M whose image does not have the ξ-direction at m, e.g., for the curves of constant speed and of constant direction at M. On the other hand, the image of the C^- has the ξ-direction at m; this is also true of the images of the streamline and the equipotential line at M, since neither has the $\mathfrak{L}_1$-direction at M. As before, this last fact can also be seen directly from $\psi_\xi = 0$, $\psi_\eta \neq 0$, etc.

Such a point M (or points) where l_1 shows an extremum and $\mathfrak{L}_1$ a cusp plays an important role; in fact, a streamline that passes through a cusp of $\mathfrak{L}_1$ separates those streamlines that do not encounter $\mathfrak{L}_1$ at all from those streamlines that do encounter it (and do so in general at two points) and consequently have cusps at $\mathfrak{L}_1$. In the characteristic plane, streamlines that intersect l_1 in the ξ-direction (and in general at two points) and those that do not intersect it at all are separated by a streamline that contacts l_1 at m in the ξ-direction. The region between the branches of $\mathfrak{L}_1$ is covered three times (see Fig. 127); otherwise the mapping is one to one (see also Fig. 118). Of course similar considerations apply to an $\mathfrak{L}_2$. All this will find its illustration in an example to be considered in detail in Sec. 20.3.

(b) *Intersection of limit lines.* We now consider a point, a, in the ξ,η-plane where $h_1 = h_2 = 0$, and consequently $\partial h_1/\partial\eta = \partial h_2/\partial\xi = 0$. In this case all four derivatives x_ξ, x_η, y_ξ, y_η vanish at a. We assume that *both* $\partial h_1/\partial\xi \neq 0$, $\partial h_2/\partial\eta \neq 0$. Hence, in the vicinity of this point we can write for $h_1(\xi,\eta) = 0$ the explicit form $\xi = g(\eta)$ with $g'(\eta) = 0$ at a. Also $h_2(\xi,\eta) = 0$ may be

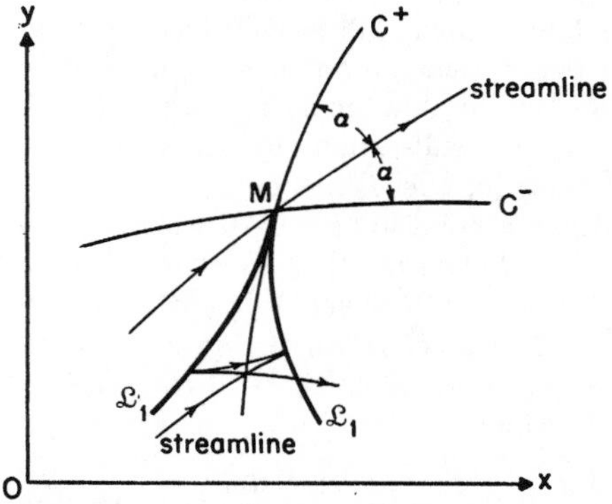

FIG. 127. Cusp of limit line.

written as $\eta = f(\xi)$, where $f'(\xi) = 0$ at a. Next, differentiating Eqs. (3) we find at the point a

$$x_{\xi\xi} = \frac{\partial h_1}{\partial \xi} \cos \phi^-, \qquad y_{\xi\xi} = \frac{\partial h_1}{\partial \xi} \sin \phi^-, \qquad x_{\eta\eta} = \frac{\partial h_2}{\partial \eta} \cos \phi^+,$$

$$y_{\eta\eta} = \frac{\partial h_2}{\partial \eta} \sin \phi^+, \qquad x_{\xi\eta} = y_{\xi\eta} = 0.$$

Consider the curve $\mathfrak{L}_1$ corresponding to $h_1(\xi,\eta) = 0$. We prove that it has a cusp at $A(x,y)$, the point corresponding to $a(\xi,\eta)$ (see Fig. 128). In fact, at A,

$$\frac{dx}{d\eta} = x_\eta + x_\xi g' = 0, \qquad \frac{dy}{d\eta} = y_\eta + y_\xi g' = 0,$$

$$\frac{d^2x}{d\eta^2} = x_{\eta\eta} + x_\xi g''.$$

The second term to the right in the last equation is zero unless $g'' = \infty$ at A; but $g'' = (-\partial^2 h_1/\partial\eta^2)/(\partial h_1/\partial\xi)$ at A, where $\partial h_1/\partial\xi \neq 0$ by hypothesis, and differentiation of (7) shows that $\partial^2 h_1/\partial\eta^2$ remains finite at A. Hence at A: $d^2x/d\eta^2 = x_{\eta\eta}$, $d^2y/d\eta^2 = y_{\eta\eta}$ and $dy/dx = y_{\eta\eta}/x_{\eta\eta} = \tan\phi^+$. We conclude that the curve $\mathfrak{L}_1$ has a cusp at A with the tangent in the C^+-direction. In exactly the same way we find that the curve $\mathfrak{L}_2$ corresponding to $h_2 = 0$ has a cusp at A with the C^--direction as tangent. It can then be seen that *the inner angle in the x,y-plane is covered four times.*

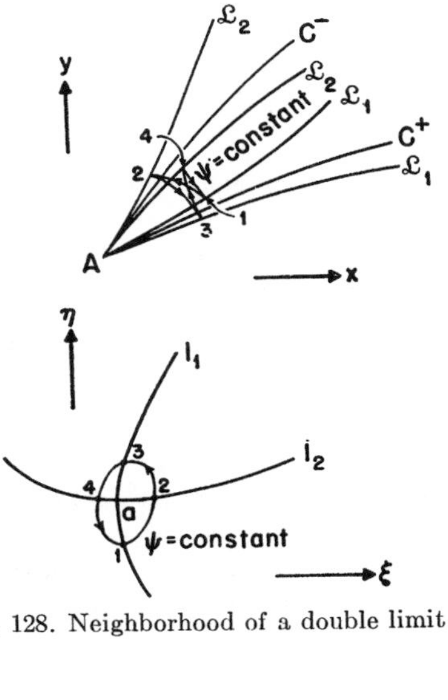

FIG. 128. Neighborhood of a double limit point.

From (4″) and the properties at the point under consideration we find that $\varphi_{\xi\eta} = \psi_{\xi\eta} = 0$, that

$$\varphi_{\xi\xi}\varphi_{\eta\eta} = q^2 \cos^2 \alpha \frac{\partial h_1}{\partial \xi}\frac{\partial h_2}{\partial \eta}, \qquad \text{and} \qquad \psi_{\xi\xi}\psi_{\eta\eta} = -\rho^2 q^2 \sin^2 \alpha \frac{\partial h_1}{\partial \xi}\frac{\partial h_2}{\partial \eta}.$$

Hence, according to whether $(\partial h_1/\partial \xi)(\partial h_2/\partial \eta) \lessgtr 0$ the point is either a saddle point (upper sign) or an extremum for φ, and the other way for ψ. For the first case, the extremum of ψ, a streamline near the point a in the hodograph will be closed, and its image in the physical plane (see Fig. 128) demonstrates the fourfold covering of the angle at A. In the next article we shall find an interesting illustration of this situation.

We do not discuss singularities of "higher order".

5. Limit singularities for $M = 1$

In our previous derivations, $\alpha \neq 90°$ and $\alpha \neq 0°$ were assumed and actually used in many conclusions. If $\alpha = 90°$, i.e., at the sonic line, a partly new situation presents itself which we shall now study with respect to limit-type singularities.[35]

For $\alpha = 90°$ the Jacobians J and D no longer exhibit the same behavior and it is now the vanishing of D which will serve as a criterion. Using (4″), we have

$$\text{(10)} \qquad \begin{aligned} D &= J \frac{\rho q}{2 \sin \alpha} \cos \alpha = \rho q h_1 h_2 \cos^2 \alpha \\ &= -\frac{\cot^2 \alpha}{\rho q} \psi_\xi \psi_\eta = \frac{\rho}{q} \varphi_\xi \varphi_\eta . \end{aligned}$$

We see that if $J < \infty$ then $D = 0$, and if $J = \infty$, D may be zero or nonzero (finite or infinite).

Writing φ_ξ, φ_η in terms of ψ_q, ψ_θ we have

$$\text{(11)} \qquad \varphi_\xi = \frac{1}{\rho}[-q\psi_q + \psi_\theta \cot \alpha], \qquad \varphi_\eta = \frac{1}{\rho}[q\psi_q + \psi_\theta \cot \alpha].$$

As $\alpha \to 90°$, $\psi_\theta \cot \alpha \to 0$, unless $\psi_\theta \to \infty$ (which we exclude).

Laying aside the case $D \to \pm\infty$ which corresponds to branch type singularities (see Sec. 6), we call an *ordinary sonic point*, in contrast to limit point or branch point, a point where $\psi_q \neq 0$; then both φ_ξ and φ_η remain nonzero. For example, in the case of spiral flow (see Sec. 17.4) the sonic line (a circle) is not a limit line. Straight computation shows that with the notation of Art. 17, where $C, k > 0$:

$$\psi_q = \frac{C\rho}{q}, \qquad \psi_\theta = k, \qquad \varphi_\xi = k\frac{\cot\alpha}{\rho} - C, \qquad \varphi_\eta = k\frac{\cot\alpha}{\rho} + C$$

$$h_1 = \frac{1}{a}\left(\frac{k}{\rho} - C\tan\alpha\right), \quad h_2 = \frac{1}{a}\left(\frac{k}{\rho} + C\tan\alpha\right), \quad b = \frac{k\rho q^3}{k^2\cot^2\alpha - C^2\rho^2}.$$

For $\alpha \to 90°$, $\psi_q \neq 0$ and hence all points are ordinary.

A new type of limit point (and limit line) which we shall call *sonic limit point* is characterized by $M = 1$, $\psi_q = 0$. At such a point either $\psi_q = 0$, $\psi_\theta \neq 0$, or $\psi_q = 0$, $\psi_\theta = 0$. The first case can clearly happen along a whole arc of curve which we then denote as a *sonic limit line*, $\mathfrak{L}_t$. In the second case the point is isolated, or as we shall see presently, it may be the common point of an $\mathfrak{L}_t$ and an $\mathfrak{L}_1$, for instance.

The simplest example of an $\mathfrak{L}_t$ appears in the radial flow (Sec. 17.4). Along the line $M = 1$, which in this case is a circle, we find, using the above formulas with $C = 0$:

$$\psi_q = 0, \qquad \psi_\theta = k, \qquad \varphi_\xi = \varphi_\eta = k\frac{\cot\alpha}{\rho},$$

$$h_1 = h_2 = \frac{k}{a\rho}, \qquad b = \frac{1}{k}\rho q^3 \tan^2\alpha.$$

There is no other limit line in this example, since both h_1 and h_2 are everywhere nonzero.*

From $\varphi_\xi = \varphi_\eta = 0$ a sonic limit line is an equipotential line. Hence, at every point of a sonic limit line, the streamline direction is perpendicular to the $\mathfrak{L}_t$, and since $\alpha = 90°$, it follows that both the C^+ and C^- are enveloped by the $\mathfrak{L}_t$ (see Fig. 129). Since at the $\mathfrak{L}_t$, $\psi_q = 0$, $\psi_\theta \neq 0$, $d\psi = \psi_q\,dq + \psi_\theta\,d\theta = \psi_\theta\,d\theta$, we see that if $d\psi = 0$ also $d\theta = 0$, and vice versa; hence the line $\theta =$ constant, the isocline, is likewise perpendicular to the $\mathfrak{L}_t$.

More generally, and in analogy to our study of the $\mathfrak{L}_1$ and $\mathfrak{L}_2$, we conclude from Eqs. (17.25), since $d\varphi = 0$, $\psi_q = 0$, $\psi_\theta \neq 0$ at the $\mathfrak{L}_t$:

$$dx = -\frac{\psi_\theta}{\rho q}\sin\theta\,d\theta, \qquad dy = \frac{\psi_\theta}{\rho q}\cos\theta\,d\theta.$$

Thence

$$\frac{dy}{dx} = -\cot\theta, \qquad \text{if } d\theta \neq 0. \tag{12}$$

* Note that in the spiral flow $h_1 \to -\infty$, $h_2 \to \infty$ along the ordinary sonic line, while in the radial flow h_1 and h_2 are finite along the sonic limit line.

Hence an element of any curve in the hodograph on which $d\theta \neq 0$ maps into an element with slope $dy/dx = -\cot\theta$ in the x,y-plane, i.e., normal to the streamline, or tangential to the $\mathfrak{L}_t$. Next, considering an exceptional element with $d\theta = 0$, and using q as parameter, we find from Eq. (17.25′) that $dx/dq = 0$, $dy/dq = 0$, but that d^2x/dq^2, d^2y/dq^2 cannot vanish simultaneously, and we conclude that the curves which at the sonic line in the hodograph have the exceptional (radial) direction will map into cusped curves. We also note that the characteristics C^+, C^- lie on different sides of the isocline (and streamline) forming an angle of 180° with each other since in the hodograph they are separated by the isocline.

We conclude from

$$b = q\frac{\partial q}{\partial s} = \frac{q^2}{D}\psi_\theta, \tag{9'}$$

that the acceleration, which has been found infinite along an $\mathfrak{L}_1$ or $\mathfrak{L}_2$, is also infinite at any sonic limit point where $\psi_\theta \neq 0$.

In the radial flow example the lines of constant direction are the radii (which are perpendicular to the $\mathfrak{L}_t$). The radii are also the streamlines for both flows, and at each point of the sonic limit circle the two streamlines may be considered to form a degenerate cusp.

A sonic limit point could be the point of intersection of an $\mathfrak{L}_t$ with an $\mathfrak{L}_1$, or with an $\mathfrak{L}_2$, or with both* (double sonic limit point). It could also be the sonic point of an $\mathfrak{L}_1$ or of an $\mathfrak{L}_2$ (an example will be found in Sec. 20.3) or of both (again double sonic limit point). If the point is the point of intersection of an $\mathfrak{L}_t$ and an $\mathfrak{L}_1$, say, then ψ_θ must also vanish there. For, since $\psi_q = 0$ along the sonic line $q = q_t$, we have from Taylor's formula applied to ψ_q that $\psi_q = O(q - q_t)$ while $\tan\alpha = O(q - q_t)^{-\frac{1}{2}}$, so that $\psi_q \tan\alpha = O(q - q_t)^{\frac{1}{2}}$; hence $\psi_\xi = q\psi_q \tan\alpha - \psi_\theta$ tends to the value of $-\psi_\theta$ at $q = q_t$ irrespective of the path of approach. But $\psi_\xi = 0$ everywhere

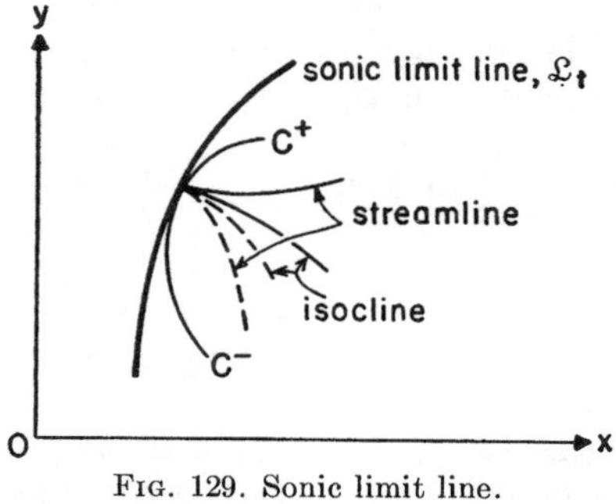

FIG. 129. Sonic limit line.

* For an example see the last of the papers quoted in Note 35.

along $\mathfrak{L}_1$; hence $\psi_\theta = 0$ at such a point of intersection. This also shows that a line $\mathfrak{L}_1$ along which $M = 1$ everywhere is not possible. The conclusion $\psi_\theta = 0$ is not valid if the point under consideration is the sonic point of an $\mathfrak{L}_1$ without lying on an $\mathfrak{L}_t$ (see examples in Secs. 20.3 and 20.5 where $\psi_\theta \neq 0$).

We review: *The sonic limit line $\mathfrak{L}_t$, characterized by $\psi_q = 0$ along a sonic line, is a (piecewise) equipotential line and a line of constant velocity; at a point where $\psi_\theta \neq 0$, it is the envelope of both families of characteristics and, more generally, of all curves whose images in the hodograph plane do not have the exceptional radial direction. All curves whose images have the radial direction at the sonic circle have cusps at the $\mathfrak{L}_t$. On the $\mathfrak{L}_t$ the acceleration becomes infinite.*

We do not study here or in the following the case $\alpha = 0°$, since lines with $\rho = 0$ will in any case appear only as boundaries of flow regions.

6. Branch lines

We now study the vanishing of the reciprocals d and j of Jacobians D and J, respectively. We consider $M \geqq 1$ (see end of Sec. 2 for $M < 1$). Now $D = \rho q h_1 h_2 \cos^2 \alpha$ cannot become infinite unless *either h_1 or h_2 becomes infinite.* Here, therefore, we have only to investigate the loci $h_1(x,y) = \pm\infty$, $h_2(x,y) = \pm\infty$; in other words, there are only branch lines B_1 and B_2, and no analogue of the $\mathfrak{L}_t$. We now consider a (locally) single-valued solution in the x,y-plane and lines (or points) along which h_1 or h_2 considered as a function of x, y becomes infinite. The discussion becomes similar to that for limit lines if we invert the Eqs. (3), which were basic in our previous investigation. Then $x_\xi = \eta_y \partial(x,y)/\partial(\xi,\eta)$ and from (3) and (4) we obtain $x_\xi = \eta_y h_1 h_2 \sin 2\alpha = h_1 \cos\phi^-$, or $\eta_y = \cos\phi^-/h_2 \sin 2\alpha$. This suggests that we introduce

$$k_1 = \frac{1}{h_1 \sin 2\alpha}, \qquad k_2 = \frac{1}{h_2 \sin 2\alpha}. \tag{13}$$

If we assume $\alpha \neq 90°, 0°$, then the k_i are equivalent to $1/h_i$ and we obtain from (3)

$$\frac{\partial\xi}{\partial x} = k_1 \sin\phi^+, \qquad \frac{\partial\eta}{\partial x} = -k_2 \sin\phi^-,$$
$$\frac{\partial\xi}{\partial y} = -k_1 \cos\phi^+, \qquad \frac{\partial\eta}{\partial y} = k_2 \cos\phi^-, \tag{14}$$

$$j = k_1 k_2 \sin 2\alpha = (h_1 h_2 \sin 2\alpha)^{-1}. \tag{14'}$$

The locus $k_1(x,y) = 0$ is called a *branch line* in the x,y-plane, briefly a B_1, and its image in the ξ,η-plane an *edge* b_1. We assume now $k_2 \neq 0$, so that $d\eta \neq 0$ on locus (then the flow in the physical plane is not a simple wave or a uniform flow). From $k_1 = 0$, we see by (14) that $d\xi = 0$; hence $\xi =$ constant. Therefore b_1 is an η-line in the ξ,η-plane, and *B_1 is a C^+ in the flow plane.* It follows as in the previous investigations that all lines in the x,y-plane, with the exception of those lines which have the ξ-direction at their point of intersection with B_1, appear in the ξ,η-plane tangent to the straight vertical η-line, b_1, the edge.[36] Among these are the lines $\varphi =$ constant and $\psi =$ constant. Hence the edge b_1 is an envelope of streamlines and equipotential lines. The ξ-direction, i.e., the C^--direction, is the exceptional direction at B_1. Assuming also $\partial k_1/\partial s_1 \neq 0$, it can be shown (see similar proofs in a previous section) that the image of a line which has the ξ-direction at its intersection with B_1, and in particular of a C^-, must have a cusp at b_1. Since the images of the C^- are the straight horizontal ξ-lines, they return at the edge (see Fig. 130). For a C^--line

$$d(\theta - \alpha) = \frac{\partial(\theta - \alpha)}{\partial \xi}\, d\xi = -\left(1 + \frac{\alpha'}{Q'}\right) d\xi,$$

and since $d\xi$ changes sign at an intersection with b_1, the same must be true

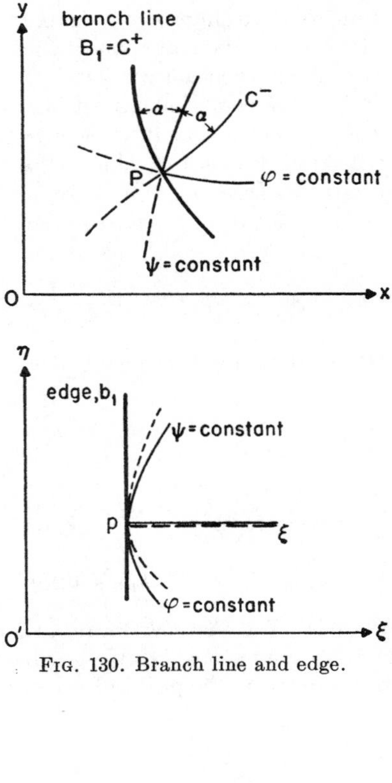

FIG. 130. Branch line and edge.

for $d(\theta - \alpha)$ unless, exceptionally, $1 + \alpha'/Q' = 0$; hence in general *a C^- has an inflection point* at its intersection with the B_1.[37]

Since the lines of constant speed and those of constant direction are lines $\eta + \xi = $ constant and $\eta - \xi = $ constant in the ξ,η-plane, they intersect the vertical edge at $\pm 45°$ and thus are not tangent to it; therefore in the flow plane they have the exceptional direction at B_1. Hence, at each point of B_1, the C^-, the line of constant speed and that of constant direction touch each other, while the streamlines and equipotential lines bisect the angles between B_1 and the inflection tangent of the C^-, and cross the branch line without singularity.

It also follows that at B_1 the direction of the vector grad q is perpendicular to the C^--direction: $\partial q/\partial s_1 = 0$ (the same holds for the vector grad θ). It is easily seen that in steady irrotational flow (also in three dimensions) the direction of grad q coincides with that of $d\mathbf{q}/dt$; hence at any point of B_1 the acceleration vector is perpendicular to the C^--direction there. The acceleration at B_1 is finite. In fact

$$b = \frac{dq}{dt} = q\,\frac{\partial q}{\partial s} = q^2 \sin\alpha \tan\alpha\,(k_1 + k_2).$$

This is finite for $k_1 = 0$, k_2 finite, and zero if k_2 also vanishes. A point where $k_1 = k_2 = 0$ is called a *double branch point*.

A branch line has physical reality. In fact, its characteristic property of dividing two flow regions in which the same velocity $\mathbf{q}$ occurs for different $\mathbf{r}$ is in no way extraordinary. One and the same (single-valued) hodograph solution $\mathbf{r} = \mathbf{r}(\mathbf{q})$, however, cannot represent such a flow; hence a series expansion of the hodograph solution must break down at the edge in the hodograph.

Branch lines appear (and have been described by M. J. Lighthill and by T. M. Cherry[38]) in flow in a symmetrical channel, which accelerates from zero velocity at one end to supersonic velocity at the other (see Sec. 25.1). For reasons of symmetry there is not one but two branch lines, B_1 and B_2, with images b_1 ($k_1 = 0$) in the η-direction, and b_2 ($k_2 = 0$) in the ξ-direction. The image of their point of intersection is a sonic double branch point, $k_1 = k_2 = 0$.

We review the main properties of a branch line obtained here: *Branch lines exist only in supersonic flow. A branch line B_1 in the flow plane, $k_1(x, y) = 0$, is a C^+-characteristic. At each point P of B_1, where $k_2 \neq 0$, $\partial k_1/\partial s_1 \neq 0$, the C^-, the curve of constant speed, and that of constant direction through P touch each other; streamlines and potential lines bisect the angles between the B_1 and the C^- at P which has there an inflection point. The acceleration is not infinite at P and has the direction perpendicular to the C^-. The image b_1*

of B_1, the edge, is a (vertical, straight) η-line. It is the envelope of streamlines and potential lines, and the images of all curves having the exceptional C^--direction at their intersection with B_1 have cusps on b_1.

We further review: In case of a limit line, the edge of the fold separates different sheets in the x,y-plane; in case of a branch line, the separating edge is in the ξ,η-plane. The ξ,η-plane in the vicinity of the critical curve in the first case, the x,y-plane in the vicinity of the branch line in the second case, are each covered once, by hypothesis. The difference in properties found for limit line and edge (critical curve and branch line), which have analogous mathematical definitions, is due to the difference between the x,y-plane and the ξ,η-plane; the values of ξ,η are directly related to the flow variables, the x,y are not.

7. Final remarks

(a) *Conditions at the sonic line.* As noted before, there is no analogue to the sonic limit line since D cannot tend toward infinity unless h_1 or h_2 or both do so. However, at $\alpha = 90°$, for a point to be a branch point, it is necessary that not only $h_1(h_2)$ but also $h_1 \cos \alpha$ ($h_2 \cos \alpha$) tends to infinity; then $k_1 \rightarrow 0$ ($k_2 \rightarrow 0$), $\varphi_\xi \rightarrow \infty$ ($\varphi_\eta \rightarrow \infty$). A *double branch point* is characterized by *both k_1 and k_2 being zero.* This result may be added to the results established in Sec. 5 for the singularities at $\alpha = 90°$.

(b) *Remarks on some particular solutions.* The simplest example with a limit line, radial flow (Sec. 17.4), shows a sonic limit line, where some of the properties established for ordinary limit lines are modified. The spiral flow (Sec. 17.4) shows an ordinary limit line, $h_1 = 0$, with no cusps. In the next article, we shall find in a flow studied by Ringleb a limit line with two cusps. This flow features a double limit point at infinity and a sonic point of a straight streamline*. A limit line with a double limit point $h_1 = h_2 = 0$, will likewise be found in the next article in the example of the compressible doublet (which likewise has a sonic limit point on a straight streamline).

For a *simple wave*, $j(x,y) = 0$ in a two-dimensional region; the whole hodograph image reduces to one η-line—in the case of a forward wave—and the transition to hodograph equations is not possible. Some features of the geometry of simple waves can be easily discussed from the present point of view.

Consider a forward wave, $Q - \theta = 2\xi =$ constant, for which therefore $d\xi = 0$ (we assume $\alpha \neq 0°, 90°$ for simplicity); Eqs. (14) then show that $k_1 \equiv 0$, $h_1 \equiv \infty$, and each cross-characteristic, C^+, is a branch line $k_1 = 0$. The two-dimensional flow region is covered by these branch lines. On the other hand the straight characteristics, C^-, have in general an en-

* Such a point has some particular features, based on $d\psi = 0, d\theta = 0$—and hence $\psi_q = 0$, $\theta_\varphi = 0$—on the straight streamline.

velope which is easily seen to have all the properties of a limit line $\mathcal{L}_2$, although there is no hodograph solution to serve as a starting point. Each straight characteristic C^- is simultaneously a line of constant speed and of constant direction and the C^+, the streamlines, and the equipotentials are all cusped at the envelope. Along this limit line the Jacobian $j(x,y)$ is indeterminate: $k_1 = 0$, $k_2 = \pm\infty$.

(c) *Remark on the (x,t)-problem.* Branch lines and limit lines exist likewise in the one-dimensional nonsteady flow of an ideal fluid considered in Chapter III. Limit lines appear as envelopes of either family of characteristics, and in particular as envelopes of the straight characteristics in simple waves. An example of a branch line was found in Sec. 12.5 and illustrated in Fig. 62. It is the $(u - a)$-characteristic BD in the physical plane, which intersects the characteristics of the other family at their inflection points.

The theory can be worked out along the lines of this article. The $(u + a)$-characteristics (i.e., the lines $dx/dt = u + a$) are the η-lines and the $(u - a)$-characteristics the ξ-lines; we know that $v - u$ and $v + u$ remain constant along these ξ-lines and η-lines, respectively. There are two types of limit lines, $\mathcal{L}_1$ and $\mathcal{L}_2$, and of branch lines, B_1 and B_2.[39]

Article 20

Chaplygin's Hodograph Method

1. Separation of variables[40]

In Art. 16 we derived the linear equations (16.32″) for the stream function $\psi(q,\theta)$ in polytropic flow:

$$(1)\quad q^2\left(1 - \frac{\kappa - 1}{2}\frac{q^2}{a_s^2}\right)\frac{\partial^2\psi}{\partial q^2} + \left(1 - \frac{\kappa + 1}{2}\frac{q^2}{a_s^2}\right)\frac{\partial^2\psi}{\partial\theta^2} + q\left(1 - \frac{\kappa - 3}{2}\frac{q^2}{a_s^2}\right)\frac{\partial\psi}{\partial q} = 0.$$

Following Chaplygin we now introduce, instead of q, a new variable τ by setting

$$(2)\quad \tau = \frac{q^2}{q_m^2} = \frac{\kappa - 1}{2}\frac{q^2}{a_s^2}.$$

The Bernoulli equation then takes the form

$$(3)\quad \tau + \left(\frac{a}{a_s}\right)^2 = 1 \qquad \text{or} \qquad a = a_s(1 - \tau)^{\frac{1}{2}},$$

and since, with $\rho_s = 1$ as before, $(a/a_s)^2 = \rho^{\kappa-1}$, the following relations hold:

$$\rho = (1-\tau)^{1/(\kappa-1)}, \qquad p = p_s(1-\tau)^{\kappa/(\kappa-1)}$$

(3′)
$$M^2 = \frac{1}{\kappa-1}\frac{2\tau}{1-\tau}, \qquad 1 - M^2 = \frac{1-h^2\tau}{1-\tau}, \qquad \tau_t = \frac{\kappa-1}{\kappa+1} = \frac{1}{h^2};$$

the subscripts s, m and t in Eqs. (2), (3), (3′) have their usual meaning. As q goes from 0 to q_m, τ varies from 0 to 1; to the sonic value $q_t = a_t$ corresponds the above value τ_t which equals $\frac{1}{6}$ for $\kappa = \gamma = 1.4$. The first-order equations (16.31) become

(4)
$$P(\tau)\frac{\partial\varphi}{\partial\tau} = \frac{\partial\psi}{\partial\theta}, \qquad Q(\tau)\frac{\partial\psi}{\partial\tau} = \frac{\partial\varphi}{\partial\theta},$$

with

(4′)
$$P(\tau) = \frac{2(\kappa-1)\tau}{(\kappa+1)\tau-(\kappa-1)}(1-\tau)^{\kappa/(\kappa-1)}, \qquad Q(\tau) = \frac{2\tau}{(1-\tau)^{1/(\kappa-1)}},^*$$

and Eq. (1) becomes

(5)
$$PQ\frac{\partial^2\psi}{\partial\tau^2} - \frac{\partial^2\psi}{\partial\theta^2} + P\frac{dQ}{d\tau}\frac{\partial\psi}{\partial\tau} = 0;$$

or, substituting from (4′) and simplifying, we obtain

(5′)
$$4\tau^2(1-\tau)\frac{\partial^2\psi}{\partial\tau^2} + \left(1 - \frac{\kappa+1}{\kappa-1}\tau\right)\frac{\partial^2\psi}{\partial\theta^2} + 4\tau\left(1 + \frac{2-\kappa}{\kappa-1}\tau\right)\frac{\partial\psi}{\partial\tau} = 0.$$

It is seen that the coefficient of $\partial^2\psi/\partial\theta^2$ changes sign from positive to negative as τ increases through τ_t.

To this last equation we apply the method of separation of variables, writing $\psi(\tau,\theta) = A(\theta)\,B(\tau)$. Upon substitution one finds from well-known considerations that A''/A must equal a constant. If we put $A''/A = -n^2$, we obtain, for $n \neq 0$, $A(\theta) = A_n(\theta) = \alpha_n e^{in\theta} + \beta_n e^{-in\theta} = \gamma_n \sin(n\theta + \delta_n)$, with α_n, β_n or γ_n, δ_n as arbitrary constants. Following Chaplygin,[41] we set

$$B(\tau) = \psi_n(\tau) = \tau^{n/2}f_n(\tau),$$

so that

(6)
$$\psi(\tau,\theta) = \psi_n(\tau)A_n(\theta) = \tau^{n/2}f_n(\tau)(\alpha_n e^{in\theta} + \beta_n e^{-in\theta}).$$

* This notation is used by several authors. The $Q(\tau)$ of (4′) has nothing to do with the angle $Q(q)$ introduced in Art. 16.

This gives for $\psi_n(\tau)$, and finally for $f_n(\tau)$, the equations

$$(7)\quad \tau^2(1-\tau)\psi_n'' + \tau\left[1+\frac{2-\kappa}{\kappa-1}\tau\right]\psi_n' - \frac{n^2}{4}\left[1-\frac{\kappa+1}{\kappa-1}\tau\right]\psi_n = 0$$

and

$$(7')\quad \tau(1-\tau)f_n'' + \left[(n+1)-\left(n-\frac{2-\kappa}{\kappa-1}\right)\tau\right]f_n' + \frac{n(n+1)}{2(\kappa-1)}f_n = 0.$$

Equation (7′) is a hypergeometric equation

$$(7'')\qquad \tau(1-\tau)f'' + [c_n - (a_n+b_n+1)\tau]f' - a_n b_n f = 0,$$

involving, however, only two parameters n and κ instead of the three parameters in (7″).

A solution of (7″) regular for $\tau = 0$ is given by the hypergeometric function $F(a_n, b_n, c_n; \tau)$, whose Taylor expansion is

$$F(a_n, b_n, c_n; \tau) = \frac{\Gamma(c_n)}{\Gamma(a_n)\Gamma(b_n)} \sum_{\nu=0}^{\infty} \frac{\Gamma(a_n+\nu)\Gamma(b_n+\nu)}{\Gamma(c_n+\nu)} \frac{\tau^\nu}{\nu!},$$

provided c_n is not a negative integer or zero.[42] Hence we obtain the following solution of (7), called Chaplygin Function or Chaplygin solution:

$$\psi_n(\tau) = \tau^{n/2}F[a_n, b_n, n+1; \tau]$$

$$(8)\qquad a_n, b_n = \frac{1}{2}\left[n - \frac{1}{\kappa-1} \pm \left(\frac{\kappa+1}{\kappa-1}n^2 + \frac{1}{(\kappa-1)^2}\right)^{\frac{1}{2}}\right]$$

$$F(a_n, b_n, n+1; \tau) = 1 + \frac{a_n b_n}{1!(n+1)}\tau + \frac{a_n(a_n+1)b_n(b_n+1)}{2!(n+1)(n+2)}\tau^2 + \cdots.$$

Here n is arbitrary except that it must not be a negative integer. It is easily seen that this series converges out to the singularity at $\tau = 1$, i.e., for $\tau \leqq 1$; F is therefore an analytic function of τ which tends to 1 as $\tau \to 0$.* Solutions of (7) corresponding to the exceptional values of n, $|n| > 1$, will not be of the form (8) (see Sec. 4). For $n = -1$ we see from (8) that either a_n or b_n vanishes; in this case $F = 1$ satisfies (7′), and $\psi_{-1}(\tau) = \tau^{-\frac{1}{2}}$ is still of the form (8). (See Sec. 3, and beginning of Sec. 4).

* Near $\tau = 1$ the following expansion may be used in general:

$$F(a,b,c;\tau) = F(a,b,c;1)\,F(a,b,a+b-c+1;1-\tau) + F(c-a,c-b,c;1)(1-\tau)^{c-a-b}F(c-a,c-b,c+1-a-b;1-\tau).$$[43]

This formula is used in Sec. 5.

It is well-known, and we mention it for later use, that the ordinary differential equation of second order (7″) has a second independent solution,

$$\tau^{-n}F(a_n - n, b_n - n, 1 - n;\tau)$$

with the a_n and b_n as in (8). Any solution of the equation (7′) is a linear combination of these two particular solutions.*

By use of the principle of superposition a solution of (5) more general than (6) is obtained in the form

$$\psi(\tau,\theta) = a\theta + \sum_{(n)} \psi_n(\tau)\,(\alpha_n e^{in\theta} + \beta_n e^{-in\theta}),\dagger$$

where the α_n, β_n are arbitrary constants and $\psi_n(\tau)$ is given by (8). The range of n will have to be set in each case; also, when there are infinitely many terms, the convergence must be investigated. To each term $\psi(\tau,\theta) = A_n(\theta)\psi_n(\tau)$, where $A_n(\theta)$ stands for $\gamma_n \sin(n\theta + \delta_n)$, a potential $\varphi(\tau,\theta)$ corresponds by (4), namely,

$$\varphi(\tau,\theta) = Q(\tau)\psi_n' \int A_n(\theta)\,d\theta \equiv \varphi_n(\tau)\int A_n(\theta)\,d\theta. \tag{6′}$$

2. Relation to incompressible flow solutions

We now want to relate these results to corresponding results for an incompressible flow. The passage to the limit from compressible flow to incompressible flow may be made by letting $q_m \to \infty$. From (2) and (3′) it is seen that this corresponds *for fixed* q *and* θ to $\tau \to 0$ and to $M \to 0$. If in (1) we let $M \to 0$, we obtain

$$q^2 \frac{\partial^2\psi}{\partial q^2} + \frac{\partial^2\psi}{\partial\theta^2} + q\frac{\partial\psi}{\partial q} = 0, \tag{9}$$

a Laplace equation in polar coordinates q, θ. The equation has the particular solutions $q^n e^{\pm in\theta}$ (for any n), and also the solution $(a + b\theta)(c + d \log q)$, where a, b, c, d are arbitrary constants. *We want to find compressible flow solutions that reduce to solutions of* (9) *as* $q_m \to \infty$. Such solutions can be defined in many ways. The following is the correspondence proposed by Chaplygin. We consider $q^n e^{\pm in\theta}$ and use the above-mentione$ fact that $\lim_{\tau\to 0} F(a_n, b_n, c_n;\tau) = 1$. We introduce now a reference speed q_1, which will be kept constant when we let q_m tend towards infinity. We may assume $q_1 = 1$ without loss of generality; if τ_1 is the corresponding value of τ,

* Two independent solutions of (7) for $n = 0$ are 1 and $\int_{\tau_1}^{\tau} Q^{-1}\,d\tau$.

† We use here $\psi(\tau,\theta)$ also for the sum of terms (6). The notation $\psi_n(\tau) = \psi_n$ is reserved for a solution of (7).

namely, $\tau_1 = 1/q_m^2$, then $(q/q_1)^2 = \tau/\tau_1$ or $q^2 = \tau/\tau_1$. Hence,

$$\lim_{q_m \to \infty} \frac{\psi_n(\tau)}{\psi_n(\tau_1)} = \lim_{q_m \to \infty} \left(\frac{\tau}{\tau_1}\right)^{n/2} = q^n,$$

and

$$\lim_{q_m \to \infty} e^{\pm in\theta} \frac{\psi_n(\tau)}{\psi_n(\tau_1)} = e^{\pm in\theta} q^n.$$

We therefore decide to associate with each term $q^n e^{\pm in\theta}$ occurring in an incompressible flow solution the term $[\psi_n(\tau)/\psi_n(\tau_1)]e^{\pm in\theta}$ which satisfies (5′). It is seen that in this correspondence the argument θ of $\mathbf{q}$ remains unchanged. If then the stream function of an incompressible flow is given in the form

$$\psi_0 = a\theta + \sum_{(n)} q^n(\alpha_n e^{in\theta} + \beta_n e^{-in\theta}), \tag{10}$$

we associate with it the "corresponding" compressible flow

$$\psi = a\theta + \sum_{(n)} \frac{\psi_n(\tau)}{\psi_n(\tau_1)} (\alpha_n e^{in\theta} + \beta_n e^{-in\theta}). \tag{11}$$

Before continuing we collect for reference solutions of (4) which correspond to the previously mentioned solutions of (9):

(a) $\psi_0 = A\theta, \qquad \varphi_0 = -A \log q$

(b) $\psi_0 = B \log q, \qquad \varphi_0 = B\theta$

(c) $\psi_0 = C_n q^n \sin(n\theta + \delta_n), \qquad \varphi_0 = -C_n q^n \cos(n\theta + \delta_n);$

and correspondingly

(a) $\psi = A\theta, \qquad \varphi = A \int_{\tau_\perp}^{\tau} P^{-1}\, d\tau$

(b) $\psi = B \int_{\tau_\perp}^{\tau} Q^{-1}\, d\tau, \qquad \varphi = B\theta$

(c) $\psi = C_n \dfrac{\psi_n(\tau)}{\psi_n(\tau_\perp)} \sin(n\theta + \delta_n), \qquad \varphi = -\dfrac{C_n}{n} Q \dfrac{\psi_n'(\tau)}{\psi_n(\tau_\perp)} \cos(n\theta + \delta_n),$

where $\psi_n(\tau)$ is given in (8). Each of the last three solutions satisfies (4).[44] Solutions (a) and (b), which correspond to $n = 0$, are the source (sink) solution and the vortex solution of Sec. 17.4 respectively.

We now complete the explanation of Chaplygin's method. Consider the two first-order equations obtained from Eq. (16.31) in the limit of incompressible flow:

$$q \frac{\partial \varphi_0}{\partial q} = -\frac{\partial \psi_0}{\partial \theta}, \qquad \frac{\partial \varphi_0}{\partial \theta} = q \frac{\partial \psi_0}{\partial q}. \tag{12}$$

From these, considered as Cauchy-Riemann equations, it is seen that $\varphi_0 + i\psi_0$ is an analytic function of $(\log q - i\theta)$, hence of $qe^{-i\theta} = q_x - iq_y = \zeta$. Chaplygin's method developed by him for the study of gaseous jets can then be described as follows. Suppose that an incompressible flow problem has been solved by the method of the complex potential. With $z = x + iy$ let $w(z)$ be the *complex potential*, denote by $dw/dz = \zeta = qe^{-i\theta}$ the *complex velocity*, and call $w_0(\zeta) = w(z) = \varphi_0(q,\theta) + i\psi_0(q,\theta)$ the *hodograph potential*. The transformation inverse to $dw/dz = \zeta(z)$, namely, $z = z(\zeta)$, exists provided that $\zeta'(z) \neq 0$.* Next, in the neighborhood of a stagnation point $\zeta = 0$—provided this is a regular point—expand $w_0(\zeta)$ into a Taylor series in ζ so that

$$\psi_0 = \mathscr{I}\left[\sum_{n=0}^{\infty} c_n \zeta^n\right] = \mathscr{I}\left[\sum_{n=0}^{\infty} c_n q^n e^{-in\theta}\right], \tag{13}$$

$\mathscr{I}$ denoting "imaginary part", and form the corresponding series

$$\psi = \mathscr{I}\left[\sum_{n=0}^{\infty} c_n \frac{\psi_n(\tau)}{\psi_n(\tau_1)} e^{-in\theta}\right].\dagger \tag{14}$$

The series (14) is, within its region of convergence, *the stream function of a compressible flow and reduces to* (13) *as* $q_m \to \infty$. Following Chaplygin we then consider (14) as an (approximate) compressible flow solution of the problem whose incompressible flow solution is given by $w_0 = \varphi_0 + i\psi_0$.

We have to keep in mind, however, that the solution (14) need not be the correct solution of the compressible flow problem. The fact that (14) reduces to (13) as $q_m \to \infty$ (and therefore in the limit of incompressible flow satisfies both the differential equation and the given boundary conditions) does not imply that (14) satisfies these boundary conditions while q_m is finite.

The method, which as we shall see works without trouble in the case of the jet problem for which it was designed, meets great difficulties if applied to other boundary-value problems.

Let us finally note that we are led to the same solutions $\psi_n(\tau)$ if (17.12) for the Legendre transform Φ, rather than the stream-function equation for ψ, is used. We then obtain an equation with the same $c_n = n + 1$, but a_n', b_n' instead of a_n, b_n, where $a_n' + b_n' = n + 1/(\kappa - 1)$, $a_n'b_n' = -n(n - 1)/2(\kappa - 1)$. Hence the hypergeometric functions appear again, however with different dependence of a_n', b_n' on the parameters n and κ.[45]

* In general the function $z(\zeta)$ as an inverse function is not single-valued. In the case of a multivalued solution $z(\zeta)$, the hodograph potential $w_0(\zeta)$ will represent only one branch of the solution.

† Note that for $\kappa = -1$ this correspondence is not the same as that given in Sec. 17.6.

3. A flow with imbedded supersonic region

In the remainder of this article we shall illustrate the above-explained correspondence principle by several applications.

We have already reconsidered the source and sink flows. In the present setup they correspond to $n = 0$. We take next $n = -1$,* and note that Eq. (7′) is then satisfied by $f_n = 1$. Hence we obtain the solution

$$(15) \qquad \psi(\tau,\theta) = A\tau^{-\frac{1}{2}} \sin\theta, \qquad \varphi(\tau,\theta) = A\tau^{-\frac{1}{2}}(1-\tau)^{-1/(\kappa-1)} \cos\theta,$$

and, of course, a similar one with $\cos\theta$ in ψ and $\sin\theta$ in ϕ.

We shall now study in some detail this simple exact solution, due to F. Ringleb, which will be found interesting from many aspects.[46] Using now q instead of τ we consider for k real the hodograph solution [satisfying Eqs. (16.31)]

$$(16) \qquad \psi = \frac{k}{q}\sin\theta, \qquad \varphi = \frac{k}{\rho q}\cos\theta$$

with

$$\frac{\partial\psi}{\partial q} = -\frac{k}{q^2}\sin\theta, \qquad \frac{\partial\varphi}{\partial q} = -\frac{k}{\rho q^2}\left(1 - \frac{q^2}{a^2}\right)\cos\theta$$

$$\frac{\partial\psi}{\partial\theta} = \frac{k}{q}\cos\theta, \qquad \frac{\partial\varphi}{\partial\theta} = -\frac{k}{\rho q}\sin\theta,$$

where $d(\rho q)/dq$ has been obtained from Eq. (8.5). We find from Eq. (17.25′)

$$\frac{\partial x}{\partial q} = \frac{k}{\rho}\left(-\frac{\cos 2\theta}{q^3} + \frac{\cos^2\theta}{a^2 q}\right), \qquad \frac{\partial x}{\partial\theta} = -\frac{k}{\rho}\frac{\sin 2\theta}{q^2}$$

$$\frac{\partial y}{\partial q} = \frac{k}{\rho}\left(-\frac{\sin 2\theta}{q^3} + \frac{\sin 2\theta}{2a^2 q}\right), \qquad \frac{\partial y}{\partial\theta} = \frac{k}{\rho}\frac{\cos 2\theta}{q^2},$$

and, as may be verified by differentiation,

$$(17) \qquad x = k\left[\frac{\cos^2\theta}{\rho q^2} + \int_{q_t}^{q}\frac{dq}{\rho q^3}\right], \qquad y = \frac{k}{2\rho q^2}\sin 2\theta,$$

where the constants are chosen for reasons of symmetry. It is seen from (16) that the streamlines in the hodograph form a system of circles through the origin with centers on the q_y-axis; their equation is

$$(16') \qquad q = C\sin\theta, \qquad (C > 0).$$

* Regarding the case $n = 1$, see p. 341.

(We make the restriction to $C > 0$, to avoid double covering of the physical plane.) This shows that for $C \leqq q_m$ the maximum speed on each streamline equals C and is reached for $\theta = 90°$, while the maximum velocity equals q_m for $C > q_m$. If θ is substituted from (16′) into (17), a parametric representation of the streamlines in the x,y-plane is obtained with q as parameter. Thus, if $T = T(q)$ denotes the integral in (17), we obtain

$$(16'') \qquad x = \frac{k}{\rho}\left(\frac{1}{q^2} - \frac{1}{C^2}\right) + kT(q), \qquad y = \frac{k}{\rho q C}\sqrt{1 - \frac{q^2}{C^2}}.$$

For comparison we consider the analogous incompressible flow, which can be derived by introducing $\rho = \rho_0$ and $a^2 = \infty$ in the preceding formulas, and obtain instead of (17)

$$x = k\frac{\cos 2\theta}{2\rho_0 q^2}, \qquad y = k\frac{\sin 2\theta}{2\rho_0 q^2}, \qquad r = \frac{k}{2\rho_0 q^2},$$

where we now assume that k is positive. Or

$$x = r\cos 2\theta, \qquad y = r \sin 2\theta, \qquad \psi = \sqrt{k\rho_0(r - x)};$$

hence for the streamlines of the incompressible flow

$$r - x = \frac{k}{C^2\rho_0} \qquad \text{or} \qquad y^2 = \frac{k}{C^2\rho_0}\left(2x + \frac{k}{C^2\rho_0}\right).$$

For various values of C the streamlines form a family of confocal parabolas in the physical plane and of circles in the hodograph (see Fig. 131). The streamline for $C \to \infty$ is the strip of the x-axis from $x = 0$ to $x = +\infty$, and the flow may be considered as *flow around this edge*. The larger the values of C, the more nearly the parabolas approach the edge. The velocity has a constant value q on concentric circles about the origin, with radius equal to $k/2\rho_0 q^2$.

The compressible flow cannot be considered as flow around the edge since *some streamlines will reach a limit line* before completing the turn. To find this line, $\mathfrak{L}$, we compute from (16) $\partial(\varphi,\psi)/\partial(q,\theta) = 0$, using Eq. (17.27), and find

$$(M^2 - 1)\left(\frac{\partial\psi}{\partial\theta}\right)^2 - q^2\left(\frac{\partial\psi}{\partial q}\right)^2 = \frac{k^2}{q^2}(M^2\cos^2\theta - 1) = 0.$$

Hence the equation of the limit line in the hodograph, i.e., of the critical curve, is simply

$$(18) \qquad \cos\theta = \pm\frac{1}{M},$$

thus consisting of two branches. Since we have assumed $C > 0$, we have

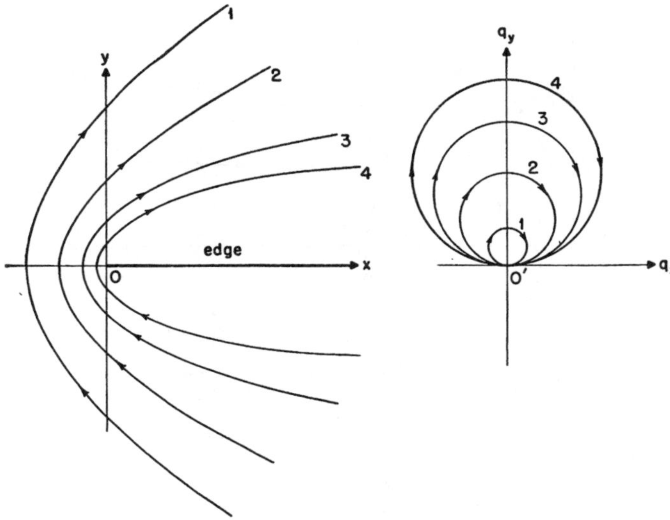

FIG. 131. Streamlines for incompressible flow around an edge.

restricted ourselves to considering the hodograph solution in the upper half-plane $q_y \geqq 0$. The branch to the right ($0 \leqq \theta \leqq 90°$) corresponds to the upper sign in (18), or $\theta = 90° - \alpha$, that to the left ($90° \leqq \theta \leqq 180°$) to the lower sign in (18), or $\theta = 90° + \alpha$. We next compute $\partial\psi/\partial\xi$ and $\partial\psi/\partial\eta$ which for $\alpha < 90°$ are equivalent to h_1 and h_2,

$$\frac{\partial\psi}{\partial\xi} = -\frac{k}{q}(\sin\theta\tan\alpha + \cos\theta),$$

$$\frac{\partial\psi}{\partial\eta} = -\frac{k}{q}(\sin\theta\tan\alpha - \cos\theta),$$

and see that on the branch where $\theta = 90° - \alpha$, $\partial\psi/\partial\xi \neq 0$ $\partial\psi/\partial\eta = 0$, while for $\theta = 90° + \alpha$, $\partial\psi/\partial\xi = 0$, $\partial\psi/\partial\eta \neq 0$. Hence the branch to the right is an l_2, that to the left an l_1. The limit line in the hodograph is tangent to both the sonic circle and the maximum circle; in the polytropic case it is an ellipse (see the second Fig. 132 graphed for $\kappa = \gamma = 1.4$). Combining (18) and (16′) we find, for the polytropic case,

$$\cos^2\theta = \frac{1}{M^2} = \frac{a^2}{q^2} = \frac{a_s^2}{q^2} - \frac{\kappa - 1}{2} = \frac{a_s^2}{C^2\sin^2\theta} - \frac{\kappa - 1}{2},$$

or

$$\sin^4\theta - \frac{\kappa + 1}{2}\sin^2\theta + \frac{a_s^2}{C^2} = 0, \tag{19}$$

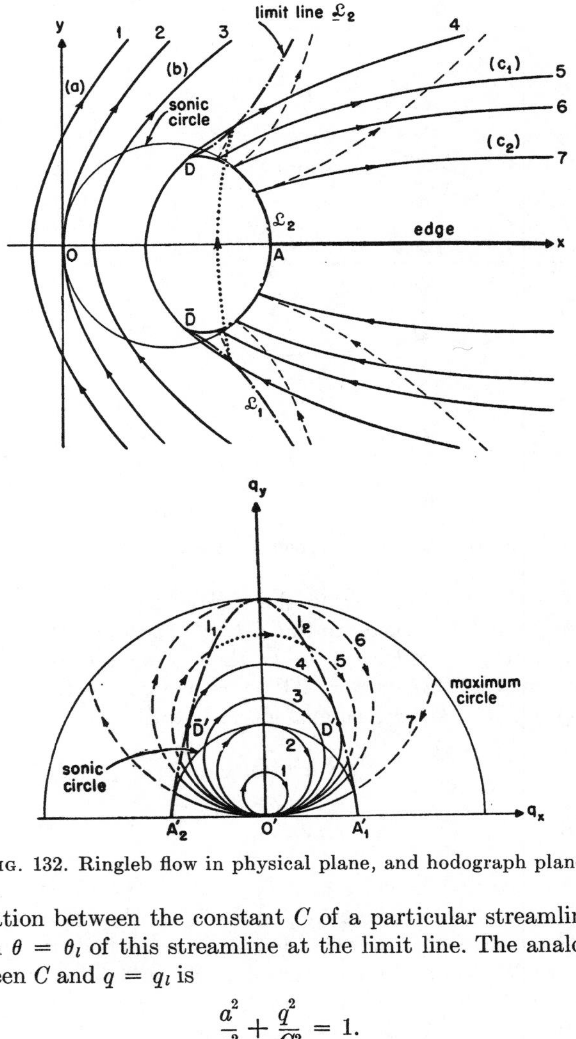

FIG. 132. Ringleb flow in physical plane, and hodograph plane.

as the relation between the constant C of a particular streamline and the inclination $\theta = \theta_l$ of this streamline at the limit line. The analogous relation between C and $q = q_l$ is

$$(19') \qquad \frac{a^2}{q^2} + \frac{q^2}{C^2} = 1.$$

Writing θ_l and q_l in Eqs. (16′) and (19) and solving we obtain

$$(19'') \qquad q_l = C \sin \theta_l, \qquad \sin^2 \theta_l = \frac{\kappa + 1}{4} \pm \frac{1}{4} \sqrt{(\kappa + 1)^2 - \frac{16 a_s^2}{C^2}}.$$

Hence the streamline corresponding to C reaches $\mathcal{L}$ only if $[(\kappa + 1)/4]^2 C^2 \geqq$

a_s^2; or with $\kappa = 1.4$, if $C \geqq \frac{5}{3}a_s = 1.67a_s$. It is thus seen that limit singularities occur only on streamlines for which

$$C \geqq \tfrac{5}{3}a_s. \tag{20}$$

To the smallest C-value, which equals $\frac{5}{3}a_s$, corresponds $\sin^2 \theta_l = (\kappa + 1)/4 = \frac{3}{5}$, $M_l^2 = 4/(3 - \kappa) = 2.5$, $M_l = 1.58$, $q_l = (2/\sqrt{\kappa + 1})a_s = 1.29\, a_s$; the largest M^2-value on this streamline is $M_{\max}^2 = 16/(3 - \kappa)^2 = \frac{25}{4}$, $M_{\max} = M_l^2 = 2.5$. Equation (19″) shows that those streamlines for which (20) holds with the inequality sign meet the limit line in the physical plane in four real points, symmetric in pairs with respect to the x-axis, the axis of symmetry; the general theory of Sec. 19.3 or direct computation shows that these are cusps of the streamlines (see first Fig. 132).

The equation of the limit line in the x,y-plane is found in parametric form by replacing in (17), θ from Eq. (18):

$$x = \frac{k}{\rho q^2 M^2} + kT, \qquad y = \pm \frac{k\sqrt{M^2 - 1}}{\rho q^2 M^2}. \tag{18′}$$

This limit line is symmetric with respect to the x-axis. It consists of two branches: one on the positive side of the x-axis, corresponding to the right branch in the hodograph, and one on the negative side, corresponding to the left one. Each branch has a cusp and then extends towards infinity. The upper branch, $\mathcal{L}_2$, goes from A to the cusp D and then to infinity; and the lower branch $\mathcal{L}_1$ is symmetric with respect to the x-axis. For $M \to \infty$: $\cos \theta = 0$, $\alpha = 0$, to which corresponds $x \to \infty$, $y \to \infty$; there $h_1 = h_2 = 0$. The point A of $\mathcal{L}$ on the x-axis is a sonic point: $M = 1$; it is the image of the two different hodograph points A_1' and A_2'. We see that at both A_1' and A_2': $\psi_q = 0$, $\psi_\theta \neq 0$ (see Art. 19.5). The point A in the physical plane is not a double limit point but the sonic point of an $\mathcal{L}_2$ and of an $\mathcal{L}_1$ which map (exceptionally) into the same point A. This point is also the sonic point of a straight streamline. The streamline direction at A coincides with that of the line of constant θ.

We have seen that a limit line generally has cusps (Sec. 19.3) which play an important role. We determine them now for this particular problem. Computing from (18′), dx/dq and dy/dq and equating both derivatives to zero, we obtain

$$q^2 = 2a^2 - aq\frac{da}{dq}, \qquad q = \frac{2a}{\sqrt{3 - \kappa}} = 1.6a, \qquad \cos \theta = \pm \frac{a}{q} = \pm\, 0.63,$$

corresponding to the points D', $\bar{D}'$ in Fig. 132.

Next we consider the curves of constant velocity. It is seen from (17) that these curves are circles, as in the incompressible case, with centers on the x-axis but no longer concentric. Thus, in particular, the sonic line in the physical plane is a circle. For the circle of velocity q, the abscissa of the center is $k[T + (1/2\rho q^2)]$ and the radius is $k/2\rho q^2$, as in the incompressible case with $\rho = \rho_0$; for $q = q_m$, the center recedes towards infinity.

From what precedes, it is seen that there are three types of streamlines in this example:

(a) Streamlines on which the speed is entirely subsonic and which resemble the streamlines of the incompressible flow around the edge, such as streamline 1 in Fig. 132.

(b) Smooth streamlines which are not entirely subsonic, such as streamlines 2 and 3. Their maximum Mach number, however, is less than 2.5. They have zero speed at infinity; the speed increases and becomes sonic when the streamlines enter the circle of constant sonic velocity; it reaches a maximum at the x-axis, and decreases again towards subsonic values. Their hodograph intersects the sonic circle but remains entirely inside the limit line. The streamline 4 corresponding to $M_{\max} = 2.5$, separates the streamlines (b) from the next group. In the hodograph this streamline is tangent to the limit line, and in the physical plane it passes through the two symmetric cusps D and $\bar{D}$ of the limit line. At these points it has infinite curvature.

(c) Streamlines on which the value of the maximum velocity is greater than $\frac{5}{3}a_s$ (with corresponding maximum Mach number greater than 2.5); they have cusps at their intersections with the limit line. There are two types of such streamlines. The first, (c_1), such as streamline 5 for instance, intersects the hodograph limit line at four points, which correspond to four cusps in the physical plane. Such a line actually has three parts; the first one (solid) extends in the hodograph from left to right through O', the second one (dashed) is outside of the hodograph limit line and the third one (dotted) back inside. The corresponding parts are shown in the physical plane. The other type, (c_2), e.g. streamline 7, intersects the hodograph limit line only twice and has accordingly two (symmetric) cusps. Streamline 6 separates these two types.

The group (b) gives an example of streamlines leading from subsonic to supersonic velocities and back again. This happens in a continuous way and without shock. We may consider any two streamlines of type (b) as walls of a channel; it is thus seen that passage through sonic speed in this type of channel flow is possible in isentropically accelerating and decelerating flow. This is an example of a smooth transonic flow.* (See also Sec. 25.3 ff.).

4. Further comments and generalizations

The solution (8), which holds for $n \neq -1, -2, \cdots$, has the property that F is analytic and $\rightarrow 1$ as $\tau \rightarrow 0$, and there is just one such solution of

* The meeting of streamlines such as 1, 2, 3, 4 with the outer branch of the limit line is only apparent.

(7′) in general. In the exceptional case $n = -1$, there is, however, a second solution which has this property. We may verify directly that

$$f = (1 - \tau)^{\kappa/(\kappa-1)}$$

satisfies (7′). Thus in addition to (15),

$$\text{(21)} \quad \psi = A\tau^{-\frac{1}{2}}(1 - \tau)^{\kappa/(\kappa-1)} \sin\theta, \qquad \varphi = A\tau^{-\frac{1}{2}}\left(1 + \frac{\kappa+1}{\kappa-1}\tau\right)\cos\theta$$

is one more solution corresponding to $n = -1$. This solution has features quite similar to (15). The flow has again a limit line with two cusps. As in (15) there are in this flow smooth streamlines along which a subsonic flow becomes sonic, then supersonic, and decelerates again to subsonic velocities. The streamline which passes through the cusps separates the regular smooth streamlines from streamlines with cusps at the limit line.

If in (15) and (21) we put $Aq_m = C$, where C remains fixed as $q_m \to \infty$, it is seen that *to these two compressible flows there corresponds the same incompressible flow* $\psi = (C/q)\sin\theta$, $\varphi = (C/q)\cos\theta$, discussed in Sec. 3.

The two functions $\psi_{-1}(\tau)$ in (15) and (21) must each satisfy Eq. (7) for $n = +1$ also. If we consider directly the case $n = 1$ we have from (8), $a_1 = 1$, $b_1 = -1/(\kappa - 1)$, and

$$\begin{aligned}\psi_1(\tau) &= \tau^{\frac{1}{2}} - \frac{1}{\kappa-1}\frac{\tau^{3/2}}{2!} + \frac{1}{\kappa-1}\left(\frac{1}{\kappa-1} - 1\right)\frac{\tau^{5/2}}{3!} - \cdots \\ &= \frac{\kappa-1}{\kappa\sqrt{\tau}}\,[1 - (1-\tau)^{\kappa/(\kappa-1)}].\end{aligned}$$

Thus $\psi_1(\tau)$ appears as the sum of two terms, which are respectively equal to the $\psi_{-1}(\tau)$ in (15) and (21).[47] For the corresponding $F(a_1, b_1, 2;\tau)$:

$$F\left(1, -\frac{1}{\kappa-1}, 2; \tau\right) = \frac{\kappa-1}{\kappa}\,[\tau^{-1} - \tau^{-1}(1-\tau)^{\kappa/(\kappa-1)}],$$

and one may easily check that the expression to the right tends toward 1 as τ tends to zero.

We turn now to another example. The flow studied in detail in Sec. 3 has been considered as the compressible counterpart of incompressible flow around an edge, i.e., a "corner" of 0° opening. It is natural to try to generalize in a similar way the well-known incompressible flow around a convex corner (see Fig. 133). It is known that the hodograph streamlines corresponding to the incompressible flow around a corner of angle $360° - \alpha$ are a family of lemniscates (actually one loop of each lemniscate), all within the angle $\alpha - 180°$; their common tangents form the hodograph streamline $\psi = 0$, which is the image of the two legs of the angle in the flow plane. These lemniscates play the role of the circles in the edge flow. Compressible counterparts of these and similar flows may be constructed by means of the method considered in this article. A few results are the following.

The principal features regarding the limit line, etc., remain unchanged. Consider, e.g., $\alpha = 270°$ (see Fig. 134). The limit line in the physical plane

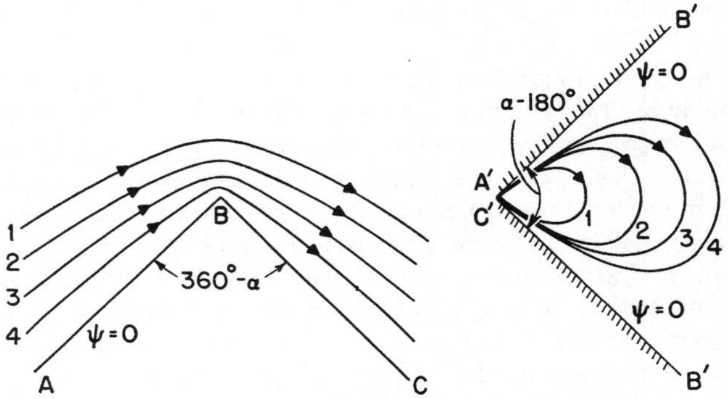

FIG. 133. Incompressible flow around a 90° corner.

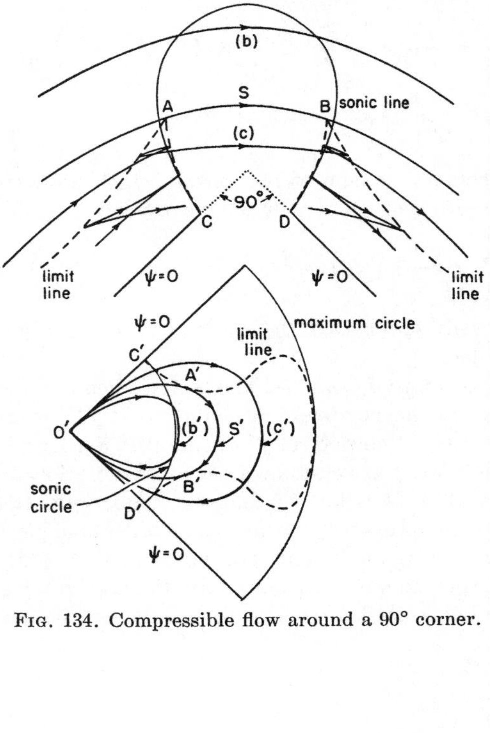

FIG. 134. Compressible flow around a 90° corner.

has again two cusps. There is again one well-defined streamline, S (the analogue of streamline 4 in the edge flow), which separates the smooth streamlines (b) which do not meet the limit line at all from those streamlines (c) which feature two or four cusps (one on each of the four branches of the limit line). The sonic line is now no longer a circle but resembles a loop of a lemniscate.

Consider now the complex potential w of the incompressible flow:

$$w = z^m, \qquad \zeta = qe^{-i\theta} = mz^{m-1}.$$

We eliminate z and obtain, with $A = m^{-m/(m-1)}$,

$$w = A\zeta^{m/(m-1)} = Aq^{m/(m-1)}\left(\cos\frac{m}{m-1}\theta - i\sin\frac{m}{m-1}\theta\right).$$

In this notation the Ringleb flow corresponds to $m = \frac{1}{2}$, and the flow of the last example to $m = \frac{2}{3}$. Let m be between $\frac{1}{2}$ and 1; the incompressible flow is a flow around a convex corner of angle $\beta = 2\pi - \alpha$, $\alpha = \pi/m$ (in radians); the angle α is between 2π and π, and β between 0 (Ringleb flow) and π.[48]

To the incompressible stream function

$$\psi_0 = Aq^{m/(m-1)}\sin\frac{m}{m-1}\theta = Aq^{\pi/(\pi-\alpha)}\sin\frac{\pi}{\pi-\alpha}\theta$$

corresponds, in general, the compressible flow

$$(8') \qquad \psi = Aq^{\pi/(\pi-\alpha)}F(a,b,c;\tau)\sin\frac{\pi}{\pi-\alpha}\theta.$$

Here $\pi/(\pi - \alpha)$ stands for the n of Eq. (8) and we have

$$a + b = n - \frac{1}{\kappa - 1} = \frac{\pi}{\pi - \alpha} - \frac{1}{\kappa - 1}$$

$$(8'') \qquad ab = -\frac{1}{2}n(n+1)\frac{1}{\kappa - 1} = -\frac{1}{2(\kappa - 1)}\frac{\pi(2\pi - \alpha)}{(\pi - \alpha)^2}$$

$$c = n + 1 = \frac{2\pi - \alpha}{\pi - \alpha}.$$

If here the angle α is of the form

$$\alpha = \frac{p+2}{p+1}\pi \qquad (p = 1, 2, 3, \cdots)$$

(as, e.g., the angle $\alpha = 270°$ of Fig. 134) then $c = (2\pi - \alpha)/(\pi - \alpha) = -p$, i.e a negative integer; it is thus seen that with $\beta = 2\pi - \alpha$, for β-values such as $\beta = \pi/2, 2\pi/3, 3\pi/4$, etc., a solution, ψ, of the preceding form (8′)

does not exist and another expansion must be used.[49] The exceptional case $\alpha = 270°$ of Fig. 134 has been computed explicitly.[50]

For angles not of the above form, e.g., for all angles $\beta < 90°$, the above solution exists; in particular for[51] $\beta = 60°$, $c = -\frac{1}{2}$, $n = -\frac{3}{2}$, and for[52] $\beta = 46.8°$ the solutions have been studied in detail. In the last case the infinite hypergeometric series reduces to a polynomial of degree four. These cases show no new feature compared with the cases $\beta = 0°, 90°$ above (Figs. 132 and 134).

The examples of the last two sections can be adapted to general elastic fluids. The restriction to polytropic flow (with the value of κ taken as 1.4 in computations) is made in order to obtain concrete results for the most important case. Each of these particular solutions can be regarded a posteriori as a solution of a boundary-value problem, e.g., by considering in each case certain streamlines as fixed boundaries of the flow.

5. Compressible doublet

We pass now to an example that is distinguished by a particularly interesting limit line. With the notation of p. 343 we consider the case $m = -1$, $n = \frac{1}{2}$, i.e., we consider a compressible analogue of the *doublet*[53]. Here

$$w = -\frac{1}{z}, \qquad \zeta = \frac{dw}{dz} = qe^{-i\theta} = \frac{1}{z^2}, \qquad \psi_0 = q^{\frac{1}{2}} \sin\frac{\theta}{2}. \tag{22}$$

The corresponding compressible stream function (for $\kappa = 1.4$) is, according to (8),

$$\begin{gathered} \psi = \sqrt{q_m}\,\tau^{\frac{1}{4}} F\left(a,b,\tfrac{3}{2};\tau\right) \sin\frac{\theta}{2} = \sqrt{q_m}\,\tau^{\frac{1}{4}} f(\tau) \sin\frac{\theta}{2}, \\ a + b = \frac{1}{2} - \frac{1}{\kappa - 1} = -2, \qquad ab = -\frac{3}{8(\kappa - 1)} = -\frac{15}{16}. \end{gathered} \tag{23}$$

For $f(\tau)$ one has the expansion

$$f(\tau) = 1 - \tfrac{5}{8}\tau + \tfrac{31}{128}\tau^2 - \cdots.$$

This converges uniformly for $\tau \leqq 1$. Near $\tau = 1$ an expansion in $1 - \tau$ may be used (see footnote p. 331).

Corresponding to this ψ the φ may be determined; the coordinates x, y are then given by (17.25′). If τ is used instead of q and (3′) is noted, the formulas (17.25′) are replaced by

$$q_m \sqrt{\tau}\, \frac{\partial x}{\partial \tau} = \frac{\partial \varphi}{\partial \tau} \cos \theta - \frac{\partial \psi}{\partial \tau} (1 - \tau)^{-1/(\kappa-1)} \sin \theta$$

and three similar ones. Integration gives

$$\begin{aligned} \sqrt{q_m}\, \tau^{1/4}(1 - \tau)^{1/(\kappa-1)}\, x &= 2\tau f'(\tau)\left(\cos\frac{\theta}{2} - \frac{1}{3}\cos\frac{3\theta}{2}\right) + f \cos\frac{\theta}{2}, \\ \sqrt{q_m}\, \tau^{1/4}(1 - \tau)^{1/(\kappa-1)}\, y &= 2\tau f'(\tau)\left(\sin\frac{\theta}{2} - \frac{1}{3}\sin\frac{3\theta}{2}\right) + f \sin\frac{\theta}{2}. \end{aligned} \tag{24}$$

We shall now discuss the singularities of this transformation. There is *no branch line*. Indeed it is seen from (23) and from the fact that both f and f' are finite, that neither $\partial\psi/\partial q$ nor $\partial\psi/\partial\theta$ can become infinite except at the point $q = 0$ for which $x \to \infty$ and $y \to \infty$; this is an isolated branch point.

The *limit singularities* are here of some interest (see Sec. 19.4). From (23)

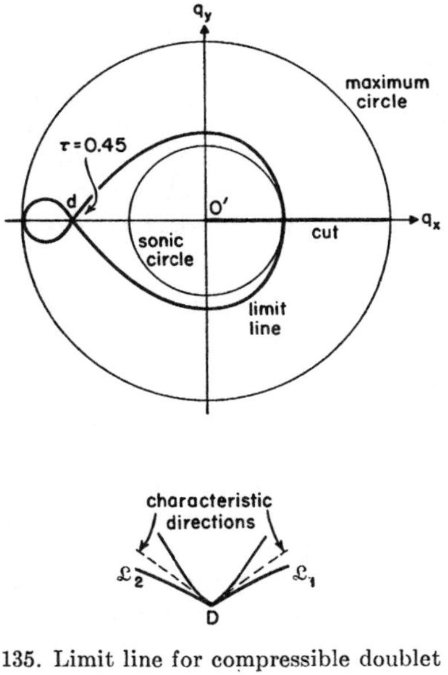

FIG. 135. Limit line for compressible doublet flow.

we compute $q\,(\partial\psi/\partial q) \pm \sqrt{M^2 - 1}\,(\partial\psi/\partial\theta)$, which are essentially the same as $\partial\psi/\partial\eta$ and $\partial\psi/\partial\xi$. This gives

$$\left(2\tau f' + \frac{1}{2} f\right) \sin\frac{\theta}{2} \pm \frac{1}{2}\sqrt{M^2 - 1}\, f \cos\frac{\theta}{2} = 0 \tag{25}$$

or

$$\pm\sqrt{M^2 - 1}\cot\frac{\theta}{2} = 4\tau\frac{f'}{f} + 1.$$

This line in the hodograph, the critical curve, has a double point d for $\theta = 180°$, where $1 + 4\tau f'/f = 0$, i.e., $\tau \approx 0.45$. At this point both $\partial\psi/\partial\xi = 0$ and $\partial\psi/\partial\eta = 0$; consequently $h_1 = h_2 = 0$, and, by (19.7), $\partial h_1/\partial\eta = \partial h_2/\partial\xi = 0$. Next we compute at the double point the second derivatives of φ and find that $\partial^2\varphi/\partial q^2 = \partial^2\varphi/\partial\theta^2 = 0$, $\partial^2\varphi/\partial q\partial\theta \neq 0$. From this we conclude as at the end of Sec. 19.4 that the point is an extremum for ψ and a saddle point for φ: *no streamline, ψ = constant, passes through this point*; the streamlines encircle it (see Fig. 128).

The critical curve is a double loop curve touching the sonic circle at $\theta = 0°$ and the maximum speed circle at $\theta = 180°$ (see Fig. 135). The first of these two points is the sonic point of a straight streamline (similar to point A in the Ringleb flow).

In the physical plane the limit line consists of the lines $h_1 = 0$ and $h_2 = 0$, which are both cusped at their common point D. The cusp tangents at D have characteristic directions. In the neighborhood of D the flow is confined to the obtuse angle (covered four times) above the cusps. This is an interesting example of a limit point of higher order, $h_1 = h_2 = 0$, where the critical curves intersect.

6. Subsonic jet

We turn now to the consideration of a *boundary-value problem*: Chaplygin's method applied to a subsonic jet.* The problem of a gas escaping through a slit between plane walls adapts itself particularly well to Chaplygin's method (Sec. 2), since this is essentially a problem in the hodograph plane.[54] Here the shape of the escaping fluid is not known beforehand: the boundaries of the jet are *free boundaries*. Since the flow is assumed steady, the boundaries, both fixed walls and free boundaries, do not change in time and are streamlines. Hence along the boundary, ψ must be piecewise constant. Along a straight wall the angle θ is a given constant; hence the hodograph image of such a line is radial through O' with known slope.

* See also comments in Sec. 15.2.

Along a free boundary the pressure is constant. The influence of gravity and other external forces being neglected, it follows from Bernoulli's equation, in either the incompressible or compressible case, that on a free boundary surface the velocity has some constant value, say q_1. Thus, the free streamlines are mapped onto arcs of circles about O' of radius q_1. Hence, for such a problem we know the boundary and the boundary values in the hodograph for the incompressible as well as for the compressible case; these boundary conditions coincide if the shape of the vessel, the total flux through the orifice and speed at the jet boundary are the same in the two cases.

The particular jet problem considered by Chaplygin is shown in the figure (see Fig. 136). There is a particularly simple disposition of walls, in which the vertical wall $\bar{A}\bar{B}$ is a continuation of AB. The distance $B\bar{B} = 2a$ is given. We assume that the half plane to the left of $AB\bar{B}\bar{A}$ is filled with fluid; the flow starts with zero velocity at infinity to the left and converges to a parallel jet, with constant velocity q_1 on the free boundaries of the jet, which are streamlines. Along the horizontal center streamline we take $\psi = 0$; and $\psi = \psi_1$, say, along ABC while $\psi = -\psi_1$ along $\bar{A}\bar{B}\bar{C}$.

The boundary ABC is mapped onto $A'B'C'$ in the hodograph plane and the same holds for $\bar{A}\bar{B}\bar{C}$ and $A'\bar{B}'C'$. All streamlines in the hodograph go from A' to C'. Along $\bar{B}'C'B'$ the velocity q equals q_1.

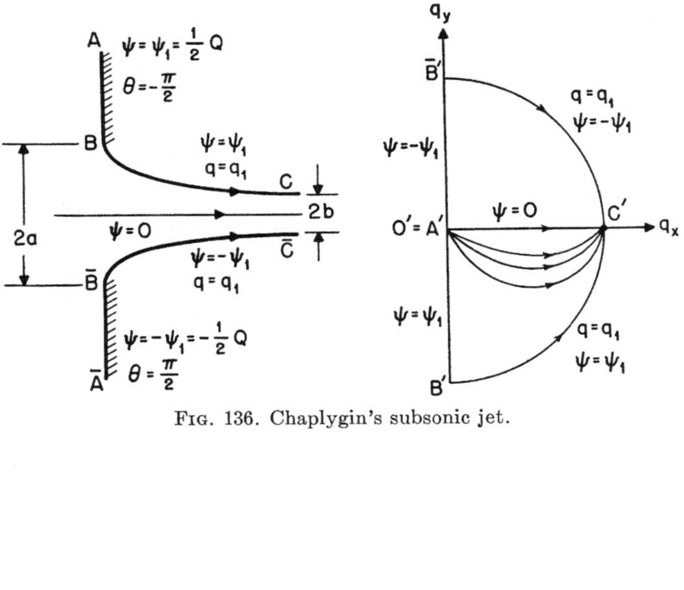

FIG. 136. Chaplygin's subsonic jet.

In this problem *the method of Helmholtz-Kirchhoff-Joukowski furnishes directly what we need for the application of Chaplygin's method, namely the (incompressible) hodograph potential* $w_0(\zeta)$. The method can be applied to much more general data than the ones considered here. In our case the result is simple and well-known. The formula which gives w_0 in terms of ζ is

$$
\begin{aligned}
w_0 &= \varphi_0 + i\psi_0 = -\frac{Q}{\pi} \log \frac{1}{2}\left(\frac{1}{\zeta} - \zeta\right) \\
&= \frac{Q}{\pi} \log 2\zeta - \frac{Q}{\pi} \log (1 - \zeta^2) = \frac{Q}{\pi}\left(\log 2\zeta + \sum_1^\infty \frac{\zeta^{2n}}{n}\right).
\end{aligned} \tag{26}
$$

Here $Q = 2\psi_1$ denotes the flux through the slit, per second, and the velocity on the free boundaries BC and $\bar{B}\bar{C}$ has been put equal to unity: $q_1 = 1$. The expansion in (26) is valid for $|\zeta| < 1$. The value of Q is connected with the asymptotic width, $2b$, of the jet: $Q = 2b\rho_0 q_1 = 2b\rho_0$ in the incompressible case. From (26)

$$
\begin{aligned}
\psi_0 &= -\frac{Q}{\pi}\left(\theta + \sum_1^\infty \frac{q^{2n}}{n} \sin 2n\theta\right), \\
\varphi_0 &= \frac{Q}{\pi}\left(\log 2q + \sum_1^\infty \frac{q^{2n}}{n} \cos 2n\theta\right).
\end{aligned} \tag{27}
$$

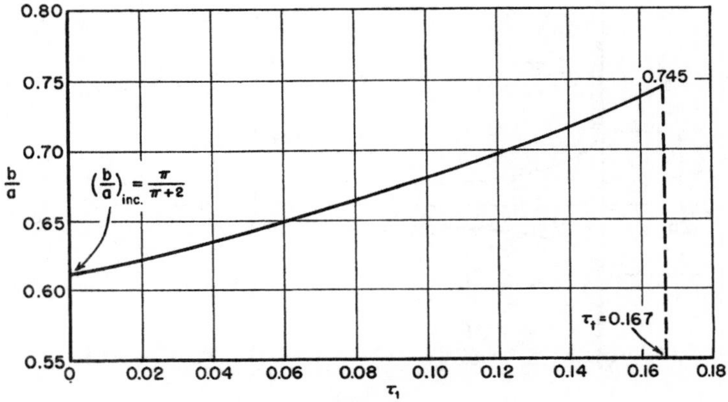

FIG. 137. Contraction ratio b/a versus τ_1.

We see that for $\theta = 0, \psi_0 = 0$. For $q = q_1 = 1$, $\theta \gtrless 0$, the series

$$\sum_1^\infty \frac{q^{2n}}{n} \sin 2n\theta = \sum_1^\infty \frac{\sin 2n\theta}{n} = \pm\frac{\pi}{2} - \theta.$$

Hence if $\theta > 0$, $\psi_0 = -(Q/\pi)(\theta + (\pi/2) - \theta) = -Q/2$ on $\bar{B}\bar{C}$, while if $\theta < 0$, $\psi_0 = -(Q/\pi)(\theta - (\pi/2) - \theta) = Q/2$ on BC, as it should be. On AB and $\bar{A}\bar{B}$, for $\theta = \mp\pi/2$, the sum of the series is zero; hence on $AB, \psi_0 = Q/2$, on $\bar{A}\bar{B}$, $\psi_0 = -Q/2$. Thus the boundary conditions are satisfied.

Now consider the *compressible jet*. The velocity q will still be constant on the free boundaries; we call it q_1 and assume it equal to one, $\tau_1 = 1/q_m^2$. According to the rule of Sec. 2 we then form

$$(28) \qquad \psi = -\frac{Q}{\pi}\left[\theta + \sum_1^\infty \frac{1}{n}\frac{\psi_n(\tau)}{\psi_n(\tau_1)} \sin 2n\theta\right].$$

We remember and may check that for $q_m \to \infty$ $(\tau_1 \to 0)$ this reduces to the ψ_0 of (27). Chaplygin has proved that the above converges together with all its necessary derivatives if $q < q_1 < q_t$, i.e., $\tau < \tau_1 < \frac{1}{6}$. Moreover, as we have seen, (28) *satisfies the boundary conditions exactly*. In fact for $\theta = \pm\pi/2$, $\psi = \mp Q/2$; for $\tau = \tau_1$, i.e., for $q = q_1$, the right side of (28) is exactly the same as the right side of (27) for $q = 1$. Hence, indeed, along ABC, $\psi = Q/2$, and along $\bar{A}\bar{B}\bar{C}$, $\psi = -Q/2$. Therefore (28) is the exact solution of the subsonic jet problem.

The velocity potential φ corresponding to ψ in (28) can be determined as previously explained; then x and y are obtained as infinite series in τ and θ; each series contains the flux Q as a factor. The final aim is the velocity distribution throughout the field of flow, i.e., τ,θ in terms of x,y. Since the problem is subsonic, there are no singularities of branch or limit type; the numerical computation, however, becomes very involved.

Some typical questions of jet flow can be dealt with directly by means of the hodograph solution. Such a question is—as in the incompressible case—to find the form of the jet (i.e., the shape of the free streamlines) and, in particular, its asymptotic width. If we call the asymptotic width $2b$ (see Fig. 136), while $2a$ is the given width of the slit, we wish to find $\lambda = b/a$. Since we know x and y in terms of τ and θ, we can find the values of x and y on the jet boundary, i.e. for $\tau = \tau_1$. In particular the expression for y is of the form: y/Q equal to a known function of τ_1, θ, and Q is proportional to $2b$. Now as $2a$ is the width of the slit, the value of y on the jet boundary, for $\theta = \pi/2$, is equal to $-a$; hence one obtains λ in terms of τ_1 (see Fig. 137 which corresponds to $\kappa = 1.4$). In the incompressible case, i.e. for $\tau_1 \to 0$, the value of λ has been found by Kirchhoff: $\lambda = \pi/(\pi + 2) = 0.611$. The other extreme appears for $\tau_1 = 0.167$ (sonic value), for which $\lambda = 0.745$. The values of λ for various τ_1, were computed by Chaplygin.

The method of this section is applicable to more general steady-flow jet problems as long as the speed q on the free streamlines is everywhere below the sonic speed. In terms of pressure, this will be so if the outside pressure p_1 in the receiver is higher than the "critical" pressure corresponding to $M = 1$. If p_1 is lower than the critical pressure, the jet is wholly or partly supersonic, and the Chaplygin method does not apply.[55]

CHAPTER V

INTEGRATION THEORY AND SHOCKS

Article 21

Development of Chaplygin's Method

1. The problem

In this article we shall describe two different approaches aimed at obtaining solutions of certain boundary-value problems. The jet problem of the preceding section was a boundary problem that could be solved exactly. The main reason for this was that the boundary conditions appeared in a natural way in the hodograph plane; in addition, the problem dealt with was entirely subsonic.

In the important problems of flow past a body and of channel flow, the boundary conditions are given in the physical plane and therefore the application of the hodograph method to these problems meets with great difficulties. In the first part of the present article we shall discuss methods due mainly to M. J. Lighthill. The work of T. M. Cherry is independent of that of Lighthill; it goes in the same direction and in certain respects further. However, it does not seem appropriate to discuss here both Lighthill's and Cherry's work, and the latter is somewhat more difficult to present in a small space. The second part of the article will deal with S. Bergman's method and some of his results. We shall also point out the mathematical relation between the two approaches.

In Secs. 20.1 and 20.2 Chaplygin's method for the construction of compressible flows was described. A compressible flow was constructed corresponding to an incompressible flow, according to definite rules. In the problem of flow past an obstacle the application of Chaplygin's method runs as follows. First, we attempt to determine the complex potential $w(z) = \varphi(x, y) + i\psi(x, y)$ of the given boundary-value problem for incompressible flow. Then, with $dw/dz = q_x - iq_y = qe^{-i\theta} = \zeta$, the complex potential $w(z)$ is expressed as a function of ζ, namely, $w(z) = w_0(\zeta) = \varphi_0(q, \theta) + i\psi_0(q, \theta)$. As explained in Sec. 2 of the preceding article $w_0(\zeta)$ is next expanded into a Taylor series in the neighborhood of a stagnation point $\zeta = 0$, and using $\tau = q^2/q_m{}^2$ we associate with the incompressible $\psi_0(q, \theta)$ a compressible stream function $\psi(\tau, \theta)$ by means of the rule contained in Eqs.

(20.10) and (20.11); we accept this $\psi(\tau, \theta)$ as an approximate solution of the original boundary-value problem on the basis of the fact that $\psi(\tau, \theta)$ tends to $\psi_0(q, \theta)$ as $q_m \to \infty$.

Two basic difficulties regarding this conception are immediately recognized. First, there is no reason to assume that $\psi(\tau,\theta)$, if reverted to the physical plane, will satisfy the original boundary conditions for $M > 0$. The streamline in the x,y-plane which in the compressible flow corresponds to the given contour, for which we have solved the incompressible problem, will not coincide with this contour; its shape will depend on a (dimensionless) parameter, e.g. on τ_1, the "velocity" of the undisturbed flow. We shall have to be satisfied if, for a certain range of values of τ_1, this curve is close to the given contour.

We may also look at this difficulty from the following point of view. We are unable to find a solution which satisfies both the equations of motion and the given boundary conditions. In the approach just mentioned, we satisfy the differential equations exactly and the boundary conditions approximately.* Another possibility, well known in applications, is to try to satisfy the boundary conditions exactly and the differential equations only approximately. This is the approach of methods in which simplified differential equations in the physical plane are substituted for the exact equations. (If these equations are linear, then the boundary-value problem can be solved in many cases, at least in principle.)

Second, the construction of $\psi(\tau,\theta)$ clearly works only in a region of convergence of both $\psi_0(q,\theta)$ and $\psi(\tau,\theta)$. Since the incompressible flow has a singularity in the hodograph, the Taylor expansion of $w_0(\zeta)$ (about a certain stagnation point) will have a limited circle of convergence; a separate power series is required for each region of convergence. These series must be analytic continuations of each other across the boundaries of these regions, and we assume here that this problem in the theory of functions can be solved. However, even if such a solution of the incompressible flow problem has been found and we associate with these series the corresponding Chaplygin series, as in Eqs. (20.10) and (20.11), it cannot be asserted that the latter need be analytic continuations of each other, and we shall see that actually they are not (Sec. 3).

The developments which we shall discuss in this article concern only this second difficulty, for which decisive results have been obtained. The first difficulty is not tackled in any of them. This holds for the investigations of Cherry, Lighthill, and others working in a similar direction, as well as for those of Bergman and his collaborators.

We shall present in the following three sections Lighthill's method and main result. In so doing it seems useful, after some preparatory work to be

* This applies even for the simplified approach discussed in Secs. 17.5 and 17.6.

given in the next section, to start with the consideration of a concrete example, namely the compressible circulation-free flow about a circular cylinder, and to explain the principle of the solution and the essential difficulties encountered, without going into too many mathematical details. The study of this problem, which involves supersonic as well as subsonic velocities, will thus contribute to an understanding of the general situation. The solution presented in the example is based throughout on series expansions of the (incompressible) hodograph potential $w_0(\zeta)$. This could be avoided in the subsonic region (see Sec. 4), but for the continuation to supersonic speeds one has at any rate to start from the series expansion of the hodograph potential.

A general solution for the subsonic region will then be discussed in Sec. 4.[1] However, we do not enter into an explanation of the general procedure recommended by Lighthill in the case of supersonic velocities.

2. Replacement of Chaplygin's factor $[\psi_n(\tau_1)]^{-1}$

We refer to Eqs. (20.13) and (20.14) with $\psi_n(\tau)$ given by Eq. (20.8). For $\tau = \tau_1$, the right side of Eq. (20.14) coincides with that of Eq. (20.13) with $q = 1$, where $q = q_1 = 1$ is the velocity at infinity and $\tau_1 = 1/q_m^2$. We have indicated earlier that if we start with one branch of the compressible flow $\psi(\tau,\theta)$, the main problem is to find its continuation over the desired region. Lighthill noticed that for this purpose Chaplygin's factor $[\psi_n(\tau_1)]^{-1}$ is inappropriate (leading to avoidable complications) and replaced it by another factor which retains the essential properties of the former and is better adapted to the problem of continuation. What are these properties? We have seen that

$$\lim_{q_m \to \infty} \psi_n(\tau)/\psi_n(\tau_1) = q^n,$$

and require accordingly, $f(n, \tau_1)$ denoting the factor in question, that

$$\lim_{q_m \to \infty} \psi_n(\tau)f(n, \tau_1) = \lim_{q_m \to \infty} \left(\frac{\tau}{\tau_1}\right)^{n/2} = q^n, \tag{1}$$

i.e., that $f(n, \tau_1)$ should behave like $\tau_1^{-n/2}$ as $q_m \to \infty$. A second requirement is that

$$\psi = \mathcal{I}\left[\sum_{n=0}^{\infty} c_n \psi_n(\tau) f(n, \tau_1) e^{-in\theta}\right], \tag{2}$$

which replaces Eq. (20.14), should have the same circle of convergence as the original ζ-series expansion (20.13) about $\zeta = 0$; and in fact should exhibit for $\tau = \tau_1$, a behavior similar to that of (20.13) for $q = 1$.

To arrive at an appropriate normalizing factor $f(n, \tau_1)$ we introduce instead of τ, the new variable, λ, the same as that in (17.17), defined by $d\lambda/dq = \sqrt{1 - M^2}/q$; it leads to a normal form of the second-order equation for $\psi(q, \theta)$. In terms of τ we have

$$(3) \qquad \frac{d\lambda}{d\tau} = \frac{d\lambda/dq}{d\tau/dq} = \frac{\sqrt{1 - M^2}}{q} \frac{q_m^2}{2q} = \frac{\sqrt{1 - M^2}}{2\tau} = \frac{1}{2\tau} \sqrt{\frac{1 - \tau/\tau_t}{1 - \tau}}.$$

We chose the limits of the integral so that $\lambda = 0$ for $\tau = \tau_t$:

$$(3') \qquad \begin{aligned} \lambda &= -\frac{1}{2} \int_\tau^{\tau_t} \frac{1}{\tau} \sqrt{\frac{1 - \tau/\tau_t}{1 - \tau}}\, d\tau \\ &= \tau_t^{-\frac{1}{2}} \tanh^{-1} \sqrt{\frac{\tau_t - \tau}{1 - \tau}} - \tanh^{-1} \sqrt{\frac{1 - \tau/\tau_t}{1 - \tau}}. \end{aligned}$$

We shall use the variable

$$(4) \qquad s = \lambda + \sigma,$$

where σ is a constant defined by the requirement that $s \approx \log (q/q_m)$ for small τ, i.e.

$$(5) \qquad \lim_{\tau \to 0} \frac{e^{2s}}{\tau} = 1 \qquad \text{or} \qquad \lim_{\tau \to 0} \frac{e^{2\lambda}}{\tau} = e^{-2\sigma}.$$

This gives with $\kappa = \gamma = 1.4$ and $h^2 = \tau_t^{-1} = 6$,

$$\sigma = -h \tanh^{-1} \frac{1}{h} - \frac{1}{2} \log (h^2 - 1) + \log 2 = -1.17.$$

It is seen that $s \to -\infty$, as $\tau \to 0$, $M \to 0$. As τ increases from 0 to τ_t, s increases from $-\infty$ to $\sigma = -1.17$.

With s (or λ) as an independent variable the stream-function equation (20.5) is transformed to

$$(6) \qquad \frac{\partial^2 \psi}{\partial s^2} + \frac{\partial^2 \psi}{\partial \theta^2} = T \frac{\partial \psi}{\partial s},$$

where T is a function of s (or τ or q), namely,[2]

$$T = -\frac{\rho}{\sqrt{1 - M^2}} \frac{d}{ds} \left(\frac{\sqrt{1 - M^2}}{\rho} \right).$$

If we apply to Eq. (6) the separation of variables $\psi = \psi_n(s) e^{in\theta}$, we obtain

$$(6') \qquad \frac{d^2 \psi_n}{ds^2} - n^2 \psi_n = T \frac{d\psi_n}{ds},$$

an equation which will be used later.

In order to obtain an equation with a term containing ψ rather than

$\partial\psi/\partial s$ as in Eq. (6), the dependent variable is changed by putting

(7) $$\psi = V(\tau)\psi^*.$$

An elementary computation gives

(8) $$\frac{\partial^2\psi^*}{\partial s^2} + \frac{\partial^2\psi^*}{\partial\theta^2} + F\psi^* = 0,$$

where, with a prime denoting differentiation with respect to s,

(9) $$\frac{2V'}{V} = T, \qquad F = \frac{V''}{V} - \frac{2V'^2}{V^3}$$

(9′) $$V = \left(\frac{\rho^2}{1 - M^2}\right)^{1/4} = \left[\frac{h^2(1 - \tau)}{1 - h^2\tau}\right]^{1/4} = \left(-\frac{P}{Q}\right)^{1/4}.*$$

The equation corresponding to Eq. (6′) is then

$$\frac{d^2\psi_n^*}{ds^2} - n^2\psi_n^* + F\psi_n^* = 0,$$

which suggests that asymptotically for large $|n|$, $\psi_n^* \sim e^{ns}$. This, together with Eq. (7), makes somewhat plausible the following result (which we give without proof). The function $\psi_n(\tau)$ is *asymptotically equal* to $e^{ns}V(\tau)$, as $|n| \to \infty$, for subsonic τ (negative integers n excluded); or, more precisely,

(10) $$\psi_n(\tau) = V(\tau)e^{ns}\left[1 + O\left(\frac{1}{n}\right)\right]$$

as $|n| \to \infty$, uniformly for $0 \leqq \tau \leqq \tau_t - \epsilon$, for complex n and $|n + m| \geqq \delta$ for all positive integers m (δ, ϵ arbitrarily small positive numbers).

We can now see that for $\tau_1 < \tau_t$ a simple and appropriate normalizing factor is

(11) $$f(n, \tau_1) = e^{-ns_1}, \qquad \text{where } s_1 = s(\tau_1).$$ [3]

In fact, from the asymptotic equality of e^{2s} and τ for small τ, as expressed in Eq. (5), we see that e^{-s} behaves like $\tau_1^{-\frac{1}{2}}$ in the limit $\tau_1 \to 0$, as was intended. Also it follows from Eq. (10) that the general term of the series in Eq. (2) will behave for large n like $V(\tau)c_n e^{n(s-s_1-i\theta)}$ and for $\tau = \tau_1$, $s = s_1$, this is indeed a behavior similar to that of the general term in Eq. (20.13) for $q = 1$.

3. Flow around a circular cylinder

To gain insight into the problem of flow around an obstacle, we consider here what is probably the simplest profile, the circle.[4] We shall give the

* Note that $V^4 = K^{-1}$, where K was introduced in (17.24′).

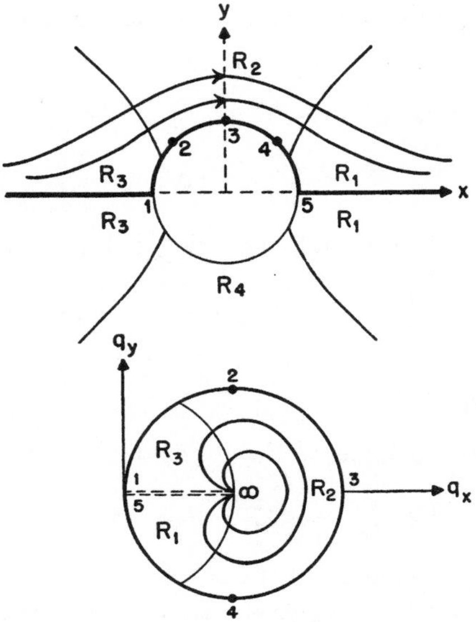

FIG. 138. Flow past circular cylinder.

main line of procedure and refer the reader for details to the literature that will be quoted.*

Let the radius of the circle be equal to one, let, as before, $q_\infty = q_1 = 1$, and the corresponding τ_1 be subsonic. The complex potential $w(z)$ of the incompressible flow is well known:

$$
(12) \qquad \begin{gathered} w(z) = z + \frac{1}{z}, \qquad \zeta = \frac{dw}{dz} = qe^{-i\theta} = 1 - \frac{1}{z^2}, \qquad z = (1 - \zeta)^{-\frac{1}{2}} \\ w(z) = w_0(\zeta) = (1 - \zeta)^{\frac{1}{2}} + (1 - \zeta)^{-\frac{1}{2}}. \end{gathered}
$$

Consider the upper half of the z-plane. It is seen (Fig. 138) that the x-axis from point 5 to ∞ and from $-\infty$ to 1 is mapped onto the cut between $\zeta = 0$ and $\zeta = 1$; the image of the profile streamline 1 2 3 4 5 appears in the hodograph as the circle of center $q_x = 1$, $q_y = 0$ through the origin, and a few more streamlines are sketched roughly indicating how the flow region in the upper z-plane outside the obstacle is mapped onto the inside of the circle in the q_x,q_y-plane.

The expansion of $w_0(\zeta)$ must be made separately for $|\zeta| < 1$ and $|\zeta| > 1$. The line $|\zeta| = 1$ is an arc of circle in the hodograph plane, and its image $|1 - 1/z^2| = 1$, or $x^2 - y^2 = \frac{1}{2}$, separates the flow region above the x-axis

* The example of this section requires only some idea of analytic continuation and the residue theorem.

into three regions R_1, R_2, R_3, with corresponding images in the hodograph.

Expanding for $|\zeta| < 1$ we must take $z = +(1 - \zeta)^{-\frac{1}{2}}$ in R_1 and $z = -(1 - \zeta)^{-\frac{1}{2}}$ in R_3 (then for $\zeta = 0$ we obtain $z = 1$ in R_1 and $z = -1$ in R_3); for $|\zeta| > 1$, $z = i\zeta^{-\frac{1}{2}}(1 - 1/\zeta)^{-\frac{1}{2}}$ in R_2.

For the regions R_1, R_2, R_3 we then obtain the expansions

$$w_1 = \frac{1}{\Gamma(\frac{1}{2})} \sum_{n=0}^{\infty} (n - 1)\Gamma(n - \tfrac{1}{2}) \frac{\zeta^n}{n!}$$

(13)
$$w_2 = \frac{i}{\Gamma(\frac{1}{2})} \sum_{n=0}^{\infty} (n + \tfrac{1}{2})\Gamma(n - \tfrac{1}{2}) \frac{\zeta^{\frac{1}{2}-n}}{n!}$$

$$w_3 = -w_1 .$$

If we consider also the lower part of the z-plane, namely, $y < 0$, the regions R_1 and R_3 are continued symmetrically below the x-axis, and we denote by R_4 the region symmetric to R_2. The hodograph image of this lower part covers a second time the same hodograph circle as in Fig. 138 and is not shown. The formulas (13) remain correct and we have to add the formula

(13′)
$$w_4 = -w_2 .$$

One may verify that each of the four series is the analytic continuation of its two neighbors. This expansion into series of $w(z) = w_0(\zeta)$ is the first step.

Next we seek a corresponding compressible flow and begin by constructing the series W_1 and W_2, corresponding to w_1 and w_2, according to Eqs. (2) and (11),

(14)
$$W_1 = \frac{1}{\Gamma(\frac{1}{2})} \sum_{n=0}^{\infty} \frac{(n - 1)\Gamma(n - \frac{1}{2})}{n!} \psi_n(\tau) e^{-n(s_1+i\theta)},$$

(15)
$$W_2 = \frac{i}{\Gamma(\frac{1}{2})} \sum_{n=0}^{\infty} \frac{(n + \frac{1}{2})\Gamma(n - \frac{1}{2})}{n!} \psi_{\frac{1}{2}-n}(\tau) e^{(n-\frac{1}{2})(s_1+i\theta)}.$$

We shall see immediately that W_2 is *not* the analytic continuation of W_1 and that it can never be so, no matter how $f(n, \tau_1)$ might have been chosen. To find the continuation of W_1 for $\tau > \tau_1$ is a mathematical problem which can be approached in various ways. A comparatively simple solution (see Note 4) is based on a generalization of the representation of the hypergeometric function $F(a, b, c; x)$ by a "Barnes Integral".

To explain this idea we first return for a moment to the incompressible flow problem. Denote by B the integral[5]

$$B = \frac{1}{2\pi i \Gamma(\frac{1}{2})} \int_{-i\infty}^{+i\infty} (\nu - 1)\Gamma(\nu - \tfrac{1}{2})\Gamma(-\nu)(-\zeta)^{\nu} \, d\nu$$

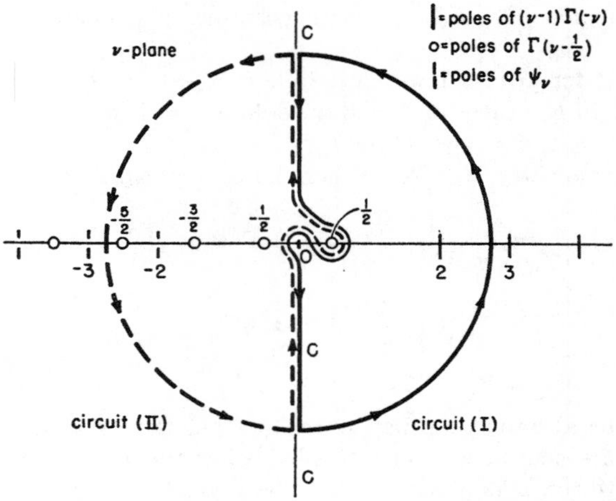

FIG. 139. Integration paths used in connection with Barnes integral.

along the path C in the complex ν-plane indicated in Fig. 139. We then apply the residue theorem to the above integral taken around an appropriate closed circuit (I) to the right of the imaginary axis, such as the one indicated in the figure. We use the fact that $\Gamma(\nu)$ has simple poles at $\nu = -n$, $n = 0, 1, 2, \cdots$ with residues $(-1)^n/n!$. Then the poles of $(\nu - 1)\Gamma(-\nu)$, which are at $n = 0, 2, 3, \cdots$ are inside (I) and those of $\Gamma(\nu - \frac{1}{2})$, which are at $\frac{1}{2} - n$, $n = 0, 1, 2, \cdots$ are outside (I). It can be shown that, as $|\nu| \to \infty$, $q < 1$, the integral around the semicircle to the right converges to zero and therefore $-B$ equals the limit of the sum of the residues at the poles which are inside (I), or

$$B = \frac{1}{\Gamma(\frac{1}{2})} \sum_{n=0}^{\infty} (n - 1)\Gamma(n - \tfrac{1}{2}) \frac{\zeta^n}{n!}, \qquad |\zeta| < 1. \tag{13''}$$

The series to the right in Eq. (13″) is identical with the expansion (13) of w_1. In this way w_1 is now expressed by the integral B, and we can use this representation to find the continuation of w_1 for $|\zeta| > 1$. We shall now explain this main point directly for the compressible flow problem.

If, with $\zeta = qe^{-i\theta}$, we replace, according to Eqs. (2) and (11), q^n by $\psi_n(\tau)e^{-ns_1}$, we are led to consider, instead of B, the integral

$$B' = \frac{1}{2\pi i\Gamma(\frac{1}{2})} \int_{-i\infty}^{+i\infty} (\nu - 1)\Gamma(\nu - \tfrac{1}{2})\Gamma(-\nu)(-1)^\nu \psi_\nu(\tau)e^{-\nu(s_1+i\theta)}\, d\nu \tag{16}$$

along the same path C as in Fig. 139. We now apply the residue theorem to the integrand in (16) and the circuit (I). We find that for $\tau < \tau_1$, (corresponding to $q < 1$) $-B'$ equals the limit of the sum of the residues of the integrand at $\nu = 0, 2, 3, \cdots$ and that this sum is exactly the series W_1. To find then the analytic continuation we integrate around the closed circuit (*II*) (indicated in the figure) to the left, where for $|\nu| \to \infty$ and $\tau > \tau_1$, the integral around the left semicircle tends to zero. Inside this circuit are now *not only the poles of* $\Gamma(\nu - \frac{1}{2})$, *but also those of* $\psi_\nu(\tau)$, considered as a function of ν for fixed τ $(0 < \tau < 1)$. The former are at $\frac{1}{2} - n$ $(n = 0, 1, 2, \cdots)$ and the limit of the corresponding sum of the residues equals W_2 if θ is negative and $-W_2$ if θ is positive.* The (simple) poles of $\psi_\nu(\tau)$ are at $-n$ $(n = 2, 3, \cdots)$, with residues $\rho_n = -nC_n\psi_n(\tau)$ where C_n depends on the constants a_n, b_n, introduced in Eq. (20.8)†. [We note that $f(n, \tau_1) = e^{-ns_1}$ does not contribute additional poles and corresponding residues as Chaplygin's original $[\psi_n(\tau_1)]^{-1}$ would.[6]] The limit of the sum of the residues at these poles is equal to a series $-L$, where

$$(17)\qquad \begin{aligned} L &= \frac{1}{\Gamma(\frac{1}{2})} \sum_{n=2}^{\infty} (-1)^{n+1}\Gamma(n + 2)\Gamma(-n - \tfrac{1}{2})C_n\psi_n(\tau)e^{n(s_1+i\theta)} \\ &= \sum_{n=2}^{\infty} \frac{2^{n+1}\Gamma(n + 2)}{1\cdot3\cdot5 \cdots (2n + 1)} C_n\psi_n(\tau)e^{n(s_1+i\theta)}. \end{aligned}$$

Hence, the *analytic continuation* of W_1 in the region corresponding to R_2 is $W_2 - L$ and we recognize that Eq. (15) alone could not give the desired continuation since the W_2 in (15) does not take care of all relevant poles. No matter what our choice of $f(n, \tau_1)$, the poles of $\psi_n(\tau)$—as a function of n—must be taken into consideration. In a similar way we obtain $-W_1 - 2L$ and $-W_2 - L$ in the regions corresponding to R_3 and R_4.

This solution is not symmetric about the y-axis. We obtain a symmetric solution if we take $W_1 + L$ as the solution in R_1; it can be seen that $L \to 0$ as $q_m \to \infty$, as we expect. The whole solution in the regions R_1, R_2, R_3, R_4 is then:

$$(18)\qquad W_1 + L, \qquad W_2, \qquad -W_1 - L, \qquad -W_2.$$

The continuation of $-W_2$ into the first region is again $W_1 + L$, demonstrating that ψ is single-valued in the physical plane.

By means of asymptotic estimates of $|\psi_n(\tau)|$ such as (10), which are different for subsonic and supersonic τ-values, it follows that W_1 converges for $\tau < \tau_1$, W_2 converges for $\tau_1 < \tau \leqq 1 - \epsilon$ (ϵ a positive number, appear-

* This corresponds to the fact that the argument of -1 in (16) must be taken as $-\pi$ and π respectively for the integrals along the semicircles to tend to zero.

$$\dagger\ C_n = \frac{\Gamma(a_n)\Gamma(n - b_n + 1)}{(n!)^2\Gamma(a_n - n)\Gamma(1 - b_n)} = \frac{e^{-2\sigma n}}{2\pi n}\left[1 + O\left(\frac{1}{n}\right)\right] \qquad (n = 2, 3, \cdots)$$

(cf. Lighthill quoted Note IV. 50). This asymptotic formula is obtained by application of Stirling's formula.

ing in the estimate of $|\psi_n(\tau)|$), and L converges for all τ. Thus, we have indeed obtained the continuation of $W_1 + L$ over the whole field of flow, including the supersonic region.

This example demonstrates both the principle of the method and its difficulties.

4. General solution for the subsonic region*

For subsonic τ a compact and explicit solution is available.[7]

In addition to the above-mentioned properties of $\psi_n(\tau)$ the main tool is the *partial fraction expansion* (with respect to n for fixed τ) of $\psi_n(\tau)$ holding for $0 \leqq \tau < \tau_t$ and complex n:

$$\psi_n(\tau) = e^{ns}\left[1 + n\sum_{m=2}^{\infty}\frac{1}{m+n}C_m e^{ms}\psi_m(\tau)\right] \tag{19}$$
$$= e^{ns}\sum_{m=0}^{\infty}\frac{n}{m+n}C_m e^{ms}\psi_m(\tau),$$

where in the last form of this expansion we define $C_0 = 1$, $C_1 = 0$, and $n/(n+m) = 1$ for $n = m = 0$. Equation (19) expresses $\psi_n(\tau)$ for complex n in terms of the $\psi_m(\tau)$, where m is a positive integer.†

From (19) we obtain, with $r_m = C_m\psi_m(\tau)e^{ms_1}$,

$$\psi_n(\tau)e^{-ns_1} = n\sum_{m=0}^{\infty}\frac{r_m}{n+m}e^{(n+m)(s-s_1)} \tag{20}$$

and since, by (10), $V(\tau) = \lim_{|n|\to\infty}\psi_n(\tau)e^{-ns}$, we also have

$$V(\tau) = \sum_{m=0}^{\infty} r_m e^{m(s-s_1)} = \sum_{m=0}^{\infty} C_m e^{ms}\psi_m(\tau). \tag{21}$$

From Eqs. (20) and (21) follows the desired result, namely, the *continuation of expansion* (2) *throughout the subsonic region.* Using (20), Eq. (2) can be written

$$\psi(\tau,\theta) = \mathcal{I}\left[\sum_{n=0}^{\infty} nc_n e^{-in\theta}\sum_{m=0}^{\infty}\frac{r_m}{m+n}e^{(m+n)(s-s_1)}\right] \tag{22}$$
$$= \mathcal{I}\left[\sum_{m=0}^{\infty} r_m e^{im\theta}\sum_{n=0}^{\infty}\frac{nc_n}{m+n}e^{(m+n)(s-s_1-i\theta)}\right]$$
$$= \mathcal{I}\left\{\sum_{m=0}^{\infty} r_m e^{im\theta}\left[\int_0^{e^{s-s_1-i\theta}}\left(\sum_{n=0}^{\infty} nc_n\zeta^{m+n-1}\right)d\zeta\right] + c_0\right\}.$$

* The considerations of this section, due to Lighthill, are not elementary in character. The main result is that the integral representation (23) is valid in the whole subsonic region.

† Equation (19) is the result of applying Mittag-Leffler's theorem to $e^{-ns}\psi_n(\tau)$.

From now on to the end of the section we write simply w rather than w_0. From (20.13),

$$\frac{dw}{d\zeta} = \sum_{n=0}^{\infty} nc_n\zeta^{n-1},$$

and Eq. (22) may be written:

$$\begin{aligned} \psi &= \mathfrak{g}\left\{\sum_{m=0}^{\infty} r_m e^{im\theta}\left[\int_0^{e^{s-s_1-i\theta}} \zeta^m \frac{dw}{d\zeta}\, d\zeta\right] + w(0)\right\} \\ &= \mathfrak{g}\left\{\sum_{m=0}^{\infty} C_m\psi_m(\tau)\, e^{m(s_1+i\theta)}\left[\int_0^{e^{s-s_1-i\theta}} \zeta^m\, dw(\zeta)\right] + w(0)\right\}. \end{aligned} \tag{23}$$

We may now verify that ψ satisfies the conditions postulated in Sec. 1. First, as $q_m \to \infty$, all $r_m \to 0$, except $r_0 = 1$; also from (5):

$$\lim_{q_m\to\infty} e^{s-s_1} = \frac{q}{q_1} = q,$$

and we see that the right side of Eq. (23) reduces to

$$\mathfrak{g}\left[\int_0^{qe^{-i\theta}} dw + w(0)\right] = \mathfrak{g}[w(\zeta)].$$

One likewise verifies directly that (23) satisfies the stream-function equation.

We shall now show how this representation of the solution is defined for all subsonic τ. From $dw/dz = w'(z) = \zeta$, we conclude that z is an analytic function of ζ in the hodograph of the incompressible flow. This, however, is not a simple plane but a Riemann surface, R (consisting of two sheets with a branch point at $\zeta = 1$ in our example of the circular cylinder), and $z(\zeta)$ is regular on R. The integral in Eq. (23) can be written

$$\int_0^{e^{s-s_1-i\theta}} \zeta^m dw(\zeta) = \int_0^{e^{s-s_1-i\theta}} \zeta^{m+1}\frac{dz}{d\zeta}\, d\zeta = \int_0^{e^{s-s_1-i\theta}} \zeta^{m+1}\, dz(\zeta). \tag{24}$$

If $z = z_0$ corresponds to $\zeta = 0$,

$$Y_m(z) = \int_{z_0}^{z} \zeta^{m+1}\, dz$$

is a regular function of z in the z-plane outside the body *if we assume that there is no circulation*. Since we know, in addition, that $z(\zeta)$ is regular on R it follows that $Z_m(\zeta) = Y_m[z(\zeta)]$ is also regular on R. The last term in (24) can be written as $Z_m(e^{s-s_1-i\theta})$. In the absence of circulation this is therefore single-valued on a Riemann surface, R^*, which corresponds to the incompressible hodograph surface R as locus of the points (τ, θ) for which $e^{s-s_1-i\theta}$ lies on R. The R^* is the hodograph of the subsonic part of the compressible flow and (23) is regular on R^* if it converges.

We now show that

$$(23') \qquad \psi = \mathcal{J}\left[\sum_{m=0}^{\infty} r_m e^{im\theta} \int_0^{e^{s-s_1-i\theta}} \zeta^{m+1}\, dz(\zeta)\right]$$

converges everywhere if the flow is purely subsonic.[8] In fact, if we integrate along a path, of length l in the physical plane, joining $\zeta = 0$ and $e^{s-s_1-i\theta}$ on R then, since at any point $s < \sigma$, its sonic value,

$$\left|\int_{(l)} \zeta^{m+1}\, dz(\zeta)\right| \leqq le^{(\sigma-s_1)(m+1)}.$$

On the other hand $r_m = C_m\psi_m(\tau)e^{ms_1}$ where, for large m (p. 359): $C_m \sim e^{-2\sigma m}/2\pi m$, and $\psi_m(\tau) \sim Ve^{ms}$, hence

$$r_m \sim \frac{V}{2\pi m}\, e^{m(s+s_1-2\sigma)}.$$

Thus we see that for large enough m a term of (23′) is comparable with $(Vle^{\sigma-s}/2\pi m)e^{-m(\sigma-s)}$, which assures the convergence.*

Thus for subsonic flow without circulation, the problem has been solved in the explicit form of the integral representation (23).

We turn now to a presentation of Bergman's method, and at the end of the article, we shall indicate the relation between Bergman's and Lighthill's methods.

5. Bergman's integration method

In an attempt to continue and improve Chaplygin's pioneer work, which, while highly successful in many ways, failed for the problem of flow past an obstacle, S. Bergman began to apply a general mathematical idea to this problem.[9] The purpose was to establish a correspondence between analytic functions of a complex variable (i.e. solutions of incompressible flow problems) and solutions of linear partial differential equations of elliptic type (such as the equation for ψ). This is achieved by means of an integral representation of the solution from which properties of the solution can be deduced by means of complex function theory. (In particular, solutions of the compressible flow equations can be obtained which are multivalued and have singularities of the type needed in the flow problems under consideration. In fact, as seen before, even in the incompressible flow around a circle the stream function cannot be represented by a single convergent series of single-valued functions).

Let P_0 be the contour of an obstacle in the physical plane; denote by $\psi_0(q,\theta)$ the stream function of the incompressible flow problem, so that $\psi_0 = 0$ on P_0, and $\Delta\psi_0 = 0$ outside P_0. Just as in the methods of Chaply-

* Actually there is uniform convergence. Also, the differentiated series converge uniformly, etc., and we may verify that ψ satisfies the stream-function equation.

gin and Lighthill, we wish to associate with ψ_0 a ψ satisfying the compressible hodograph equation, which is not too different from ψ_0 as long as a representative Mach number is small, and which reduces to ψ_0 in the limit $q_m \to \infty$. Of course ψ will not vanish on P_0 : it can however be assumed and has been proved under certain circumstances that $\psi = 0$ on a streamline P close to P_0.[10]

First we rewrite Eq. (20.5), in the same way as before, by introducing instead of q the new variable λ of Eq. (3′) and obtain Eq. (6) with either s or λ as independent variable. Next we introduce ψ^* by Eq. (7) or by

$$\psi^* = \alpha\psi, \qquad \alpha = V^{-1}, \qquad T = -\frac{2\alpha'}{\alpha}, \tag{25}$$

where primes denote differentiation with respect to λ, and we obtain Eq. (8), where in the polytropic case,

$$F = -\frac{\alpha''}{\alpha} = \frac{(\kappa + 1)M^4}{16(1 - M^2)^3}[16 - 4(3 - 2\kappa)M^2 - (3\kappa - 1)M^4]. \tag{26}$$

Since both F and λ are given in terms of M, F is given in terms of λ.[11]

We try to integrate Eq. (8) by setting

$$\psi^* = g_0(\lambda, \theta) + \sum_{n=1}^{\infty} G_n(\lambda)g_n(\lambda, \theta) \tag{27}$$

(note that this is not a separation of variables).[12] Each g_n in Eq. (27) is a *harmonic function* of λ, θ. Using the symbol $\Delta = \partial^2/\partial\lambda^2 + \partial^2/\partial\theta^2$ and substituting, we obtain with $G_0 = 1$

$$\begin{aligned} \Delta\psi^* + F\psi^* &= \sum_{n=0}^{\infty} [\Delta(G_n g_n) + FG_n g_n] \\ &= \sum_{n=0}^{\infty} \left(G_n'' g_n + 2G_n' \frac{\partial g_n}{\partial\lambda} + G_n\,\Delta g_n + FG_n g_n\right) \\ &= \sum_{n=1}^{\infty} \left[(G_n'' + FG_n)\, g_n + 2G_n' \frac{\partial g_n}{\partial\lambda}\right] + Fg_0. \end{aligned} \tag{28}$$

We shall see that Eq. (8) is satisfied if we put

$$\frac{\partial g_n}{\partial\lambda} = -\frac{1}{2} g_{n-1} \qquad (n = 1, 2, \cdots) \tag{29}$$

$$G_{n+1}' = G_n'' + FG_n \qquad (n = 0, 1, \cdots) \tag{29′}$$

where g_0 is arbitrary and $G_0 = 1$. The right side of Eq. (28) becomes:

$$\sum_{n=1}^{\infty} [g_n G_{n+1}' - g_{n-1}G_n'] + Fg_0.$$

This series equals the limit for $n \to \infty$ of $g_n G'_{n+1} - g_0 G'_1$. This is so far a formal computation. It will be shown later, using a majorant method, that the series in (27) and the necessary derivatives converge uniformly in a certain region and that not only $g_n G_n$ but also $g_n G'_{n+1} \to 0$ as $n \to \infty$. Thus the right-hand side of Eq. (28) reduces to $-g_0 G'_1 + F g_0 = 0$, since, because of Eq. (29′) for $n = 0$, $G'_1 = F G_0 = F$.

The G_n are determined by (29′) if we add the condition

$$G_n(-\infty) = 0 \qquad (n = 1, 2 \ldots), \tag{29''}$$

and it is seen that the sequence $G_0, G_1, G_2, \cdots$ depends only on $F(\lambda)$. Hence it is uniquely determined for a given (p, ρ)-relation, it can be computed once and for all, and can be tabulated.

On the other hand, the sequence $g_0, g_1, g_2, \cdots$ depends on the arbitrary function g_0. We introduce now the complex variable

$$Z = \Lambda - i\theta \qquad \text{with} \qquad \Lambda = \lambda + \sigma + \log q_m \tag{30}$$

where $\sigma = -1.17$, as defined in Eqs. (4) and (5). Since $\tau = q^2/q_m^2$, the second Eq. (5) can be written

$$\lim_{q_m \to \infty} (\lambda + \log q_m) = \log q - \sigma, \tag{31}$$

and it follows that with $\zeta = qe^{-i\theta}$ denoting again the complex velocity

$$\lim_{q_m \to \infty} Z = \log q - i\theta = \log \zeta. \tag{31'}$$

The variable Z is the same as Lighthill's variable $s - s_1 - i\theta$, and it will serve a similar purpose.[13]

For a fixed q_m the $g_n(\lambda, \theta)$ of Eq. (27) which are harmonic in (λ, θ) are also harmonic in (Λ, θ). We then define a sequence of functions of Z: $f_0(Z), f_1(Z), \cdots$ where $f_0(Z)$ is an arbitrary analytic function of Z and where $f'_n = -\frac{1}{2} f_{n-1}$ $(n = 1, 2, \cdots)$ by putting

$$f_n(Z) = -\frac{1}{2} \int_0^Z f_{n-1}(t)\, dt \qquad (n = 1, 2, \cdots), \tag{32}$$

which implies $f_n(0) = 0$ for $n = 1, 2, \cdots$. Let

$$g_n(\lambda, \theta) = \mathcal{I}[f_n(Z)] \qquad (n = 0, 1, \cdots). \tag{33}$$

Then from Eq. (32), since $\mathcal{I}[f'_n(Z)] = \partial g_n / \partial \lambda$,

$$\frac{\partial g_n}{\partial \lambda} = \mathcal{I}[f'_n(Z)] = -\mathcal{I}\left[\frac{1}{2} f_{n-1}(Z)\right] = -\frac{1}{2} g_{n-1},$$

and (29) is satisfied. Substituting successively into Eq. (32) and finally writing the n-tuple integral as a single integral we obtain

$$(32') \qquad f_n(Z) = \frac{(-1)^n}{2^n(n-1)!}\int_0^Z f_0(t)(Z-t)^{n-1}\,dt \quad (n = 1, 2, \cdots),$$

so that, with $df_0/dt = f_0' \neq 0$, and assuming from now on that $f_0(0) = 0$:

$$(32'') \qquad f_n(Z) = \frac{(-1)^n}{2^n n!}\int_0^Z f_0'(t)(Z-t)^n\,dt \quad (n = 0, 1, \cdots).$$

This last formula may be verified by applying partial integration to the integral in (32′). For $n = 0$ the right side of (32″) reduces to $f_0(Z) - f_0(0) = f_0(Z)$. Thus, on account of Eq. (33)

$$(33') \qquad g_n(\lambda, \theta) = \frac{(-1)^n}{2^n n!}\mathscr{I}\left[\int_0^Z f_0'(t)(Z-t)^n\,dt\right] \qquad (n = 0, 1, 2, \cdots),$$

Substituting these into Eq. (27) we obtain

$$(34) \qquad \psi^*(\lambda, \theta) = \sum_{n=0}^{\infty}\frac{(-1)^n}{2^n n!}G_n(\lambda)\,\mathscr{I}\left[\int_0^Z f_0'(t)(Z-t)^n\,dt\right].$$

The similarity of (34) and (23) is obvious.

If we introduce the function of three variables

$$(35) \qquad G(t; \lambda, \theta) = \sum_{n=0}^{\infty}\frac{1}{2^n n!}G_n(\lambda)(t-Z)^n,$$

we can write (34) in the form

$$(36) \qquad \psi^*(\lambda, \theta) = \mathscr{I}\left[\int_0^Z G(t; \lambda, \theta)\,df_0(t)\right],$$

where $df_0(t) = f_0'(t)dt$. Also, using (25),

$$(37) \qquad \psi(\lambda, \theta) = V\mathscr{I}\left[\int_0^Z G(t; \lambda, \theta)\,df_0(t)\right].$$

The right side of Eq. (37) thus *transforms the arbitrary analytic function* $f_0(Z)$ *into a solution* $\psi(\lambda, \theta)$ *of the second-order equation* (6).

With respect to the legitimacy of these operations, we note that to justify the interchange of differentiation and summation which led to (34) we need the uniform convergence of this series and its derivatives in a λ,θ-region. The interchange of summation and integration which leads from (34) to (36) is not essential for our final result. To justify it we need uniform convergence in a t-region for fixed λ,θ. Both are covered by the considerations in the next section.

Next, let us make an appropriate choice of the arbitrary function $f_0(t)$

which occurs in these various formulas. Denote by $w_0(\zeta)$ with $\zeta = qe^{-i\theta}$ the hodograph potential of an incompressible problem. Then $\mathfrak{g}[w_0(\zeta)] = \psi_0(q, \theta)$ is the incompressible stream function of this problem, and we choose

$$f_0(t) = w_0(e^t), \qquad w_0(1) = 0. \tag{38}$$

Then, consider the passage to the limit $q_m \to \infty$ for fixed (q, θ). For greater clarity we write $\psi(q, \theta)$ or $\psi^*(q, \theta)$ instead of $\psi(\lambda, \theta)$ or $\psi^*(\lambda, \theta)$. We assume (it will be proved presently) that

$$\lim_{q_m \to \infty} G(t; \lambda, \theta) = 1.$$

(In a formal way, this is seen from (35) and (29″) and $G_0 = 1$). Note also that in (37) $V \to 1$ as $q_m \to \infty$. We have then from (36) and (37), using (31′) and (38)

$$\lim_{q_m \to \infty} \psi(q, \theta) = \lim_{q_m \to \infty} \psi^*(q, \theta) = \mathfrak{g}[w_0(\zeta)] = \psi_0(q, \theta), \tag{39}$$

as was intended.

6. Convergence

We now sketch the investigation of the *uniform convergence of* (35) *and of its* λ,θ*-derivatives as far as the subsonic range is concerned.* We consider functions $F(\lambda)$ such that for each $\epsilon < 0$, $\lambda < \epsilon$, C an appropriate constant, and

$$\mathfrak{F} = \frac{C}{(\epsilon - \lambda)^2}, \tag{40}$$

we have

$$| F | \leqq \mathfrak{F}, \qquad \text{and} \qquad \frac{d^\kappa F}{d\lambda^\kappa} \leqq \frac{d^\kappa \mathfrak{F}}{d\lambda^\kappa} \qquad (\kappa = 1, 2, \cdots). \tag{41}$$

We will write briefly $F \ll \mathfrak{F}$ for the inequalities (41) and we say that $\mathfrak{F}$ dominates F or that $\mathfrak{F}$ is a majorant function for F.

If (41) holds we can find majorant functions $P_n(\lambda)$ for the G_n, as follows. We define $P_n(\lambda)$ by

$$\begin{gathered} P'_{n+1} = P''_n + C(\epsilon - \lambda)^{-2} P_n \qquad (n = 0, 1, \cdots) \\ P_0 = 1, \qquad P_n(-\infty) = 0. \end{gathered} \tag{42}$$

Then,

$$G_n \ll P_n. \tag{43}$$

In fact, for $n = 0$, we have $P_0 = G_0$, hence $G_0 \ll P_0$ in the sense of the above definition. Now suppose $G_n \ll P_n$, then from formulas (29′), (41), (42) we see that $G_{n+1} \ll P_{n+1}$.

It is easily verified that $P_n(\lambda)$ can be given explicitly as

$$P_n = n!\,\mu_n(\epsilon - \lambda)^{-n} \tag{42′}$$

where with

$$\alpha_1 = \tfrac{1}{2} - (\tfrac{1}{4} - C)^{\frac{1}{2}}, \qquad \alpha_2 = \tfrac{1}{2} + (\tfrac{1}{4} - C)^{\frac{1}{2}}$$

$$\mu_0 = 1, \qquad \mu_{n+1} = \mu_n(n^2 + n + C)(n + 1)^{-2} \tag{44}$$

$$= \mu_n(n + \alpha_1)(n + \alpha_2)(n + 1)^{-2}.$$

This recurrence formula for the μ_n coincides with the recurrence formula for the coefficients of the hypergeometric series $H(\alpha_1, \alpha_2, 1; x)$ and we may write

$$H(\alpha_1, \alpha_2, 1; x) = \sum_{n=0}^{\infty} \mu_n x^n. \tag{44′}$$

Since neither α_1 nor α_2 is a negative integer, the series does not terminate. It converges for $|x| < 1$, diverges for $|x| > 1$, and converges uniformly with all its derivatives for $|x| \leqq b < 1$, where b is an arbitrary positive number less than 1.

We can now estimate the $G(t; \lambda, \theta)$ of Eq. (35). Substituting (42) and (42′) into (35), we obtain

$$G \ll \sum_{n=0}^{\infty} P_n(\lambda) 2^{-n} \frac{1}{n!} |Z - t|^n$$

$$= \sum_{n=0}^{\infty} \mu_n n!\, 2^{-n}(\epsilon - \lambda)^{-n} \frac{1}{n!} |Z - t|^n$$

$$= \sum_{n=0}^{\infty} \mu_n \left(\frac{1}{2} \frac{1}{|\epsilon - \lambda|} |Z - t|\right)^n.$$

Thus, using (44′) we have

$$G(t; \lambda, \theta) \ll H\left(\alpha_1, \alpha_2, 1; \frac{1}{2} \frac{|Z - t|}{|\epsilon - \lambda|}\right). \tag{45}$$

Therefore G, as well as all its derivatives with respect to λ and θ, converges uniformly and absolutely in a region where

$$\frac{1}{2} \frac{|Z - t|}{|\epsilon - \lambda|} \leqq b < 1. \tag{46}$$

This justifies all transformations previously performed in Sec. 5 and proves that the ψ^* as given in (36) satisfies Eq. (8).

Now let $q_m \to \infty$. Since the convergence of the series (35) is also uniform with respect to q_m we conclude that $G(t; \lambda, \theta) \to 1$, as $q_m \to \infty$. We used this fact in establishing the result (39).

Finally, we want to express condition (46) in a geometric form. If the integration of $G(t; \lambda, \theta)$ from $t = 0$ to $t = Z$ is along the straight line from 0 to Z the maximum of $| Z - t |$ is $| Z |$ and (46) will be satisfied if $| Z | < 2 | \epsilon - \lambda |$, or

$$\Lambda^2 + \theta^2 < 4(\epsilon - \lambda)^2. \tag{46'}$$

The boundary of this region (see Fig. 140) is a hyperbola which as $\epsilon \to 0$, becomes

$$3\Lambda^2 - \theta^2 - 8c\Lambda + 4c^2 = 0, \qquad \text{where } c = \sigma + \log q_m .$$

We see that the larger q_m is, i.e., the more the effects of compressibility vanish, the larger is the part of the Λ,θ-plane which we obtain as region of validity and as $q_m \to \infty$, the whole $\log q$, θ-plane is obtained.[14] The uni-

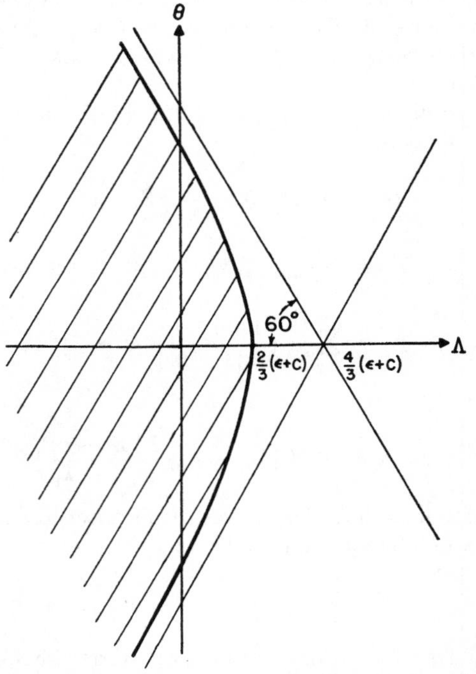

FIG. 140. Region of convergence for Bergman's method.

form convergence of (35) and hence of $\psi^* = \sum G_n g_n$ is assured in this region, provided $f_0(Z)$ itself is bounded there.

The above simple proof must be slightly modified for the polytropic F, where the inequalities (41) do not hold directly. This can be done in several ways, conserving the essential idea of the proof and retaining the result (46) with respect to the region of convergence.[15]

7. Integral transformation

In the preceding section we constructed a solution of the stream-function equation depending on an arbitrary analytic function $f_0(t)$ in the convenient form of the integral (37) with the generating function $G(t; \lambda, \theta)$ defined by Eq. (35). It seems obvious that this particular generator G is not the only one by means of which such a solution can be found. We shall now characterize such generating functions by means of a differential equation and side conditions which they must satisfy.

Following Bergman we consider the partial differential equation

$$L^+(v) \equiv \frac{\partial^2 v}{\partial z_1\, \partial z_2} + cv = 0 \tag{47}$$

where z_1, z_2 are complex variables and c is an analytic function of z_1, z_2. Let the function of three independent complex variables $K(z_1, z_2, t)$ be an analytic function of its variables in some suitable domain B. We assume that for all t in B, K satisfies (47) and the boundary condition

$$\frac{\partial K}{\partial z_2} = 0 \qquad \text{for} \qquad t = z_1 . \tag{48}$$

Then, with $f(t)$ an arbitrary analytic function of t we form

$$v(z_1, z_2) = \int_a^{z_1} K(z_1, z_2, t) f'(t)\, dt \tag{49}$$

and prove that v is a solution of Eq. (47). Indeed, we have

$$\frac{\partial^2 v}{\partial z_1\, \partial z_2} = \int_a^{z_1} \frac{\partial^2 K}{\partial z_1\, \partial z_2} f'(t)\, dt + f'(z_1) \left(\frac{\partial K}{\partial z_2}\right)_{t=z_1} .$$

On account of (48) the second term on the right-hand side vanishes. Since c is independent of t and K satisfies (47), we obtain

$$L^+(v) = \int_a^{z_1} \left(\frac{\partial^2 K}{\partial z_1\, \partial z_2} + cK\right) f'(t)\, dt = 0. \tag{50}$$

To establish the connection between Eqs. (47) and (8) we extend the variables Λ, θ of the preceding section into the complex domain and call these complex variables Λ_1, θ_1. Then $z_1 = \Lambda_1 - i\theta_1$ and $z_2 = \Lambda_1 + i\theta_1$ are independent complex variables, and

$$\frac{\partial}{\partial z_1} = \frac{1}{2}\frac{\partial}{\partial \Lambda_1} - \frac{1}{2i}\frac{\partial}{\partial \theta_1}, \qquad \frac{\partial}{\partial z_2} = \frac{1}{2}\frac{\partial}{\partial \Lambda_1} + \frac{1}{2i}\frac{\partial}{\partial \theta_1};$$

if we then put for the c in (47): $c(z_1, z_2) = \frac{1}{4} F(\Lambda_1)$ we see that

$$L^+(v) = \frac{1}{4}\left(\frac{\partial^2 v}{\partial \Lambda_1^2} + \frac{\partial^2 v}{\partial \theta_1^2} + Fv\right).$$

If in Λ_1 and θ_1 the imaginary parts reduce to zero, the previously independent variables z_1 and z_2 reduce to the conjugate complex variables $z_1 = Z = \Lambda - i\theta$, as introduced in Eq. (30), and $z_2 = \bar{Z} = \Lambda + i\theta$. It then follows from the fact that $L^+(K) = 0$ for all z_1, z_2 in B that $K(Z, \bar{Z}, t)$ satisfies

$$L(K) = \frac{\partial^2 K}{\partial \lambda^2} + \frac{\partial^2 K}{\partial \theta^2} + F(\lambda)K = 0. \tag{51}$$

We next restrict attention to generators K for which

$$\lim_{q_m \to \infty} K(Z, \bar{Z}, t) = 1, \tag{52}$$

uniformly in t for fixed q, θ. For the $f(t)$ in (49) we choose, as in (38),

$$f(t) = w_0(e^t), \tag{53}$$

where $w_0(\zeta)$ denotes a hodograph potential, as in previous sections. If one puts as an abbreviation $dw_0/dt = w_0'(t)$ so that $w_0'(e^t)e^t\, dt = dw_0(e^t)$, it follows from the above construction that

$$u(Z, \bar{Z}) = \int_a^Z K(Z, \bar{Z}, t)\, dw_0(e^t) \tag{54}$$

satisfies Eq. (8), namely

$$L(u) = \frac{\partial^2 u}{\partial \lambda^2} + \frac{\partial^2 u}{\partial \theta^2} + Fu = 0. \tag{51'}$$

Then it follows from Eqs. (54), (52), and (31′) that:

$$\lim_{q_m \to \infty} u(Z, \bar{Z}) = \int_a^{\log \zeta} dw_0(e^t) = w_0(\zeta) + \text{constant}.$$

It is thus seen that for $\psi^*(q, \theta) = \mathcal{I}[u(Z, \bar{Z})]$, with $u(Z, \bar{Z})$ defined by (54),

$$\lim_{q_m \to \infty} \psi^*(q, \theta) = \lim_{q_m \to \infty} \psi(q, \theta) = \psi_0(q, \theta) + \text{constant}, \tag{55}$$

where $\psi_0(q, \theta) = \mathscr{I}[w_0(\zeta)]$ and $\psi = V\psi^*$ as in Eq. (7).

If the generator $K(z_1, z_2, t)$, *where* z_1, z_2, t *are independent complex variables, satisfies Eq.* (47), *namely* $L^+(K) = 0$, *and the side condition* (48), *then* $v(z_1, z_2)$ *defined by* (49) *satisfies the same equation, namely* $L^+(v) = 0$. *If then we put* $c = F/4$ *and restrict the complex variables* z_1, z_2 *to* $Z = \Lambda - i\theta$, $\bar{Z} = \Lambda + i\theta$ *then* $K(Z, \bar{Z}, t)$ *and* $u(Z, \bar{Z})$ *will both satisfy Eq.* (8). *If finally, we assume for* K *the condition* (52) *and choose the previously arbitrary* $f(t)$ *as in* (53), *the imaginary part of* $u(Z, \bar{Z})$ *will reduce to the incompressible stream function as* $q_m \to \infty$.

The discussion was made in the domain of complex Λ_1, θ_1 on account of its simplicity [Eq. (47) being hyperbolic, (48) being a condition along a characteristic, etc.]. However, it is easily checked that the generator K need only satisfy—for real Λ, θ—Eqs. (51), (52), and the side condition

$$\frac{\partial K}{\partial \lambda} - i\frac{\partial K}{\partial \theta} = 0, \quad \text{for} \quad t = \lambda - i\theta. \tag{48'}$$

A generating function satisfying all these conditions is the one given in (35), where we now write $G(t; \lambda, \theta) = K(Z, \bar{Z}, t)$ or

$$K(Z, \bar{Z}, t) = \sum_{n=0}^{\infty} \frac{1}{2^n n!} G_n(\lambda)(t - Z)^n, \tag{56}$$

where $G_n' = G_{n-1}'' + FG_{n-1}$. We find by straightforward computation that the right side of (56) satisfies Eq. (51). Hence, *not only the* $\psi^*(q, \theta)$ *defined by Eq.* (36) *but also the generator* $G(t; \lambda, \theta)$ *of Eq.* (35) *satisfies Eq.* (8). We have seen before that (52) holds for the present K of (56), and we can verify that (48′) holds.[16]

As a second example consider Lighthill's generating function, in Eq. (23). Here the notation is different: $Z = s - s_1 - i\theta$, $t = \log \zeta$, or $e^t = \zeta$; the w in Eq. (23) is the hodograph potential w_0 of the present notation, and with $\alpha = V^{-1}$ we consider the generator

$$K(Z, \bar{Z}, t) = \alpha \sum_{m=0}^{\infty} C_m \psi_m(\tau) e^{m(s_1 + i\theta + t)}. \tag{57}$$

This function satisfies the differential equation since this is so for each term separately. Also condition (48) holds true. We have, in fact,

$$\frac{\partial K}{\partial s} - i\frac{\partial K}{\partial \theta} = \sum_{m=0}^{\infty} C_m e^{m(s_1 + i\theta + t)} \left[\frac{d}{ds}(\alpha\psi_m) + m\alpha\psi_m\right]$$
$$= \frac{d}{ds}\left[\alpha \sum_{m=0}^{\infty} C_m e^{ms} \psi_m(\tau)\right] \qquad \text{for } t = Z.$$

According to Eq. (21), this equals $(d/ds)(\alpha V) = 0$, since $V = 1/\alpha$; that condition (52) holds was verified before in Sec. 4.

We thus possess a general principle which allows us to *transform an analytic function*, a solution of an incompressible problem, *into a solution of the stream-function equation*, by means of a generator K, and we possess two examples of such generators.

In order to be useful such a generator must have certain properties: the new compressible flow pattern should be similar to the incompressible pattern as long as some representative Mach number is small. The generator should be explicitly given, e.g., by means of a well-converging expansion, and convergence should hold in as large a region as possible. Lighthill's and Bergman's operators have these properties, though, each with different emphasis in these respects.

8. Relation of the two methods[17]

We shall now consider in more detail the relation between the two methods. We take as a starting point Lighthill's formula (23)

$$\psi = \mathscr{I}\left[\sum_{m=0}^{\infty} r_m e^{im\theta} \int_0^{e^{s-s_1-i\theta}} \zeta^m \frac{dw}{d\zeta}\, d\zeta\right]$$

and rewrite it as

$$\psi = \mathscr{I}\left[\sum_{m=0}^{\infty} R_m(s) e^{-m(s-s_1-i\theta)} \int_0^{e^{s-s_1-i\theta}} \zeta^m\, dw(\zeta)\right],$$

where $R_m(s) = r_m e^{m(s-s_1)}$. The term

$$e^{-m(s-s_1-i\theta)} \int_0^{e^{s-s_1-i\theta}} \zeta^m\, dw(\zeta)$$

is an analytic function of $s - i\theta$ and therefore

$$\mathscr{I}\left[e^{-m(s-s_1-i\theta)} \int_0^{e^{s-s_1-i\theta}} \zeta^m\, dw\right] = h_m(s, \theta) \tag{58}$$

is harmonic in s and θ. Thus (23) can be expressed as:

$$\psi(s, \theta) = \sum_{m=0}^{\infty} R_m(s) h_m(s, \theta) \tag{59}$$

or with $\psi^* = \alpha\psi$

$$\psi^* = \sum_{m=0}^{\infty} H_m(s) h_m(s, \theta), \tag{60}$$

where

$$H_m(s) = \alpha(s) R_m(s). \tag{61}$$

From (21), remembering that $\alpha = 1/V$, we obtain

$$\sum_{m=0}^{\infty} H_m(s) = 1, \tag{62}$$

and see that there is an identically valid relation between the $H_m(s)$.

Thus, comparing Eq. (60) with Eq. (27) and remembering that, except for a constant, s is the same as λ, we see that the same equation is solved by the two authors by the same type of expansion and that (60) leads necessarily to the identity (28) with h_n and H_n taking the place of g_n and G_n.

Now, however, the difference becomes manifest: the requirement for the g_n is different from that for the h_n. Bergman chooses in (29)

$$\frac{\partial g_n}{\partial s} = -\tfrac{1}{2}g_{n-1},$$

while we find from (58)

$$\frac{\partial h_n}{\partial s} = -nh_n + \mathcal{g}[w'(e^{s-s_1-i\theta})].$$

Consequently from (28) the following formula for the H_n obtains:

$$\sum_{n=0}^{\infty} (H_n'' - 2nH_n' + FH_n)h_n = -\mathcal{g}[w'(e^{s-s_1-i\theta})]\cdot 2 \sum_{n=0}^{\infty} H_n'(s). \tag{63}$$

Here, according to (62), the right side is zero and (63) will be satisfied if

$$H_n'' - 2nH_n' + FH_n = 0. \tag{64}$$

Now $H_n = \alpha(s)R_n(s) = C_n\alpha(s)\psi_n(s)e^{ns}$; using $F = -(\alpha''/\alpha)$ we find

$$\begin{aligned} H_n'' - 2nH_n' + FH_n &= \alpha[R_n'' - (T + 2n)R_n' + nTR_n] \\ &= \alpha C_n e^{ns}(\psi_n'' - T\psi_n' - n^2\psi_n) = 0 \end{aligned}$$

because of (6′). Hence (64) holds. Equation (64), namely, $H_n'' - 2nH_n' + FH_n = 0$ $(n = 0, 1, 2, \cdots)$ compares with Eq. (29′), namely, $G_n'' - G_{n+1}' + FG_n = 0$ $(n = 1, 2, \cdots)$, $G_0 = 1$.

It is an advantage of Lighthill's method that the functions $H_n(s)$ are connected with the hypergeometric functions, whose properties have been dealt with extensively. As seen in Sec. 4 the convergence is guaranteed in the whole subsonic domain.

Bergman's $G_n(s)$, defined by the recurrence (29′), are not connected with a well-known system of functions. This is a consequence of his definition of the g_n which in turn constitutes an asset of his method. His recurrence formula (29) for the g_n is, in fact, the recurrence formula for harmonic functions $Re[(\lambda + i\theta)^n]$ with proper normalization. Due to this correspondence the operator defined in Eq. (37), which transforms an analytic function of a complex variable into a solution of Eq. (8), preserves many properties of the former.[18]

Article 22

Shock Theory

1. Nonexistence of solutions

Just as in the case of one-dimensional nonsteady flow, see Sec. 14.1, it can be shown that the differential equations governing the steady plane flow of an ideal fluid admit no solution satisfying certain boundary conditions which can be enforced by simple physical arrangements. The equations in question are the x- and y-components of Newton's equation (1.I), with gravity omitted:

$$\rho \frac{dq_x}{dt} = -\frac{\partial p}{\partial x}, \qquad \rho \frac{dq_y}{dt} = -\frac{\partial p}{\partial y} \tag{1}$$

where $d/dt = q_x\partial/\partial x + q_y\partial/\partial y$, together with the equation of continuity (16.1):

$$\frac{\partial}{\partial x}(\rho q_x) + \frac{\partial}{\partial y}(\rho q_y) = 0, \tag{2}$$

and the specifying condition: $dS/dt = 0$ (see Sec. 2.3).

Assume that between the points A and B on the y-axis (see Fig. 141) the fluid is passing uniformly with supersonic velocity parallel to the x-axis:

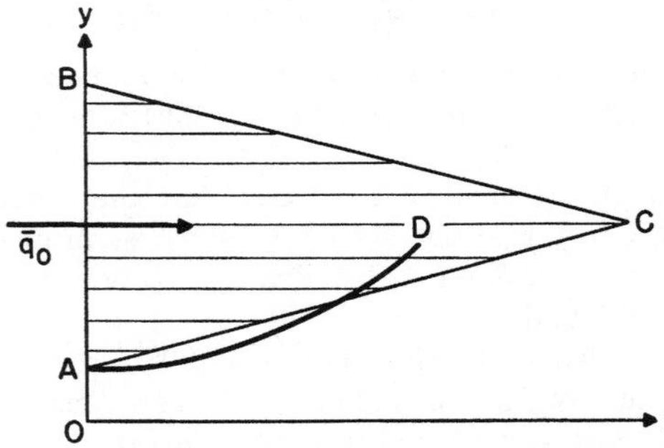

FIG. 141. Enforceable boundary conditions for which no inviscid solution exists.

$q_x = q_0$, $q_y = 0$, $p = p_0$, and $a = a_0 < q_0$, where q_0, p_0, and a_0 are constants. Through the points A and B the two straight lines inclined at angles $\pm$ arc sin (a_0/q_0) are drawn, and the horizontal streamlines, representing the uniform flow $q_x = q_0$, $q_y = 0$, and $p = p_0$ within the triangle ABC, are inserted. Then this solution satisfies the differential equations given above as well as the boundary conditions on $x = 0$. Now the fluid is elastic by virtue of the entropy being the same for all particles as they cross AB. Moreover, see Sec. 6.5, any solution of these equations satisfying the boundary conditions must be irrotational, since the Bernoulli function is prescribed by these boundary conditions to be constant throughout the corresponding flow. But according to the theory of an elastic inviscid and nonconducting fluid in irrotational motion, given in Sec. 16.4, the solution is uniquely determined in the characteristic triangle ABC by the conditions along AB. This means that the differential equations of an inviscid and nonconducting fluid have no solution in ABC consistent with the given boundary conditions other than the solution $q_x = q_0$, $q_y = 0$ everywhere in ABC.

It is certainly possible to subject an actual fluid mass passing between A and B to some additional conditions. The argument is based on two observable properties of fluids in supersonic motion.[19] First, a fluid mass moving uniformly can be deflected without appreciably disturbing conditions upstream, provided the deflection is gradual enough. Secondly, if a given deflection is gradual enough for one Mach number, then it is also for all larger Mach numbers. These facts will be reflected in the theory presented in the last two sections of this article.

Suppose that in the present example, Fig. 141, the fluid is bounded from below by a wall which has one end at the point A and is horizontal at that end. Provided that it does not slope upward too abruptly beyond A, such a wall will not disturb the existing conditions between A and B. In Fig. 141 the wall is represented by a curve AD, and this curve must necessarily be a streamline for the fluid particles passing through A. Now as the Mach number M_0 increases, the lines AC and BC become more nearly horizontal. Thus, however slightly the curve AD slopes upward, the point D falls within the triangle ABC for sufficiently large M_0, and a contradiction arises. The inclination θ of the streamline at D must be that determined by the tangent to the curve AD at D, and not $\theta = 0$ determined by the solution given above.

As in Art. 14 the way to overcome this contradiction is to take into account viscosity and/or heat conduction. In a fluid of low viscosity the noticeable influence of these on the flow pattern is assumed to be localized in extremely thin layers within which occur rapid changes in the state variables. Throughout the rest of the flow the equations of ideal fluid theory are supposed to describe the actual (viscous) flow adequately. A flow pattern of this combined type is observed in cases, such as the example above, for which there exists no solution based only on the theory of ideal fluids.[20]

A workable theory of steady plane flows containing such thin layers or *shocks* can be based on the following principles. *It is assumed that the original differential equations for the motion of an ideal fluid, namely the equation of continuity, Newton's equation, and the specifying condition, are valid at all points of the x,y-plane with the exception of certain "shock lines", whose positions are a priori unknown; across these lines the state variables are discontinuous, the sudden changes being governed by rules derived from the theory of viscous and/or heat-conducting fluids.*

We must now determine the form these shock conditions take in the case of steady plane flow. A particular example of such a flow of a viscous fluid can be constructed from the steady one-dimensional flow presented in Secs. 11.3 and 11.4. Suppose the latter flow is viewed from a coordinate system moving in the negative y-direction with constant speed v. This amounts to superimposing on the original flow a constant speed v in the y-direction. Then along any line parallel to the x-axis in the resulting motion a rapid change from the state $q_x = u_1$, $q_y = v_1$, p_1, ρ_1 to another state $q_x = u_2$, $q_y = v_2$, p_2, ρ_2 takes place, in which these eight variables satisfy the three relations (11.22), (11.23) and (11.24):

$$\rho_1 u_1 = \rho_2 u_2 = m, \tag{3a}$$

$$p_1 + m u_1 = p_2 + m u_2, \tag{3b}$$

$$\frac{u_1^2}{2} + \frac{\gamma}{\gamma - 1}\frac{p_1}{\rho_1} = \frac{u_2^2}{2} + \frac{\gamma}{\gamma - 1}\frac{p_2}{\rho_2}, \tag{3c}$$ [21]

as well as the extra condition

$$v_1 = v_2 = v. \tag{3d}$$

If q_1^2 is written for $u_1^2 + v_1^2$ and q_2^2 for $u_2^2 + v_2^2$ then (3c) may be replaced by

$$\frac{q_1^2}{2} + \frac{\gamma}{\gamma - 1}\frac{p_1}{\rho_1} = \frac{q_2^2}{2} + \frac{\gamma}{\gamma - 1}\frac{p_2}{\rho_2}, \tag{3c$'$}$$

by virtue of (3d). In the limit $\mu_0 \to 0$ the thickness of the transition region tends to zero [see (11.47)] and Eqs. (3) then relate the initial and final values of an abrupt transition.

Thus Eqs. (3) are the shock conditions for a particular kind of steady plane motion. It will now be shown that these equations are the transition conditions for the most general case of steady plane flow.

2. The oblique shock conditions for a perfect gas

First it is necessary to obtain the equations governing the steady plane flow of a viscous fluid. The equation of continuity is the same as for an inviscid fluid, namely Eq. (2). For a viscous fluid the three components of Newton's equation are given by Eqs. (3.8), and these must be specialized to the case of steady plane motion with gravity neglected.

For a volume element $dx\,dy\,dz$ (see Fig. 6 of Art. 3) in such a motion, no shearing stresses due to viscosity can exist on the two faces parallel to the x,y-plane, since adjacent particles on opposite sides of these faces move with the same velocity. Consequently

$$\tau_{zx} = \tau_{zy} = \sigma_z' = 0, \tag{4}$$

and hence also

$$\tau_{yz} = \tau_{xz} = 0. \tag{5}$$

There remain only the normal stresses σ_x', σ_y' and the shearing stresses

$$\tau_{xy} = \tau_{yx} = \tau \qquad \text{(say)}. \tag{6}$$

Thus with gravity neglected Eqs. (3.8) reduce to

$$\rho \frac{dq_x}{dt} = -\frac{\partial p}{\partial x} + \frac{\partial \sigma_x'}{\partial x} + \frac{\partial \tau}{\partial y}, \tag{7}$$

$$\rho \frac{dq_y}{dt} = -\frac{\partial p}{\partial y} + \frac{\partial \tau}{\partial x} + \frac{\partial \sigma_y'}{\partial y}, \tag{8}$$

where $d/dt = q_x\partial/\partial x + q_y\partial/\partial y$.

As the specifying condition we suppose that the motion is adiabatic except for heat conduction, i.e. simply adiabatic (see Sec. 1.5). Under these conditions the energy equation is (3.24), and on restricting the latter to the case of steady plane flow we have

$$\frac{d}{dt}\left(\frac{q^2}{2} + U\right) + \frac{w + w'}{\rho} = \frac{1}{\rho}\left[\frac{\partial}{\partial x}\left(k\frac{\partial T}{\partial x}\right) + \frac{\partial}{\partial y}\left(k\frac{\partial T}{\partial y}\right)\right],$$

when gravity is omitted. Here w is defined by (2.5) and w' by (3.10), which in the case of steady plane flow reduce to

$$w = \frac{\partial}{\partial x}(pq_x) + \frac{\partial}{\partial y}(pq_y),$$

$$-w' = \frac{\partial}{\partial x}(q_x\sigma_x' + q_y\tau) + \frac{\partial}{\partial y}(q_x\tau + q_y\sigma_y'),$$

by virtue of Eqs. (4), (5) and (6); hence

$$w + w' = \frac{\partial}{\partial x}[q_x(p - \sigma_x') - q_y\tau] + \frac{\partial}{\partial y}[-q_x\tau + q_y(p - \sigma_y')].$$

Finally, if it is assumed that the fluid is a perfect gas, so that (1.6):

$$p = gR\rho T \tag{9}$$

holds and $U = c_vT = p/(\gamma - 1)\rho$ according to (2.13), then the specifying condition is

$$\begin{aligned}\rho\frac{d}{dt}\left[\frac{q^2}{2} + \frac{1}{\gamma - 1}\frac{p}{\rho}\right] &+ \frac{\partial}{\partial x}[q_x(p - \sigma_x') - q_y\tau] \\ &+ \frac{\partial}{\partial y}[-q_x\tau + q_y(p - \sigma_y')] = \frac{\partial}{\partial x}\left(k\frac{\partial T}{\partial x}\right) + \frac{\partial}{\partial y}\left(k\frac{\partial T}{\partial y}\right).\end{aligned} \tag{10}$$

Equations (2) and (7) through (10) are five equations for the five unknowns q_x, q_y, p, ρ, and T, provided σ_x', σ_y' and τ are expressed in terms of these variables (see last paragraph of Sec. 3.3). If (2) is multiplied by q_x and the result added to (7), then the latter is replaced by

$$\frac{\partial}{\partial x}[\rho q_x^2 + p - \sigma_x'] + \frac{\partial}{\partial y}[\rho q_xq_y - \tau] = 0; \tag{11}$$

similarly if (2), multiplied by q_y, is added to (8), it becomes

$$\frac{\partial}{\partial x}[\rho q_xq_y - \tau] + \frac{\partial}{\partial y}[\rho q_y^2 + p - \sigma_y'] = 0. \tag{12}$$

Again, on adding (2), this time multiplied by $[q^2/2 + p/(\gamma - 1)\rho]$ to the specifying condition (10) we obtain

$$\begin{aligned}&\frac{\partial}{\partial x}\left[\rho q_x\left(\frac{q^2}{2} + \frac{1}{\gamma - 1}\frac{p}{\rho}\right) + q_x(p - \sigma_x') - q_y\tau - k\frac{\partial T}{\partial x}\right] \\ &+ \frac{\partial}{\partial y}\left[\rho q_y\left(\frac{q^2}{2} + \frac{1}{\gamma - 1}\frac{p}{\rho}\right) - q_x\tau + q_y(p - \sigma_y') - k\frac{\partial T}{\partial y}\right] = 0.\end{aligned} \tag{13}$$

Finally if T is expressed in terms of p and ρ from the equation of state (9), then (2), (11), (12), and (13) become four equations for q_x, q_y, p, and ρ as functions of x and y.

Each equation is of the form

$$\frac{\partial A}{\partial x} + \frac{\partial B}{\partial y} = 0,$$

and when integrated over the interval x_1 to x_2 supplies a relation of the form

$$A(x_2) - A(x_1) + \int_{x_1}^{x_2}\frac{\partial B}{\partial y}\,dx = 0. \tag{14}$$

We now consider solutions for which the y-derivatives remain bounded as μ_0 and k tend to zero. Then (14) holds also for the limit flow, with the integral tending to zero as x_2 approaches x_1. Thus in particular, if the x-direction is taken to be the normal to the shock line at the point under consideration, and 1, 2 refer to adjacent points on opposite sides of the curve, see Fig. 142, the difference $A_2 - A_1$ must vanish. Introducing successively for A the four expressions from (2), (11), (12), and (13) we obtain the four conditions

$$(15)\qquad \begin{gathered} [\rho q_x]_1^2 = 0, \qquad [\rho q_x^{\,2} + p - \sigma_x']_1^2 = 0, \qquad [\rho q_x q_y - \tau]_1^2 = 0, \\ \left[\rho q_x\left(\frac{q^2}{2} + \frac{1}{\gamma - 1}\frac{p}{\rho}\right) + q_x(p - \sigma_x') - q_y\tau - k\frac{\partial T}{\partial x}\right]_1^2 = 0. \end{gathered}$$

In order to compare these with Eqs. (3a) through (3d) we first replace the values of q_x and q_y on the two sides 1 and 2 of the shock by u_1, v_1 and u_2, v_2, respectively. Then the first of Eqs. (15) is exactly (3a). Now since we assume that the fluid behaves like an ideal fluid on either side of the shock, the viscous stresses σ_x', τ and the heat flux $k(\partial T/\partial x)$ must vanish on the two sides of the shock. Thus the second of Eqs. (15) gives (3b) immediately, while the third gives

$$\rho_2 u_2 v_2 - \rho_1 u_1 v_1 = m(v_2 - v_1) = 0.$$

Therefore, since $m \neq 0$ (i.e., particles cross the shock line), (3d) must hold. Under the same condition the fourth of Eqs. (15) yields (3c′).

Thus we have shown that the four equations (3a) through (3d) represent necessary conditions relating the initial and final values of an abrupt transition in steady plane flow. One restriction must still be added to the conditions. The flow studied in Art. 11, which in the limit supplies a special case, at least, of such a transition (namely one with v constant), is not re-

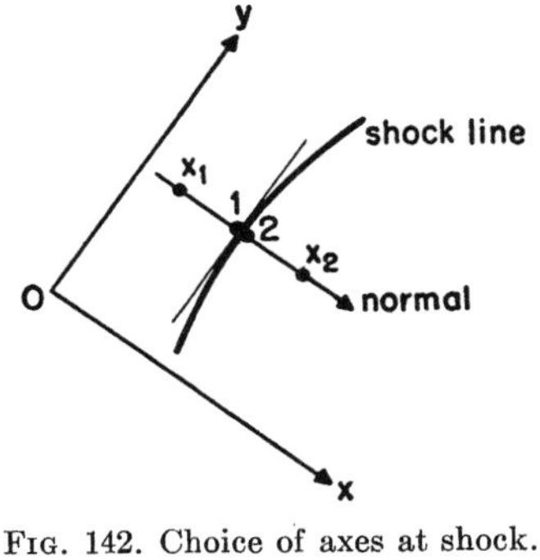

FIG. 142. Choice of axes at shock.

versible: it always goes from lower to higher values of Θ [see (11.13)] and therefore of T or p/ρ.[22] Thus the program outlined in the last section may be formulated more precisely as follows.

We consider flow patterns in the x,y-plane which satisfy the differential equations of ideal fluid theory everywhere except on certain curves (of unknown shape); *across these "shock lines" the tangential component of velocity v is continuous, while p, ρ, and the normal component u have discontinuities which satisfy the three conditions* (3a) *through* (3c) *and the inequality*

$$\frac{p_2}{\rho_2} \geqq \frac{p_1}{\rho_1}, \tag{16}$$

where for any particle state 1 *precedes state* 2.[23]

It is not correct to refer to these flow patterns as "discontinuous solutions of the equations of ideal fluid flow", see Sec. 15.2. Rather they should be called asymptotic solutions of the equations of viscous flow.

3. Analysis of the shock conditions

The shock conditions consist of the four equations (3a) through (3d) and the inequality (16). The main features of the conditions are that the tangential components of velocity v_1, v_2 enter only into (3d), and that the normal components u_1, u_2, the pressures p_1, p_2, and the densities ρ_1, ρ_2 satisfy exactly the same equations as the relative velocities u_1', u_2', the pressures p_1, p_2, and the densities ρ_1, ρ_2 do in the case of nonsteady one-dimensional flow, see Eqs. (14.2). Moreover the inequalities (14.9) and (16) are the same. Thus to every result which is a consequence of Eqs. (14.2) and the inequality (14.9) there corresponds a result in the present case, obtained from it by changing u_1', u_2' to u_1, u_2. This remark saves us a certain amount of work, since results corresponding to those obtained in Secs. 14.3 and 14.4 may be given immediately.

For this purpose it is convenient to introduce the Mach numbers M_{1n}, M_{2n} corresponding to the normal velocities u_1, u_2; these *normal Mach numbers* are related to M_1, M_2 by the equations

$$M_{1n} = M_1 \sin \sigma_1, \qquad M_{2n} = M_2 \sin \sigma_2,$$

where σ_1 and σ_2 are the angles (between 0° and 180°) at which the shock line is inclined to the stream direction for states 1 and 2 respectively, see Fig. 143a. They replace the M_1', M_2' of Secs. 14.3 and 14.4.

(*a*) Consider first some limiting cases which satisfy the shock conditions (3a) through (3d). If $u_1 = u_2 = 0$, then $m = 0$, and Eq. (3b) gives $p_1 = p_2$. For this case the third condition (3c) is satisfied for an arbitrary value of $\rho_1 = \rho_2$. No particle crosses the line of discontinuity and hence this case is not usually included in the concept of shock. Another limiting case is $u_1 =$

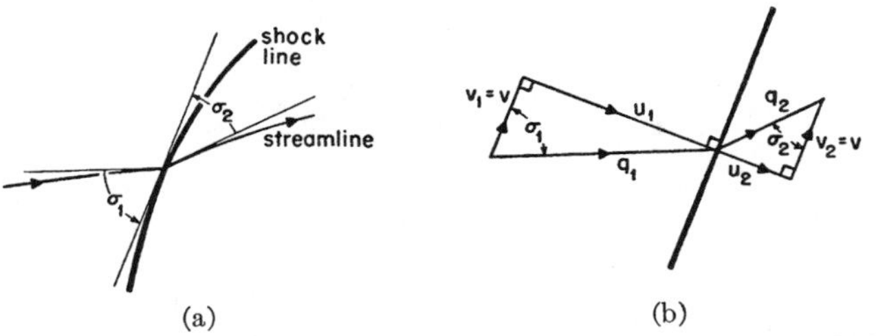

FIG. 143. Deflection of streamline at shock. (a) The angles σ_1 and σ_2, (b) Velocity components.

$u_2 \neq 0$. Again from (3b) it follows that $p_1 = p_2$, while the third condition leads to $\rho_1 = \rho_2$. With the fourth condition (3d) it follows that no actual discontinuity occurs, and this case will be referred to as *zero shock*. The same conclusion follows if we know only that $p_1 = p_2$, or that $\rho_1 = \rho_2$, provided the particles actually cross the shock line.

(*b*) *A physically possible shock* (i.e., a rapid transition governed by viscous fluid theory) *is always a "compression shock"*; *pressure, density and temperature increase, while the normal component of the velocity decreases and the tangential component remains unchanged.*

Note that this implies that the speed q decreases and that *the streamline is deflected towards the shock line on passing through the shock* (Fig. 143).

(*c*) The normal Mach numbers M_{1n}, M_{2n} are given in terms of the pressure ratio p_2/p_1 by the analogue of (14.13), namely

$$M_{1n}^2 = \frac{\gamma + 1}{2\gamma}\frac{p_2}{p_1} + \frac{\gamma - 1}{2\gamma}, \qquad M_{2n}^2 = \frac{\gamma + 1}{2\gamma}\frac{p_1}{p_2} + \frac{\gamma - 1}{2\gamma}.$$

In the preceding section we assigned to the state in front of the shock the subscript 1, and to that behind the shock the subscript 2. Hence $p_2/p_1 \geqq 1$, so that M_{1n}^2 cannot be less than 1 and M_{2n}^2 cannot be greater than 1. Only in the case of zero shock can either be equal to 1. In addition, as in (14.14),

$$\frac{\gamma - 1}{2\gamma} \leqq M_{2n}^2 \leqq M_{1n}^2 \leqq \infty .$$

For a physically possible shock, the component of velocity normal to the shock front is supersonic before and subsonic after the shock. The density ratio ρ_2/ρ_1 *cannot exceed* h^2(=6 *in air*) *and the square of the normal Mach number after the shock cannot be less than* $(\gamma - 1)/2\gamma$ (=1/7), *the extreme values corresponding to an infinite pressure ratio* $p_2/p_1 = \infty$. Note that since $q \geqq u$ the Mach number M_1 cannot be less than 1, nor can M_2 be less than $\sqrt{(\gamma - 1)/2\gamma}$. In addition since (9) gives $a_2 > a_1$ we have $M_2^2 = M_{2n}^2 + v^2/a_2^2 \lessgtr M_{1n}^2 + v^2/a_1^2 = M_1^2$. Thus *the speed* q_1 *is supersonic, whereas* q_2 *may be subsonic or supersonic with*

$$\frac{\gamma - 1}{2\gamma} \leqslant M_2^2 \leqslant M_1^2 \leqslant \infty .$$

From the fact that u_1 is supersonic it follows that $\sin \sigma_1 = u_1/q_1 \geq a_1/q_1 = \sin \alpha_1$. Similarly since u_2 is subsonic we have $\sin \sigma_2 \leq \sin \alpha_2$, whenever $M_2 \geq 1$. Hence *the acute angle between streamline and shock is no smaller than the Mach angle α_1 before the shock, and, when $M_2 \geqslant 1$, no larger than the Mach angle α_2 after the shock*. Moreover since $M_2^2 \sin^2 \sigma_2 = M_{2n}^2 \geqslant (\gamma - 1)/2\gamma$, *the angle after the shock is never smaller than arc sin $\sqrt{2\gamma}/M_2\sqrt{\gamma - 1}$.*

(*d*) On crossing a shock line a fluid particle experiences an increase of entropy (and therefore of p/ρ^γ), the amount of increase depending on the values of p and ρ before and after the shock transition. On the other hand, the theory of steady plane flow developed in Arts. 16 through 21, as applied to a perfect fluid, is based on the assumptions that a relation $p/\rho^\gamma =$ constant holds throughout the fluid and that the flow is irrotational. In other words, *the results derived in the six preceding articles need not hold in the region of the x,y-plane beyond a shock line*. The case in which all particles undergo the same change in entropy is the exception. For then the flow behind the shock will again be isentropic and, as we shall see in Sec. 24.1, irrotational.

However, it may be recalled from Sec. 14.3 that the change in entropy is in most cases small. In fact it is of the third order in $(\xi - 1)$, $\xi = \rho_2/\rho_1$, so that for shocks of moderate strength it is not a bad approximation to assume that the value of p/ρ^γ does not change. Hence the conclusions drawn in Arts. 16 through 21 are approximately valid behind weak shocks.[24]

(*e*) The shock conditions (3a), (3b), (3c) are three relations between the two sets of variables u_1, p_1, ρ_1 and u_2, p_2, ρ_2. The remaining condition (3d) states that the tangential velocity components v_1, v_2 must be equal. Each of the four conditions is unaltered when the subscripts 1 and 2 are interchanged. From the first three relations u_2, p_2, ρ_2 may be expressed in terms of u_1, p_1, ρ_1; thus in analogy to (14.20):

$$(17) \qquad \begin{aligned} u_2 &= \frac{1}{\gamma + 1}\left[2\gamma \frac{p_1}{m} + (\gamma - 1)u_1\right], \\ p_2 &= \frac{1}{\gamma + 1}[-(\gamma - 1)p_1 + 2mu_1], \\ \rho_2 &= \frac{(\gamma + 1)m^2}{2\gamma p_1 + (\gamma - 1)mu_1}. \end{aligned}$$

(*f*) If H_n is the sum of the velocity head, corresponding to the normal component u, and the pressure head, i.e. $gH_n = (u^2/2) + \gamma p/(\gamma - 1)\rho$, then

$$H_{1n} = H_{2n} = h^2 \frac{u_1 u_2}{2g}. \tag{18}$$

According to condition (3d) this implies that

$$H_1 = H_2, \tag{19}$$

where H is the "total head" which occurs in the Bernoulli equation. Thus we see that *the total head of a particle is unaltered on crossing the shock line.* This conclusion could have been read off directly from (3c′).

We have assumed in Sec. 2 that the motion on either side of the shock is strictly adiabatic. Now we have seen in Sec. 2.5 that for such a steady motion gH is a constant on any streamline. This constant is $q_m^2/2 = h^2 q_t^2/2$, where q_m is the maximum speed and q_t the sonic speed. Hence the *maximum speed and the sonic speed on any given streamline are the same before and after the shock*. Moreover since $gH = gH_n + v^2/2$, Eq. (18) may be written

$$u_1 u_2 = \frac{1}{h^2}(q_m^2 - v^2) = q_t^2 - \frac{v^2}{h^2}. \tag{20}$$

If the shock is a *normal shock*, i.e., $v = 0$ and $\sigma_1 = \sigma_2 = 90°$, then this last relation reduces to

$$q_1 q_2 = q_t^2.$$

In this case the sonic speed on a streamline is the geometric mean of the particle speeds before and after the shock.[25] Since $q_1 \geqq q_2$ it follows that *for a normal shock q_1 is always supersonic and q_2 subsonic.* This conclusion is also covered by the first result in (*c*), since the normal component of velocity is now the speed itself.

(*g*) If the dimensionless parameters

$$\frac{p_2}{p_1} = \eta, \qquad \frac{u_2}{u_1} = \frac{\rho_1}{\rho_2} = \frac{1}{\xi} = \zeta$$

are introduced, then the points $P_1(\zeta, \eta)$ and $P_2(1/\zeta, 1/\eta)$ lie on the hyperbola of Fig. 75, and represent the given shock. The Mach numbers M_{1n} and M_{2n} appear in the slopes $-\gamma M_{1n}^2$ and $-\gamma M_{2n}^2$ of the lines AP_1 and AP_2, respectively; these lines form equal angles with OA.

The five quantities η, ζ, ξ, M_{1n}^2 and M_{2n}^2 are determined when any one of them is given. This follows from the fact that any one of the five fixes the corresponding points P_1 and P_2 on the hyperbola, and once these points are known each of the five can be read off from the figure by means of a length or a slope. Algebraic relations between them are given by Eqs. (14.24) through (14.27) with M' replaced each time by M_n. Thus in particular from Eqs. (14.25)

$$(21)\qquad \begin{cases} \eta = \dfrac{p_2}{p_1} = \dfrac{2\gamma}{\gamma+1} M_{1n}^2 - \dfrac{1}{h^2}, \\[2ex] \dfrac{1}{\xi} = \zeta = \dfrac{u_2}{u_1} = \dfrac{\rho_1}{\rho_2} = \dfrac{2}{\gamma+1}\dfrac{1}{M_{1n}^2} + \dfrac{1}{h^2}, \end{cases}$$

and from (14.27)

$$(22)\qquad \left(\frac{2\gamma}{\gamma-1} M_{1n}^2 - 1\right)\left(\frac{2\gamma}{\gamma-1} M_{2n}^2 - 1\right) = h^4.$$

The Hugoniot equation, given by (14.26):

$$(23)\qquad h^2(\eta - \xi) = \xi\eta - 1, \qquad \xi = \frac{\rho_2}{\rho_1}, \qquad \eta = \frac{p_2}{p_1},$$

is unaltered.[26]

With the same change, Fig. 76 gives the graphs of ξ, M_{1n}^2, M_{2n}^2 versus η, for $\eta \geqq 1$.

4. Representation of a shock in the hodograph plane

At the beginning of Art. 8 we saw that in steady flow the state of a moving particle at any moment is determined by its velocity $\mathbf{q}$. It was assumed that a (p, ρ)-relation holds on the streamline traced by the particle and that the Bernoulli constant (Sec. 2.5) is known. For plane flow the velocity is represented by a point in the hodograph plane, and the continuous motion of a particle by a curve in this plane.

When a particle passes through a shock front its representative point (q_x, q_y) in the hodograph plane undergoes a sudden jump. We shall now study the conditions governing this discontinuous transition. For this purpose, see Fig. 143b, it is better to use the speed q and the angle σ (between the streamline and shock front) to characterize the velocity, rather than the velocity components $u = q \sin \sigma$ and $v = q \cos \sigma$.

We assume that the position of the shock is a priori unknown, but that the state before the shock, i.e., the set of values p_1, ρ_1, q_1, and θ_1 is given. Now when p_1, ρ_1, q_1 are known the shock conditions (3a) through (3d) may be considered as four equations governing the five quantities p_2, ρ_2, q_2, σ_1, σ_2. Since there are five unknowns and only four equations, there is a single infinity of solutions. But to each set of values q_2, σ_1, σ_2 there corresponds just one velocity vector behind the shock; its magnitude is q_2 and its direction is $\theta_1 + \sigma_1 - \sigma_2$. Thus *the end points of shock transitions from a fixed initial point lie on a curve in the hodograph plane.* On this curve we may take σ_1 as parameter.

Let the points Q_1, Q_2 represent the velocity vectors $\mathbf{q}_1$, $\mathbf{q}_2$ in the hodograph plane, and take Cartesian axes $O'U$, $O'V$ with $O'U$ along $O'Q_1$, as in Fig. 144. The condition (3d) then implies that Q_1Q_2 is perpendicular to the line $O'S$ which represents the direction of the shock. Thus the coordinates of the point Q_2 are given by

$$U = v \cos \sigma_1 + u_2 \sin \sigma_1 , \qquad V = (q_1 - U) \cot \sigma_1 . \tag{24}$$

The quantities u_2 and v occurring in these equations may be expressed in terms of the given q_1 and the parameter σ_1 by means of the relation (20) and the equation $v = q_1 \cos \sigma_1$. Thus

$$u_2 = \frac{1}{h^2 u_1} (q_m^{\ 2} - v^2) = \frac{q_m^{\ 2} - q_1^{\ 2} \cos^2 \sigma_1}{h^2 q_1 \sin \sigma_1}.$$

The first of Eqs. (24) now becomes

$$U = (U_B - U_A) \cos^2 \sigma_1 + U_A , \tag{25}$$

where

$$U_A = \frac{q_m^{\ 2}}{h^2 q_1}, \qquad U_B = U_A + q_1 \left(1 - \frac{1}{h^2}\right). \tag{26}$$

From the second of Eqs. (24) we have $\cot^2 \sigma_1 = V^2/(q_1 - U)^2$, and from (25) $\cot^2 \sigma_1 = \cos^2 \sigma_1/(1 - \cos^2 \sigma_1) = (U - U_A)/(U_B - U)$. On equating these two expressions for $\cot^2 \sigma_1$ we find that the locus of the point Q_2 is the Cartesian leaf (Folium of Descartes)

$$\frac{V^2}{(q_1 - U)^2} = \frac{U - U_A}{U_B - U}; \tag{27}$$

this curve is sketched in Fig. 145a.

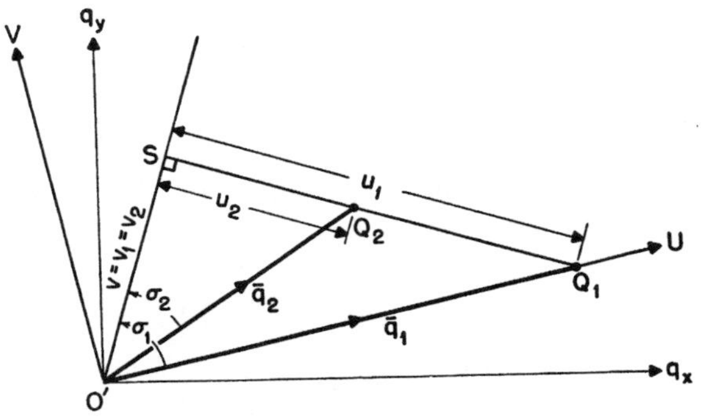

FIG. 144. U,V-axes in the hodograph plane.

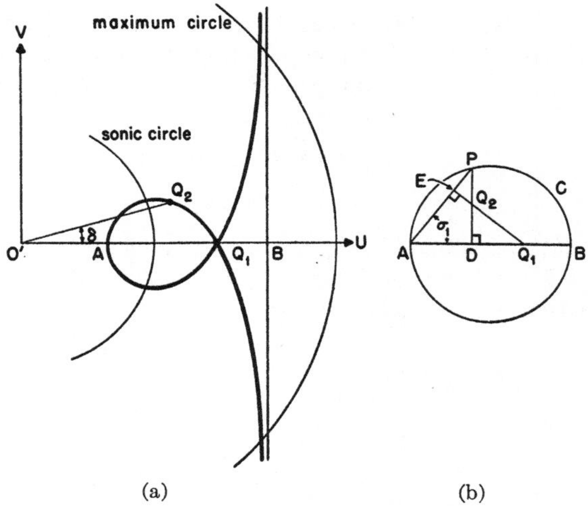

FIG. 145. Folium of Descartes and its construction.

The fact that $U_A \geqq q_1 \leqq U_B$ (or that the points A, Q_1, B lie in the order shown) follows immediately from the conditions that q_1 be supersonic and less than q_m. Thus since the sonic speed q_t is q_m/h we have $q_1^2 \geqq q_m^2/h^2$ or $q_1 \geqq q_m^2/h^2 q_1 = U_A$; from $q_1^2 \leqq q_m^2$ we have $q_1^2 \leqq q_m^2/h^2 + q_1^2(1 - 1/h^2)$ or $q_1 \leqq (q_m^2/h^2 q_1) + q_1(1 - 1/h^2) = U_B$. Furthermore, it is easily verified from (26) that the points A and Q_1 are inverse with respect to the sonic circle $U^2 + V^2 = q_m^2/h^2$. The folium is symmetric about the U-axis, has a double point at Q_1, and a double asymptote $U = U_B$. This diagram is referred to as the *shock polar*, and was introduced by A. Busemann.[27]

Equation (27), defining the shock polar, contains only distances measured from the points A, B, Q_1, and the line joining these points. It is possible to give a geometrical construction of the shock polar based only on these three points. Let C be the circle on AB as diameter, and P any point of C (Fig. 145b). Then the point of intersection, Q_2, of the line PD perpendicular to AQ_1B, and the line Q_1E perpendicular to AP, lies on the shock polar. As P traces out the circle C, the point Q_2 traces out the polar. For the similar triangles Q_2DQ_1 and ADP yield $Q_2D:Q_1D = AD:PD$, and for the circle C we have $(PD)^2 = AD \cdot DB$. Hence

$$\left(\frac{Q_2D}{Q_1D}\right)^2 = \left(\frac{AD}{PD}\right)^2 = \frac{AD}{DB},$$

which is an alternative way of writing (27).

The slope and curvature of each branch of the shock polar at the point Q_1 are of particular interest. The first two derivatives of V with respect to

U are easily computed from (27); on the branch with negative slope at Q_1, their values at this point are $dV/dU = -\sqrt{(q_1 - U_A)/(U_B - q_1)}$ and $d^2V/dU^2 = -(U_B - U_A)/\sqrt{(q_1 - U_A)(U_B - q_1)^3}$. The corresponding radius of curvature is therefore

$$R = \sqrt{(q_1 - U_A)(U_B - U_A)}. \tag{28}$$

From (26) and $q_m^2 = 2gH_1$, the slope is found to be $-\sqrt{M_1^2 - 1} = -\cot \alpha_1$ where α_1 is the Mach angle of the incident flow. Thus this branch of the shock polar is inclined at an angle $-(90° - \alpha_1)$ to the velocity vector $\mathbf{q}_1$ and hence is tangent at Q_1 to the characteristic epicycloid Γ^- through that point, see (16.37). Similarly, the other branch of the shock polar is tangent at Q_1 to the Γ^+-epicycloid through Q_1. Now the shock direction $O'S$ in Fig. 144 is perpendicular to Q_1Q_2S which in the limit $Q_2 \to Q_1$ becomes the tangent at Q_1. Hence *for very weak shocks* (q_2/q_1 close to 1) *the angle between the shock front and the incident streamline is approximately the Mach angle* α_1.[28]

As Q_2 approaches Q_1 in Fig. 145b, the line AP approaches the normal direction to the negatively inclined branch of the shock polar at Q_1. Moreover, in the limit D coincides with Q_1, and the length of the segment AP is then, according to (28), the radius of curvature R. Therefore the center of curvature $\hat{Z}$ is the intersection of the line through Q_1 parallel to the limit position of AP and the tangent to the circle C at A. Thus, the projection of $\hat{Z}$ on $O'Q_1$ is A, the inverse of Q_1 with respect to the sonic circle. From the construction given in Sec. 16.5 for the center of curvature Z of the corresponding characteristic epicycloid, it can be shown that Z has the same property. Hence $\hat{Z} = Z$ and *the two curves actually osculate at* Q_1.

The Cartesian leaf is determined by the values $\mathbf{q}_1$ and q_m alone. Each point Q_2 on it represents a transition satisfying conditions (3a) through (3d). However, these equations do not determine which of the points Q_1, Q_2 represents the state in front of the shock. But we have seen in the preceding section that in a physically possible transition the speed must decrease. Thus Q_2 can represent conditions behind the shock only if it lies on the closed loop of the folium. In the following, the term *shock polar* will refer only to this closed loop. The remainder of the folium inside the maximum circle represents those states from which the state Q_1 can be reached by transition through a shock.

5. Shock diagram and pressure hills

Apart from size and orientation in the q_x, q_y-plane the shock polar is determined by the ratio q_1/q_m or, according to (16.10) and (16.11), by the Mach number M_1 of the incident stream. Hence *all polars corresponding to the same value of* M_1 *are similar*. This remark enables us to confine our at-

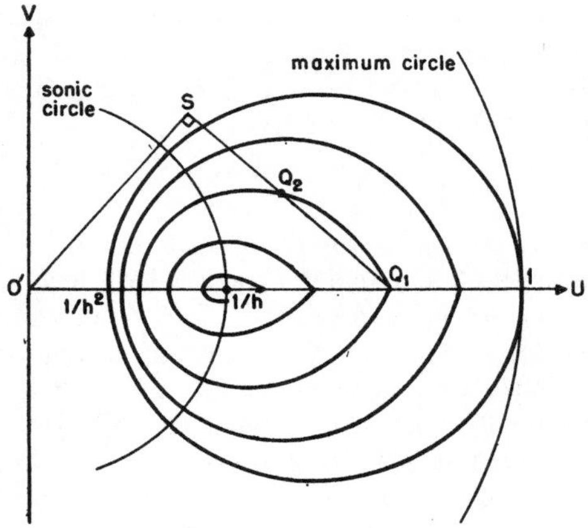

FIG. 146. The shock diagram.

tention for the rest of this section to those polars which correspond to a fixed direction of $\mathbf{q}_1$, and $q_m = 1$. Alternatively the velocity $\mathbf{q}$ in the following may be considered to represent the dimensionless velocity $\mathbf{q}/q_m$ of any particular case.

For $q_m = 1$ the sonic speed q_t is $1/h$, so that q_1 may vary between $1/h$ and 1. As q_1 increases from $1/h$ to 1, U_A decreases from $1/h$ to $1/h^2$, and in the U, V-plane the corresponding shock polar expands from an infinitesimal circuit around the point $(1/h, 0)$ to the circle with the points $(1/h^2, 0)$ and $(1, 0)$ at the ends of a diameter, see Fig. 146. All shock polars lie between these extreme curves and to each point Q_2 within the circle there corresponds one possible shock: $\overline{O'Q_2}$ is the velocity vector $\mathbf{q}_2$, the corner Q_1 of the polar through Q_2 gives the velocity vector $\mathbf{q}_1 = \overline{O'Q_1}$, and the perpendicular to Q_1Q_2 determines the shock direction $O'S$. The graph of this family of polars is called the *shock diagram*.

So far the pressure and density have entered only through the Bernoulli constant $q_m^2/2$ which is the same on either side of the shock line and equal to each member of (3c′). As we have seen in (8.24), this constant is a multiple of p_s/ρ_s and therefore of the stagnation temperature T_s. Thus the stagnation temperature has the same value on either side of the shock. However, the stagnation pressure p_s (and hence the stagnation density ρ_s) does change on crossing the shock.

For any given stagnation pressure p_s, the pressure itself may be introduced by erecting the pressure hill (see Sec. 8.2) on the U,V-plane of the shock polar, i.e., on the hodograph plane. For the polytropic case this hill is given by (16.12). On setting $\kappa = \gamma$ and $q_m = 1$ this equation becomes

$$p = p_s[1 - q^2]^{\gamma/(\gamma-1)}, \qquad q^2 = U^2 + V^2. \tag{29}$$

The hill, which is a surface of revolution, is shown in Fig. 147a. Its base is the circle $q = q_m = 1$ and its summit lies on the p-axis at the point $p = p_s$.

Consider now the intersections of the plane perpendicular to the U,V-plane whose trace in that plane is Q_1Q_2, with the family of pressure hills obtained by varying p_s. For any two such curves the ordinates are constant multiples of each other. In particular select the two intersections which correspond to $p_s = p_{s1}$, p_{s2}, the stagnation pressures on the two sides of the shock (Fig. 147b). Let P_1 and P_2 be the points on these two curves corresponding to the pressures p_1 and p_2 (i.e., those points whose projections are Q_1 and Q_2 respectively). Then the line P_1P_2 is tangent to both curves. For when the velocity vector $\mathbf{q}$ corresponds to points on the line Q_1Q_2 in Fig. 147a, it may be resolved into a constant component v along $O'S$ and a variable component u perpendicular to this direction. Now (29) is derived from the Bernoulli equation, of which (2.21′) is the differential form. Hence

$$dp = -\rho q\, dq = -\rho[u\, du + v\, dv] = -\rho u\, du$$

on either of the curves in Fig. 147b. Thus the slope of the first curve at P_1 is $(dp/du)_1 = -\rho_1 u_1$ and that of the second at P_2 is $(dp/du)_2 = -\rho_2 u_2$. These are equal by the first shock condition (3a). Moreover, the slope of the line P_1P_2 is $(p_1 - p_2)/(u_1 - u_2)$ which, according to the second shock condition (3b), is the same as the two previous slopes. Hence P_1P_2 is the common tangent of the two curves.

As Q_2 moves along a given polar, P_2 moves in the tangent plane at P_1 to the pressure hill through P_1, this hill being determined by p_{s1}. For any given position of Q_2, the point P_2 is the intersection of the vertical line through Q_2 with this tangent plane. Once P_2 is determined, there is a unique pressure hill of the family (29) passing through P_2, or in other words a

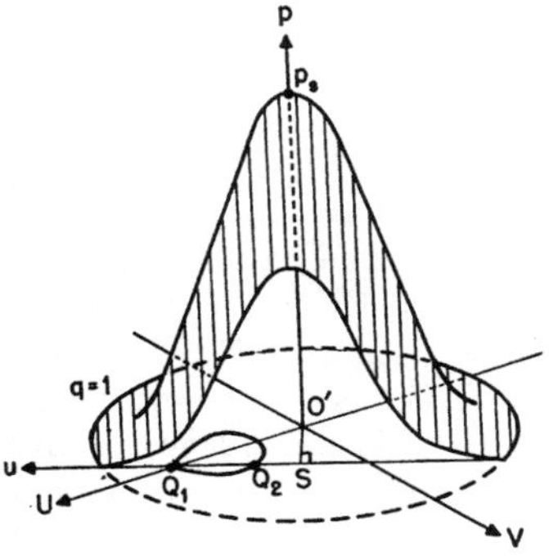

(a)

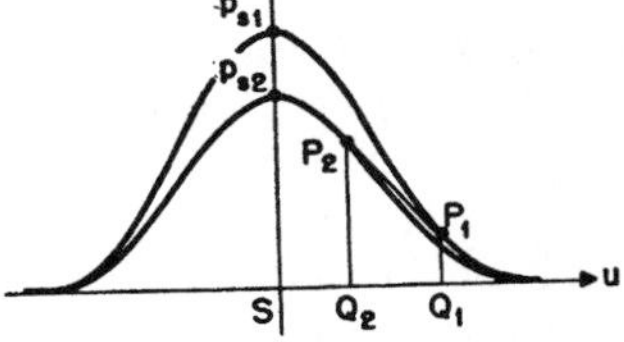

(b)

FIG. 147. Section of the pressure hill. (a) Vertical section with trace Q_1Q_2, (b) Common tangent.

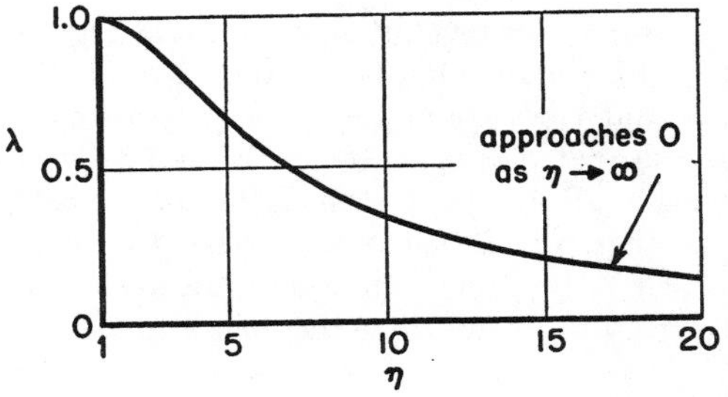

FIG. 148. Graph of stagnation pressure ratio versus pressure ratio.

unique value of p_{s2}.[29] We conclude this section by obtaining explicit formulas for the ratio p_{s2}/p_{s1}.

Since T_s is the same on the two sides of the shock and the flow is adiabatic there, we have

$$\left(\frac{p_{s2}}{p_{s1}}\right)^{\gamma-1} = \left(\frac{p_{s1}}{\rho_{s1}^{\gamma}}\right)\left(\frac{\rho_{s2}^{\gamma}}{p_{s2}}\right) = \left(\frac{p_1}{\rho_1^{\gamma}}\right)\left(\frac{\rho_2^{\gamma}}{p_2}\right).$$

Hence

$$\lambda = \frac{p_{s1}}{p_{s2}} = \eta^{-1/(\gamma-1)}\xi^{\gamma/(\gamma-1)},$$

where, as in Sec. 3, $\eta = p_2/p_1$ and $\xi = \rho_2/\rho_1$.*

Equations (21) through (23) now allow us to express λ in terms of ξ, η, ζ, M_{1n} or M_{2n} alone. In particular, we have

$$\lambda = \eta^{-1/(\gamma-1)}\left[\frac{h^2\eta + 1}{\eta + h^2}\right]^{\gamma/(\gamma-1)}, \tag{30}$$

and Fig. 148 gives the graph of λ versus η.

6. The deflection of a streamline by a shock

The deflection δ of a streamline on crossing a shock is given by the angle $Q_2O'Q_1$ in Fig. 144, i.e., $\sigma_1 - \sigma_2$. As Q_2 moves along the polar in Fig. 145a from Q_1 to A this angle increases from zero to a maximum, and then decreases to zero again. Now we have seen that the shock polar is determined except for size and orientation by the Mach number M_1 of the incident stream. Thus we conclude: *For each value of M_1, there is a maximum deflection which can be produced by means of a shock.* This is analogous to what was found in Sec. 18.3 for the deflection of a stream by means of a simple wave.

On a given polar there is a relation between the deflection δ and the in-

* From Eq. (1.7), $\lambda = \exp[(S_1 - S_2)/gR]$ so that it also represents the change in entropy.

clination σ_1 of the shock to the incident streamline. Thus from Eqs. (24), (25), and Fig. 145 we find

$$\tan \delta = \frac{V}{U} = \left(\frac{q_1}{U} - 1\right) \cot \sigma_1$$

$$= \left[\frac{(q_1 - U_A) + (q_1 - U_B) \cot^2\sigma_1}{U_A + U_B \cot^2\sigma_1}\right] \cot \sigma_1 .$$

If we write

$$\text{(31)} \qquad \begin{cases} \tau_1 = \cot \sigma_1 , \quad \epsilon = \tan \delta, \\ c = 1 - \dfrac{U_B}{q_1} = \dfrac{1}{h^2}\left(1 - \dfrac{q_m^2}{q_1^2}\right) = -\dfrac{2}{(\gamma + 1)M_1^2}, \\ d = 1 - \dfrac{U_A}{q_1} = 1 - \dfrac{q_m^2}{h^2 q_1^2} = \dfrac{2(M_1^2 - 1)}{(\gamma + 1)M_1^2}, \end{cases}$$

then this formula becomes

$$\text{(32)} \qquad \epsilon = \frac{(c\tau_1^2 + d)\tau_1}{(1 - c)\tau_1^2 + (1 - d)} .$$

For each value of M_1 there is a corresponding curve in the τ_1, ϵ-plane relating the deflection δ to the shock inclination σ_1, see Fig. 149. As M_1 increases from 1 to ∞, c increases monotonically from $-2/(\gamma + 1)$ to 0 and d from 0 to $2/(\gamma + 1)$. Thus for fixed positive τ_1, Eq. (32) shows that

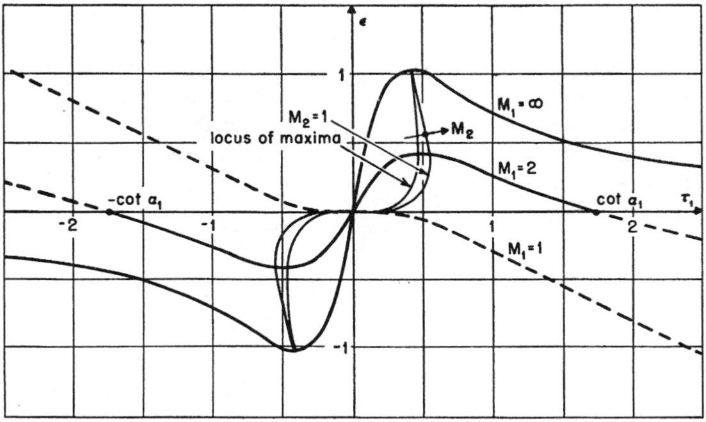

FIG. 149. Relation between deflection and shock angle for fixed incident Mach number.

ϵ increases monotonically with M_1 from

$$M_1 = 1: \qquad \epsilon = -\frac{2\tau_1^3}{(\gamma + 3)\tau_1^2 + (\gamma + 1)},$$

to

$$M_1 = \infty: \qquad \epsilon = \frac{2\tau_1}{(\gamma + 1)\tau_1^2 + (\gamma - 1)},$$

and vice versa for $\tau_1 < 0$. Every curve (32) which corresponds to a physically possible transition lies between these last two. Each such curve intersects the τ_1-axis at the origin and at the two points

$$\tau_1 = \pm\sqrt{-d/c} = \pm\sqrt{M_1^2 - 1} = \pm\cot \alpha_1 ,$$

where α_1 is the Mach angle corresponding to M_1. The origin corresponds to the point A on the shock polar, and the other two points to the double point Q_1. Hence for physically possible shocks our attention must be restricted to the range

$$(33) \qquad | \tau_1 | \leqq \cot \alpha_1 ;$$

the end points correspond to zero shock, for which $\sigma_1 = \alpha_1$, $180° - \alpha_1$ as was seen before. Thus, as we found in Sec. 3, for each M_1 the smallest possible angle between the shock line and incident streamline is the Mach angle $\alpha_1 = \text{arc sin } 1/M_1$; the largest possible angle is 90°. The latter corresponds to $\tau_1 = 0$ or normal shock. In both cases the resulting deflection is zero.

The symmetry allows attention to be focussed on the range $0 \leqq \tau_1 \leqq \cot \alpha_2$, which corresponds to the top half of the shock polar. It is clear from Fig. 149 that for each value of M_1, ϵ has a maximum in this range. This is in agreement with the statement about δ which was made at the beginning of this section. The value of τ_1 for which this maximum occurs is found from (32) by setting $d\epsilon/d\tau_1$ equal to zero. It is therefore a root of the equation $c(1 - c)\tau_1^4 + (3c - 2cd - d)\tau_1^2 + d(1 - d) = 0$ or

$$A\tau_1^4 + B\tau_1^2 + C = 0,$$

where

$$A = (\gamma + 1)M_1^2 + 2,$$

$$B = (\gamma + 1)M_1^4 + 2(\gamma - 1)M_1^2 + 4,$$

$$C = -(M_1^2 - 1)[(\gamma - 1)M_1^2 + 2].$$

The roots of this equation which correspond to real τ_1 are given in terms of M_1 by

$$\tau_1^2 = \cot^2\sigma_1 = -\frac{[(\gamma+1)M_1^4 + 2(\gamma-1)M_1^2 + 4]}{2[(\gamma+1)M_1^2+2]} + \frac{M_1^2\sqrt{(\gamma+1)[(\gamma+1)M_1^4 + 8(\gamma-1)M_1^2+16]}}{2[(\gamma+1)M_1^2+2]} \tag{34}$$

and the root in question is the positive square root of this expression. A somewhat simpler expression is obtained for the sine of the angle σ_1 at which maximum deflection occurs:

$$\sin^2\sigma_1 = \frac{[(\gamma+1)M_1^2 - 4]}{4\gamma M_1^2} + \frac{\sqrt{(\gamma+1)[(\gamma+1)M_1^4 + 8(\gamma-1)M_1^2+16]}}{4\gamma M_1^2}. \tag{35}$$

For each given M_1 the maximum possible deflection, $\delta_{\max}$, of the streamline may be determined by first computing the position σ_1 of the shock from (34) or (35), and substituting the result in (32) to obtain the corresponding $\epsilon = \tan\delta$. A graph of $\delta_{\max}$ versus M_1 is given in Fig. 150, and a few corresponding values in Table VI (Sec. 7).

It is clear (see Fig. 149) that *the maximum deflection increases monotonically with* M_1. Letting $M_1 \to \infty$ we have from (34), $\tau_1^2 \to 1/h^2$ and from (32), $\epsilon \to (h^2-1)/2h = 1/\sqrt{\gamma^2-1}$. Thus *a streamline cannot be deflected by more than arc sin* $(1/\gamma)$ *by means of a shock, whatever the incident Mach number may be.* For $\gamma = 7/5$ this angle is 45° 35′.

The formula (32) can also be used to find $\tau_2 = \cot\sigma_2$ in terms of τ_1. Thus since $\delta = \sigma_1 - \sigma_2$ it yields

$$\epsilon = \frac{\tau_2 - \tau_1}{1+\tau_1\tau_2} = \frac{(c\tau_1^2+d)\tau_1}{(1-c)\tau_1^2+(1-d)},$$

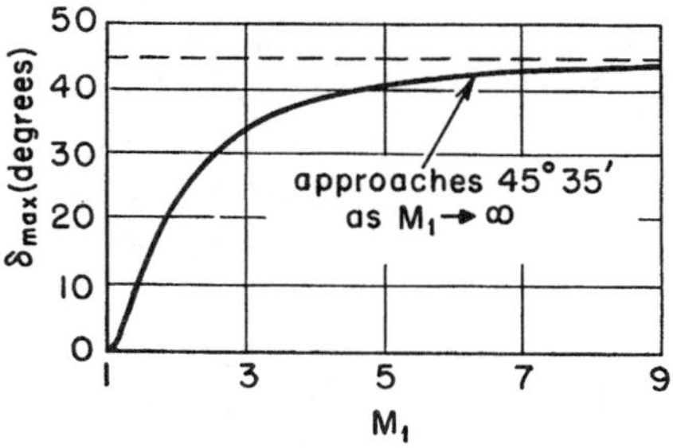

FIG. 150. Graph of maximum deflection angle versus incident Mach number.

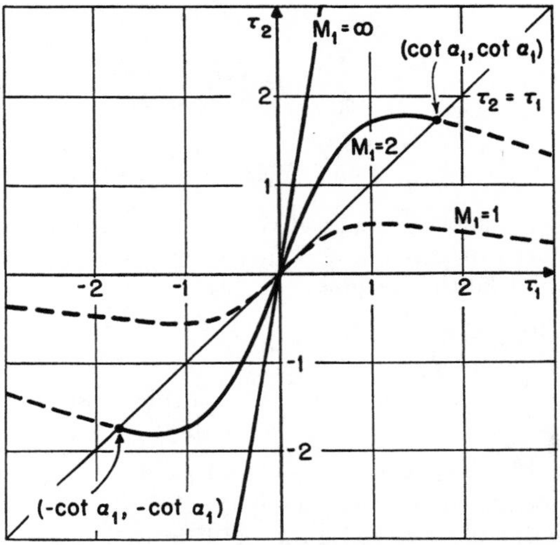

FIG. 151. Relation between shock angles for fixed incident Mach number.

so that

$$\tau_2 = \frac{(\tau_1^2 + 1)\tau_1}{-c\tau_1^4 + (1 - c - d)\tau_1^2 + (1 - d)} = \frac{\tau_1}{c_1\tau_1^2 + d_1}, \tag{36}$$

where

$$c_1 = -c = \frac{2}{(\gamma + 1)M_1^2}, \qquad d_1 = 1 - d = \frac{(\gamma - 1)M_1^2 + 2}{(\gamma + 1)M_1^2}.$$

For each value of M_1 there is a corresponding curve in the τ_1, τ_2-plane relating the inclinations σ_1, σ_2 of the streamline to the shock before and after the transition, see Fig. 151. As M_1 increases from 1 to ∞, c_1 decreases monotonically from $2/(\gamma + 1)$ to 0, and d_1 from 1 to $(\gamma - 1)/(\gamma + 1)$. Thus for each positive τ_1, (36) shows that τ_2 increases monotonically with M_1 from

$$M_1 = 1: \qquad \tau_2 = \frac{(\gamma + 1)\tau_1}{2\tau_1^2 + (\gamma + 1)},$$

to

$$M_1 = \infty: \qquad \tau_2 = \frac{(\gamma + 1)}{(\gamma - 1)}\tau_1,$$

and vice versa for $\tau_1 < 0$. Every curve (36) which corresponds to a physically possible transition lies between these last two. Each such curve intersects the line $\tau_2 = \tau_1$ at the origin and the two points $(\pm\cot \alpha_1, \pm\cot \alpha_1)$. Only the segment of the curve which lies between these last two points is of interest since for a physically possible shock the inequality (33) must be satisfied.

The formula (36) was obtained by algebraic manipulation of Eqs. (3). Now interchange of the subscripts 1 and 2 leaves these equations unaltered. It follows therefore that interchange of these subscripts in (36) will yield a new formula, which is also a consequence of Eqs. (3). Thus we obtain

$$\tau_1 = \frac{\tau_2}{c_2\tau_2^2 + d_2}, \tag{37}$$

where

$$c_2 = \frac{2}{(\gamma + 1)M_2^2}, \qquad d_2 = \frac{(\gamma - 1)M_2^2 + 2}{(\gamma + 1)M_2^2}.$$

This may be considered as an equation determining M_2 at any point of the τ_1, τ_2-plane.

Consider the family of curves in the τ_1, τ_2-plane given by (37) with unrestricted parameter M_2. This family is the reflection in the line $\tau_2 = \tau_1$ of the family given by Eq. (36) with the parameter M_1 unrestricted. Now we have just seen that only those points lying between the lines $\tau_2 = \tau_1$ and $\tau_2 = (\gamma + 1)\tau_1/(\gamma - 1)$ correspond to physically possible shocks. Accordingly, only that segment of each of the curves (37) which lies in this wedge-shaped region is of physical interest. The bounds on σ_2 given in Sec. 3 can now be verified by considering the values of τ_2 at the ends of this segment.

In some problems it is the pressure ratio η across the shock, and not the Mach number M_1 in front, which is known (see Sec. 23.4). The deviation of η from 1 is a measure of the strength of the shock front, as also are the deviations of the equivalent quantities ξ, ζ, M_{1n} and M_{2n}. From Fig. 143b we have

$$\frac{\tau_1}{\tau_2} = \frac{\cot \sigma_1}{\cot \sigma_2} = \frac{v/u_1}{v/u_2} = \frac{u_2}{u_1} = \zeta,$$

which provides a relation betweeen σ_1, σ_2 and the strength represented by ζ. Now since $\sigma_2 = \sigma_1 - \delta$, we may replace τ_2 in this equation by $(\tau_1 + \epsilon)/(1 - \epsilon\tau_1)$. Then on solving for ϵ we have

$$\epsilon = \frac{(1 - \zeta)\tau_1}{\zeta + \tau_1^2} = \frac{\tau_1}{s(\tau_1^2 + 1) - 1}, \tag{38}$$

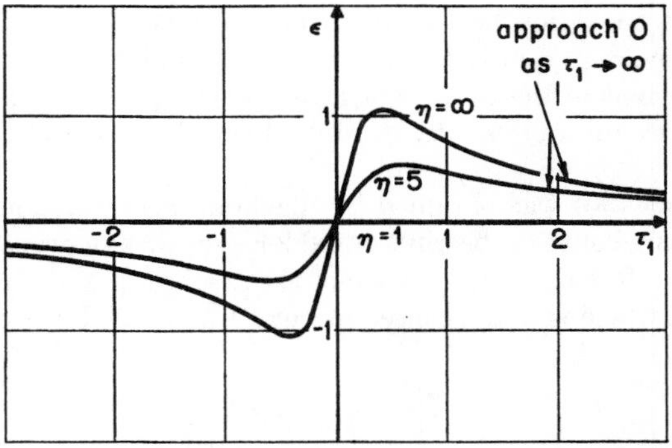

FIG. 152. Relation between deflection and shock angle for fixed shock strength.

where

$$s = \frac{1}{1-\zeta} = \frac{\xi}{\xi - 1} = \frac{h^2\eta + 1}{(h^2 - 1)(\eta - 1)}.$$

For each shock strength s, (38) represents a curve in the ϵ, τ_1-plane, see Fig. 152. As η increases from 1 to ∞, s decreases monotonically from ∞ to $h^2/(h^2 - 1) = (\gamma + 1)/2$. Thus for fixed positive τ_1, (38) shows that ϵ increases monotonically with η from

$$\eta = 1 : \qquad \epsilon = 0,$$

to

$$\eta = \infty : \qquad \epsilon = \frac{2\tau_1}{(\gamma + 1)\tau_1^2 + (\gamma - 1)}.$$

All points lying between these two curves correspond to physically possible transitions. Notice that the curve $\eta = \infty$ is the same as $M_1 = \infty$ in Fig. 149.

Equation (38) can also be obtained by setting $M_1^2 = M_{1n}^2 (\tau_1^2 + 1)$ in the coefficients c and d of Eq. (32) and using (21).

7. Strong and weak shocks

We have seen that for each given supersonic state p_1, ρ_1, q_1, θ_1 there is a simple infinity of states p_2, ρ_2, q_2, θ_2 which satisfy the shock conditions. Thus at least one further condition is required in order to fix the state 2. In many problems (see Secs. 23.2 and 23.3) this condition consists in prescribing the deflection $\delta = \theta_2 - \theta_1$ of the streamline. We assume, for definiteness, that the given value of δ is positive and also, of course, that it is no larger than δ_{max}.

When M_1 and δ are known, (32) becomes a cubic equation for τ_1. This cubic has just two roots lying in the range given by the inequality (33), (see Fig. 149). If $\delta = \delta_{\max}$ these roots are coincident and there is just one possible inclination σ_1 of the shock. In this case the state 2 is uniquely determined. If however $\delta < \delta_{\max}$ these roots are distinct and there are two possible inclinations for the shock. To each there corresponds a different state 2 behind the shock. The transition corresponding to the larger root (smaller σ_1) will be called the *weak shock* for the given deflection, and that corresponding to the smaller root (larger σ_1), the *strong shock*.[30] To justify these names we notice that as τ_1 increases, $M_{1n}^2 = M_1^2/(\tau_1^2 + 1)$ decreases, and therefore the pressure ratio p_2/p_1 decreases, see (21). In Fig. 145a the weak shock is given by the position of Q_2 nearer to Q_1 and the strong shock by the position nearer to A. As $\delta \to 0$ the strong one tends to the normal shock and the weak to the zero shock.

In general the flow behind the strong shock is subsonic and that behind the weak shock supersonic, see Fig. 149. However this distinction fails if δ is too close to $\delta_{\max}$. To prove this it is sufficient to show that for any $M_1 > 1$ the positive value, δ_{son}, of δ for which the flow is sonic behind the shock (i.e., $M_2 = 1$), is not equal to $\delta_{\max}$. We shall do this by showing that the value of τ_1 corresponding to δ_{son} is in fact always larger than that corresponding to $\delta_{\max}$. Thus we shall also have proved that *the flow behind the strong shock is always subsonic*, see Fig. 149.

According to (37), when $M_2 = 1$ the angles σ_1 and σ_2 are related by

$$\tau_1 = \frac{(\gamma + 1)\tau_2}{2\tau_2^2 + (\gamma + 1)}.$$

But τ_2 can always be expressed in terms of τ_1 and M_1 by means of (36). Thus on eliminating τ_2 we find

$$2\tau_1^2 + (\gamma + 1)(c_1\tau_1^2 + d_1)^2 - (\gamma + 1)(c_1\tau_1^2 + d_1) = 0.$$

Hence for each $M_1 \geqq 1$, the value of τ_1 for which $\delta = \delta_{\text{son}}$ is a root of the equation

$$\bar{A}\tau_1^4 + \bar{B}\tau_1^2 + \bar{C} = 0,$$

where

$$\begin{aligned} \bar{A} &= 2, \\ \bar{B} &= (\gamma + 1)M_1^4 - (3 - \gamma)M_1^2 + 4, \\ \bar{C} &= -(M_1^2 - 1)[(\gamma - 1)M_1^2 + 2]. \end{aligned}$$

The roots of this equation which correspond to real τ_1 are given by

$$\tau_1^2 = \cot^2 \sigma_1 = -\frac{1}{4}[(\gamma + 1)M_1^4 - (3 - \gamma)M_1^2 + 4] \tag{39}$$
$$+ \frac{M_1^2}{4}\sqrt{(\gamma + 1)[(\gamma + 1)M_1^4 - 2(3 - \gamma)M_1^2 + (\gamma + 9)]},$$

and the root in question is the positive square root of this last expression. Equation (39) should be compared with (34), which gives the value of τ_1 for which $\delta = \delta_{max}$. Also the formula

$$\sin^2 \sigma_1 = \frac{[(\gamma + 1)M_1^2 - (3 - \gamma)]}{4\gamma M_1^2}$$
$$+ \frac{\sqrt{(\gamma + 1)[(\gamma + 1)M_1^4 - 2(3 - \gamma) M_1^2 + (\gamma + 9)]}}{4\gamma M_1^2}$$

compares with (35).

To show that the value of τ_1 given by (39) is always larger than that given by (34) we consider the behavior of the functions $f(x) = Ax^2 + Bx + C$ and $g(x) = \bar{A}x^2 + \bar{B}x + \bar{C}$ for $x > 0$. Clearly $f(x)$ and $g(x)$ are both positive for sufficiently large x, and $f(x) - g(x) = (\gamma + 1)M_1^2x(x + 1)$ is positive for $x > 0$. Hence $g(x)$ is negative when x takes on the positive root of $f(x) = 0$, but eventually becomes positive. Thus the positive root of $g(x) = 0$, given by (39), is larger than that of $f(x) = 0$, given by (34).

Table VI gives δ_{max}, δ_{son} and the corresponding values of σ_1 for various values of M_1 and $\gamma = 7/5$. The corresponding curves in the τ_1,ϵ-plane are indicated in Fig. 149. The difference between the two values of σ_1 never exceeds 4°30′. Similarly $\delta_{max} - \delta_{son}$ is never greater than 30′.

As an example we consider a uniform stream with $M_1 = 2$ which is to be deflected through an angle $\delta = 10°$ by means of a straight shock line OS through a given point O, see Fig. 153. This value of δ is less than both

TABLE VI

THE DEFLECTIONS δ_{max} AND δ_{son} WITH CORRESPONDING VALUES OF σ_1 ($\gamma = 7/5$)

M_1	Maximum Deflection		$M_2 = 1$	
	σ_1	δ_{max}	σ_1	δ_{son}
1.0	90°	0°	90°	0°
1.5	66°36′	12° 6′	62°15′	11°41′
2.0	64°40′	22°58′	61°29′	22°43′
2.5	64°48′	29°48′	62°39′	29°40′
3.0	65°15′	34° 4′	63°46′	34° 1′
3.5	65°41′	36°52′	64°37′	36°50′
4.0	66° 3′	38°47′	65°15′	38°45′
∞	67°48′	45°35′	67°48′	45°35′

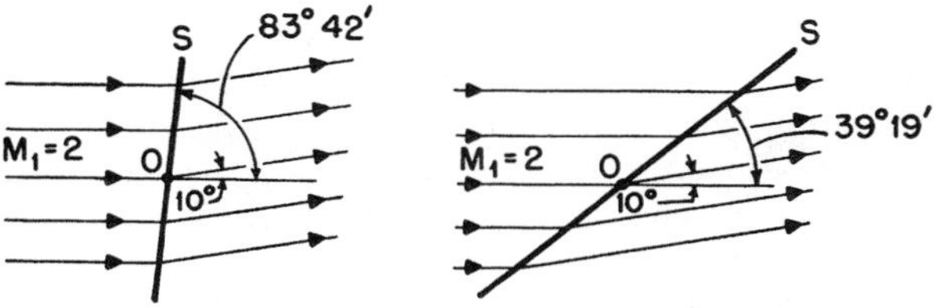

FIG. 153. Deflection of a uniform stream by strong and weak shocks.

δ_{max} and δ_{son} for the given M_1. From (32) we find that the strong shock has an inclination of $\sigma_1 = 83°42'$ to the incident stream and the weak shock an inclination of $\sigma_1 = 39°19'$. These are indicated in the figure. The corresponding values of M_{1n} are 1.9879 and 1.2671, and hence from Eqs. (21) and (22) we obtain

$$M_{2n} = 0.5794 \quad \text{and} \quad 0.8032\,,$$

$$\frac{p_2}{p_1} = 4.4438 \quad \text{and} \quad 1.7066\,,$$

$$\frac{\rho_2}{\rho_1} = 2.6487 \quad \text{and} \quad 1.4584\,,$$

The two values of σ_2 are $83°42' - 10° = 73°42'$ and $39°19' - 10° = 29°19'$, and the corresponding values of M_2 are 0.6037 and 1.6405.

Article 23

Examples Involving Shocks

1. Comparison of deflections caused by shocks and simple waves

We first derive a result concerning the similarity of transitions through shocks and simple waves, which is useful in numerical problems involving these phenomena. Consider a compression through a simple wave in which an incident stream with $M_1 = 2$ is deflected by 10°. From Sec. 18.2 we find for the state 2 after the deflection: $p_2/p_1 = 1.7052$, $\rho_2/\rho_1 = 1.4640$ and $M_2 = 1.6514$. These values are almost the same as those given at the end of the last article for a deflection of the same amount through a weak shock. This can be understood from the following considerations.

We wish to compare the transition through a weak shock front from a fixed state 1 to a variable state 2, with that through a simple wave. For either of the transitions we may consider the state 2 to be a function of the variable deflection $\delta = \theta_2 - \theta_1$. Three similar properties of these two types of transition have been found.

(a) To the second order of the density difference $\rho_2 - \rho_1$, the quantity p/ρ^γ remains constant across a shock. Through a simple wave it remains exactly constant.

(b) The same relation, (22.3c′) with fixed state 1, holds between p_2, ρ_2, and q_2 for the two phenomena.

(c) If a weak shock for which $\delta < 0$ is compared with a forward wave, and a weak shock for which $\delta > 0$ with a backward wave, then the coefficients of the first two terms in the expansion of $q_2 - q_1$ in powers of $\theta_2 - \theta_1$ are the same for the two phenomena in each case.

The first property was discussed in Sec. 14.3 and mentioned again in Sec. 22.3; the second follows from the fact that (22.3c′) is both a shock condition and an expression of Bernoulli's equation for the simple wave. The third is a restatement of the osculatory property of the shock polar and the two epicycloids through its corner given in Sec. 22.4.

It follows from (a) and (b) that the expansion property (c) holds also for $p_2 - p_1$ and $\rho_2 - \rho_1$. Furthermore if $F(p, \rho, q, \theta)$ is any function of the state variables p, ρ, q, and θ which has a Taylor series at p_1, ρ_1, q_1, θ_1, then the same expansion property holds for $F_2 - F_1$. Thus, for instance, if the transition takes place through a weak shock for which $\delta < 0$ or through a forward wave, we may write

(1) $$F_2 - F_1 = F_1'(\theta_2 - \theta_1) + \tfrac{1}{2}F_1''(\theta_2 - \theta_1)^2 + O(\theta_2 - \theta_1)^3,$$

the two phenomena differing only in the term $O(\theta_2 - \theta_1)^3$. Here F' and F'' are the first and second derivatives of F with respect to θ taken along a Γ^+-characteristic; thus

$$F' = \frac{dp}{d\theta}\frac{\partial F}{\partial p} + \frac{d\rho}{d\theta}\frac{\partial F}{\partial \rho} + \frac{dq}{d\theta}\frac{\partial F}{\partial q} + \frac{\partial F}{\partial \theta},$$

where $dp/d\theta$, $d\rho/d\theta$, and $dq/d\theta$ signify the rates of change of pressure, density and speed with polar angle along a Γ^+-characteristic in the hodograph plane. Hence

(2) $$F' = q \tan \alpha \left[-\rho q \left(\frac{\partial F}{\partial p} + \frac{1}{a^2}\frac{\partial F}{\partial \rho} \right) + \frac{\partial F}{\partial q} \right] + \frac{\partial F}{\partial \theta},$$

since from Bernoulli's equation $dp = -\rho q\, dq$ and on a Γ^+-characteristic $dq = q \tan \alpha\, d\theta$. Likewise, if the transition takes place through a weak shock for which $\delta > 0$ or through a backward wave

(3) $$F_2 - F_1 = {}'F_1(\theta_2 - \theta_1) + \tfrac{1}{2}{}''F_1(\theta_2 - \theta_1)^2 + O(\theta_2 - \theta_1)^3$$

where now the differentiation is along a Γ^--characteristic, i.e.,

$$(4) \qquad {}'F = q \tan \alpha \left[\rho q \left(\frac{\partial F}{\partial p} + \frac{1}{a^2} \frac{\partial F}{\partial \rho} \right) - \frac{\partial F}{\partial q} \right] + \frac{\partial F}{\partial \theta}.$$

The function F may, of course, contain p_1, ρ_1, q_1, θ_1 as parameters. Also, from (2) and (4) it follows that if F (as a function of p, ρ, q, θ) is even in $\theta - \theta_1$ then ${}'F_1 = -F_1'$ and ${}''F_1 = F_1''$; this result is a consequence of the symmetry of the pair of Γ-characteristics through the point $\mathbf{q}_1$, in the hodograph, since this ensures that F varies with $\theta - \theta_1$ on Γ^+ in the same way as with $\theta_1 - \theta$ on Γ^-.

For example, consider the component of velocity after the transition in the direction of the initial velocity $\mathbf{q}_1$, which in Sec. 22.3 was designated by U. Then $F = U = q \cos(\theta - \theta_1)$ is an even function of the deflection $\theta - \theta_1$. Applying the differentiation in (2) to this function we have

$$F' = q\,[\tan\alpha \cos(\theta - \theta_1) - \sin(\theta - \theta_1)],$$

$$F'' = 2q \tan \alpha \left[\left(1 - \frac{\gamma + 1}{\sin^2 2\alpha}\right) \tan\alpha \cos(\theta - \theta_1) - \sin(\theta - \theta_1) \right].$$

On putting $\theta = \theta_1$ in these equations and substituting the results into (1), we obtain

$$\frac{U_2 - q_1}{q_1} = a(\theta_2 - \theta_1) + b(\theta_2 - \theta_1)^2 + O(\theta_2 - \theta_1)^3,$$

where

$$a = \tan \alpha_1 = \frac{1}{\sqrt{M_1^2 - 1}},$$

$$b = \left(1 - \frac{\gamma + 1}{\sin^2 2\alpha}\right) \tan^2\alpha_1 = \frac{(3 - \gamma)M_1^4 - 8M_1^2 + 4}{4(M_1^2 - 1)^2}.$$

This formula applies if the transition is through a shock for which $\delta < 0$ or a forward wave. For the other case, the only difference is that a is to be replaced by $-a$.

Suppose that now we restrict our attention to forward waves and shocks for which the deflection is negative. Let the final state be denoted by 2,1 when the transition is a shock, and by 2,2 when it is a simple wave. Now (1) gives $F_{2,2}$ as a function of θ_2 along the Γ^+-characteristic through (q_1, θ_1). Hence, since this characteristic is also the Γ^+-characteristic through $(q_{2,2}, \theta_{2,2})$ we may differentiate (1) to give

$$(5) \qquad \begin{aligned} F_{2,2}' &= F_1' + F_1''(\theta_2 - \theta_1) + O(\theta_2 - \theta_1)^2, \\ F_{2,2}'' &= F_1'' + O(\theta_2 - \theta_1). \end{aligned}$$

But $F'_{2,1}$, $F''_{2,1}$ differ from $F'_{2,2}$, $F''_{2,2}$, respectively, by terms $O(\theta_2 - \theta_1)^3$ since F' and F'' are themselves functions of the state variables, see (2), and the states 2,1 and 2,2 differ by this amount. Hence Eqs. (5) will hold also for $F'_{2,1}$ and $F''_{2,1}$.

If now the flow passes through a second transition to a third state 3, we have

$$F_3 - F_2 = F'_2(\theta_3 - \theta_2) + \tfrac{1}{2}F''_2(\theta_3 - \theta_2)^2 + O(\theta_3 - \theta_2)^3, \tag{6}$$

where, if the first transition is a shock $F'_2 = F'_{2,1}$, $F''_2 = F''_{2,1}$, and if it is a simple wave $F'_2 = F'_{2,2}$, $F''_2 = F''_{2,2}$. In either case

$$\begin{aligned} F'_2 &= F'_1 + F''_1(\theta_2 - \theta_1) + O(\theta_2 - \theta_1)^2, \\ F''_2 &= F''_1 + O(\theta_2 - \theta_1). \end{aligned} \tag{7}$$

From Eqs. (1), (6), and (7) we now obtain

$$\begin{aligned} F_3 - F_1 &= (F_3 - F_2) + (F_2 - F_1) \\ &= F'_1(\theta_3 - \theta_1) + \tfrac{1}{2}F''_1(\theta_3 - \theta_1)^2 + O(\theta_3 - \theta_2, \theta_2 - \theta_1)^3. \end{aligned}$$

Clearly the same argument may be extended to any number of transitions.

If a fixed initial state 1 is connected to a variable final state 2 by a series of transitions, each of which is either a weak shock with negative deflection or a forward wave, then for any function $F(p, \rho, q, \theta)$:[31]

$$F_2 - F_1 = F'_1(\theta_2 - \theta_1) + \tfrac{1}{2}F''_1(\theta_2 - \theta_1)^2 + O(\Delta^3), \tag{8}$$

where Δ is the biggest deflection (regardless of sign) caused by the transitions, and a prime denotes the differentiation in (2). If each of the transitions is either a weak shock with positive deflection or a backward wave, then a similar formula holds with F'_1, F''_1 replaced by $'F_1$, $''F_1$ respectively, a prime now denoting the differentiation in (3). Moreover, if F is an even function of $\theta - \theta_1$, then $F'_1 = -'F_1$ and $F''_1 = ''F_1$.

2. Supersonic flow along a partially inclined wall

We consider now a limiting case of the problem discussed in Sec. 22.1.

First the point A is chosen as the origin and the point B removed to infinity. The fluid is therefore assumed to be passing horizontally across the positive y-axis at constant pressure p_0 and density ρ_0, and with uniform supersonic speed q_0. These boundary conditions, namely

$$a = a_0, \qquad q = q_0 > a_0, \qquad \theta = 0 \qquad \text{on } x = 0 \text{ for all } y > 0,$$

are realized if the coordinate system is moved in the negative x-direction with speed q_0 into the fluid at rest.

Secondly, we consider a wall which slopes upwards at an angle δ_0 for a distance l to the point D, and then runs horizontally, see Fig. 154. The wall introduces the additional boundary conditions

$$(9)\qquad \begin{aligned} &\theta = \delta_0 > 0 && \text{on} && y = x \tan \delta_0 && \text{for} && 0 < x < l \cos \delta_0 , \\ &\theta = 0 && \text{on} && y = l \sin \delta_0 && \text{for} && x > l \cos \delta_0 . \end{aligned}$$

This represents the limiting case of walls which have small curvature except in the neighborhood of two points A and D, where the curvature becomes very large; such walls were considered in Sec. 22.1.

The boundary conditions on $x = 0$ determine a unique continuous solution above the line AC which is inclined at an angle $\alpha_0 = \text{arc sin } (a_0/q_0)$ to the horizontal. This solution is represented in Fig. 154 by the set of horizontal streamlines, and the figure is drawn under the assumption that δ_0 is larger than the Mach angle α_0. In this case it is immediately evident that this continuous solution is inconsistent with the boundary conditions on AD. Thus a flow pattern which includes a shock line must be sought.

In Sec. 22.7 it was shown that provided δ_0 is less than a certain maximum (depending on M_0) there are two possible positions for a straight shock AS (Fig. 155) which deflects the incident flow abruptly into the direction $\theta = \delta_0$. We shall not consider here a shock with subsonic flow behind it. Thus our discussion will be limited to values of $\delta_0 \leqq \delta_{\text{son}}$ and furthermore to the weak shock for such a value.[32]

In the hodograph plane (Fig. 155) the point P_0 with polar coordinates, $q = q_0$, $\theta = 0$ represents the whole region of the physical plane in which q and θ remain constant. The end point P_1 of the shock transition lies on or outside the sonic circle on the upper half of the shock polar with corner at P_0, and has the polar angle $\theta_1 = \delta_0$. The straight shock front AS in the physical plane is perpendicular to P_0P_1, and after passing through it the fluid particles move parallel to the wall AD.

The deflected streamlines may be given a second deflection, equal and opposite to the first, by means of a simple wave centered at D, see Sec. 18.3. Two such waves are possible, a forward and a backward one, corresponding to the segments of Γ^+- and Γ^--characteristics, $P_1P_2^*$ and P_1P_2, joining

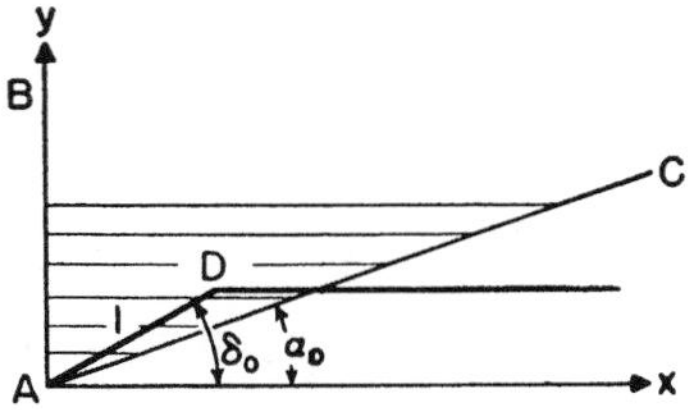

FIG. 154. Boundary conditions for which no continuous solution exists.

P_1 to the q_x-axis. It is clear, however, that if P_1 lies too near to the point of intersection, I, of the polar with the sonic circle (i.e., δ_0 is too close to δ_{son}), then the Γ^+-characteristic through P_1 will not reach the q_x-axis and the point P_2^* will not exist. On the other hand the Γ^--characteristic through P_1 will always intersect the q_x-axis. For it is easily proved that this property holds for any sonic or supersonic point on the upper half of the circular limit polar Λ, corresponding to $q_0 \to q_m$. Hence it holds for any sonic or supersonic point lying above the q_x-axis and within Λ.

We therefore select the arc P_1P_2 of the Γ^--characteristic corresponding to the backward wave. The first of Fig. 155 shows the centered rarefaction wave between the characteristic DS (perpendicular to the tangent at P_1) and the characteristic DW (perpendicular to the tangent at P_2). All particles cross DW horizontally and here the curved streamlines turn into straight lines parallel to the x-axis.

In Sec. 22.3 we found that the angle between a nonzero shock and the streamline behind it is smaller than the Mach angle downstream. Hence the shock line AS, and the characteristic DS eventually intersect, as indicated in the figure. Beyond the point S the solution outlined above is no longer valid, and the character of the flow is different. Since such a change can only take place across a characteristic, the flow pattern is bounded by the cross-characteristic ST of the simple wave, and its rectilinear extension TU.

The problem of determining the solution beyond $VSTU$ is more difficult.

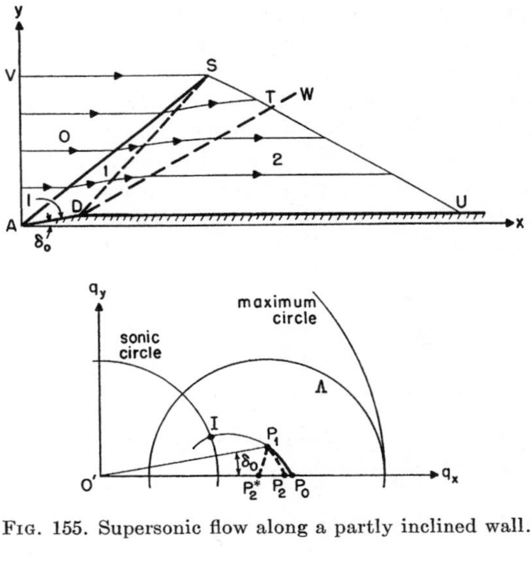

FIG. 155. Supersonic flow along a partly inclined wall.

It can be anticipated that a curved extension of the shock line AS will appear, but neither its shape nor the flow pattern behind it can be expressed in terms of known functions. In such cases one has to resort to numerical methods of integration based, for instance, on the use of characteristics of the equations of ideal (nonpotential) fluid flow and the shock conditions. The unknown shock line is determined step by step along with the flow on either side. The extensive computations involved can be carried out on high-speed computing machines.[33] (The basic theory of such flows will be given in Secs. 24.1 and 24.2; see also the discussion of Riemann's problem in Secs. 15.4 and 15.5.)

The full solution in the region $VSTUDAV$ is shown in Fig. 155. The region is divided into four subregions with uniform horizontal flows in the triangles AVS and DTU, uniform inclined flow in the triangle ASD, and a centered simple wave in SDT.

Analytically the flow regions are determined as follows. The inclination σ_0 of the shock AS to the x-axis is found by replacing the subscript 1 by 0 in (22.32) and determining the larger positive root $\tau_0 = \cot \sigma_0$ of the cubic equation in τ_0 which results on setting ϵ equal to $\tan \delta_0$. Once σ_0 is found, the state 1 within the triangle ASD is determined from Eqs. (22.21) and (22.22), with the subscripts 1, 2 replaced by 0, 1. Thus with $M_{0n} = M_0 \sin \sigma_0$ we compute M_{1n}, p_1/p_0, and ρ_1/ρ_0. The Mach number M_1 behind the shock is then given by $M_1 = M_{1n} \operatorname{cosec} \sigma_1 = M_{1n} \operatorname{cosec} (\sigma_0 - \delta_0)$.

The simple wave extends from the line DS, making an angle arc sin $(1/M_1)$ with the direction AD, to the line DW, making an angle arc sin $(1/M_2)$ with DU, where M_2 is the Mach number for which

$$Q_2 = Q_1 + \delta_0 .$$

The equation of the streamlines within the wave [see (18.9)] is

$$r\left(\cos \frac{\phi - 2\eta}{h}\right)^{h^2} = \text{constant}; \qquad 2\eta = Q_1 + \delta_0 ,$$

where r, ϕ are polar coordinates with origin at D. The state 2 in the triangle DTU is known once M_2 has been determined.

The distance of S from D is given by $DS/\sin \sigma_1 = l/\sin (\alpha_1 - \sigma_1)$; the equation of the cross-characteristic ST of the simple wave [see (18.11)] is therefore

$$r^2 \sin \frac{\phi - 2\eta}{h} \left(\cos \frac{\phi - 2\eta}{h}\right)^{h^2} = (DS)^2 \sin \frac{\alpha_1 + \delta_0 - 2\eta}{h} \left(\cos \frac{\alpha_1 + \delta_0 - 2\eta}{h}\right)^{h^2}, \tag{10}$$

since ϕ is measured from the direction DU. The distance DT is determined from (10) by setting $\phi = \alpha_2$ and solving for r. Finally the line TU has slope $-\text{arc sin } (1/M_2)$.

As an example we consider the case $M_0 = 2$, $\delta_0 = 10°$. In Sec. 22.7 we found that for this Mach number the weak shock was inclined at an angle of 39°19′. The corresponding value of M_{0n} was 1.2671 and behind the shock we had

$$M_{1n} = 0.8032, \qquad p_1 = 1.7066p_0\,, \qquad \rho_1 = 1.4584\rho_0\,.$$

The corresponding value of M_1 was 1.6405.

For this value of M_1 we find $Q_1 = 106.0581°$, so that $Q_2 = 116.0581°$, to which corresponds $M_2 = 1.9884$ (see Table V in Sec. 18.2). The inclination of DS to AD is therefore 37°34′, and that of DW to DU is 30°12′. For these values of M_1 and M_2 we have from Eqs. (8.15) and (8.16): $p_1/p_s = 0.2215$, $p_2/p_s = 0.1301$, $\rho_1/\rho_s = 0.3407$, and $\rho_2/\rho_s = 0.2330$ (see Table I in Art. 8).* Hence

$$p_2 = 0.5875p_1 = 1.0026p_0\,,$$

$$\rho_2 = 0.6839\rho_1 = 0.9975\rho_0\,.$$

This example was used in constructing Fig. 155.

It should be noticed that M_2, p_2, ρ_2 differ from M_0, p_0, ρ_0 by well under 1 per cent. This is in agreement with the result obtained in the last section concerning successive transitions through shocks with positive deflection and backward waves. In the notation of that section $\theta_1 = \theta_2 = 0$, $\Delta = 10°$.

The above arguments hold whether δ_0 is greater or smaller than α_0 (provided $\delta_0 \leqq \delta_{son}$ for the given M_0). It remains to be checked that even when the point D lies below the Mach line AC in Fig. 154 (as it does in the example given) no continuous solution exists which satisfies the boundary conditions (9). The argument is completely analogous to that given at the end of Sec. 14.6 for the corresponding one-dimensional case.

3. Supersonic flow past a straight line profile: contact discontinuity[34]

In this example the fluid is supposed to be moving horizontally across the whole y-axis at constant pressure p_0 and density ρ_0, and with supersonic speed q_0:

$$a = a_0\,, \qquad q = q_0 > a_0\,, \qquad \theta = 0 \qquad \text{on } x = 0 \text{ for all } y.$$

The presence of the straight line profile AB in Fig. 156, of length l and inclination δ_0, introduces the additional boundary condition

$$\theta = \delta_0 > 0 \qquad \text{on} \qquad y = x \tan \delta_0\,, \qquad \text{for} \qquad 0 < x < l \cos \delta_0\,.$$

* This accuracy, which is not given by Tables I and V, is required for Sec. 3, where the present example is used.

This problem has features in common with the last one, and in particular it may be seen that there is no continuous solution on the upper side of the profile. As before we assume that $\delta_0 \leqq \delta_{son}$ for the given M_0, and use a straight weak shock AS_1 to deflect the stream abruptly into the direction AB. Conditions before and after the shock are represented by hodograph points P_0 and P_1, Fig. 156, with P_0P_1 an arc of the shock polar with corner at P_0. The deflection on the lower side of the profile is effected by means of a simple wave centered at A, and in order not to violate the boundary conditions on the y-axis, this must be a forward wave. The image of this wave is the arc P_0P_2 of the Γ^+-epicycloid through P_0, and the lines AW_2, AS_2 are perpendicular to the tangents of this epicycloid at P_0 and P_2, respectively. It is of course assumed that δ_0 is sufficiently small for the epicycloid to meet $O'P_1$ (see Sec. 18.3). The uniform flow on the lower side of the profile is then represented by the point P_2.

The flow above the profile may be joined to the flow below by as-

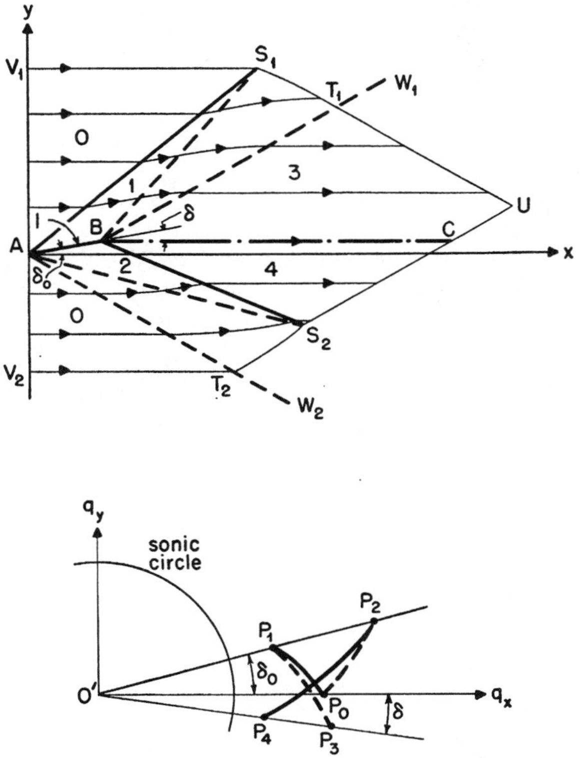

FIG. 156. Supersonic flow about an inclined straight line profile.

suming that a straight shock forms along an appropriate line BS_2 and a centered backward wave S_1BW_1 beyond BS_1, the C^+-characteristic through B.

The uniform state of particles after passing through this simple wave is represented by a point P_3 lying on the Γ^--characteristic through P_1, and that after passing through the shock by a point P_4 lying on the shock polar with corner at P_2. The two points P_3, P_4 must satisfy two conditions: they must lie on the same ray through the origin O', in order that the direction of motion of all particles be the same behond W_1BS_2, and secondly their pressure values p must be the same. We now restrict the discussion to cases where P_4 lies on or outside the sonic circle, so that the flow behind the shock BS_2 is sonic or supersonic. Under suitable conditions it is possible to satisfy the two requirements above by choosing the inclination of the shock line BS_2 and the extent of the wave S_1BW_1 appropriately.

The values of the density, however, will not necessarily be the same for particles coming from above and below the profile. Thus the dividing streamline BC through B will be a discontinuity line analogous to the one described in Sec. 15.2 for the one-dimensional case. As there, it will be called a *contact discontinuity*. It is a characteristic of the ideal fluid equations (see Sec. 9.6).

As we noted in Sec. 2, the shock line AS_1 meets the characteristic through B above the profile. Similarly the characteristic AS_2 intersects the shock line through B below the profile. Thus, as in the preceding example, the flow regions are limited, namely by the cross-characteristics S_1T_1, S_2T_2 of the simple waves, and their extensions T_1U and S_2CU. The problem of determining the flow pattern outside the region $V_1S_1T_1UCS_2T_2V_2$ is more difficult and as before numerical methods of integration must be employed.[35]

The flow patterns up to the states 1 and 2 may be computed as in the last section. It is then easy, in principle, to determine the rest of the flow pattern. Let $-\delta$ (to be found) be the inclination of the straight dividing streamline BC to the profile AB. Then, according to (18.3) which governs the transition through the backward wave, and (8.16):

$$\delta = Q_3 - Q_1,\qquad p_3 = p_1\left[\frac{(\gamma-1)M_1^2+2}{(\gamma-1)M_3^2+2}\right]^{\gamma/(\gamma-1)}. \tag{11}$$

Moreover by Eqs. (22.32) and (22.21), governing the shock transition across BS_2,

$$\delta = -\text{arc tan}\,\frac{(c\tau_2^2+d)\tau_2}{(1-c)\tau_2^2+(1-d)},\qquad p_4 = p_2\left[\frac{2\gamma M_2^2}{(\gamma+1)(1+\tau_2^2)}-\frac{1}{h^2}\right] \tag{12}$$

where r_2 (negative) is the cotangent of the angle ABS_2. Here c, d must be computed from (22.31) with M_1 replaced by M_2. The remaining problem is to determine those values of M_3 and τ_2 for which Eqs. (11) and (12) yield the same value of δ and give $p_3 = p_4$.

If M_3 and τ_2 are allowed to vary in Eqs. (11) and (12) respectively, then the corresponding points (δ, p_3) and (δ, p_4) trace out two curves in the δ,p-plane, see Fig. 157, whose intersection gives the common values of deflection and pressure.[36] For each δ_0 these curves have the general shape shown in the figure, as may be seen by considering (11) as M_3 increases from M_1 to ∞, and (12) as τ_2 increases from $-\cot \alpha_2$ to 0. It is easily shown that for δ_0 sufficiently small (depending on M_0) the point A_1 will lie between A_2 and A_2^*, so that the curves have just one point of intersection, and the latter will be close enough to A_2 for it to correspond to sonic or supersonic state 4. For in the limit $\delta_0 \to 0$ the distance $A_2A_2^* = 2\gamma p_2(M_2^2 - 1)/(\gamma + 1) \to 2\gamma p_0(M_0^2 - 1)/(\gamma + 1) > 0$ and $A_1 \to A_2$ which is then the point of intersection of the curves, with $M_4 = M_2 > 1$. In addition A_1 lies above A_2 for $\delta_0 > 0$ since $p_1 > p_0 > p_2$. The result now follows from the continuity of all functions involved with respect to δ_0.

Once this point of intersection is known, M_3, which gives the inclination $\alpha_3 = \arcsin (1/M_3)$ of BW_1 to BC, and τ_2, which determines the position of BS_2, may easily be found from Eqs. (11) and (12). The streamlines in the backward wave region S_1BT_1 are then given by

$$r\left[\cos \frac{\phi - 2\eta}{h}\right]^{h^2} = \text{constant}; \qquad 2\eta = Q_1 + \delta_0,$$

see (18.9), where r, ϕ are polar coordinates with pole at B.

In practice one would obtain this intersection by taking various values for δ in Eqs. (11) and (12), computing the corresponding values of p_3 and p_4, and then interpolating for equality. A first approximation is obtained by noticing that on the lower side of the profile the transitions are a forward wave and a shock with negative deflection. Hence from (8), with $F = p$, we have

$$(13) \qquad p_4 - p_0 = a(\delta_0 - \delta) + b(\delta_0 - \delta)^2 + O(\delta_0, \delta)^3,$$

where a, b are certain constants depending on the state 0 but not on δ_0 or δ. Similarly

$$(14) \qquad p_3 - p_0 = -a(\delta_0 - \delta) + b(\delta_0 - \delta)^2 + O(\delta_0, \delta)^3$$

for the transitions on the upper side. Thus

$$p_4 - p_3 = 2a(\delta_0 - \delta) + O(\delta_0, \delta)^3,$$

so that the first approximation $\delta = \delta_0$ (BC horizontal) will make p_3 and p_4 differ by at most $O(\delta_0^3)$.[37]

As an example we consider the case $M_0 = 2$, $\delta_0 = 10°$. The flow over the upper side of the profile for the first approximation $\delta = \delta_0 = 10°$ was discussed in the preceding section. In particular (the suffix 2 of that section being replaced by 3) we found

$$p_3 = 1.0026p_0\,, \qquad M_3 = 1.9884.$$

Similarly for the flow on the lower side of the profile we find, with $\delta = 10°$,

$$p_4 = 0.9995p_0\,, \qquad M_4 = 1.9862.$$

Thus the approximation $\delta = 10°$ gives a higher value of p_3 than of p_4. The true value of δ is therefore slightly larger, see Fig. 157. For $\delta = 10°6'$ the corresponding values are

$$p_3 = 0.9970p_0\,, \qquad M_3 = 1.9920,$$

$$p_4 = 1.0052p_0\,, \qquad M_4 = 1.9822,$$

where now p_4 exceeds p_3. By linear interpolation we find the second approximation $\delta = 10°2'$. Any desired accuracy can be obtained by repeating this process with more accurate pressure values. Thus in this example even though the stream is initially deflected *upwards*, the dividing streamline BC is inclined *downwards*.

The corresponding values of the Mach numbers are $M_3 = 1.9894$ and $M_4 = 1.9850$, and the common value of the pressures $p_3 = p_4 = 1.0011p_0$. The reader should compute for himself the values of p_2, ρ_2, M_2, ρ_3, ρ_4 and determine the positions of the characteristic BT_1 and the shock line BS_2. It will be found that the density ratio ρ_4/ρ_3 is very close to unity so that the discontinuity across BC is in fact insignificant, even though we started

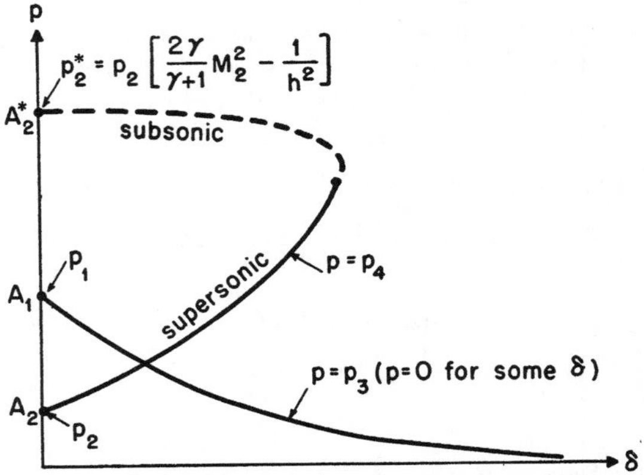

FIG. 157. Pressure-deflection figures for flows behind profile.

with a sizable deflection $\delta_0 = 10°$. The numerical values of the present example were used in constructing Fig. 156.

The pressure force on the profile is

$$(p_1 - p_2)l = (1.7066 - 0.5480)p_0 l = 1.1586 p_0 l,$$

and it is directed downwards perpendicular to AB. Notice that no knowledge of the states 3 and 4 is required to determine this force. In general we may write in analogy to Eqs. (13) and (14)

$$p_2 - p_0 = a\delta_0 + b\delta_0^2 + O(\delta_0^3),$$

$$p_1 - p_0 = -a\delta_0 + b\delta_0^2 + O(\delta_0^3),$$

so that $p_1 - p_2 = -2a\delta_0 + O(\delta_0^3)$. According to (2) the constant a is equal to $-\rho_0 q_0^2 \tan \alpha_0 = -\gamma M_0^2 p_0/\sqrt{M_0^2 - 1}$. Hence the pressure force on the profile is

$$\frac{2\gamma M_0^2 \delta_0}{\sqrt{M_0^2 - 1}} p_0 l + O(\delta_0^3).$$

For the above example the first term in this expression gives the value $1.1285 p_0 l$, which is about 3 per cent too small.

4. Behavior of a shock at a wall (oblique shock reflection)[38]

In this section we shall discuss another example on the basis of the deflection theory developed in the preceding article. The present example complements that of the head-on reflection of a shock at a wall which was treated in Sec. 15.1.

Assume (see Fig. 158) that a fixed plane wall is set at an oblique angle to the direction of motion of a one-dimensional shock front whose strength is constant and in front of which the gas is at rest. After the shock has passed, the gas moves towards the wall, whereas at the wall it must move parallel to it. The problem is to find a flow pattern behind the shock which satisfies both these boundary conditions.

Since the velocity of the shock is constant the line of intersection of shock and wall moves parallel to itself along the wall with constant velocity. In a system of coordinates moving with this velocity, the shock is effectively at rest and the wall is moving tangential to itself. If the z-axis is now taken along the line of intersection, then since the boundary conditions are independent of both z and t, we may treat the problem as one of steady flow in the x,y-plane.

In this plane (see Fig. 159) the straight line OS represents the shock front incident to the wall $WO\overline{W}$ at an angle ω. The gas in front of the shock is now moving along the direction of the wall, and the streamlines are vertical straight lines which continue below OS in a direction inclined

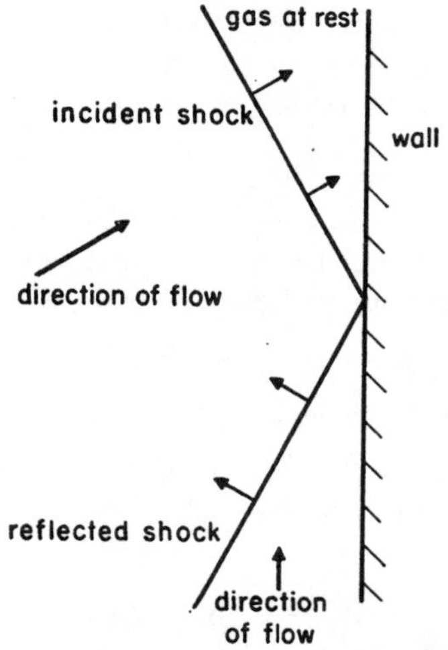

FIG. 158. Oblique reflection of a shock at a fixed wall.

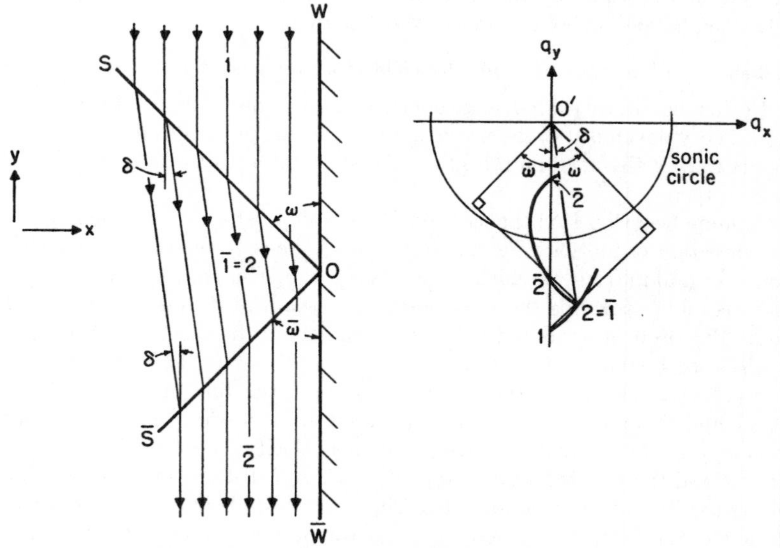

FIG. 159. Shock reflection viewed from moving coordinate system.

towards the wall. At the wall the velocity must be vertical, and hence the streamline through O, the point at which the shock front meets the wall, must remain vertical. Under suitable conditions a solution satisfying this requirement is obtained by assuming another shock line $O\bar{S}$, making an angle $\bar{\omega}$ with the downward vertical, across which the streamlines again change direction so as to become vertical once more. Then according to Sec. 22.7 there are two possible positions for this second shock and hence two values of $\bar{\omega}$ for each suitable ω. In contrast to the example in the last section we shall in the present case be able to specify precisely the conditions under which such a solution is valid.

An observer at rest with respect to the (now moving) wall first sees the shock OS moving obliquely towards the wall, and then the shock $O\bar{S}$ moving obliquely away from the wall. This phenomenon is known as *oblique shock reflection*. The term "reflection" suggests the symmetry condition $\bar{\omega} = \omega$. We shall see however that this condition can be satisfied only for one particular value of ω. For very weak shocks it can be satisfied approximately for all values of ω.

In the hodograph plane (Fig. 159), the fluid state above the first shock is represented by a point 1 on the q_y-axis. The point 2, corresponding to the state after the first shock, lies at the intersection of the shock polar with its corner at 1 and the line through 1 making an angle $90° - \omega$ with the q_y-axis. The point 2 will also be denoted by $\bar{1}$ to indicate that it represents flow in front of the second shock. The point $\bar{2}$ (representing the state after the second shock) must lie on the shock polar with corner at $\bar{1}$ and in addition on the q_y-axis, since we require the streamlines to be vertical again after the second shock. There are in general two possible positions for the point $\bar{2}$, one corresponding to a weak reflected shock and the other to a strong reflected shock. The figure shows the $\bar{\omega}$ corresponding to the weak shock.

To fix the state 1, we note that the shock OS is characterized by its strength, which may be represented by $\eta = p_2/p_1$, and its angle of incidence ω to the wall. Then Eq. (22.21), with $M_{1n} = M_1 \sin \omega$, determines M_1 and hence the point 1, so that the construction of the last paragraph can be effected. Our task is now to compute $\bar{\omega}$ and the pressure ratio $\bar{\eta}$ across the second shock as functions of ω and η. To this end we initially replace η by δ, the deflection of the streamlines caused by either shock in the x,y-plane. Equation (22.38) gives the relationship between δ, η and ω.

We now derive two equations, one connecting M_2, δ, ω and the other $M_{\bar{1}}$, δ, $\bar{\omega}$. By elimination of $M_2 = M_{\bar{1}}$ we then obtain a relation between $\bar{\omega}$, ω and δ. The equation in M_2, δ, and ω is obtained from Eq. (22.37) by writing $\tau_1 = \cot \omega$ and

$$\tau_2 = \cot(\omega - \delta) = \frac{\tau_1 + \epsilon}{1 - \epsilon\tau_1}, \qquad \epsilon = \tan \delta.$$

On solving for M_2, we have

$$(15) \qquad \frac{M_2^2}{(h^2 - 1)(\epsilon^2 + 1)} = \frac{\tau_1(\tau_1^2 + 1)}{(1 - \epsilon\tau_1)[\epsilon\tau_1^2 + (h^2 - 1)\tau_1 + h^2\epsilon]}.$$

The second equation is obtained from this one by changing the subscripts 1 and 2 to $\bar{2}$ and $\bar{1}$ respectively, and at the same time ϵ to $-\epsilon$:

$$\text{(16)} \qquad \frac{M_{\bar{1}}^{2}}{(h^2 - 1)(\epsilon^2 + 1)} = -\frac{\tau_{\bar{2}}(\tau_{\bar{2}}^{2} + 1)}{(1 + \epsilon\tau_{\bar{2}})[\epsilon\tau_{\bar{2}}^{2} - (h^2 - 1)\tau_{\bar{2}} + h^2\epsilon]},$$

where clearly $\tau_{\bar{2}} = \cot \bar{\omega}$.

The right-hand sides of Eqs. (15) and (16) become identical on setting $\tau_{\bar{2}} = -\tau_1$. This possibility must be excluded since it corresponds to an equal and opposite transition immediately following the first, and such a transition would violate the inequality condition (22.16) for shocks. On equating these two right-hand members we obtain a cubic equation for $\tau_{\bar{2}}$ as a function of τ_1 and ϵ. One root of this equation is $\tau_{\bar{2}} = -\tau_1$ and is to be neglected; the two remaining roots satisfy

$$\text{(17)} \qquad A\tau_{\bar{2}}^{2} + B\tau_{\bar{2}} + C = 0, \qquad \tau_{\bar{2}} = \cot \bar{\omega},$$

where

$$A = (h^2 - 2)\epsilon\tau_1^{2} + (h^2 - 1)(\epsilon^2 - 1)\tau_1 - h^2\epsilon,$$
$$B = -(h^2 - 1)\tau_1[(\epsilon^2 - 1)\tau_1 - 2\epsilon],$$
$$C = -h^2\epsilon(\tau_1^{2} + 1).$$

For each given pair of values ω, δ the two corresponding values of $\bar{\omega}$ are determined by this last equation.

In order to express the coefficients A, B, C in (17) in terms of the original parameters ω and η, we eliminate ϵ from them by means of Eq. (22.38):

$$\text{(18)} \qquad \epsilon = \frac{\tau_1}{s(\tau_1^{2} + 1) - 1}, \qquad s = \frac{h^2\eta + 1}{(h^2 - 1)(\eta - 1)}.$$

If a common factor $-\tau_1(\tau_1^{2} + 1)/[s(\tau_1^{2} + 1) - 1]^2$ of the three coefficients is ignored, they may then be written:

$$\text{(17')} \qquad \begin{aligned} A &= s[(h^2 - 1)s - (h^2 - 2)](\tau_1^{2} + 1) - 1, \\ B &= -(h^2 - 1)\tau_1[s^2(\tau_1^{2} + 1) - 1], \\ C &= h^2[s(\tau_1^{2} + 1) - 1]. \end{aligned}$$

At one extreme we have $\eta = 1$ or $s = \infty$, for which (17) reduces to

$$\text{(19)} \qquad \tau_{\bar{2}}^{2} - \tau_1\tau_{\bar{2}} = 0, \qquad \tau_{\bar{2}} = 0 \quad \text{or} \quad \tau_1.$$

At the other we have $\eta = \infty$ or $s = h^2/(h^2 - 1)$ for which it reduces to

$$\text{(20)} \qquad A_\infty\tau_{\bar{2}}^{2} + B_\infty\tau_{\bar{2}} + C_\infty = 0,$$

where

$$A_\infty = 2h^2\tau_1^2 + (h^2 + 1),$$
$$B_\infty = -\tau_1[h^4\tau_1^2 + (2h^2 - 1)],$$
$$C_\infty = h^2[h^2\tau_1^2 + 1].$$

The graphs of $\bar{\omega}$ versus ω corresponding to Eqs. (19) and (20) for $\gamma = 7/5$ are indicated by broken lines in Fig. 160. They intersect at the points: $\omega = \bar{\omega} = 0°$; $\omega = 0°$, $\bar{\omega} = 90°$; $\omega = \bar{\omega} = \omega_0$ where

$$\omega_0 = \text{arc cot}\sqrt{(\gamma + 1)/(3 - \gamma)} = 39°14'.$$

For each value of s between the extreme values, the corresponding graph of $\bar{\omega}$ versus ω passes through these three points and lies in the region enclosed by the broken lines. Two such curves, for $\eta = 1.25$ and $\eta = 5$, are drawn in the figure.

For each value of s there is a maximum value of ω (minimum of τ_1) beyond which (17) has imaginary roots, and a solution of the type considered does not exist. The condition for this maximum is the vanishing of the discriminant $B^2 - 4AC$, and with $z = (\tau_1^2 + 1)$ it becomes

$$Dz^3 - Ez^2 + Fz - G = 0,$$

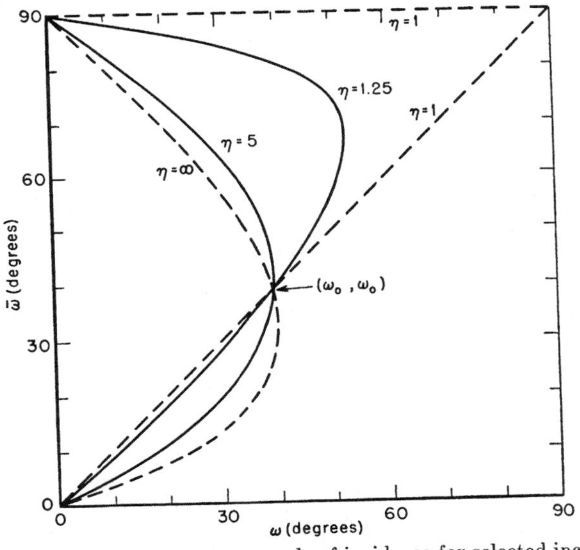

FIG. 160. Angle of reflection versus angle of incidence for selected incident shock strengths.

where

$$D = (h^2 - 1)^2 s^4,$$
$$E = (h^2 - 1)^2 s^4 + 4h^2(h^2 - 1)s^3 - 2(h^4 - 2h^2 - 1)s^2,$$
$$F = 2(h^2 - 1)(3h^2 - 1)s^2 - 4h^2(h^2 - 3)s + (h^2 - 1)^2,$$
$$G = (h^2 + 1)^2.$$

These maximum values of ω are plotted versus p_1/p_2 $(= 1/\eta)$ in Fig. 161 for $\gamma = \frac{7}{5}$. To any point which lies on or below this curve there corresponds a solution of the kind indicated in this section.

5. Properties of the reflection

To find the pressure ratio $\bar{\eta} = p_{\bar{2}}/p_{\bar{1}}$ across the reflected shock we eliminate ϵ between (18) and its analogue for the second shock, namely

$$\epsilon = \frac{\tau_{\bar{2}}}{\bar{s}(\tau_{\bar{2}}^2 + 1) + 1}, \qquad \bar{s} = \frac{h^2 + \bar{\eta}}{(h^2 - 1)(\bar{\eta} - 1)}.$$

This last equation is obtained from (18) by changing the subscripts 1 and 2 to $\bar{2}$ and $\bar{1}$ respectively, and at the same time ϵ to $-\epsilon$. We then obtain

$$\bar{s} = \frac{\tau_{\bar{2}}(\tau_1^2 + 1)s - (\tau_1 + \tau_{\bar{2}})}{\tau_1(\tau_{\bar{2}}^2 + 1)}, \tag{21}$$

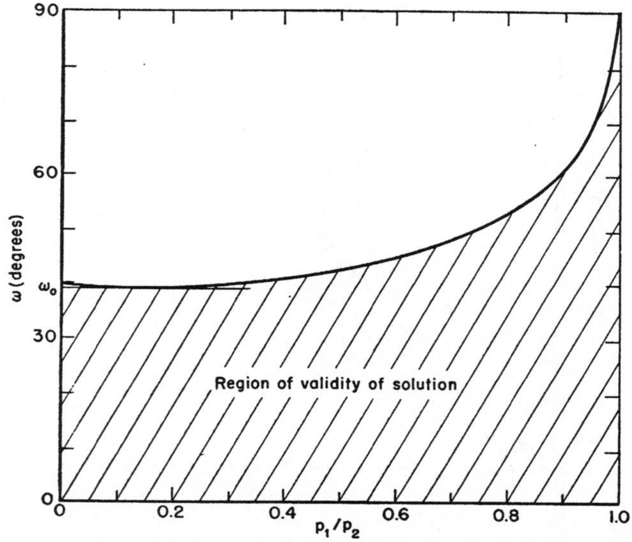

FIG. 161. Incident angle and pressure ratios for which reflection is possible.

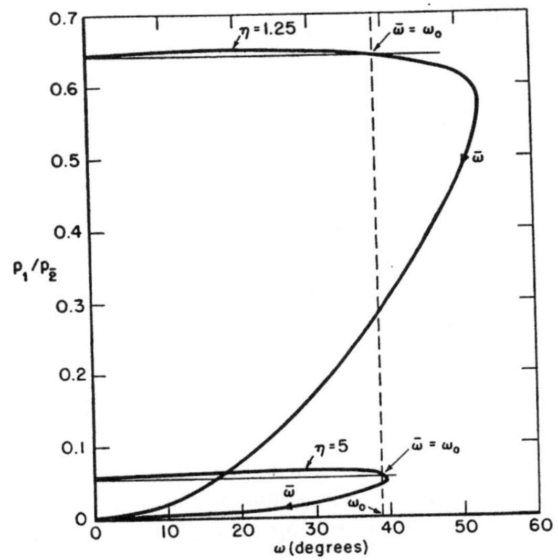

FIG. 162. Total pressure ratio versus incidence angle for $\eta = 1.25, 5$.

which expresses $\bar{s}$ in terms of τ_1, s and the appropriate root $\tau_{\bar{2}}$ of (17). In Fig. 162 the graphs of $p_1/p_{\bar{2}} = 1/\eta\bar{\eta}$ versus ω are plotted for the two values $\eta = 1.25$ and $\eta = 5$, with the direction of increasing $\bar{\omega}$ indicated by arrowheads. They show that as $\bar{\omega}$ increases, the total compression $p_{\bar{2}}/p_1$ across the two shocks initially decreases slightly and then increases indefinitely.

For fixed ω, the Taylor developments of the larger root of (17) through terms of the first order in $(\eta - 1)$, see (17′), and of $\bar{\eta}$, see (21), through terms of the second, give

$$\tau_{\bar{2}} \sim \tau_1 + K(\eta - 1), \qquad K = \frac{(h^2 - 1)\tau_1^2 - h^2}{\tau_1(h^2 + 1)},$$

$$\bar{\eta} \sim \eta + L(\eta - 1)^2, \qquad L = -\frac{(\tau_1^2 - 1)(\tau_1^2 + h^2)}{(h^2 + 1)\tau_1^2(\tau_1^2 + 1)}.$$

These formulas apply to the weak reflected shock. According to Eq. (22.23) the corresponding development for the density ratio is

$$\bar{\xi} \sim \xi + \frac{L}{\gamma}(\eta - 1)^2.$$

To an observer at rest with respect to the wall, the incident and reflected shocks will have velocities of propagation c and $\bar{c}$ which satisfy

$$\bar{c} \operatorname{cosec} \bar{\omega} = c \operatorname{cosec} \omega, \tag{22}$$

this being the condition that the point of intersection of the two shocks should always lie at the wall. Thus $\bar{c}/c = (\tau_1^2 + 1)^{\frac{1}{2}}/(\tau_2^2 + 1)^{\frac{1}{2}}$ has a Taylor development

$$\frac{\bar{c}}{c} \sim 1 - \frac{\tau_1 K}{(\tau_1^2 + 1)} (\eta - 1).$$

The equivalent formulas in the case of a strong reflected shock are left for the reader to derive.

The principal results concerning oblique shock reflection are as follows: *The weak as well as the strong reflected shock can have an appreciably different inclination to the wall from that of the incident shock, and considerable increases of pressure and density may be caused by the reflection.* The former implies also that *the velocity of propagation of the reflected shock may be appreciably different from that of the incident shock*, see Eq. (22). *For each η there is just one value of ω for which the reflected shock can have the same inclination to the wall as the incident shock, and this value is independent of η.*

The results in Sec. 15.1 concerning the head-on reflection of a shock can be recaptured by considering a limiting case. Thus as $\omega \to 0$, i.e. $\tau_1 \to \infty$, we see from (17′) that $A = O(\tau_1^2)$, $B = O(\tau_1^3)$, $C = O(\tau_1^2)$, so that the larger root of (17) tends to infinity with τ_1 such that

$$\frac{\tau_2}{\tau_1} \to \lim_{\tau_1 \to \infty} \left(-\frac{B}{A} \right) = \frac{(h^2 - 1)s}{(h^2 - 1)s - (h^2 - 2)}. \tag{23}$$

In addition $\epsilon \to 0$, according to (18), in such a way that $\epsilon\tau_1$ tends to $1/s$. Hence from (15), or (16), we see that

$$M_2^2\epsilon^2 \to \frac{(h^2 - 1)}{(s - 1)[(h^2 - 1)s + 1]}. \tag{24}$$

To an observer who is at rest with respect to the wall, the flow behind the incident shock will be perpendicular to the shock, and its Mach number, $\bar{M}_2$ say, different from M_2. However, since he moves in the y-direction, components of velocity perpendicular to the wall are unaltered. Hence $\bar{M}_2 \cos \omega = M_2 \sin \delta$, so that $\bar{M}_2$ and $M_2\epsilon$ tend to the same limit. If we denote this limit by $\bar{M}_2$ itself, we find from (24)

$$s = \frac{(h^2 - 2)}{2(h^2 - 1)} + S, \qquad S = \sqrt{\left(\frac{\gamma + 1}{4}\right)^2 + \frac{1}{\bar{M}_2^2}}. \tag{25}$$

From (22), (23), and (25) it follows that

$$\frac{\bar{c}}{c} \to \frac{2(h^2 - 1)S + (h^2 - 2)}{2(h^2 - 1)S - (h^2 - 2)}.$$

This result agrees with Eq. (15.3′) when $\bar{c}$ is replaced by $-c'$ in the limit, since clearly $\bar{M}_2$ is the M_2 of Sec. 15.1. We leave it for the reader to verify that the limiting pressure and density ratios are given by Eqs. (15.4′) and (15.5′).

One interesting result can be deduced without knowing these limit ratios explicitly. From (21) we find that in the limit

$$\bar{s} = s - \frac{h^2 - 2}{h^2 - 1} = s - \frac{3 - \gamma}{2}.$$

This is also the form taken by Eq. (21) when $\tau_1 = \tau_{\bar{2}} = h/\sqrt{h^2 - 2}$. Since s and $\bar{s}$ are simple functions of η and $\bar{\eta}$ respectively, this means that for each η *the pressure ratio across the reflected shock is the same for the oblique reflection* $\omega = \bar{\omega} = \omega_0$ *as it is for head-on reflection*. For $\bar{\omega} < \omega_0$ (see Fig. 162), $p_{\bar{2}}/p_1$ is less than that for head-on reflection, while for $\bar{\omega} > \omega_0$ it is greater.

6. Intersection of two shocks

A simple example of the interaction of shocks occurs when two shock lines intersect. Assume that fluid crossing the segment AB of the y-axis (Fig. 163) is in a state of uniform supersonic motion: $q = q_0$, $\theta = 0$, $p = p_0$, $\rho = \rho_0$, $a = a_0 < q_0$. Consider two shocks located on oppositely inclined lines AC and BC, the first causing the state $q_0,0,p_0,\rho_0$ to change to q_1,θ_1,p_1,ρ_1 where $\theta_1 > 0$, and the second causing it to change to q_2,θ_2,p_2,ρ_2 where $\theta_2 < 0$.[39] At the line $x = x_0$ passing through C, the particles below C have speed q_1, direction θ_1, pressure p_1 and density ρ_1, while those above C have q_2, θ_2, p_2, ρ_2. In the hodograph plane, Fig. 163, these two states are represented by two points 1, 2 lying on the shock polar with corner 0 at $(q_0, 0)$, with chords 01 and 02 perpendicular to AC and BC respectively.

Under suitable conditions, we can find a flow pattern for $x > x_0$ which satisfies all requirements by assuming that two new shock fronts (reflected

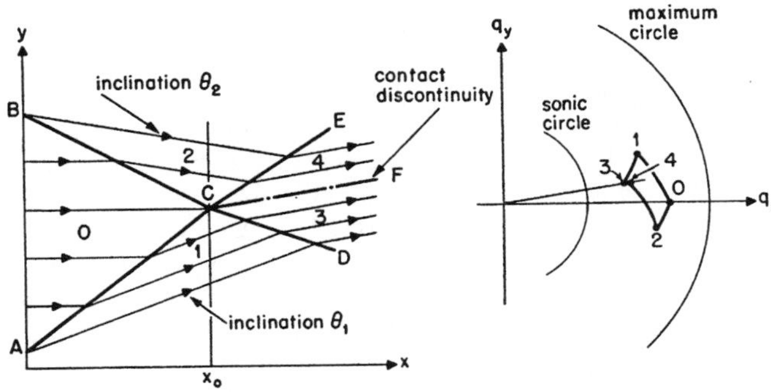

FIG. 163. Intersection of two shocks crossing a uniform stream.

shocks) form along appropriate lines CD and CE through C. The two states of the particles after passing through these second shock fronts will be represented by two points 3 and 4; the point 3 lies on the shock polar with corner at 1 (the line 13 being perpendicular to CD), and 4 on the shock polar with corner at 2 (the line 24 being perpendicular to CE). The two points 3 and 4 must satisfy the same two conditions as in Sec. 3, and again the dividing streamline CF through C will, in general, be a contact discontinuity.

It is easy, in principle, to find the values of q, θ, p, ρ after the second shocks, when the states 1 and 2 are given. We simply apply Eqs. (22.21) and (22.32) to the transitions across CD and CE. Using p and θ to denote the final values of the pressure and stream inclination, we obtain

$$\begin{aligned} \theta &= \theta_1 + \arctan \frac{(c\tau_1^2 + d)\tau_1}{(1 - c)\tau_1^2 + (1 - d)}, \\ p &= p_1 \left[\frac{2\gamma M_1^2}{(\gamma + 1)(\tau_1^2 + 1)} - \frac{1}{h^2} \right] \end{aligned} \tag{26}$$

for the transition across CD, and

$$\begin{aligned} \theta &= \theta_2 + \arctan \frac{(c'\tau_2^2 + d')\tau_2}{(1 - c')\tau_2^2 + (1 - d')}, \\ p &= p_2 \left[\frac{2\gamma M_2^2}{(\gamma + 1)(\tau_2^2 + 1)} - \frac{1}{h^2} \right] \end{aligned} \tag{27}$$

for that across CE. In these equations τ_1 and τ_2 are the cotangents of the inclinations of CD and CE to the streamlines in ACD and BCE respectively, the first being negative and second positive; the primes in Eq. (27) indicate that M_1, in the definitions (22.31) of c and d must be replaced by M_2.

By eliminating τ_1 between the two equations (26) and τ_2 between the two equations (27), we obtain two relations between θ and p, one for the transition across CD and the other for that across CE. The angle θ is easily eliminated between these last two equations to give a single equation for p. Once p has been determined, the second equation in (16) yields τ_1, and the second in (17) τ_2; the first in each pair will then give the common inclination θ of the streamlines between CD and CE. In addition the density ratios ρ_3/ρ_1 and ρ_4/ρ_2 across the two reflected shocks can be determined from (22.23), since the corresponding pressure ratios p/p_1 and p/p_2 are known. Hence the two densities ρ_3 and ρ_4 below and above the discontinuity line CF can be computed. Finally the Mach numbers on either side of this line can be found from (22.22).

In practice, the common values of θ and p in (26) and (27) are found in the same way as for the example discussed in Sec. 3. A first approximation to θ, which is accurate when all four shocks are very weak, is obtained by using the results of Sec. 1. Thus although the theorem at the end of that section does not apply here (the transitions concerned being of different kinds) we can adapt the theory to the present example.

We note that if F does not depend on θ explicitly then neither does $'F$ nor F', and these two derivatives differ only in sign: $'F = -F'$, see Eqs. (2) and (4). Consider the flow below C. Then for the transition of positive deflection across the shock CA, we have according to Eq. (3):

$$F_1 - F_0 = -F_0'\theta_1 + \tfrac{1}{2}F_0''\theta_1^2 + O(\theta_1^3), \tag{28}$$

and for the second transition, of negative deflection, across the shock CD, Eq. (1) gives

$$F_3 - F_1 = F_1'(\theta - \theta_1) + \tfrac{1}{2}F_1''(\theta - \theta_1)^2 + O(\theta - \theta_1)^3. \tag{29}$$

Now in (28) we may replace F by F' and F'' successively to obtain

$$F_1' = F_0' - F_0''\theta_1 + O(\theta_1^2),$$
$$F_1'' = F_0'' + O(\theta_1).$$

Hence on substituting into (29) and adding (28) to the result we find

$$F_3 - F_0 = F_0'(\theta - 2\theta_1) + \tfrac{1}{2}F_0''(\theta - 2\theta_1)^2 + O(\theta - \theta_1, \theta_1)^3, \tag{30}$$

and similarly

$$F_4 - F_0 = -F_0'(\theta - 2\theta_2) + \tfrac{1}{2}F_0''(\theta - 2\theta_2)^2 + O(\theta - \theta_2, \theta_2)^3. \tag{31}$$

In particular we may take $F = p$. Then on equating the right-hand sides of (30) and (31) we see that for the pressure to be the same above and below the discontinuity line CF, the inclination of that line must satisfy $\theta - 2\theta_1 = -(\theta - 2\theta_2)$ or

$$\theta = \theta_1 + \theta_2, \tag{32}$$

to the second order in the deflections θ_1, θ_2. Another way of stating this result is to say that the *deflections caused by diagonally opposite shocks are approximately equal.*

As an example we take $M_0 = 3$, $\theta_1 = 20°$ and $\theta_2 = -10°$ with $\gamma = 7/5$. Then in the usual way we find for weak shocks CA and CB: $M_1 = 1.9941$, $p_1/p_0 = 3.7713$ and $M_2 = 2.5050$, $p_2/p_0 = 2.0545$. With the first approximation $\theta = 10°$ we obtain $p/p_1 = 1.7050$ across CD and $p/p_2 = 3.2158$ across CE. Thus

$$\theta = 10°: \qquad \frac{p}{p_0} = \begin{matrix} 6.4301 & \text{across } CD, \\ 6.6069 & \text{across } CE. \end{matrix}$$

This indicates that the correct value of θ is somewhat smaller. We therefore try

$$\theta = 9°40': \quad \frac{p}{p_0} = \begin{matrix} 6.5400 & \text{across } CD, \\ 6.4946 & \text{across } CE. \end{matrix}$$

By linear interpolation we obtain the second approximation $\theta = 9°44'$ and this gives $p/p_0 = 6.52$, correct to two decimal places, across both CD and CE. The corresponding density ratios are $\rho_3/\rho_0 = 3.56$ and $\rho_4/\rho_0 = 3.61$ so that

$$\frac{\rho_4}{\rho_3} = 1.016,$$

which differs very little from unity. Similarly the difference between $M_3 = 1.63$ and $M_4 = 1.66$ is very slight, so that the discontinuity is insignificant even though we started with large deflections θ_1, θ_2. These values were used in constructing Fig. 163; the apparently equal spacing of the streamlines throughout DCE illustrates the weakness of the discontinuity CF.

Finally, we must consider the conditions under which the present solution is valid. If the two shocks AC, BC have equal strengths, so that they also have equal and opposite inclinations, then the solution will be symmetric about the horizontal line through C. This dividing streamline may be replaced by a fixed wall, and then we have the oblique reflection problem of the last section. Thus, in this case, it is sufficient that the strength η of the shock AC and its inclination ω to the horizontal determine a point in Fig. 161 lying on or below the curve. We have seen that when the point lies below the curve there are two possible flow patterns. Now consider the case of unequal shocks. Their strengths and corresponding inclinations determine two distinct points in Fig. 161. A continuity argument similar to the one given in Sec. 3 shows that if one of these points lies below the curve and the second lies sufficiently close to the first, then there are again two solutions of the present kind. This may be seen by considering the curves in the θ,p-plane given by (26) and (27) when τ_1 and τ_2 are allowed to vary (see the curve $p = p_4$ in Fig. 157). The same figure also shows that whenever there is one such flow pattern there is in general a second. One of these corresponds to a pair of strong reflected shocks, and the other to a pair of weak reflected shocks. The approximation (32) applies to the latter. In particular we can always obtain these flow patterns for sufficiently weak shocks AC, BC.

This gives an indication of some conditions under which our solution is valid. On the other hand it is definitely not valid when either of the shocks AC, BC is the strong shock for the deflection it produces. For the flow behind a strong shock is always subsonic (see Sec. 22.7) whereas the flows in front of the reflected shocks must be supersonic.

Article 24

Nonisentropic Flow

1. Strictly adiabatic flow of an inviscid fluid

In all the examples discussed in the preceding article, solutions could be given in terms of regions of uniform flow or simple waves, separated by straight shock lines or straight characteristics. It is clear that problems with more general boundary conditions cannot be solved in this way. For instance, in the example discussed in Sec. 23.2 the solution was obtained in a limited region only. The state of the fluid on the boundary of this region forms a set of boundary conditions on the flow beyond, for which we anticipated that a curved shock line would be required. In this section we shall investigate the nature of the inviscid flow behind a curved shock line.

In Arts. 16 through 21 the theory of steady plane inviscid flow was developed under the assumptions that the fluid was elastic and the motion irrotational. Both these assumptions are realized when the motion is isentropic and the Bernoulli function H is constant throughout the flow. Such is the case for the strictly adiabatic flow behind a straight shock when the incident flow is uniform. For the total head H is the same for all particles behind the shock by (22.19), and since the values of p_2, ρ_2 are the same at all points of the shock, see Eq. (22.17), so also is the entropy. However, if the shock is curved the flow after it will not be isentropic since the values of p_2, ρ_2 depend, for each particle, on the slope of the shock line at the point where the particle reaches the transition, see Eq. (22.17). On the other hand the total head is still constant behind the shock. There is no longer an over-all relation between p and ρ after the shock since different particles have different values of entropy. The condition of strictly adiabatic flow for the region behind the shock leads, as was seen in Sec. 1.5, only to the condition

$$\frac{dS}{dt} = 0, \tag{1}$$

where S, the entropy or any given function of the entropy, is a known function of p and ρ.

To study the inviscid flow behind a shock—or any inviscid flow for which p is a given function of ρ only for each particle—we must first find an equation to replace (6.17). This equation was obtained from Newton's equation (1.I) by writing grad P for (grad $p)/\rho$ and using certain vector identities. This can only be done for an elastic fluid and we therefore seek another way of transforming (grad p)/ρ. According to Eqs. (2.11) and (2.23), which relate the three functions $I(p, \rho)$, $U(p, \rho)$, and $S(p, \rho)$, we have

$$dI = dU - \frac{p}{\rho^2}\,d\rho + \frac{dp}{\rho} = T\,dS + \frac{dp}{\rho}.$$

Hence for the actual changes occurring in the flow, we have

$$\frac{1}{\rho}\,\text{grad}\,p = \text{grad}\,I - T\,\text{grad}\,S,$$

which provides the necessary transformation.

In Eq. (6.17) grad P must now be replaced by grad $I - T$ grad S. The equation then reads

$$\frac{\partial \mathbf{q}}{\partial t} + (\text{curl}\,\mathbf{q} \times \mathbf{q}) = T\,\text{grad}\,S - \text{grad}\,g\tilde{H}, \tag{2}$$

where

$$\tilde{H} = \frac{q^2}{2g} + h + \frac{I}{g}. \tag{2'}$$

For steady flow this reduces to

$$\text{curl}\,\mathbf{q} \times \mathbf{q} = T\,\text{grad}\,S - \text{grad}\,g\tilde{H}.\text{[40]} \tag{3}$$

Note that in the derivation of this equation we have not used the specifying condition (1). It is a consequence of Newton's equation and the First Law of Thermodynamics. We shall now investigate Eqs. (2) and (3) quite generally. Later the discussion will be restricted to steady plane flow and in particular to the problem formulated at the beginning of this section.

In strictly adiabatic steady motion Eq. (1) reads $\mathbf{q}\cdot\text{grad}\,S = 0$. Thus the vectors curl $\mathbf{q} \times \mathbf{q}$ and grad S are each perpendicular to $\mathbf{q}$, and hence, see (3), the same is true for grad $g\tilde{H}$. In this case the derivative of $\tilde{H}$ along a streamline is zero, and $\tilde{H}$ is therefore constant along the line.[41] Now the total head H differs from $\tilde{H}$ only in having I replaced by P, and as we saw at the end of Sec. 2.5 the latter differ, in strictly adiabatic motion, by at most a constant for each particle. Hence H remains constant along a streamline and we recapture Bernoulli's equation (2.20′).

For isentropic motion S = constant, and there is an over-all (p, ρ)-relation. Then P, which differs from I by at most a constant for each particle, will differ from it by at most an over-all constant. In (2) we may then put grad $S = 0$ and replace $\tilde{H}$ by H, so as to recover Eq. (6.17), which is valid for an arbitrary elastic fluid. The latter equation has already been discussed in Sec. 6.5.

Since $\tilde{H}$ may be used in place of H whenever the motion is strictly adiabatic, it may be called the *total head* or *Bernoulli function* in this case.

We now consider steady flow which is not necessarily strictly adiabatic, and the vanishing first of grad $\tilde{H}$ and then of curl $\mathbf{q}$ in Eq. (3). If $\tilde{H}$ is constant throughout the flow, then the equation becomes

$$\text{(4)} \qquad \text{curl}\,\mathbf{q} \times \mathbf{q} = T\,\text{grad}\,S.$$

This means that grad S is perpendicular to both $\mathbf{q}$ and curl $\mathbf{q}$. Thus: *In the steady motion of an inviscid fluid throughout which $\tilde{H}$ is constant, the surfaces on which the entropy has constant values are composed of streamlines and vortex lines.* An important consequence of (4) is the following. If also curl $\mathbf{q} = 0$ throughout the flow then grad $S = 0$, and we can state: *In the steady irrotational motion of an inviscid fluid throughout which $\tilde{H}$ is constant, all particles have the same entropy.* In Sec. 6.5 we saw that the converse is not necessarily true.

Returning to Eq. (3), we now suppose that the motion is irrotational: curl $\mathbf{q} = 0$. Then T grad S = grad $g\tilde{H}$, so that

$$\text{curl}\,(T\,\text{grad}\,S) = \text{grad}\,T \times \text{grad}\,S = 0.$$

Thus either the flow is isothermal: grad $T = 0$, or it is isentropic: grad $S = 0$, or the surfaces on which T has constant values coincide with those on which S has constant values. In strictly adiabatic flow the first and third alternatives imply that both S and T remain constant on streamlines. Since according to Eq. (2.12) S and T are not functionally related, this means that the pressure p and density ρ also remain constant on the streamlines. The second alternative implies, by (3), that $\tilde{H}$ is also constant throughout the flow. *In any irrotational, strictly adiabatic, steady motion of an inviscid fluid, either the particles carry constant values of p and ρ, or the flow is isentropic with constant total head.*[42]

For a perfect gas $I = U + p/\rho = \gamma p/(\gamma - 1)\rho$, according to Eqs. (2.13) and (2.23). Hence Eq. (22.19) should more strictly read $\tilde{H}_1 = \tilde{H}_2$, since there is not necessarily any connection between the expressions for P on the two sides of the shock. Thus if in the steady motion in front of a plane shock $\tilde{H}$ = constant, as in the case when the motion is uniform, then $\tilde{H}$ = constant after the shock for strictly adiabatic flow. If in addition S is not constant after the shock, as occurs in the case mentioned when the shock is not straight, then the motion is rotational after the shock. This follows from the second result above. *For uniform incident flow the strictly adiabatic, steady, plane motion behind a curved shock is rotational.*[43]

In strictly adiabatic steady flow both $\tilde{H}$ and S are constant on each streamline, and we may assume that their variation from streamline to streamline is determined by the boundary conditions in any particular problem. If the motion is not only steady but also plane, we may introduce (as in Sec. 16.2) a stream function $\psi(x, y)$ to satisfy the equation of continuity:

$$\rho q_x = \frac{\partial \psi}{\partial y}, \qquad \rho q_y = -\frac{\partial \psi}{\partial x}. \tag{5}$$

Now we have seen that (5) implies that ψ remains constant along streamlines. Thus S and $g\tilde{H}$ are functions of ψ alone, say $F(\psi)$ and $G(\psi)$, which are determined by the boundary conditions. Also for plane motion, the vortex vector curl $\mathbf{q}$ has just a z-component, ω say. Equation (3) therefore reduces to a scalar equation along the normal to the streamline, namely

$$\begin{aligned} \omega &= \frac{1}{q}\left(T\frac{\partial S}{\partial n} - g\frac{\partial \tilde{H}}{\partial n}\right), \\ &= \frac{1}{q}\left(T\frac{dS}{d\psi} - g\frac{d\tilde{H}}{d\psi}\right)\frac{\partial \psi}{\partial n}. \end{aligned}$$

Thus finally we have*

$$\omega = \rho(TF' - G'), \tag{6}$$

where primes signifiy differentiation with respect to ψ, and we have used Eq. (16.19): $\rho q = \partial\psi/\partial n$.

This result, essentially due to L. Crocco,[44] has the following interpretation. A material filament initially perpendicular to the x,y-plane remains perpendicular and is therefore at each instant a vortex filament. By continuity its cross-sectional area is inversely proportional to ρ, and hence its vorticity is proportional to ω/ρ. Thus from (6) we may state: *In a strictly adiabatic steady plane flow the vorticity varies linearly with temperature on any streamline.* For isentropic flow, $F' = 0$, the vorticity is constant, in agreement with Helmholtz' second vortex theorem (Sec. 6.4).

2. Equation for the stream function

We may now deduce an equation for ψ to replace (16.21) when the flow is strictly adiabatic but not necessarily elastic. Referring to Eq. (16.20) we see that $\partial p/\partial n$ can no longer be replaced by $a^2\partial\rho/\partial n$ since there is no over-all (p, ρ)-relation. Instead we must proceed as in Sec. 15.6. Thus from $S(p, \rho) = F(\psi)$ we derive

$$\frac{\partial S}{\partial p}\frac{\partial p}{\partial n} + \frac{\partial S}{\partial \rho}\frac{\partial \rho}{\partial n} = F'(\psi)\frac{\partial \psi}{\partial n} = \rho q F',$$

* For a perfect gas, this shows that the mean rotation ω is a linear function of p and ρ on each streamline.

so that

$$\frac{\partial p}{\partial n} = a^2 \frac{\partial \rho}{\partial n} + \frac{\rho q F'}{\partial S/\partial p}, \tag{7}$$

where as before

$$a^2 = -\frac{\partial S/\partial \rho}{\partial S/\partial p} = \frac{dp/dt}{d\rho/dt}, \tag{8}$$

according to (1).

A change must also be made in (16.20′). The term $\partial q_n/\partial s$ cannot be replaced by $\partial q/\partial n$ since the irrotationality condition (16.7) no longer holds. Now since $\partial q_n/\partial s - \partial q/\partial n = \omega$, we must write

$$\frac{\partial q_n}{\partial s} = \frac{\partial q}{\partial n} + \omega,$$

and use the expression in (6) for ω.

When these two changes are made, the final equation for ψ becomes

$$(M^2 - 1)\frac{\partial^2 \psi}{\partial s^2} - \frac{\partial^2 \psi}{\partial n^2} = \rho F'\left(\rho T + \frac{M^2}{\partial S/\partial p}\right) - \rho^2 G'.$$

As in Sec. 16.2 the left-hand side of this equation may be written in Cartesian form to give

$$\left(1 - \frac{q_x^2}{a^2}\right)\frac{\partial^2 \psi}{\partial x^2} - \frac{2q_x q_y}{a^2}\frac{\partial^2 \psi}{\partial x\, \partial y} + \left(1 - \frac{q_y^2}{a^2}\right)\frac{\partial^2 \psi}{\partial y^2} = \rho^2 G' - \rho F'\left(\rho T + \frac{M^2}{\partial S/\partial p}\right). \tag{9}$$

The left-hand member of (9) is identical with that of Eq. (16.21) but there the right member was zero.

The right-hand side of (9) includes F' and G', which are known functions of ψ, as well as T, a^2 and $\partial S/\partial p$, given functions of p and ρ which, by virtue of $S(p, \rho) = F(\psi)$, can be expressed as functions of ρ and ψ. Thus we still have to express ρ, q_x, and q_y as functions of ψ and its derivatives. From the definition of $\tilde{H}$ we have, on neglecting the gravity term, the Bernoulli equation

$$\tfrac{1}{2}q^2 + I = g\tilde{H} = G(\psi), \tag{10}$$

where I is a known function of p and ρ which can be expressed as a function of ρ and ψ from $S(p, \rho) = F(\psi)$. This equation, together with (5), serves to determine ρ, q_x, and q_y as functions of ψ, $\partial\psi/\partial x$ and $\partial\psi/\partial y$. Thus: *Eq.* (9)

is a planar nonhomogeneous differential equation of the second order in* ψ. It plays a role similar to that of Eq. (15.23) in the one-dimensional nonsteady case.

As an illustration of these calculations, consider the case of a perfect gas, for which the entropy S is†

$$S = \frac{gR}{(\gamma - 1)} \log \frac{p}{\rho^\gamma}, \tag{11}$$

see Eq. (1.7). Then

$$\frac{\partial S}{\partial p} = \frac{gR}{(\gamma - 1)p}, \qquad \frac{\partial S}{\partial \rho} = -\frac{\gamma g R}{(\gamma - 1)\rho}, \qquad a^2 = \frac{\gamma p}{\rho},$$

and in this case the right-hand member of (9) is

$$\rho^2 \left[G' - \frac{a^2 + (\gamma - 1)q^2}{\gamma g R} F' \right]. \tag{12}$$

In addition $I = \gamma p/(\gamma - 1)\rho = a^2/(\gamma - 1)$, where in terms of ρ and ψ

$$a^2 = \gamma \rho^{\gamma - 1} \exp\,[(\gamma - 1)F/gR]. \tag{13}$$

Hence (10) becomes

$$\frac{1}{2} q^2 + \frac{\gamma}{\gamma - 1} \rho^{\gamma - 1} \exp[(\gamma - 1)F/gR] = G. \tag{14}$$

In order to determine ρ, q_x, q_y as functions of ψ, $\partial\psi/\partial x$, $\partial\psi/\partial y$, we first express ρ in terms of q from (5):

$$\rho = \frac{1}{q} \left[\left(\frac{\partial \psi}{\partial x}\right)^2 + \left(\frac{\partial \psi}{\partial y}\right)^2 \right]^{\frac{1}{2}}, \tag{15}$$

and use this to eliminate ρ from (14). The resulting equation for q as a function of ψ, $\partial\psi/\partial x$, $\partial\psi/\partial y$ is

$$Aq^2 + \frac{B}{q^{\gamma - 1}} = 1, \tag{16}$$

where

$$A = \frac{1}{2G(\psi)}, \qquad B = \frac{\gamma \exp\,[(\gamma - 1)F(\psi)/gR]}{(\gamma - 1)G(\psi)} \left[\left(\frac{\partial \psi}{\partial x}\right)^2 + \left(\frac{\partial \psi}{\partial y}\right)^2 \right]^{(\gamma - 1)/2}.$$

Once q is determined from this equation, the density follows from (15) and then the velocity components from (5) and the speed of sound from (13).

We return now to the general equation (9) and show how the correspond-

* That is, linear in the derivatives of highest order, see Sec. 9.4.

† Here we must use the entropy S itself, and not a function of it as in Sec. 15.6.

ing equation for the rotational motion of an elastic fluid can be deduced from it. Let the (p, ρ)-relation be written in the form $S(p, \rho) = 0$, and consider $S(p, \rho)$ to be the entropy function of the gas.* Then since the motion is isentropic, (9) will hold with $F' = 0$:

$$\left(1 - \frac{q_x^2}{a^2}\right)\frac{\partial^2\psi}{\partial x^2} - 2\frac{q_x q_y}{a^2}\frac{\partial^2\psi}{\partial x\,\partial y} + \left(1 - \frac{q_y^2}{a^2}\right)\frac{\partial^2\psi}{\partial y^2} = \rho^2 G'. \tag{17}$$

For isentropic motion I may be replaced by P in the Bernoulli equation (10), so that it becomes

$$\tfrac{1}{2}q^2 + P = gH = G(\psi). \tag{18}$$

This equation, together with (5), serves to determine ρ, q_x and q_y as functions of ψ, $\partial\psi/\partial x$, $\partial\psi/\partial y$. We see that when P is expressed as a function of ρ, the latter satisfies

$$\frac{C}{\rho^2} + P(\rho) = D, \tag{18$'$}$$

where

$$C = \frac{1}{2}\left[\left(\frac{\partial\psi}{\partial x}\right)^2 + \left(\frac{\partial\psi}{\partial y}\right)^2\right], \qquad D = G(\psi).$$

Once ρ is determined, q_x and q_y follow from (5) and a^2 from its expression as a function of ρ. In particular, if the motion is irrotational then $G =$ constant in (18), and (17) reduces to (16.21). Even in this case the coefficients are not known explicitly as functions of ψ and its derivatives [see remark after (16.21)].† This completes the discussion of Sec. 16.2 and extends it to the case when irrotationality is not granted.

An important conclusion which we can draw from (9) is the following. Since the characteristics of the differential equation depend only on the second-order terms, they are the same for (9) as for (16.21). Thus, *real characteristic lines in the x,y-plane exist only for supersonic motion, $M \geqq 1$, and then they are lines with inclinations $\theta \pm \alpha$, where $\alpha =$* arc sin (a/q) *and a is the same function of p, ρ as before.* Further conclusions analogous to those of Art. 16, however, cannot be drawn. The interchange of dependent and independent variables, the use of the hodograph, etc., are no longer of avail, since (9) is nonhomogeneous.

The above conclusions concerning the characteristics also follow by considering Eqs. (22.1), (22.2), and (1) as a system of four homogeneous

* The corresponding temperature and internal energy functions $T(p, \rho)$ and $U(p, \rho)$ are solutions of Eqs. (2.11) and (2.12).

† Equation (18′) shows that ρ and a are then functions of $(\partial\psi/\partial x)^2 + (\partial\psi/\partial y)^2$ alone, and the coefficients are in fact independent of ψ itself. (See Note IV. 1.)

first-order equations for q_x, q_y, p, and ρ. For this purpose Eq. (1) is written out as

$$\frac{dp}{dt} - a^2 \frac{d\rho}{dt} = 0, \tag{1'}$$

see (8). Following Sec. 9.6 we find that the characteristics of the system are the streamlines and the two families of Mach lines. For the streamlines there are two compatibility relations, namely, (1′) and the Bernoulli equation (10). Across the streamlines ρ and q may have discontinuities (see Note II. 31), as we have already seen in Sec. 23.3. The compatibility relations on the Mach lines are

$$dp \pm \rho q^2 \tan \alpha \, d\theta = 0, \qquad \text{along a } C^{\pm}.$$

However, we did not find the streamlines appearing as characteristics of Eq. (9). The reason is that we have assumed in the present section that S and $\tilde{H}$ are prescribed functions of ψ. This leads to two new equations for the derivatives of the state variables normal to a streamline which, in conjunction with the above-mentioned system, are sufficient to determine these derivatives from given values on the streamline.

It was shown in Sec. 22.3 that the actual change in entropy across a shock is in most cases very small. Thus if the motion before the shock is isentropic with constant total head, if the shock is not too strong and if, at the same time, the variation of slope along the shock line is not large, the derivative G' will be zero and F' insignificant.[45] Under these conditions one may, as a rule, consider the flow after the shock to be of the same type as that before the shock.

3. Substitution principle. Modified stream function

For the important case of flow behind a curved shock line with uniform incident stream, the entropy S varies from streamline to streamline but the Bernoulli function $\tilde{H}$ is constant. We shall now show that any strictly adiabatic flow of a perfect gas can be replaced by one with this property. Moreover the substitute flow has the same streamline pattern and pressure distribution as the original.

Consider the equation of continuity and the two components of Newton's equation in natural coordinates (Sec. 16.1):

$$\frac{\partial}{\partial s}(\rho q) + \rho q \frac{\partial \theta}{\partial n} = 0,$$

$$\rho \frac{\partial q}{\partial s} = -\frac{\partial p}{\partial s}, \qquad \rho q^2 \frac{\partial \theta}{\partial s} = -\frac{\partial p}{\partial n}.$$

These equations are satisfied by the set of variables q, θ, p, ρ in the x,y-plane corresponding to the adiabatic flow of a given gas. When these variables are replaced by λq, θ, p, ρ/λ^2, respectively, the equations are still satisfied provided that λ is a function which remains constant along streamlines: $\partial\lambda/\partial s = 0$. For the second flow has the same streamlines, so that the s- and n-directions are the same as before. Note however that the second set of variables need not necessarily correspond to a strictly adiabatic flow of the same gas. For the fact that $S(p, \rho)$ is constant along streamlines does not imply that $S(p, \rho/\lambda^2)$ has the same property. Of course, another entropy function can always be found which will have this property for the second flow, i.e., the second flow may always be considered the strictly adiabatic flow of a suitable gas.

We therefore have the following result: *To each given strictly adiabatic flow in the x,y-plane there corresponds an infinity of such flows with the same streamline pattern and pressure distribution. These flows are obtained from the original by multiplying the velocity vector* $\mathbf{q}$ *at each point by* λ *and the density* ρ *by* $1/\lambda^2$, *where* λ *is any function whose level lines are the streamlines.* Moreover since $a^2 = dp/d\rho$, where differentiation is along a streamline, the velocity of sound is also multiplied by λ. The Mach number at a point is therefore unchanged by this transformation, so that all the flows are subsonic (or supersonic) in the same region. This is known as the *substitution principle.*[46]

We have just seen that in general the character of the gas changes under this substitution principle; normally the entropy function cannot be the same in the two cases. An exception is the perfect gas, for which $S(p, \rho)$ is given by (11). For if p/ρ^γ remains constant along streamlines then so also does $p\lambda^{2\gamma}/\rho^\gamma$. In particular, by choosing λ proportional to $\rho^{\frac{1}{2}}/p^{1/2\gamma}$ we make the second flow isentropic, since then $p\lambda^{2\gamma}/\rho^\gamma$ takes the same constant value on all streamlines. Similarly the function $g\tilde{H} = \gamma p/(\gamma - 1)\rho + q^2/2$ is multiplied by λ^2 in the substitute flow. Hence, in particular, if λ is chosen inversely proportional to $\tilde{H}^{\frac{1}{2}}$, the corresponding function in the substitute flow will be constant throughout. Thus: *If the original flow is that of a certain perfect gas then the substitute flows are also*; *in addition one of them is isentropic and another has constant total head.*

For strictly adiabatic flow of a perfect gas in which $\tilde{H}$ is constant throughout, the equations of the preceding section can be simplified. There is no loss in generality when we take this constant value of $\tilde{H}$ to be $1/2g$ (this is equivalent to choosing units so that $q_m = 1$). Then (16) becomes

$$q^2(1 - q^2)^{2/(\gamma-1)} = \left(\frac{\partial\Psi}{\partial x}\right)^2 + \left(\frac{\partial\Psi}{\partial y}\right)^2, \tag{19}$$

where

$$\Psi = \left(\frac{2\gamma}{\gamma - 1}\right)^{1/(\gamma-1)} \int^{\psi} \exp\,[F(\psi)/gR]\; d\psi. \tag{20}$$

Thus q is a function of $(\partial\Psi/\partial x)^2 + (\partial\Psi/\partial y)^2$ alone, and it then follows from (10) that the same is true for a^2:

$$a^2 = \frac{\gamma - 1}{2}(1 - q^2). \tag{21}$$

From (5) and (15) we find

$$\begin{aligned} q_x &= (1 - q^2)^{-1/(\gamma-1)} \frac{\partial\Psi}{\partial y}, \\ q_y &= -(1 - q^2)^{-1/(\gamma-1)} \frac{\partial\Psi}{\partial x}. \end{aligned} \tag{22}$$

Note that the derivatives of Ψ, unlike those of ψ, are determined by the velocity components alone. The modified stream function Ψ was introduced by L. Crocco.[47] It has the property of remaining constant on streamlines, since it is a function of ψ alone; unlike ψ, however, it is discontinuous across shock lines.

To find the equation satisfied by Ψ we compute the left member of (9) with ψ replaced by Ψ. Thus

$$\begin{aligned} &\left(1 - \frac{q_x^2}{a^2}\right)\frac{\partial^2\Psi}{\partial x^2} - 2\frac{q_x q_y}{a^2}\frac{\partial^2\Psi}{\partial x\,\partial y} + \left(1 - \frac{q_y^2}{a^2}\right)\frac{\partial^2\Psi}{\partial y^2} \\ &\quad = \left[\left(1 - \frac{q_x^2}{a^2}\right)\frac{\partial^2\psi}{\partial x^2} - 2\frac{q_x q_y}{a^2}\frac{\partial^2\psi}{\partial x\,\partial y} + \left(1 - \frac{q_y^2}{a^2}\right)\frac{\partial^2\psi}{\partial y^2}\right]\frac{d\Psi}{d\psi} \\ &\quad\quad + \left[\left(1 - \frac{q_x^2}{a^2}\right)\left(\frac{\partial\psi}{\partial x}\right)^2 - 2\frac{q_x q_y}{a^2}\frac{\partial\psi}{\partial x}\frac{\partial\psi}{\partial y} + \left(1 - \frac{q_y^2}{a^2}\right)\left(\frac{\partial\psi}{\partial y}\right)^2\right]\frac{d^2\Psi}{d\psi^2}, \\ &\quad = -\frac{\rho^2 F'}{\gamma g R}[a^2 + (\gamma - 1)q^2]\frac{d\Psi}{d\psi} + \rho^2 q^2 \frac{d^2\Psi}{d\psi^2}, \end{aligned}$$

where in the second step we have used Eqs. (5), (9), and (12). Now from (20) we have $d^2\Psi/d\psi^2 = (F'/gR)\, d\Psi/d\psi$, so that this last expression becomes

$$\frac{(q^2 - a^2)}{\gamma g R}\rho^2 F' \frac{d\Psi}{d\psi}.$$

Finally from (15) and (19) we find

$$\rho = (1 - q^2)^{1/(\gamma-1)} \frac{d\psi}{d\Psi},$$

so that Ψ satisfies the equation

$$\begin{aligned} &\left(1 - \frac{q_x^2}{a^2}\right)\frac{\partial^2\Psi}{\partial x^2} - 2\frac{q_x q_y}{a^2}\frac{\partial^2\Psi}{\partial x\,\partial y} + \left(1 - \frac{q_y^2}{a^2}\right)\frac{\partial^2\Psi}{\partial y^2} \\ &\quad = \frac{(q^2 - a^2)}{\gamma g R}(1 - q^2)^{2/(\gamma-1)}\frac{dF}{d\Psi}. \end{aligned}$$

All coefficients on the left-hand side of this equation depend on $\partial\Psi/\partial x$ and $\partial\Psi/\partial y$ alone [see (21) and (22)]. Apart from $dF/d\Psi$, which is a given function of Ψ, the right member depends only on $(\partial\Psi/\partial x)^2 + (\partial\Psi/\partial y)^2$.

4. A second approach[48]

It is possible to give a different approach to the problem of strictly adiabatic rotational flow. This parallels, to some extent, the discussion of adiabatic one-dimensional flow given in Sec. 15.7.

If q_x and q_y times the equation of continuity (22.2) are added to the first and second components (22.1) of Newton's equation respectively, we obtain

$$\text{(23)}\qquad \begin{aligned} \frac{\partial}{\partial x}(p + \rho q_x^2) + \frac{\partial}{\partial y}(\rho q_x q_y) &= 0, \\ \frac{\partial}{\partial x}(\rho q_x q_y) + \frac{\partial}{\partial y}(p + \rho q_y^2) &= 0. \end{aligned}$$

These two equations allow us to introduce two new functions $\tilde{\xi}(x,y)$ and $\tilde{\eta}(x,y)$ such that

$$\text{(24)}\qquad \begin{aligned} d\tilde{\xi} &= -\rho q_x q_y\,dx + (p + \rho q_x^2)\,dy, \\ d\tilde{\eta} &= -(p + \rho q_y^2)\,dx + \rho q_x q_y\,dy, \end{aligned}$$

just as the equation of continuity permits the introduction of $\psi(x,y)$ such that

$$\text{(25)}\qquad d\psi = -\rho q_y\,dx + \rho q_x\,dy$$

[compare (5)]. Substituting from this into Eq. (24), we have

$$\text{(26)}\qquad d\tilde{\xi} = q_x\,d\psi + p\,dy, \qquad d\tilde{\eta} = q_y\,d\psi - p\,dx.^*$$

We have seen that in strictly adiabatic flow the entropy $S(p, \rho)$ is a function of ψ alone. This function, $F(\psi)$ say, is supposed to be determined by the boundary conditions. Prescription of F therefore provides a relation between p, ρ, ψ throughout the flow. If any two of these three variables are selected as new independent variables in place of x and y, then the third may be considered a known function of these two for any given problem. Moreover, (42) can be rewritten as

$$\text{(27)}\qquad d\xi = q_x\,d\psi - y\,dp, \qquad d\eta = q_y\,d\psi + x\,dp,$$

where $\xi = \tilde{\xi} - py$ and $\eta = \tilde{\eta} + px$. We are thus led to select ψ, p as new

* For an element of streamline $d\psi = 0$, and $d\tilde{\xi}$, $d\tilde{\eta}$ are the x- and y-components of the pressure force across the element. Hence the total changes in $\tilde{\xi}$ and $\tilde{\eta}$ along a fixed boundary give the components of the force exerted by the fluid.

independent variables in place of x, y and to replace $\bar{\xi}$, $\bar{\eta}$ by ξ, η. In the new variables, we have

$$x = \partial\eta/\partial p, \qquad y = -\partial\xi/\partial p, \tag{28}$$

$$q_x = \partial\xi/\partial\psi, \qquad q_y = \partial\eta/\partial\psi, \tag{29}$$

and Eq. (25) yields

$$\begin{aligned} q_x \frac{\partial y}{\partial \psi} - q_y \frac{\partial x}{\partial \psi} &= \frac{1}{\rho}, \\ q_x \frac{\partial y}{\partial p} - q_y \frac{\partial x}{\partial p} &= 0. \end{aligned} \tag{30}$$

Substitution from (28) in (30) gives

$$\frac{\partial \xi}{\partial \psi} \frac{\partial^2 \xi}{\partial \psi \partial p} + \frac{\partial \eta}{\partial \psi} \frac{\partial^2 \eta}{\partial \psi \partial p} + \frac{1}{\rho} = 0, \tag{31}$$

$$\frac{\partial \xi}{\partial \psi} \frac{\partial^2 \xi}{\partial p^2} + \frac{\partial \eta}{\partial \psi} \frac{\partial^2 \eta}{\partial p^2} = 0, \tag{32}$$

as simultaneous equations for ξ, η as functions of ψ, p. In these equations ρ is considered to be a known function of ψ and p; once a suitable solution of them has been determined, the position coordinates x, y and the velocity components q_x, q_y are given as functions of ψ, p by (28) and (29), respectively.

Equation (32) has an immediate interpretation since it is equivalent to the second of Eqs. (30). It expresses the fact that the lines in the physical plane on which ψ is constant have the slope of the velocity vector, i.e. $(\partial y/\partial p)/(\partial x/\partial p) = q_y/q_x$. In order to interpret (31) we note that the left-hand side may be written

$$\frac{1}{2} \frac{\partial}{\partial p} \left[\left(\frac{\partial \xi}{\partial \psi} \right)^2 + \left(\frac{\partial \eta}{\partial \psi} \right)^2 \right].$$

Thus the equation may be integrated to give

$$\frac{1}{2} \left[\left(\frac{\partial \xi}{\partial \psi} \right)^2 + \left(\frac{\partial \eta}{\partial \psi} \right)^2 \right] + P(\psi, p) = G(\psi), \tag{33}$$

where $P(\psi, p) = \int^p dp/\rho$ is g times the pressure head introduced in Sec. 2.5, the integral being taken with ψ held fixed, and $G(\psi)$ is an arbitrary function. Now according to (29)

$$\left(\frac{\partial \xi}{\partial \psi} \right)^2 + \left(\frac{\partial \eta}{\partial \psi} \right)^2 = q^2, \tag{34}$$

so that (33) is the Bernoulli equation with $G(\psi)/g$ the value of the total head $\tilde{H}$ on each streamline. As before we assume that this function is determined by the boundary conditions.

Equation (33) may be written in the form

$$\frac{1}{2}q^2 = G(\psi) - P(\psi, p), \tag{35}$$

where the right member is a function of ψ and p determined by the boundary conditions. Thus the Bernoulli equation expresses the fact that the speed q is a known function of ψ and p for any given problem. Having determined this function we may replace (31) by (34).

On differentiating (35) with respect to p we find

$$\frac{1}{\rho} = -q\frac{\partial q}{\partial p}. \tag{36}$$

Now a^2 is the rate of change of p with respect to ρ as a particle moves along a streamline, so that

$$\frac{\partial \rho}{\partial p} = \frac{1}{a^2},$$

since this derivative is taken with ψ constant. Thus on differentiating (36) we obtain

$$M^2 - 1 = \rho^2 q^3 \frac{\partial^2 q}{\partial p^2}. \tag{37}$$

The motion is therefore subsonic or supersonic according as $\partial^2 q/\partial p^2$ is negative or positive. (Cf. Sec. 8.1)

If (32) and (34) are solved simultaneously for $\partial\eta/\partial\psi$ and $\partial^2\eta/\partial p^2$, and η is eliminated from the resulting equations by equating $\partial^3\eta/\partial p^2\partial\psi$ and $\partial^3\eta/\partial\psi\partial p^2$, we obtain a nonplanar second-order differential equation for ξ. This equation is of Monge-Ampère type, but unlike that in Sec. 15.7 it is very complicated. Since (32) and (34) are symmetric in ξ and η, the same Monge-Ampère equation is obtained for η when ξ is eliminated from them.

A simpler equation is obtained if ξ and η are expressed in terms of the stream direction $\theta(\psi, p)$. According to (29) we may write

$$\frac{\partial \xi}{\partial \psi} = q\cos\theta, \qquad \frac{\partial \eta}{\partial \psi} = q\sin\theta, \tag{38}$$

and then (34) is automatically satisfied. By differentiating (32) twice with respect to ψ we obtain three equations from which ξ and η may be eliminated by means of (38). The resulting equation for θ is

$$A\frac{\partial^2\theta}{\partial\psi^2} + 2B\frac{\partial^2\theta}{\partial\psi\,\partial p} + C\frac{\partial^2\theta}{\partial p^2} = F,$$

where

$$A = \frac{\partial^2 q}{\partial p^2} - q\left(\frac{\partial \theta}{\partial p}\right)^2, \qquad B = q\,\frac{\partial \theta}{\partial \psi}\,\frac{\partial \theta}{\partial p}, \qquad C = -q\left(\frac{\partial \theta}{\partial \psi}\right)^2,$$

$$F = \frac{\partial \theta}{\partial \psi}\left[\frac{\partial^3 q}{\partial \psi\,\partial p^2} - \frac{\partial q}{\partial \psi}\left(\frac{\partial \theta}{\partial p}\right)^2 + 2\,\frac{\partial q}{\partial p}\,\frac{\partial \theta}{\partial \psi}\,\frac{\partial \theta}{\partial p}\right].$$

This equation, like (9) for ψ in the physical plane, is planar and nonhomogeneous. Its advantage in comparison with (9) is that the coefficients A, B, C, and F are known explicitly as functions of ψ, p, $\partial\theta/\partial\psi$, $\partial\theta/\partial p$ once the distributions of total head and entropy from streamline to streamline are given by the boundary conditions. For, by (35), the function $q(\psi,p)$ appearing in these coefficients is then known. In Eq. (9) there appear ρ, q_x, q_y which must be determined as functions of ψ and its derivatives by means of (5) and (10), so that in general they are not known explicitly.

5. The sufficiency of the shock conditions[49]

In Sec.14.2 it was shown that the shock conditions (14.2) and (14.9) are necessary conditions relating the initial and final states of an abrupt transition which is the limit of a one-dimensional nonsteady viscous flow. In Sec. 22.2 the same was shown for the shock conditions (22.3) and (22.16) in the case of steady plane flow. We shall now show that these conditions are, in a definite sense, also sufficient. To this end we first discuss the one-dimensional non steady case, and introduce a more convenient system of coordinates in the x,t-plane.

We consider a given line $\mathcal{S}$: $x = f(t)$, which is nowhere parallel to the x-axis, but otherwise arbitrary, and introduce a curvilinear coordinate system whose coordinate lines consist of the parallels to the x-axis and the curves obtained by translating $\mathcal{S}$ in the x-direction (see Fig. 164). For one coordinate, α, at a point P we take the displacement in the x-direction from $\mathcal{S}$, and for the other, β, we take the time t:

$$\alpha = x - f(t), \qquad \beta = t. \tag{39}$$

Thus

$$\frac{\partial}{\partial x} = \frac{\partial}{\partial \alpha}, \qquad \frac{\partial}{\partial t} = \frac{\partial}{\partial \beta} - c(\beta)\,\frac{\partial}{\partial \alpha}, \tag{40}$$

where $c(\beta) = df(\beta)/d\beta$ is the slope of $\alpha =$ constant at P (measured from the t-axis).

The differential equations governing the simply adiabatic flow of a per-

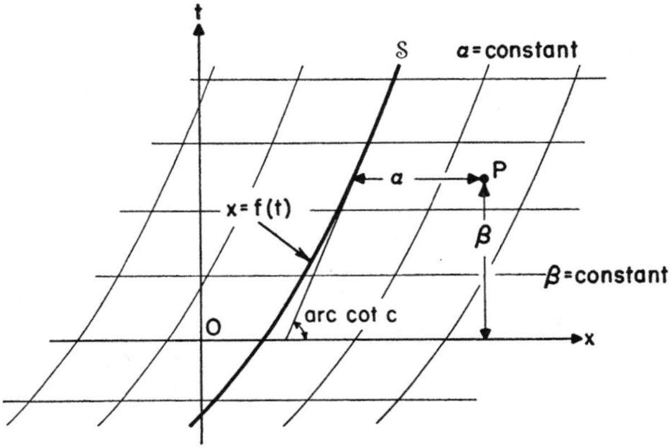

FIG. 164. Curvilinear coordinates in the x,t-plane.

fect viscous fluid in the x,t-plane are (11.2), (11.3), and (11.4′). If Eqs. (40) are introduced into them we obtain

$$\frac{\partial}{\partial \alpha}[\rho(u - c)] + \frac{\partial \rho}{\partial \beta} = 0, \tag{41a}$$

$$\rho(u - c)\frac{\partial u}{\partial \alpha} + \frac{\partial}{\partial \alpha}(p - \sigma_x') + \rho\frac{\partial u}{\partial \beta} = 0, \tag{41b}$$

$$\begin{aligned}\rho(u - c)\frac{\partial}{\partial \alpha}\left[\frac{u^2}{2} + \frac{gR}{\gamma - 1}T\right] &+ \frac{\partial}{\partial \alpha}[u(p - \sigma_x') - K] \\ &+ \rho\frac{\partial}{\partial \beta}\left[\frac{u^2}{2} + \frac{gR}{\gamma - 1}T\right] = 0,\end{aligned} \tag{41c}$$

where $K = k\,\partial T/\partial \alpha$ and, according to (11.6), $\sigma_x' = \mu_0\,\partial u/\partial \alpha$. These equations are similar to (14.4), (14.5), and (14.6). The derivative $\partial'/\partial t$ appearing in the latter is equivalent to $\partial/\partial \beta$. It was not necessary to introduce there the new coordinate system explicitly.

We now magnify the α-coordinate in the ratio $1:\mu_0$, i.e., we introduce a new independent variable $s = \alpha/\mu_0$, assuming for convenience that μ_0 is constant. Under this change of variable Eqs. (41) become

$$\frac{\partial}{\partial s}[\rho(u - c)] = -\mu_0\frac{\partial \rho}{\partial \beta}, \tag{42a}$$

$$\rho(u - c)\frac{\partial u}{\partial s} + \frac{\partial}{\partial s}(p - \sigma_x') = -\mu_0\rho\frac{\partial u}{\partial \beta}, \tag{42b}$$

$$\rho(u - c)\frac{\partial}{\partial s}\left[\frac{u^2}{2} + \frac{gR}{\gamma - 1}T\right] + \frac{\partial}{\partial s}[u(p - \sigma_x') - K] = -\mu_0\rho\frac{\partial}{\partial \beta}\left[\frac{u^2}{2} + \frac{gR}{\gamma - 1}T\right], \tag{42c}$$

where

$$\sigma_x' = \frac{\partial u}{\partial s}, \qquad K = \frac{k}{\mu_0}\frac{\partial T}{\partial s}. \tag{43}$$

The right-hand side of each of Eqs. (42) has the factor μ_0 which, as we saw in Sec. 11.5, is extremely small for air (and other gases). Let now the right-hand sides be replaced by zeros, and consider a solution of the resulting equations in which u, p, ρ, and T tend to finite values at both $s = -\infty$ and $s = +\infty$ (each state depending on β). We shall use subscripts 1 and 2 for these states, assigning 1 to $s = -\infty$ and 2 to $s = +\infty$ if $u - c$ is positive, and vice versa if $u - c$ is negative. Such a solution will be denoted by $\mathbf{T}(s, \beta)$ where the single symbol stands for the set of state variables. Then $\mathbf{T}(s, \beta)$ is called a *shock transition solution of the original Eqs.* (42), *with respect to the line* $\mathcal{S}$ *and the states* 1 *and* 2. It depends implicitly on the viscosity coefficient μ_0 through the variable s which is $1/\mu_0$ times the displacement from $\mathcal{S}$.

If the right-hand members of Eqs. (42) are replaced by zeros, the resulting equations no longer contain derivatives with respect to β, and are in fact equivalent to Eqs. (11.8), governing the one-dimensional steady flow of a perfect viscous gas, when $\mu_0 s = \alpha$ stands for x, $u' = u - c$ for u, and partial for ordinary derivatives. This is easily checked once it is remembered that c does not depend on s. Hence the solution of these equations follows the same lines as for Eqs. (11.8), except of course that now the integration constants m, C_1, C_2 must be considered functions of β.

With reference to Eqs. (11.8) we found that when the system possesses a solution in which the particles pass from a state 1 at $x = -\infty$ to a state 2 at $x = +\infty$, these states satisfy the shock conditions

$$\rho_1 u_1 = \rho_2 u_2 = m, \tag{44a}$$

$$p_1 + \rho_1 u_1^2 = p_2 + \rho_2 u_2^2 = C_1 m, \tag{44b}$$

$$\frac{1}{2}u_1^2 + \frac{\gamma}{\gamma - 1}\frac{p_1}{\rho_1} = \frac{1}{2}u_2^2 + \frac{\gamma}{\gamma - 1}\frac{p_2}{\rho_2} = C_2, \tag{44c}$$

with $m > 0$, and also $T_2 \geqq T_1$.

We shall now verify the converse: if u_1, p_1, ρ_1 and u_2, p_2, ρ_2 are two sets of values satisfying (44), with $m > 0$, and the condition $T_2 \geqq T_1$, then there exist solutions of (11.8) in which the particles pass from state 1 at $x = -\infty$ to state 2 at $x = +\infty$. For let (44) now be the definitions of the constants m, C_1, C_2, and rewrite (44b) and (44c) in terms of $v = u^2/2C_1^2$ and $\Theta = p/C_1^2\rho$:

$$\frac{\Theta_1}{\sqrt{2v_1}} + \sqrt{2v_1} = \frac{\Theta_2}{\sqrt{2v_2}} + \sqrt{2v_2} = 1,$$

$$v_1 + \frac{\gamma}{\gamma - 1}\Theta_1 = v_2 + \frac{\gamma}{\gamma - 1}\Theta_2 = \frac{C_2}{C_1^2}.$$

These last equations are precisely Eqs. (11.21). In the notation of Sec. 11.4 they express the fact that (v_1, Θ_1) and (v_2, Θ_2) are the points of intersection of the parabolas $\lambda = \pm\infty$ and $\lambda = -1$ in the v, Θ-plane, with the point 1 to the right of the point 2 by virtue of the condition $T_2 \geqq T_1$. Suppose now that $v(x)$, $\Theta(x)$ is any solution of Eqs. (11.14). Then $u(x)$, $T(x)$ as defined by Eqs. (11.13) will satisfy Eqs. (11.12). If now we define $\rho(x) = m/u(x)$ and $p(x) = gR\rho(x)T(x)$, then the four functions $u(x)$, $p(x)$, $\rho(x)$, $T(x)$ will satisfy Eqs. (11.9), (11.10), and (11.11) and hence Eqs. (11.8). But we have already seen in Sec. 11.4 that for $m > 0$ there are solutions of Eqs. (11.14) in which v, Θ pass from state 1 to state 2 as x increases from $-\infty$ to $+\infty$. Any two such solutions are obtained from each other by translating the origin of x. Hence there are solutions of the system (11.8) in which u, p, ρ pass from their values in the state 1 at $x = -\infty$ to their values in the state 2 at $x = +\infty$. The graphs of u, p, and ρ versus x for any two solutions are, respectively, translations of each other in the x-direction.

For $m < 0$ the same result holds with the particles passing from a state 1 at $x = +\infty$ to a state 2 at $x = -\infty$. For $m = 0$ the gas is at rest with uniform pressure and density. We now combine these results and interpret them in terms of the original system of Eqs. (42). *Let u_1, p_1, ρ_1 and u_2, p_2, ρ_2 be two sets of values of u, p, ρ depending on the time t, and $\mathcal{S}$ an arbitrary line in the x,t-plane, whose slope is denoted by $dx/dt = c(t)$. Then the shock conditions* (14.2) *and* (14.9) *are not only necessary but also sufficient conditions for there to exist shock transition solutions* $\mathbf{T}(s, \beta)$, *with respect to $\mathcal{S}$ and these two states, of the equations governing the one-dimensional nonsteady flow of a perfect viscous gas in simply adiabatic motion.* For any two such solutions the graphs of u, p, and ρ versus x at each time t are translations of each other in the x-direction.

A similar result holds in the case of steady plane flow. We consider a curve $\mathcal{S}$ in the x,y-plane and this time introduce an orthogonal system of coordinates in which one set of coordinate lines are the normals of $\mathcal{S}$, see Fig. 165. A coordinate line of the other set then cuts these normals at a constant distance from $\mathcal{S}$. Thus for one coordinate, α, at a point P we may take distance from P to $\mathcal{S}$ along the normal, and for the other, β, arc length along $\mathcal{S}$ from a fixed point P_0 to the foot of this normal. The equations governing the steady plane flow of a perfect viscous gas in simply adiabatic motion are then written in terms of these new coordinates and the corresponding components of velocity u and v; a viscosity coefficient μ_0 is introduced when the two-dimensional analogue of the viscosity assumption (11.6) is used to express the stresses in terms of the velocity gradients. In order to define shock transition solutions of this system of equations the normal coordinate α is magnified in the ratio $1:\mu_0$ and then every term having a factor μ_0 is replaced by zero. Three of the equations in the resulting system are found to be equivalent to Eqs. (11.8); the fourth expresses the fact that the component of velocity, v, in the β-direction (i.e., parallel to $\mathcal{S}$) remains constant

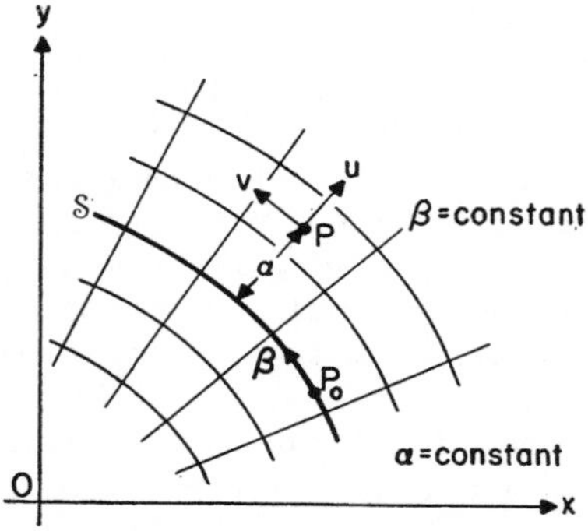

FIG. 165. Orthogonal curvilinear coordinates in the x,y-plane.

along an α-line. From this it is easily seen that *the shock conditions, Eqs.* (22.3) *and* (22.16), *are not only necessary but also sufficient conditions for the existence of shock transition solutions, with respect to* $\mathcal{S}$ *and two sets of values of* u, v, p, ρ *(depending on* β*), of the equations governing the steady plane flow of a perfect viscous gas in simply adiabatic motion.*

6. Asymptotic solutions of the equations of viscous flow

In order to understand the role of these shock transition solutions it is necessary to investigate more carefully the connection between the viscous solution of a problem and its solution by means of the principles established in Secs. 14.2 and 22.2. Suppose that a specific problem, in the x,t-plane, say, has been solved by means of the equations governing the simply adiabatic flow of a perfect viscous gas, for all small values of $\mu_0 \neq 0$. We denote this family of solutions by $\mathbf{S}(x, t; \mu_0)$, where as before the single symbol stands for the set of state variables. When x and t are replaced in it by $\mu_0 s + f(\beta)$

and β, respectively, in accordance with (39), this same family will be denoted by $\mathbf{S}(s, \beta; \mu_0)$. The dependence of the family on k is not indicated since for simplicity we may suppose that μ_0/k, which is proportional to the Prandtl number P, has the same constant value for all μ_0. Suppose also that the same problem is solved according to the principle given in Sec. 14.2; the asymptotic solution $\mathbf{S}_0(x, t)$ satisfies the equations for strictly adiabatic flow of a perfect inviscid gas at all points of the x,t-plane concerned, except for lines $\mathcal{S}$, across which the discontinuities in u, p and ρ satisfy (14.2) and (14.9).

On the basis of the discussion in Sec. 14.1 we expect that for sufficiently small μ_0, the viscous solution $\mathbf{S}$ lies close to $\mathbf{S}_0$ except in the neighborhood of $\mathcal{S}$, where it changes rapidly in the normal direction, the total change being approximately equal to the jump in $\mathbf{S}_0$ across $\mathcal{S}$ [governed by Eqs. (14.2)]. Within the transition region $\mathbf{S}(x, t; \mu_0)$ has derivatives which are unbounded as $\mu_0 \to 0$. It is plausible however that derivatives of $\mathbf{S}(s, \beta; \mu_0)$ are bounded. Then the right members of Eqs. (42) are of order μ_0, and we may expect $\mathbf{S}$ to lie close to a shock transition solution $\mathbf{T}(s, \beta)$ of (42) in this region. Now from Art. 11 we know that any fixed proportion of the total change in $\mathbf{T}(s, \beta)$ takes place in a distance of order μ_0 [see the thickness estimate (11.54), where according to (11.26) L_0 is proportional to μ_0]. Hence for successively larger proportions to be realized the interval must be of lower order than μ_0, and this will be the case for $\mathbf{S}$ also.

We now formulate this conjecture more precisely: *Let $d(\mu_0)$ be a distance which tends to zero more slowly than μ_0*, i.e., $\mu_0/d(\mu_0) \to 0$. *Then by taking μ_0 sufficiently small we can ensure, with any preassigned accuracy, that*

(a) *At points displaced more than $d(\mu_0)$ from $\mathcal{S}$, the viscous solution $\mathbf{S}(x, t; \mu_0)$ approximates $\mathbf{S}_0(x, t)$, and*

(b) *At points displaced less than $d(\mu_0)$ from $\mathcal{S}$, the viscous solution $\mathbf{S}(s, \beta; \mu_0)$ approximates $\mathbf{T}(s, \beta)$, a shock transition solution with respect to $\mathcal{S}$ and the two sets of values of u, p, ρ attained on $\mathcal{S}$ by $\mathbf{S}_0$.*

This of course has not been demonstrated in general. In the preceding section we have however proved the existence of shock transition solutions postulated by (b). In other words, it has been shown that (b) is not self-contradictory; in addition we have made it plausible. The exact solution of Sec. 11.4 provides us with the only concrete example known so far for which this conjecture can be verified.[50] The problem consists in finding a flow for which u, p, ρ take on prescribed constant values u_1, p_1, ρ_1 and u_2, p_2, ρ_2 at $x = -\infty$ and $x = +\infty$, respectively, for all t, these values satisfying the shock conditions (44) with $m > 0$, and also $T_2 \geqslant T_1$. In addition we require that $u = \frac{1}{2}(u_1 + u_2)$ at $x = 0$, say, in order to fix $\mathbf{S}(x, t; \mu_0)$. The solution $\mathbf{S}_0(x, t)$ consists of constant state values 1 to the left, and constant state values 2 to the right of $\mathcal{S}$: $x = 0$. The shock transition solution $\mathbf{T}(s, \beta)$ is

$\mathsf{S}(s, \beta; \mu_0)$ itself. This example does not depend on time, but one which does can be derived from it by superimposing a constant velocity c on the whole flow.

The situation in the case of steady plane flow is essentially the same. Its exposition is complicated however by the occurrence of a second type of rapid transition region known as the boundary layer.

These results apply equally well when μ_0 and k are functions of T, provided suitable bounds are put on their variations.

Article 25

Transonic Flow

1. On some additional boundary-value problems

In Arts. 17 and 20 we collected various examples of flows obtained by means of the hodograph method. Some of them, source flow, vortex flow, spiral flow, etc., were primarily examples of solutions of the basic equations and there was no attempt to satisfy boundary conditions. (Of course it was possible to find an a posteriori interpretation, e.g. for the flow between two streamlines considering the streamlines as walls.) An outstanding example of an exact solution of a given boundary-value problem is Chaplygin's jet problem. Various investigators have followed along similar lines.

In Art. 21 we explained the general methods of Bergman and Lighthill, which enable us to generate stream functions reducing, as $q_m \to \infty$, to the stream function of a given incompressible boundary-value problem. We shall discuss here some further results concerning flow around a profile and add some remarks on channel flow.

(*a*) *Remarks on flows past a profile.* Even in the problem of flow without circulation past a circular cylinder the elegant solutions by means of an integral formula such as found in (21.23) or (21.37) are lost, if we study the flow beyond $M = 1$.

In Sec. 21.3 we started with series expansions of the hodograph potential $w_0(\zeta)$; in the case of a circle three such series were needed to cover the ζ-plane. We also found that the compressible flow could not be correctly obtained by "translating" each of these series into its compressible counterpart, by means of the method invented by Chaplygin for the jet problem, or by an equivalent method using a different factor $f(n, \tau_1)$. The series obtained in this manner were not analytic continuations of each other; a correct procedure was to start with an appropriate branch of the compressible flow and

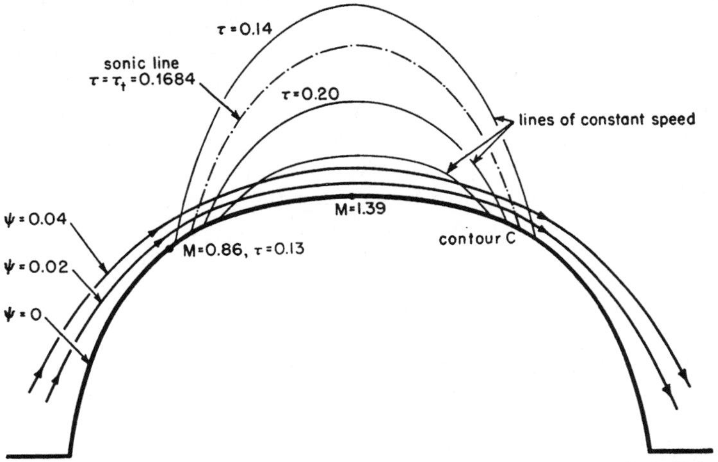

FIG. 166. Streamlines and lines of constant speed for flow around a circle-like profile derived from incompressible flow around a circle by T. M. Cherry. $M^\infty = 0.51$, $\gamma = 1.405$.

to solve the mathematical problem of analytic continuation. (This was done in Sec. 21.3 by means of an integral representation; other procedures of analytic continuation are also available.*)

Flows which for $q_m \rightarrow \infty$ reduce to flow around a circle, C_0, have been considered by several authors and by means of more or less different methods. Cherry[51] has carried out such a solution in detail. He assumes a free stream Mach number $M^\infty = M_1 = 0.51$ ($\tau^\infty = \tau_1 = 0.05$) and the flow is continued beyond the sonic line, $M = 1$ ($\tau_t = 0.17$), the maximum Mach number of the flow being $M = 1.39$, $\tau = 0.28$ (Fig. 166). The figure shows that for small values of M, the contour C (the streamline $\psi = 0$) is very close to a circle C_0; at about $M = 0.86$ ($\tau = 0.13$) the contour C starts to deviate from the circle by exhibiting a smaller ordinate than the circle. The flow obtained is thus a flow past a symmetric oval C, which stays close to a circle.

This flow around C does not encounter a limit line. If, however, the continuation of the solution inside C—which has no physical meaning—is computed, a limit line appears at about $\psi = -0.06$, with cusp well off the contour $\psi = 0$. Cherry has also obtained,[52] with the same condition at infinity, the flow past a slightly cambered cylinder (contour $\tilde{C}$ is no longer symmetric to the x-axis, but slightly thicker above and thinner below the x-axis as compared to C). In this case, the maximum Mach number rises

* See e.g. papers cited in Note 52.

to $M = 1.56$ and the cusp of the limit line comes quite close to $\tilde{C}$: at the cusp the Mach number is approximately 1.34. Of course, if the cusp were to break through the contour, there would no longer be a physically valid solution. For $M^\infty = 0.6$ a flow (with C_0 a circle) has been computed showing a limit line which has actually penetrated the curve from the inside, while for parts of the contour where $M \leqq 0.5$ the curve obtained differs little from a circle and no other limit line appears in the flow. It should be kept in mind that we consider here for one and the same given contour P_0 a family of flows (with the free-stream Mach number M^∞ as parameter); each of these reduces to flow about P_0 as $q_m \to \infty$; however, the shape of P, i.e. of the profile determined, depends upon M^∞, so that in varying M^∞ we do *not* study various flows about the same profile. (This is also important for the evaluation of the so-called limit line conjecture. See Sec. 4.)

Cherry has also shown how to proceed in the problem of flow past a circle if circulation is present. Lighthill's method in this case meets with certain theoretical difficulties.[53]

Lighthill, as well as Cherry, has indicated a general method for continuing the flow around a contour into the supersonic region. (For solutions in the subsonic region see Secs. 21.3 and 21.4). As in the example of the circle these methods are based on series expansions of $w_0(\zeta)$ and systematic procedures are given for finding the analytic continuation starting with an appropriate branch of the hodograph flow. Considerable practical difficulties are encountered.

Apart from these difficulties let us reconsider the basis of our investigations (cf. also beginning of Sec. 21.1). We started out to find a compressible flow past a given closed contour P_0; in other words, we wanted a solution of the compressible flow equations with P_0 as a streamline. We obtained, however, a family of flows depending on M^∞ as parameter, which, only as some representative quantity tends to a limit, has the given P_0 as a streamline. The shape actually obtained for the contour P in the x,y-plane depends on the value of M^∞, and varies with it. One can conclude from the results of Bergman, Cherry, and Lighthill that for values of M^∞ in a certain interval $0 < M^\infty < M_1$, a purely subsonic flow around a closed profile $P(M^\infty) \equiv P_M$ in the physical plane obtains. One can even assert that for values of M^∞ in an interval beyond M_1, say $M_1 < M^\infty < M_2 < 1$ there exists a flow past a P_M, with an imbedded supersonic region. This value M_2 may however be very close to M_1. The size of the supersonic region as well as the value of the maximum Mach number in that region depends on the value of M^∞ [in the interval (M_1, M_2)]. On the other hand, when a hodograph solution is constructed, even by a completely correct analytic continuation, so long as we are not certain of the value of M_2 (and therefore whether the chosen M^∞ is less than it), we cannot be sure that the final profile P_M in the x,y-plane will be closed, without double point, etc.

In practice, if one works out a solution to the end, actually determining the physical-plane flow and the curve P_M (as in Fig. 166), and if this P_M is closed, is not reached by a limit line, and is sufficiently close to P_0, then of course one will have found an approximate solution of the original problem.

(b) *Remarks on channel flow.* Consider a channel (we use the words duct and nozzle in the same sense), symmetric with respect to a straight axis, or center line, which we take as the x-axis. In the so-called de Laval nozzle, there is a minimum cross section at the *throat*, at $x = 0$, say; on each side of the throat the cross sections increase symmetrically. The contour is thus formed by two lines converging through the "entry" section to the throat, then diverging from the throat through the "exit" section. Two main types of flow can be distinguished, which we describe in the simplified one-dimensional or "hydraulic" way, where it is assumed (a) that we may neglect the deviation of $\mathbf{q}$ from the horizontal direction, and (b) that over a cross section normal to the center line the speed q and consequently pressure, density, etc., are the same. Then the flow may be either symmetrical with respect to the throat with subsonic velocities on both sides of it, and subsonic or sonic speed at the throat, or it may be asymmetrical with subsonic speed on one side of the throat and supersonic speed on the other (at the throat the speed is sonic).

In the symmetric type the flow starts from a state of high pressure with velocity zero at $x = -\infty$. It then accelerates while expanding throughout the converging entry section, reaching its maximum velocity at the throat. Then it decelerates while being compressed and takes on zero velocity again at infinity. This is the simplified one-dimensional description. Actually, the velocity is not the same throughout a cross section. For given contour of a channel it is possible that the speed remains subsonic throughout, even across the whole throat section; or, for the same contour, it may remain subsonic along the x-axis, the axis of the channel, while two supersonic "pockets" symmetric to both x- and y-axes form next to the wall, with the greatest speed reached at the wall.[54] Figure 167 shows curves of constant speed, the shaded regions representing the supersonic enclosures.

A nonsymmetric type of channel flow occurs when the ratio of entrance pressure to exit pressure is above a certain limit. The velocity rises from zero at $x = -\infty$ to a supersonic value to the right of the throat, while the gas is expanding.[55] (Notice the previously mentioned fact that subsonic flow is compressed and supersonic flow expanded in a diverging section.) The lines of constant speed are now quite different from those in the preceding case, as indicated in Fig. 168. At the wall the sonic speed is reached upstream of the throat, and at the center line downstream of it.

This flow exhibits a singularity so far not encountered in our examples of plane flow, namely a branch line. This singularity, typical of asymmetric channel flow, has been investigated by Lighthill and by Cherry. The phenomenon may be qualitatively understood if we follow in the hodograph the velocities along a transonic streamline.[56] With the axis of the channel as the x-axis, $q_y = 0$ everywhere along the center line of the channel so that the q_x-axis in the hodograph is the streamline $\psi = 0$ (Fig. 169). Consider a streamline in the upper half of the channel, $y > 0$; in the subsonic region $q_y < 0$ on this streamline; but on this same streamline in the supersonic region eventually $q_y > 0$. Thus, each streamline intersects the line $\psi = 0$ at a second point, in addition to the hodograph origin from which all streamlines start. These streamlines intersect each other and they have an envelope. We know from the general theory, Art. 19, that this envelope is

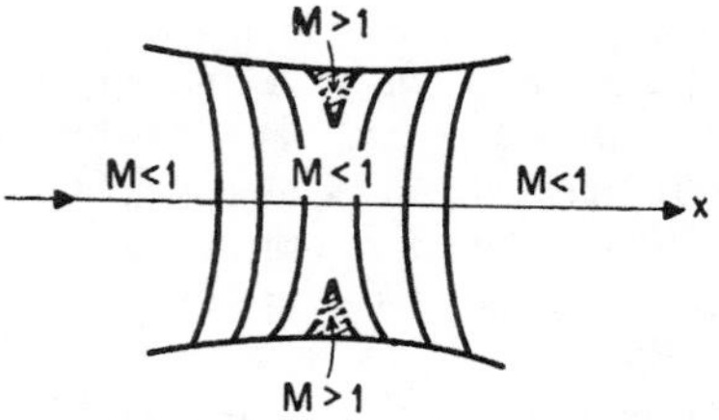

FIG. 167. Symmetric channel flow with supersonic enclosures (shaded); lines of constant speed are shown.

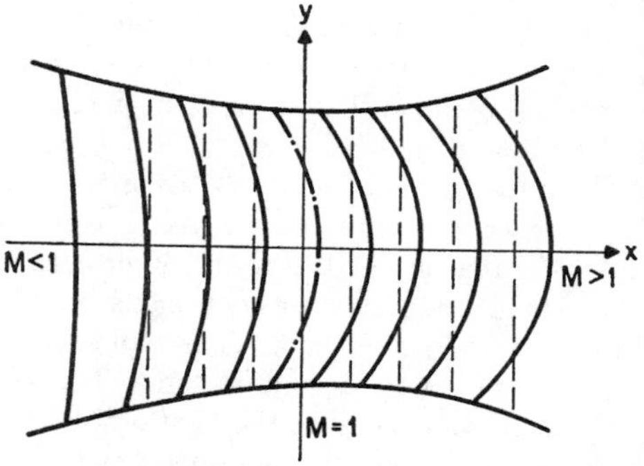

FIG. 168. Asymmetric channel flow with lines of constant speed. Sonic speed reached at wall upstream of the throat and at axis downstream of it. Vertical lines correspond to one-dimensional theory.

a characteristic Γ^+. A symmetric family of streamlines exists with the symmetric Γ^- as envelope. These two characteristics form an edge and its image in the x,y-plane is the branch line. The region (in the hodograph) between the Γ^+ and Γ^- is covered three times (Fig. 170). Through a point P in the region with positive q_y passes (a) a streamline which has cut the Γ^- (the ξ-line in Fig. 170) before reaching P, and then contacts the Γ^+; (b) one which crosses the Γ^+, reaches P and contacts the Γ^-; and (c) one which crosses the Γ^-, contacts the Γ^+ and then reaches P.

Thus it is not possible to use q, θ as independent variables in an analytic treatment; however φ and ψ may replace them, and we may then expand $q(\varphi, \psi)$ and $\theta(\varphi, \psi)$ in series in the neighborhood of a sonic point and substi-

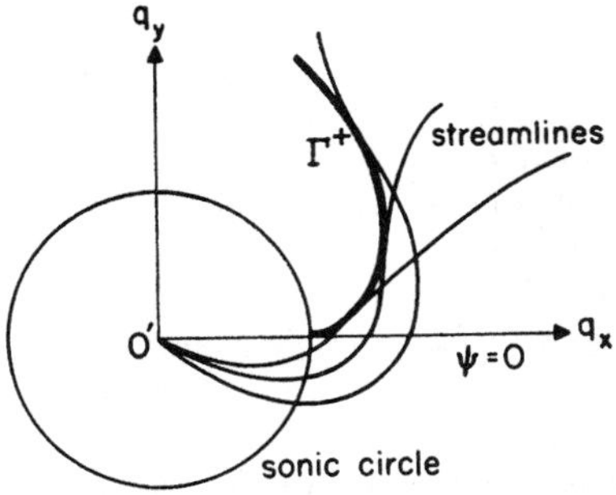

FIG. 169. Hodograph with streamlines and branch line Γ^+ in transonic channel flow.

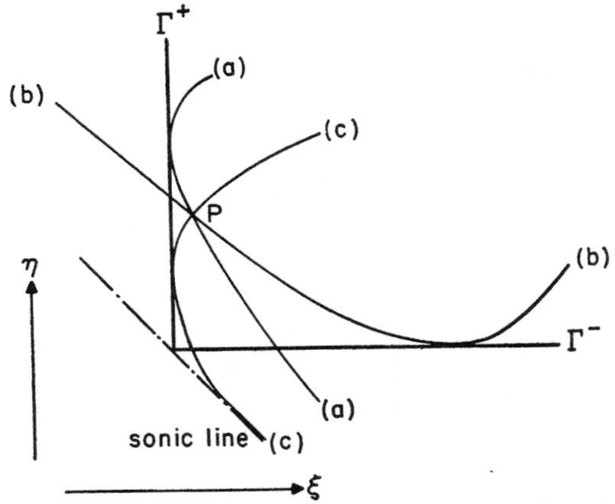

FIG. 170. Triply-covered region between branch lines in characteristic ξ,η-plane for transonic channel flow.

tute these expansions in a system obtained by interchanging variables in (16.31). Figure 171 shows the region near the sonic point on the axis in the physical plane, as investigated by Lighthill and Cherry. The branch line consists of the characteristics C_b^+ and C_b^- curving toward the right from the point of contact with the sonic line. The continuation C^+ of C_b^+ which (with an inflection point) goes down towards the left, is not a branch line [i.e. the Jacobian $\partial(q, \theta)/\partial(x, y) \neq 0$], and the same holds for the symmetric continuation C^- of the C_b^-. To each point P on the branch line C_b^+ belongs a q and a θ (of course subject to the compatibility relation), and there is some point Q which has the same q and θ, therefore satisfying the same compatibility relation (Fig. 171 shows the line of constant q and that of constant θ through P and Q); the same arguments apply to the C_b^- and C^-. To the left of the C^+ (C^-), the correspondence between x,y and q,θ is one to one. In the hodograph the C_b^+ and C^+ are both mapped into one Γ^+, the C_b^- and C^- into one Γ^-, which appear respectively as one η-line, and one ξ-line, in the ξ,η-plane. (It may be useful to visualize the mapping into the hodograph of a closed curve, such as a circle with its center at the sonic point of the x-axis. Outside the two branches of the edge the mapping is one to one. The image of the curve we are considering crosses the Γ^+, continues between the two branches towards the Γ^-, goes back again to the Γ^+, then down again crossing the Γ^-, and continues outside the two branches, forming a closed curve.)

In regard to the problem of transonic channel flow as compared with the situation considered in Art. 21, the essential difference is the following. There we could try to construct a flow along the lines of the corresponding incompressible flow, a flow which had the incompressible flow

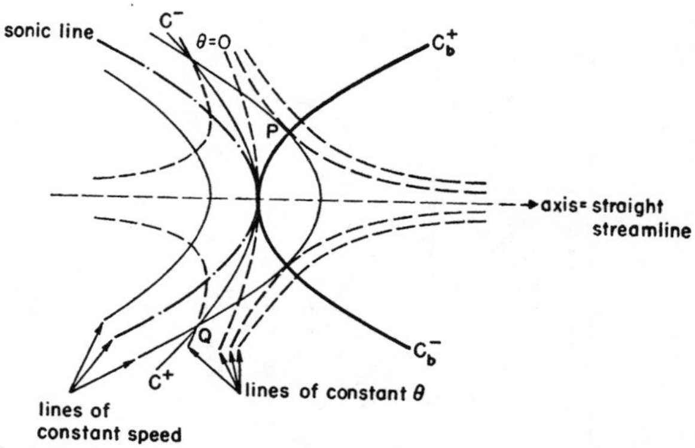

Fig. 171. Sonic point of straight streamline as double branch point in transonic channel flow.

as its limit. In the present problem, however, there is no limit case to guide us. Cherry has successfully overcome this difficulty by constructing a flow which exhibits typical channel-flow properties as explained above. For this work we refer to the original paper.[57]

2. Problem of existence of flow past a profile[58]

In preceding articles we have pointed out repeatedly that the nonlinearity of our basic equations was a source of essential difficulties. Such difficulties occur even if the problem is known to be entirely subsonic (elliptic) or entirely supersonic (hyperbolic). A second difficulty, the possibility of a partly elliptic, partly hyperbolic problem, is not restricted to nonlinear equations. This may be seen in the example of the linear Tricomi equation, $y(\partial^2u/\partial x^2) + (\partial^2u/\partial y^2) = 0$, which is elliptic for $y > 0$ and hyperbolic for $y < 0$. (Chaplygin's equation (17.24) can be approximated by Tricomi's equation for values of σ close to zero, i.e., values of q close to q_t.) However, since this equation is linear we can indicate in advance the regions of elliptic and of hyperbolic behavior and the transition line, namely $y = 0$.[59]

The deep-seated complication typified by the mixed flow of a compressible fluid lies in the combination of transition and nonlinearity which makes the hyperbolic or elliptic character depend on the solution considered, so that the transition line also varies, depending on the solution. Furthermore, a particular solution is to be singled out by appropriate boundary conditions; but the boundary conditions appropriate to an elliptic problem are quite different from those for a hyperbolic problem. Thus we anticipate a new situation which has not yet been clarified, and which is at the basis of the difficulty of indicating correct boundary conditions for mixed problems, such as channel flow, flow past a profile, etc.

To fix the ideas, consider the flow past a profile and specifically the following situation:[60] a steady uniform flow parallel to the x-axis is disturbed by the presence of a body which has, in the x,y-plane, a (convex) contour P with continuously changing tangent and curvature; we suppose given a (p, ρ)-relation and a scale factor q_m or a_s relative to which the velocity q^∞ of the undisturbed flow is subsonic.* Then, in analogy to the corresponding boundary-value problem for incompressible flow past a smooth profile, we might ask: Does there exist a smooth potential flow (that is, a flow without viscosity and heat conduction, and also without shock discontinuities) past the given profile P which coincides at infinity with the given undisturbed flow? Specifically, does there exist a potential function $\varphi(x,y)$ with continuous first and second derivatives which satisfies Eq. (16.14) everywhere outside the profile P together with the conditions $\partial\varphi/\partial n = 0$

* We consider here q_m as fixed and q^∞ (or M^∞) as a varying parameter.

along P (where $\partial/\partial n$ denotes differentiation in the direction normal to P) and, as $z = x + iy \to \infty$, the condition $\partial\varphi/\partial x \to q^\infty$, $\partial\varphi/\partial y \to 0$?

Let $q_{\max}$ denote the greatest velocity attained in the compressible flow under consideration; then with q_m now fixed, $q_{\max}$ depends on both q^∞ and P and satisfies $q_{\max} \leqq q_m$. If $q_{\max} < q_t$, then the entire flow is subsonic; however, if $q_{\max}$ is supersonic, the flow is necessarily mixed, i.e., transonic, since the velocity is subsonic at infinity, by hypothesis, and also at stagnation points on the profile.

In the first case, that of purely subsonic flow, a potential flow always exists, just as in the case of the analogous incompressible problem. This was shown independently by L. Bers and by M. Shiffman.[61] More precisely they have shown that to a given profile P of the type described above there corresponds a subsonic velocity $q_1 < q_t$ such that, for $q^\infty < q_1$, there is a unique solution of the boundary-value problem formulated above; moreover, that this solution is purely subsonic: $q_{\max} < q_t$, and, as q^∞ varies over the open interval from 0 to q_1, the maximum velocity $q_{\max}$ varies over the range from 0 to q_t. (If the profile P is symmetric with respect to the x-axis it can even be shown that $q_{\max}$ is a monotonic function of q^∞.) If $q^\infty = q_1$, then the velocity must be sonic at some point of the profile and if $q^\infty > q_1$, local supersonic speeds cannot be excluded; both the existence and uniqueness proofs fail.

At present our main concern is the case of a subsonic $q^\infty > q_1$. We ask whether a solution exists, in the form of a mixed flow including supersonic regions (where there may occur the type of discontinuities which we have found possible for hyperbolic problems) which is still a potential flow. More precisely, we ask whether, as q^∞ varies over an appropriate interval beyond q_1, there exist mixed flows past a given smooth convex profile P, in which the supersonic regions form enclosures, or "pockets" adjacent to the profile. (See Fig. 175, disregarding for the moment the lines OT_1 and OT_2. See also Fig. 166.) This last requirement is based upon previous examples and on the fact that, in subsonic flow, the maximum velocity is attained on the profile.

3. Apparent conflict between mathematical evidence and experiment.

Certainly, smooth transonic flows can exist; in fact, we have studied several concrete examples of this type. In the absence of a mathematical existence proof, we begin with a survey of the properties of these known solutions, including also the somewhat analogous case of channel flow in which the walls of the channel replace the profile.*

* However, in contrast to the problem of the profile the available space in the channel is limited.

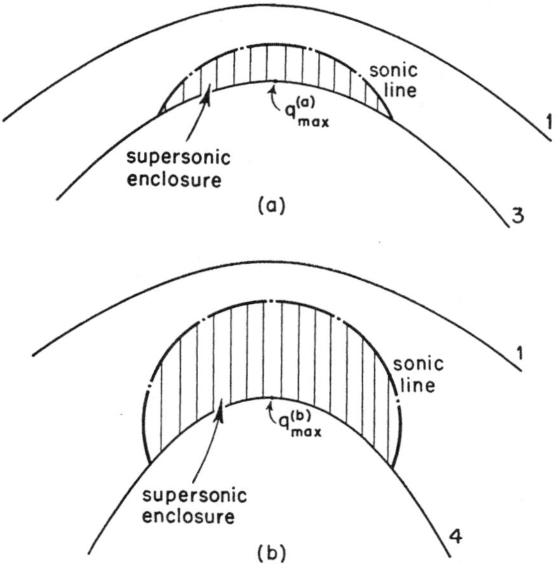

FIG. 172. Two transonic nozzles derived from Ringleb's flow.

In Ringleb's flow, Fig. 132, we may consider two streamlines as fixed walls, as in Fig. 172. We obtain in this way examples of smooth transonic channel flow; each nozzle shows a supersonic enclosure and there is a specific maximum speed obtained at one of the walls. For Fig. 172: $q_{\max}^{(a)}$ is obtained at the vertex of streamline 3, while $q_{\max}^{(b)}$ appears at the vertex of streamline 4, and $q_{\max}^{(b)} > q_{\max}^{(a)}$; on all streamlines $q^{\infty} = 0$.* Actually we obtain in this way a whole family of smooth transonic flows with varying $q_{\max}$, the corresponding maximum value of M being well above one. We may choose as walls of our channels, smooth streamlines not reached by the limit line (as in Fig. 172a). Between such smooth walls the theoretical flow accelerates continuously from a subsonic through sonic to supersonic speed and smoothly decelerates again.

An exact transonic flow solution, not of the pocket type, but rather of the type of the nonsymmetric subsonic-supersonic channel flow, is the mixed spiral flow between two streamlines (Sec. 17.4). We may also mention the vortex flow between two concentric circles, separated by the circle where $M = 1$. In this flow there is no limit line.

In Sec. 21.3 we constructed in detail a flow past a profile similar to a

* Note that in this case q^{∞} (or M^{∞}) would not be an appropriate parameter.

circle and in Sec. 1 we discussed the work of Cherry and his students, who computed such flows with supersonic enclosures and also flows past profiles similar to other prescribed shapes (see Note 52). We also know that to a given profile P there exists a speed q_1, such that for $q^\infty < q_1$ only a purely subsonic flow past the given P results.

From general considerations and the results of Lighthill and Cherry we can add that, for each given P_0, there exists a subsonic $q_2 > q_1$ such that for any q^∞ in the interval between q_1 and q_2 a transonic flow with supersonic enclosures is obtained about some smooth profile, $P(q^\infty) \equiv P_q$. However, in all these examples of flows defined so as to reduce to flow past a given profile P_0, as $q_m \to \infty$, the actual profile P_q, found by the indirect hodograph method, changes in shape with q^∞. Hence, these results do not afford examples of flows past a given fixed profile P for varying free stream speeds. (In the channel interpretation of Ringleb's flow the effect of changing profile is also observed, inasmuch as the shapes of the walls are changed as we consider regions of increasing $q_{\max}$).

We mention in this connection interesting results found by S. Tomotika and K. Tamada by means of a procedure specially adapted to the region of transonic flow.[62] The profiles they obtain vary with q^∞ (or M^∞); they found, however, that for varying (subsonic) values of M^∞ ($M^\infty = 0.717$, 0.745 and 0.752) the resulting profiles P_M were almost identical. One may thus consider their flows as an approximation to flow about a fixed profile for varying values of the parameter.

Finally, flows past *a given profile P*, as well as various types of channel flow, have been computed by some of the usual approximation methods. We shall give a few results obtained by such methods inasmuch as they are of interest for the present problem. G. I. Taylor has computed a symmetric nozzle flow with supersonic pockets by means of a series expansion in the physical plane.[63] Another method, of which the idea is due to Prandtl, consists in an iteration procedure with the "linearized flow" (Prandtl-Glauert method) as first approximation.[64] Arranged, by H. Görtler, in the form of an expansion in powers of a thickness parameter, it likewise produced smooth transonic flow patterns past given profiles; no obstacle seems to appear in the computations. Finally, H. Emmons successfully applied the classical method of replacing the original differential equations in the physical plane by difference equations and solving the resulting system of a finite number of algebraic equations by means of an iteration procedure (a method denoted today as "relaxation method"). Again, the numerical work could be carried through and some examples of potential flows with bounded supersonic regions were obtained.[65]

Thus we see: Exact transonic flows (spiral flow, Ringleb flow, etc.) exist for certain specific contours; flows with supersonic enclosures have likewise been found (by means of hodograph methods) past contours determined *a posteriori*. Approximate solutions past *given* contours have been found by various direct methods.

There seems, however, to be little experimental confirmation for the existence of such smooth mixed potential flows with imbedded supersonic regions. Smooth *acceleration* from subsonic to supersonic flow has been observed. But the observed *deceleration* from supersonic to subsonic flow is not such as might be expected from the mathematical evidence. In general, the deceleration takes place in a nonisentropic way by means of a shock, although deceleration with no observed shocks has also been reported for values of q^∞ not much larger than q_1.[66] There is no mathematical counterpart in the work discussed above (based on potential flow, which is reversible) to the observed stormy deceleration, and to the general lack of smoothness in observed transonic flows.. Thus there seems to emerge a definite discrepancy between certain theoretical results and the observations.

However, the experimental observations are made on a given *fixed* profile with some appropriate parameter, say M^∞, being varied so as to generate a family of flows past the same profile*; whereas in our theoretical examples the profile P_M changes along with M^∞. Comparing observations on flows about a fixed profile P with mathematical results regarding flows about a sequence of varying profiles P_M presupposes the unproved assumption that the essential features are the same for the two different situations. It is possible that the difference between the two situations may be of minor importance, in view of results such as those of Tomotika and Tamada where almost identical profiles P_M were found for different values of M^∞. However, as long as we have no more definite information on this point we cannot strictly speak of a discrepancy between observations and theory, since they do not refer to the same situation.

4. Limit-line conjecture

The general lack of experimental counterpart to the computed examples of smooth transonic flows points to a discrepancy between theory and observation, although it is true that in the computations we do not strictly reproduce the experimental situation. On the other hand, we note that some remarkable singularities are found in the mathematical study of flows, namely, limit lines. Can this furnish an explanation?

Comments and suggestions of v. Kármán, Tsien, Kuo, and others, prompted to a considerable extent by Ringleb's example, deal with this possibility. After enumerating some of the singular properties of limit lines (infinite acceleration, infinite pressure gradient, streamlines "turning back"

* Similar comments apply to channel flow.

at the limit line and causing "a quite impossible flow pattern", etc.) v. Kármán concludes that of the fundamental physical assumptions, continuity and irrotationality, which underly our analysis, one must be untenable. "Since the continuity cannot be violated, it must be assumed that the flow becomes rotational." This cannot happen in an inviscid steady compressible flow free of shocks. Thus the appearance of a limit line is considered by v. Kármán as the (mathematical) criterion for the (physical) breakdown[67] of steady isentropic inviscid (irrotational) flow.*

This idea, which is at the basis of the *limit-line conjecture*, has not been formulated as a precise statement but rather as a general conception which postulates for the problem under consideration *a mutual dependence between observed physical shocks and computed limit lines in the flow field.* (Other explanations based on quite different ideas have also been proposed by the same authors, in particular by Tsien.) Historically, the limit-line conjecture is strongly connected with the actual investigation of the mathematical properties of such lines; since these properties turned out to be of a very singular nature implying the breakdown of the smooth mathematical potential flow, a relation suggested itself between this breakdown and the physical phenomenon of shocks. It was then thought that the converse was also true, namely, that a physical shock implied a mathematical "shock-solution", exhibiting a limit line.

In more precise terms the conjecture may be described as follows. In the case of flow past a fixed profile, we know that if the Mach number at infinity is less than a certain M_1 then all velocities are subsonic. It is then generally assumed that there is a value $M_2 > M_1$ (which may be very close to M_1) such that for M^∞ between M_1 and M_2 smooth mixed flows result which under certain circumstances should be of the "pocket type".† It is assumed on one hand that the physical flow, for these values of M^∞, is without shocks and on the other hand, that the Jacobian $I = \partial(x, y)/\partial(q, \theta)$ (or an equivalent determinant) is different from zero everywhere in the flow field. (For subsonic velocities this Jacobian is negative, except perhaps at isolated points.) If then M^∞ is further increased to $M^\infty = M_2$, shocks will become manifest and at the same time, in the mathematical description of the problem, $I = 0$ will be found along some infinitesimally small arc of curve. This limit line will become more pronounced if M^∞ is still further increased.

* Ringleb says that at the limit line there occurs a *Stroemungsstoss* and he calls the limit line *Stosslinie*, "Stoss" being the German word for shock.

† It seems that this last point is generally assumed by analogy with the similar fact which holds for flows past varying profiles and on account of results obtained by approximation methods (Sec. 3). At any rate, this assumption should not be considered as part of the "limit-line conjecture".

We shall see at the end of this section that a careful consideration of the various examples which were discussed in the preceding sections in fact does not support the limit-line conjecture. However, historically, the decisive criticism was of a purely analytical nature. K. O. Friedrichs has investigated the mathematical question (implied in the above-described conjecture):[68] Is it possible that $I < 0$ everywhere in the flow field for $M^\infty < M_2$, but that for $M^\infty = M_2$ we find $I = 0$ somewhere in the supersonic region (i.e., in the interior of the pocket or on the sonic line or on the contour of the body)? His answer, derived under rather strong assumptions regarding the nature of the flow, is negative. (Friedrichs' work was preceded by results in the same direction due to A. A. Nikolskii and G. I. Taganov[69] which are of interest in several respects.) His result and proof were improved by A. R. Manwell and later by C. S. Morawetz and I. I. Kolodner[70] who were able to reduce Friedrichs' assumptions. He had assumed solutions $x(q,\theta)$ $y(q,\theta)$ where x and y were analytic functions of q, θ depending continuously on M^∞. In the last-named paper the authors require only, in addition to the continuous dependence of the solution on M^∞, that the streamfunction $\psi(q,\theta)$ have continuous second derivatives. The result is based on a lemma which can be applied to either of the two situations in which we are interested: the case of flow past a fixed profile studied for varying values of M^∞, and the case where the profile itself varies with M^∞. Their main results are essentially as follows. (a) Consider a potential flow past a given fixed profile with bounded curvature. Assume that the flow depends continuously on M^∞ and that, for a certain value of M^∞, the flow is mixed and has supersonic pockets. Then no limit line can appear in this supersonic pocket, either in its interior or on the boundary. (b) If by means of the hodograph method a set of profiles P_M in the physical plane is constructed, depending continuously on M^∞ and if for an $M^\infty = M_2$ a limit line appears, the corresponding profile P_M can no longer be everywhere of bounded curvature.

We now ask: What is the relation between the mathematical results reviewed in the preceding section and the above results of Friedrichs and other authors? Regarding all these theoretical results, we can see that inasmuch as they have the same subject they clearly support each other. We see that (a) can offer neither support nor contradiction since we have no mathematical example of a family of transonic flows past a fixed profile. Further, let us confront statement (b) with what the study of hodograph examples has taught us. We studied flows past profiles P_M which vary as a parameter M^∞ is varied in such a way that an originally subsonic flow (where $M^\infty < M_1$) becomes transonic for $M^\infty > M_1$. For these flows past profiles of bounded curvature, the limit line was found *inside* the profile (see Sec. 1), i.e., *outside* the flow region. As M^∞ was further increased, in a range $M_1 < M^\infty < M_2$, a cusp of the limit line approached and finally reached the *changing* profile P_M which was thereby distorted; in particular at a common point of contour and limit line there is infinite curvature of the contour (as seen in Art. 19). All this is in complete agreement with the results of Friedrichs and his followers: limit lines do not appear in the flow field, i.e., *outside* of the profile or on it, and if a limit line meets the profile its curvature cannot be everywhere bounded.

On the other hand, we may wish to compare the mathematical results including those of Friedrichs and others with the experimental evidence, and in particular, to check the limit-line conjecture. The statement (a) of Kolodner and Morawetz which relates to a fixed profile may be compared with observations on fixed smooth profiles; we know that shocks have been observed in connection with deceleration but, according to (a) for a contour with bounded curvature a limit line cannot appear in the flow field or on the contour. Thus, this comparison points against the conjecture "that the ultimate collapse of potential transonic flow is due to a limit line in the flow field".

This is also confirmed by a careful consideration of the examples described in Sec. 1. There *are* transonic flows past smooth profiles, showing supersonic enclosures with Mach numbers so far beyond one that experiments would produce shocks—and no limit line appears in the flow field. Hence (insofar as the comparison between these mathematical results regarding varying profiles and observations on fixed profiles is valid) the limit-line conjecture does not work here. Also, at any rate, wherever limit lines were obtained, they did not and could not show a relation to the above-mentioned *asymmetric* behavior where shocks appear associated primarily with transonic deceleration rather than acceleration. We repeat: the results of our computations and our examples (Sec. 17.4, Art. 20, Secs. 21.3 and 25.1) are in agreement and mutual confirmation with the mathematical results reported in the preceding pages. The limit-line conjecture even in the general sense of some kind of parallelism between physical shocks and mathematical limit lines does not seem to hit the real problem.

5. The local approach

The situation described in Secs. 2–4, with its contrast between physical observations and mathematical results, would be the more disturbing if we were in possession of a mathematical existence and uniqueness theorem regarding the problem explained at the beginning of Sec. 2. Since this is not so, efforts have been made in the other direction with the aim of proving, or at least of making it plausible, that the known examples of flows are not typical but rather exceptional.

It will be useful for what follows to introduce the concept of a "well set" or "correctly set" problem, in the sense of J. Hadamard.[71] A boundary-value problem for a partial differential equation is said to be *correctly set* or correctly posed if a solution exists, is unique, and depends continuously on the given boundary data. The precise nature of the continuous dependence on the data must of course be specified in each problem, as well as the class of functions among which one is looking for a solution. As an example of a correctly set problem, we mention the existence theorem for subsonic flow, explained in Sec. 2 where the solution can be shown to vary continuously with q^∞ and with P.

Returning to our present subject, we shall show that in the neighborhood of a profile P past which a transonic potential flow with supersonic pockets exists, other profiles can be found for which no such neighboring flow exists, although they satisfy all assumptions made at the beginning of Sec. 2.

We have seen in Sec. 18.3, Fig. 118, that in a specific instance we destroy smooth flow by introducing an arbitrarily small concave arc into the given contour. Convexity of the profile was, however, one of the hypotheses of our problem (Sec. 2). But it can be shown that a flow of the type we consider becomes impossible even if a part of the contour, within a supersonic pocket, is straightened along an arbitrarily small segment.

This interesting result was first proved by A. A. Nikolskii and G. I. Taganov (see Note 69); we give a brief but complete proof, deriving first a simple and basic inequality. If we put $M = 1$ in the first of Eqs. (16.7) we obtain $\partial\theta/\partial n = 0$ or $\partial\theta/\partial y = \tan\theta\,(\partial\theta/\partial x)$; then, the second Eq. (16.7) may be written as

$$\frac{\partial q}{\partial x}\sin\theta - \frac{\partial q}{\partial y}\cos\theta + \frac{q}{\cos\theta}\frac{\partial\theta}{\partial x} = 0.$$

Now take the y-direction normal to the sonic line S, and pointing toward subsonic velocities, and the x-direction tangent to S so that we turn from positive x to positive y by $+90°$; then, we write $\partial/\partial\sigma$ for $\partial/\partial x$ and $\partial/\partial\nu$ for $\partial/\partial y$, and θ_1 for the angle between sonic line and flow direction. Since $\partial q/\partial\sigma = 0$ along S, we obtain immediately

$$\frac{\partial q}{\partial\nu} = \frac{q}{\cos^2\theta_1}\frac{\partial\theta}{\partial\sigma}. \tag{1}$$

From $\partial q/\partial\nu \leqq 0$, it follows that $\partial\theta/\partial\sigma \leqq 0$ and we obtain the "monotonicity law": *If a point moves along the sonic line S so that the subsonic region is to the left, then the polar angle θ of the velocity vector* $\mathbf{q}$ *at the point cannot increase.*

We now apply this result to our present problem. We consider a piece of the contour along which the velocities are supersonic, i.e. in the pocket, and two points A, B on it. Through each of them we draw both characteristics and obtain the points of intersection of these characteristics with the sonic line, e.g., in the order $A_1\, B_1\, A_2\, B_2$.* Consider the hodograph image of the arc AB and of the four points of intersection. Because of the monotonicity law the images A_1', B_1', A_2', B_2' must lie on the sonic circle in the hodograph in the same order as their originals in the physical plane.

Next the contour is deformed in such a way that, within the pocket a straight segment A_*B_* is inserted. This can be done in such a manner that any number of derivatives of the function which determines the contour remain continuous. Along A_*B_* we have $\partial\theta/\partial s = 0$, and hence $\partial q/\partial n = 0$. Consider first the case that the speed q increases (or decreases) along A_*B_* or that there is a subsegment of A_*B_* along which q is monotonic. Then the hodograph image $A_*'B_*'$ of the (sub)segment A_*B_* will be a segment of a radius through the origin O'. Clearly the points of intersection of the four epicycloids through A_*', B_*' with the sonic circle will be in the order (omitting the stars) $A_1'B_1'B_2'A_2'$ (or $B_1'A_1'A_2'B_2'$), and hence not in the order required by the monotonicity law. Thus there is a contradiction.[72]

Now assume that q is constant along A_*B_*. The hodograph image is a single point A_*' and clearly the above contradiction cannot be derived. However, we can conclude that adjacent to the straight segment along which q and θ are constant, there is a small triangle in the supersonic pocket, bounded by that segment and two intersecting straight characteristics through its end points, in which $\mathbf{q}$ = constant. We call this triangle A_*B_*D. Adjacent to the straight characteristic A_*D there must then be in the pocket a simple wave W^-, say, with straight characteristics C^+, and cross-characteristics C^-. This, however, leads to a contradiction, since we have seen (Art. 18, p. 296) that the distance between two cross-characteristics measured along the straight characteristics tends to infinity as the Mach angle α tends to 90°. Thus it is impossible that a simple wave contained in the finite pocket extends all the way to the sonic line. On the other hand, as is easily seen, it is also impossible that before the sonic line is reached, the C^- pass out of the simple wave region. Hence we again reach a contradiction.

We have thus proved that a flattened segment of a contour within a supersonic pocket is incompatible with our assumptions. Hence in the neighborhood of admissible profiles there are certainly profiles for which no neighboring solution exists. Therefore (unless flattened profiles are excluded from the considerations) the original problem is not a correctly set one.†

* This order can be ensured by taking AB sufficiently small; for all arcs within such an AB the same order will then appear. Another possible order is $A_1\, A_2\, B_1\, B_2$.

† We must in principle admit the possibility that by flattening a piece of contour the whole flow pattern changes abruptly so that the supersonic pocket moves away and no longer contains the flattened piece. Such an abrupt change of flow correspond-

On the other hand take the classical problem of a Laplace potential flow. It is certainly true that in the infinitesimal neighborhood of an admissible contour (i.e., one for which the incompressible flow problem past the contour has a solution) there are contours with corners, inadmissible because of infinite velocities. Nevertheless, there is a significant difference between the two situations: the compressible flow seems to be much more "sensitive" than the Laplace potential flow to a variation of the contour. A contour may be flattened without introducing a corner, i.e., a flattened contour can still have continuous curvature and actually as many continuous derivatives as we please; in this case there would be no trouble for Laplace's equation. Interest in these considerations thus lies in the observation that even so slight a discontinuity as that introduced by the flattening can make the contour "inadmissible".[73] This sensitivity may be considered as pointing towards an explanation of why smooth transonic potential flow of the type considered is rarely observed in nature.

Frankl, Guderley, and Busemann[74] have discussed the problem of possible general lack of neighboring flows by means of suggestive arguments, partly physical, which make it plausible that the slighest irregularity of the contour leads to breakdown of potential flow.

6. Conjectures on existence and uniqueness in the large

We recall the problem that was taken as point of departure, and ask ourselves why we actually thought that an existence theorem might hold for transonic flow. The answer is obviously that it was formulated in analogy to the classical incompressible flow problem (a linear problem) and is supported by the theorem for compressible subsonic flow (a nonlinear problem). However, certain results holding for linear but mixed problems point towards a negative rather than a positive answer; the corresponding conjectures concerning our problem are particularly suggestive. Only a few hints can be given here.

In 1923 Tricomi studied the equation of mixed type

$$y\frac{\partial^2 u}{\partial x^2} + \frac{\partial^2 u}{\partial y^2} = 0 \tag{2}$$

mentioned at the beginning of Sec. 2. More generally, writing $u_{xx} = \partial^2 u/\partial x^2$, etc., the equation

$$A(x, y)u_{xx} + B(x, y)u_{xy} + C(x, y)u_{yy} = F(x, y, u, u_x, u_y) \tag{2'}$$

is an equation of mixed type if the function $\Delta(x, y) = B^2 - 4AC$ changes sign across a curve without vanishing identically.

---

ing to an arbitrarily delicate change in contour would imply that the problem is not correctly set.

Returning to Eq. (2) we shall show why a correctly posed mixed boundary-value problem may differ essentially from the analogous elliptic problem (incompressible or compressible subsonic flow past a profile is elliptic). Consider (Fig. 173) a region bounded by an arc C_0 in the elliptic half plane, $y > 0$, and two characteristics S_1T, S_2T. The equation of the characteristics is easily found, since (see Art. 9, p. 108) $y\,dy^2 + dx^2 = 0$ or $dx = \pm(-y)^{\frac{1}{2}}\,dy$, whence

$$(x - c)^2 + \tfrac{4}{9}y^3 = 0,$$

with c an arbitrary constant. These characteristics, real only for $y \leqq 0$, are semicubic parabolas. It has been proved by Tricomi that a solution u of (2) is determined in the region above by boundary values along C_0 and along one of the characteristics alone, say S_1T, while values along S_2T cannot be prescribed arbitrarily.[75]

For us the following extension (due to Frankl) of this problem is of importance. Consider (Fig. 174) a region bounded by an arc C_0 between S_1 and S_2 in the elliptic region, by two arcs of characteristics issuing from an arbitrary point O on the transition line and by two arbitrary noncharacteristic arcs issuing from S_1 and S_2, the latter intersecting the characteristics through O in T_1 and T_2 respectively (the "arbitrary" arcs must lie as shown in the figure, with S_1T_1 intersecting every characteristic of the family containing OT_1 only once and similarly for S_2T_2). The figure also shows the characteristics T_1R and T_2R, as well as the characteristics through S_1 and S_2. The value of u is given along $T_1S_1S_2T_2$ (where we go from S_1 to S_2 along C_0), but not along any arc T_1T_2 in OT_1T_2R, nor along OT_1 and OT_2.[76] The function u is then determined firstly in $OT_1S_1S_2T_2O$, and then in the quadrangle T_1OT_2R (and actually beyond this region, in the whole characteristic triangle bounded by the horizontal line S_1S_2 and the two characteristics S_1T and S_2T issuing from S_1 and S_2, respectively; this, however, is not needed for the following). But if u is uniquely determined in the region $RT_1S_1S_2T_2R$ by the values along $T_1S_1S_2T_2$, then obviously

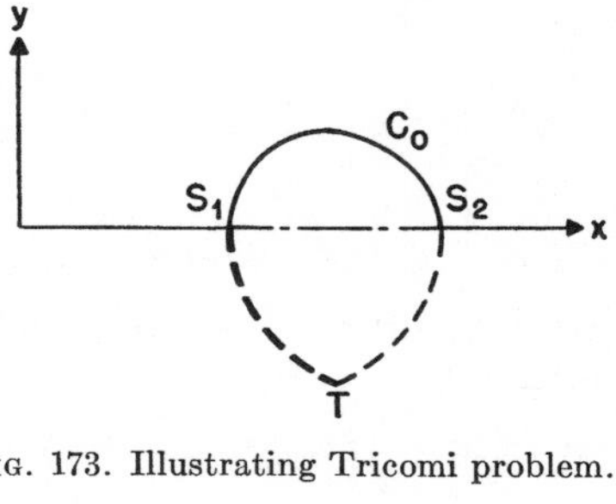

FIG. 173. Illustrating Tricomi problem.

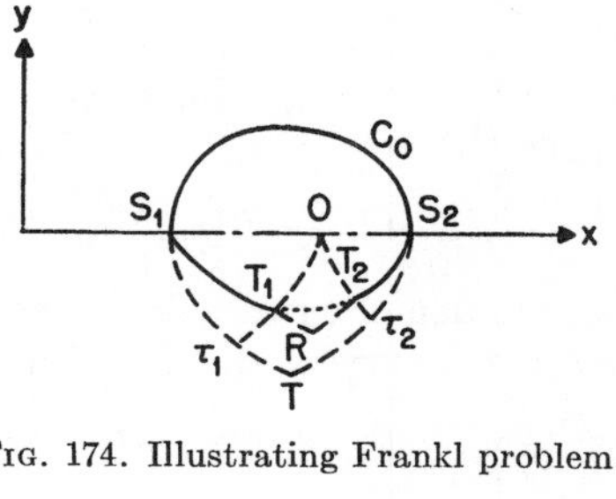

FIG. 174. Illustrating Frankl problem.

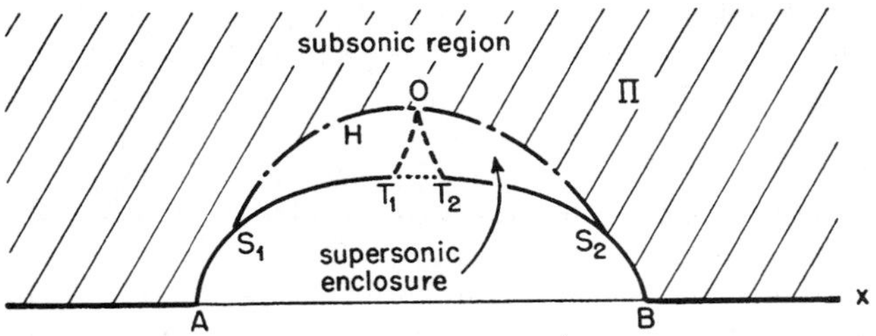

FIG. 175. Illustrating conjecture that data cannot be prescribed everywhere on arbitrary contour.

values of u *cannot be given arbitrarily* along T_1T_2. Thus, the problem in which u is prescribed along a closed contour which lies partly in the elliptic, partly in the hyperbolic region is incorrectly posed.

Following Busemann, Frankl, and particularly Guderley, we now formulate a similar problem of flow past a profile with supersonic enclosure (Fig. 175). We consider only the upper part of the profile which, for convenience, is supposed to be symmetric; let Π denote the upper half plane outside the contour, H the supersonic part of Π between S_1S_2 and the sonic line, which is here the transition line S_1OS_2. The flow is subsonic (elliptic) in the part of Π extending to infinity outside the sonic line, and supersonic (hyperbolic) in $S_1T_1T_2S_2OS_1$. This problem differs from the Frankl problem in the fact that the elliptic region, the subsonic part, extends to infinity and in the fact that the differential equation for ψ is nonlinear; consequently, we do not know a priori the position of the sonic (transition) line, nor of T_1 and T_2, as one does in Frankl's problem. (The equation for φ is likewise nonlinear; to use it we would need an analogue of the Frankl problem with $\partial\varphi/\partial n$ rather than φ given along the boundary.) If we could assume that a similar uniqueness* theorem holds for this much more difficult problem we could conclude as before: Assume that a solution exists in Π; for this solution, $\psi = 0$ along AT_1T_2B; it is however uniquely determined by the boundary value $\psi = 0$ along AT_1 and BT_2 alone. Hence, if the arc T_1T_2 of the contour is deformed (no matter how slightly) so that it is no longer part of the level line $\psi = 0$ of the solution, then no solution will exist for the deformed contour.† This would mean that the problem we took as point of departure is not correctly set in the sense explained at the beginning of the preceding section.

Thus under the assumption that in the more general mixed and nonlinear case and with the elliptic region extending to infinity, a result similar to

* Note that only a uniqueness theorem is required in this case.

† Note that neither this conclusion nor that regarding the nonexistence for flattened profiles applies if the contours are assumed analytic.

that for the Frankl problem holds, then less than the whole contour is sufficient to determine a solution of our problem.[77] Such a conjecture is very plausible. It is hard to imagine that the greater complication of the differential equation would change an incorrectly set problem into a correctly set one. The conclusion would be that smooth potential flow with supersonic pockets past an arbitrarily given profile of the type defined in Sec. 2 does not exist in general.[78]

In the light of this conjecture let us once more recall the procedure used in constructing solutions in the hodograph which were then transformed to the physical plane. First we determined a branch of a stream function which satisfied a given condition (that it reduces to a given function ψ_0 as $q_m \to \infty$) and then we computed analytic continuations of this branch. Now the solutions which have actually been obtained in this way are of an artificial regularity (they are for instance analytic functions in the hyperbolic region, which is a very unusual situation), so that we avoided typical features of hyperbolic problems and obtained only certain regular solutions. We did not solve a boundary-value problem but applied a quite different method of construction of flows, which yielded, for a given M^∞ and P_0, a smooth transonic flow past a specific contour P_M, known only a posteriori. Combining these considerations with the experimental evidence, etc., we might reason as follows. The contours P_M constructed so far by the hodograph method are artificial and the corresponding transonic flows can be considered as exceptional. The set of *all* contours P_M (depending on P_0 and on M^∞) obtainable by the hodograph method, if the original P_0 are of a reasonable generality, has not been identified. However, the contours P_M (and corresponding flows) resulting in this way will form a fairly restricted set and an arbitrarily given P will, in general, not be a P_M. Also, shapes actually used in experiments may not have the properties necessary to make them belong to the "exceptional" profiles.

In summarizing the content of Secs. 2 to 6 we must first of all admit that a complete mathematical theory which can be successfully confronted with observations is lacking; what we have are important but rather isolated bits of information. We know examples of exact solutions—obtained by the hodograph method—but they are not solutions of the boundary problem in question; and we know of (supposed) numerical solutions of boundary-value problems for given contours (Sec. 3). We saw the agreement between these mathematical results and the (mathematical) limit-line statements of Friedrichs and others (Sec. 4). Our theory does not yield a mathematical counterpart to the observed nonisentropic deceleration (Sec. 3). A hint in the direction of an explanation of these discrepancies is supplied by the proved sensitivity of solutions as seen clearly in a particular instance (Sec. 5 and Note 73). And an even stronger suggestion:

If certain recent results are valid under physically relevant conditions, then the key for resolving the various difficulties and contradictions is provided by the fact that the boundary-value problem which we took as point of departure is incorrectly set. This, as well as other points discussed in this article must, for the time being, be left undecided.

NOTES AND ADDENDA

CHAPTER I

Article 1

1. The principle of the text is NEWTON's (1642–1727) *lex secunda*, the second of three axioms formulated in the first pages of his *Philosophiae Naturalis Principia Mathematica*. First, second and third editions: London 1687, 1713, 1726. (Translated by A. Motte, London, 1729; revised by F. Cajory, Berkeley, 1934). The three axioms are at the very beginning, preceded only by the "definitions". The famous second axiom reads: "Lex II. Mutationes motus proportionalem esse motrici impressae et fieri secundam lineam rectam qua vis illa imprimitur." [Law II. The change of motion is proportional to the motive force impressed, and is made in the direction of the right line in which that force is impressed.]

For today's student of mechanics the three axioms form the starting point. They represent at the same time the completion, perfection and generalization of ideas due mainly to Galileo (1564–1642) and Huygens (1629–1695), but lead far beyond the achievements of these great predecessors. An enlightening physical and logical analysis can be found in E. MACH's *Mechanik* [10]*.

2. This rule was used implicitly by L. Euler (1707–1783) and by J. L. Lagrange (1736–1813): L. EULER, "Sectio secunda de principiis motus fluidorum", *Novi Commentarii Acad. Sci. Petrop.* **14**(1769), pp. 270–386, and "Principes généraux du mouvement des fluides", *Hist. Acad. Berlin* **11** (1755), pp. 274–315, which appeared in 1770 and 1757 respectively; J. L. LAGRANGE, "Mémoire sur la théorie du mouvement des fluides", *Oeuvres*, Vol. 4, Paris: Gauthier-Villars, 1869, pp. 695–748; the paper appeared first in 1783.

For details on this subject see TRUESDELL [11], p. 42, and [13], p. xc.

3. L. EULER, cit. Note 2. The paper in question is the second of three basic papers which all appeared in 1757 and constitute a treatise on fluid mechanics (see [13], p. LXXXIV ff.).

In this paper the concept of velocity field is explicitly formulated (though indications appear in earlier works of Euler and others), and in the first and second of these three papers the central concept of the pressure field in hydrodynamics is fully explained. (In this connection, see also Note 10, regarding J. Bernoulli.) The origin of many of the fundamental ideas developed in this work is found in L. EULER, "Principia motus fluidorum" (completed 1752), *Novi Commentarii Acad. Sci. Petrop.* **6** (1756–1757), pp. 271–311, which appeared in 1761 but was completed prior to the above (see [13], p. LXII).

4. If the state of motion is given, for all t, for each point (x,y,z) as in Eq. (1), we may call this the "spatial" description, while the Lagrangian equations give the history of each particle, the "material" description. The spatial description was par-

* We shall refer to the list of reference books by numbers in square brackets.

Article 1

tially formulated in 1749 by J. L. D'ALEMBERT (1717–1783) in his *Essai d'une nouvelle thêorie de la rêsitance des fluides*, Paris: David l'ainé, 1752, and generalized by EULER (1757), cit. Note 2, who later gave also the material description. Several authors, e.g. H. LAMB [15], have pointed out that the usual terminology of Eulerian and Lagrangian equations is unjustified.

5. In so far as the continuity equation expresses conservation of mass, its origin may be seen in Newton's *lex secunda* (see (a), p. 1 in our text). The differential equation itself in various special cases (plane flow, axially symmetric flow) was first obtained by D'ALEMBERT, cit. Note 4. The general equation for "spatial" as well as for "material" variables is due to EULER, see the two papers discussed in Note 3. (Compare also TRUESDELL [11], p. 50, on the d'Alembert-Euler continuity equation.)

6. The author's presentation, based on equations (I), (II), and (III), as developed in Arts. 1–3, differs from that in most modern textbooks where the physical point of view is emphasized. See for example L. HOWARTH [24], Vol. 1, Chapters I and II. For the author's point of view compare also R. v. MISES, "On some topics in the fundamentals of fluid flow theory", *Proc. First Natl. Congr. Appl. Mech., Chicago* (1950), pp. 667–671.

7. The theory based on Eq. (5c) is dealt with in our book in Secs. 17.5 and 17.6.

8. The equation of state for a perfect gas in equilibrium, connected with the names of Boyle (1660), Mariotte, Amontons, Gay Lussac, and Charles, has been widely known since 1800. In precisely the modern form, it was used freely by Euler, but did not appear again in the hydrodynamical literature until used by Kirchhoff (1824–1887).

In some presentations no distinction is made between the terms "perfect" and "ideal". In our book, the term "ideal" is used for "inviscid and nonconducting". The term "perfect" is defined in Eq. (1.6). Incidentally, the term "elastic" introduced on p. 7 is of very long standing in fluid dynamical literature.

9. In this work, the term "isentropic" is used if the entropy is the same everywhere and at all times. The term "strictly adiabatic" (see p. 9) applies to the case where the entropy is constant for each particle but varies from particle to particle. To assist the reader, we mention that L. HOWARTH [24] calls this latter case "isentropic" and uses "homentropic" for the case of entropy constant throughout.

Article 2

10. Equation (2.20) which we derived here from Newton's equation (1.I) (compare also p. 18) is generally attributed to Daniel Bernoulli (1700–1782): D. BERNOULLI, *Hydrodynamica, sive de viribus et motibus fluidorum commentarii*, Strassburg, 1738 (hence, some years before Euler's general equation, cit. Note 3). Much credit is due also to his father John Bernoulli (1667–1748): JOHN BERNOULLI, "Hydraulica nunc primum detecta ac demonstrata directe ex fundamentis pure mechanicis, Anno 1732", *Opera Omnia*, Vol. 4, Lausanne and Geneva: M. M. Bousquet, 1742, pp. 387–493. The equation for steady flow of an incompressible fluid, discovered but imperfectly derived by D. Bernoulli, is extended to nonsteady flow and given a satisfactory derivation by J. Bernoulli. The equation was generalized by Euler in the papers quoted in Note 3; Euler also gave explicitly the equation for a streamline. In the same papers appears the integral $\int dp/\rho$. J. Bernoulli originated the concept of hydraulic pressure which was later generalized by Euler (Note 3). Regarding the quoted works of the Bernoullis see [13] p. LXXXIV ff., and the chapter on "Bernoullian Theorems", TRUESDELL [11], p. 125 ff.

11. Flows throughout which H is constant are often called "isoenergetic". (The

Article 2

term "homenergic" is used by L. HOWARTH [24].) Compare in this connection Sec. 24.1, and Note V. 41.

Article 3

12. The fundamental concept of *stress* which is at the basis of today's mechanics of continua is due to A. L. Cauchy (1789–1857); it generalizes Euler's hydrostatic pressure: A. L. CAUCHY, "Recherches sur l'équilibre et le mouvement intérieur des corps solides ou fluides, élastiques ou non élastiques", *Bull. soc. philomath. Paris* (1823), pp. 9–13, a brief summary; and in full detail: "De la pression ou tension dans un corps solide", *Oeuvres Complètes*, Ser. 2, Vol. 7, Paris: Gauthiers-Villars, 1889, pp. 60–78; other papers by Cauchy on this subject are cited in TRUESDELL [12], p. 264, as 1827.3 and 1828.1. The basic properties of the stress tensor Σ, its symmetry, transformation formulas, etc., are easy to find on the first pages of most textbooks on elasticity. The formal theory of Σ is of course the same no matter whether elasticity theory, the theory of viscous flow, or theory of plastic flow is considered, while the physical assumptions are quite different in the various fields (cf. the title of Cauchy's 1823 paper, cited above). On these fundamentals cf. also R. v. MISES, "Über die bisherigen Ansätze in der klassischen Mechanik der Kontinua", *Verhandl. 3. intern. Congr. tech. Mech. Stockholm* **2** (1930), pp. 3–13.

13. The concept of fluid friction being proportional to the relative gliding of neighboing layers goes back to Newton, second book of the *Principia*, in the "Hypotheses" preceding Proposition LI. L. NAVIER, ["Mémoire sur les lois des mouvements des fluides", *Mém. acad. sci. Paris* **6** (1822), p. 389 ff.] developed a corpuscular theory that led to the still accepted system of partial differential equations for incompressible viscous fluids. A different corpuscular theory leading to the accepted equations without restriction to the incompressible case was given by Poisson. Saint Venant (1843) and particularly G. G. Stokes (1845) derived the same equations on the basis of the general stress concept for a continuously distributed mass: G. G. STOKES, "On the theories of the internal friction of fluids in motion, and of the equilibrium and motion of elastic solids", *Trans. Cambridge Phil. Soc.* **8** (1845), p. 287 ff. (Compare also v. MISES [16], p. 615; LAMB [15], p. 652; BUSEMANN [19], p. 351.) In the present chapter the usual simple form of the dependence of the viscous forces upon the variables p, ρ, $\mathbf{q}$ is not assumed, see p. 26 ff. and also end of Sec. 3. Actually viscous flow is not studied in our book. Compare however Arts. 11, 14, and 22.

14. The tensor symbol, grad Σ (in the sense of W. Gibbs), can be understood by considering it as the product of the symbolic vector "gradient" (see p. 3) multiplied by the tensor Σ according to the rules of matrix multiplication. This exhibits the vector character of grad Σ.

15. The invariant character of $-w'$ may be seen in the following way. The three expressions in parentheses in Eq. (10) are the components of the vector $\mathbf{u} = \Sigma' \cdot \mathbf{q}$, this product being computed by matrix multiplication. The right side of Eq. (10) is then the divergence of this vector $\mathbf{u}$, viz., $-w' = \text{div}\,(\Sigma' \cdot \mathbf{q})$.

16. θ can vanish only if either D or Σ' vanishes (cf. Note 17).

17. To see the invariant character of θ, we form by matrix multiplication the tensor $a = \Sigma' \cdot D$; then $\theta = a_{xx} + a_{yy} + a_{zz}$. Such an expression, the "trace" of an $n \times n$ matrix, is invariant; cf. Eq. (4).

18. As in Note 15: $-w' = \text{div}\,(\Sigma' \cdot \mathbf{q}) \equiv \text{div}\,\mathbf{u}$. Next, (2.27) is applied to $\int_V \text{div}\,\mathbf{u}\,dV$, and it is then easily verified that $u_n = \mathbf{t}'_n \cdot \mathbf{q}$, as in (20).

19. This general form of specifying equation was introduced and discussed by v. MISES, cit. Note 6. Our Eq. (8′) is replaced there by $d\mathbf{q}/dt + (1/\rho)\,\text{grad}\,p = \boldsymbol{F}$. In

Article 3

the "ideal" case of our text (see Note 8) A, B, C, $\boldsymbol{F}$, are given functions of the nine variables $\mathbf{r}$, t, $\mathbf{q}$, p, ρ not involving derivatives. In a more general case certain first- and second-order derivatives of $\mathbf{q}$, p, ρ are also admitted in A, B, C, $\boldsymbol{F}$.

Article 4

20. In this book magnitudes corresponding to a state of rest are in general denoted by the subscript "s", meaning "stagnation" as in "stagnation point", "stagnation pressure", etc. Here, however, the subscript "0" (zero) is used. This is because the subscript essentially serves to denote *undisturbed* flow, to which a perturbation is applied, and this undisturbed state is not necessarily a state of rest (see Art. 5). In the case of acoustics, the unperturbed state denotes a gas (air) at rest; in aerodynamical applications, however, we rather consider the small perturbation of a parallel flow. At the basis of small perturbation theory is, likewise, the wave equation. The notation in Art. 4 is adapted to that in Art. 5.

Small perturbation theory, which plays an important role in aerodynamics, is not considered further in this book. We mention the following references: G. N. WARD, *Linearized Theory of High-Speed Flow*, London and New York: Cambridge Univ. Press, 1955; the condensed monograph, S. GOLDSTEIN, "Linearized Theory of Supersonic Flow", prepared by S. I. PAI, *Inst. for Fluid Dynamics and Appl. Math., Univ. of Maryland, Lecture Ser. No.* **2** (1950), I. IMAI, "Approximation methods in compressible fluid dynamics", *ibid., Tech. Note* **BN-95**(1957), and the article by W. R. SEARS, "Small Perturbation Theory", in [31], pp. 61–121. See also Notes III.46, V. 24, 64.

21. In the famous paper, J. L. D'ALEMBERT, "Recherches sur la courbe que forme une corde tendue mise en vibration", *Kgl. Akad. Wiss. Berlin* **3** (1747), p. 214 ff. the author treats the string as a continuous medium. This theory was refounded and thoroughly exploited by Euler. In another treatment Euler and Lagrange imagined the string made up of a finite number of equally spaced particles and performed the passage to the limit, see J. L. LAGRANGE, "Recherches sur la nature et la propagation du son", *Miscellanea Taurinensia* **1** (1759), pp. 1–112.

22. The first attempt at a mathematical theory of sound was made by Newton. By a very devious argument he reached the formula $a_0 = \sqrt{p_0/\rho_0}$, thus obtaining the same result as if he had assumed the motion isothermal. In the 1687 edition of the *Principia*, he obtained the value 968 ft/sec; in the 1713 edition he got 979 ft/sec. The treatment based on the wave equation with $p = K\rho$ is due to d'Alembert and Euler. The possibility of a computation based on a different (p, ρ)-relation was suggested occasionally in the 18th century. Physical arguments in favor of $p = K\rho^\gamma$—with the corresponding value $a_0 = \sqrt{\gamma p_0/\rho_0}$—were first advanced by J. B. BIOT, (1774–1867) ["Sur la théorie du son", *J. phys.* **55** (1802), pp. 173–182] who acknowledged the assistance of LAPLACE (1749–1827) [the result was later included in P. S. LAPLACE, *Traité de mécanique céleste*, Vol. 5, Paris: Duprat (1825)]. The modern explanation based on the concept of an adiabatic process did not become possible, of course, until after the creation of the mechanical theory of heat in the 19th century.

The mathematicians d'Alembert, D. Bernoulli, Euler, and Lagrange discovered a large part of the theory of production and propagation of sound. The theory was highly developed in the 19th century by S. D. Poisson (1781–1840), G. G. Stokes (1819–1903) and particularly by H. v. Helmholtz (1821–1894), G. Kirchhoff (1824–1887), and Lord Rayleigh (1842–1919). H. v. HELMHOLTZ, "Die Lehre von den Tonempfindungen als physiologische Grundlage für die Theorie der Musik", published 1862, Braunschweig: F. Vieweg, 1913. (English translation: *On the Sensations of Tone as a*

Article 4

Physiological Basis for the Theory of Music, reprinted 1954, New York: Dover). LORD RAYLEIGH's *Theory of Sound* [28], is a work which still offers unexhausted treasures. (Compare also the instructive historical introduction to Rayleigh's work by R. B. Lindsay).

23. S. D. POISSON, "Sur le mouvement des fluides élastiques dans les tuyaux cylindriques, et sur les théories des instruments à vent", *Mém. acad. sci. Paris*, Sér. 2, **2** (1817), pp. 305–402. Poisson's method, as given in our text, is the one adopted by LORD RAYLEIGH [28], Vol. II, Sec. 273 ff.

24. The method used in this section is Hadamard's "méthode de déscente", see J. HADAMARD, *Lectures on Cauchy's Problem in Linear Partial Differential Equations*, Paris: Hermann, 1932, p. 49, (reprinted, New York: Dover, 1952).

A very interesting paper on the wave equation with discussion of the basic difference between odd and even n (in our case $n = 1$ and $n = 3$ versus $n = 2$) is M. RIESZ, "L'intégrale de Riemann-Liouville et le problème de Cauchy", *Acta Math.* **81** (1949), pp. 1–223. A shorter paper with almost equal title was presented by the same author at the *Conférences de la réunion internationale des mathématiciens*, Paris, 1937 (proceedings published by Gauthier-Villars, 1939).

Some useful particular solutions of the two-dimensional wave equation are mentioned e.g. in [31], article on "Small Perturbation Theory" by W. R. SEARS, pp. 118–119 (see Note 20). See also our comments on the two-dimensional generalized wave equation (7.45) at the end of Art. 7.

Article 5

25. ERNST MACH (1838–1916), using the interferometer method invented by him and his son LUDWIG MACH [Über ein Interferenzrefraktometer", *Sitzber. Akad. Wiss. Wien* ,Abt. II, **98** (1889), p. 1318] observed projectiles flying at a speed faster than sound: E. MACH and L. SALCHER, "Photographische Fixierung der durch Projektile in der Luft eingeleiteten Vorgänge", *Sitzber. Akad.Wiss. Wien*, Abt. II, **95** (1887), pp. 764–780; E. MACH and L. MACH, "Weitere ballistisch photographische Versuche", *Sitsber. Akad. Wiss. Wien*, Abt. IIa, **98** (1889), pp. 1310–1326. Compare also [24], Vol. II, Chapter XI, "Visualization and Photography of Fluid Motion", p. 578 ff. Mach's method was preceded by the famous *schlieren* method of A. TOEPLER, *Beobachtungen nach einer neuen optischen Methode*, Bonn: Cohen, 1864. The excellent monograph on Ernst Mach: H. HENNING, *Ernst Mach als Philosoph, Physiker und Psychologe*, Leipzig: A. BARTH, 1915, contains a bibliography of Mach's papers and books to 1912.

Mach gave not only the experimental method and results, but also the essential theoretical facts regarding what were later called "Mach cone", "Mach angle", and "Mach lines". L. Prandtl (1907) and T. Meyer (1908) were probably the first to use these terms.

CHAPTER II

Article 6

1. With respect to this article and to part of the next we refer the reader to the monograph [11], which contains a wealth of information—known results complemented by original research and abundant historical material. See also the same author's "Vorticity and the thermodynamic state in a gas flow", *Mém. sci. math.* **119** (1952).

Article 6

2. The terms "circuit" and "circulation" go back to W. Thomson, Lord Kelvin (1824–1907): W. Thomson, "On vortex motion", *Trans. Roy. Soc. Edinburgh* **25**(1869), pp. 217–260; also, W. Thomson and P. G. Tait, *Treatise on Natural Philosophy*, London and New York: Cambridge Univ. Press, which first appeared in 1867. New edition in two parts, Part I (1879), Part II (1883); and both parts, Cambridge University Press, 1912.

3. The components of curl $\mathbf{q}$ were introduced by d'Alembert, Euler, and Lagrange, and used freely but purely formally in eighteenth century studies on hydrodynamics. Cauchy proved in 1841 that the transformation rules for the components of curl $\mathbf{q}$ are the same as those for vector components: A. L. Cauchy, "Mémoire sur les dilatations, les condensations, et les rotations produites par un changement de forme dans un système de points matériels", *Oeuvres Complètes*, Ser. 2, Vol. 12, Paris: Gauthier-Villars 1916, pp. 343–377.

4. The important integral transformation (6) was actually discovered by Kelvin (1850) as evidenced by a letter from Kelvin to Stokes. The theorem was found independently by Hankel: H. Hankel, *Zur allgemeinen Theorie der Bewegung der Flüssigkeiten*, Göttingen, 1861.

5. Interpretations of curl $\mathbf{q}$ have been given by Stokes and, in several different ways, by Cauchy, cit. Note 3 and Note I.12. See further discussion in Truesdell [11], pp. 59–65; see also the presentation in Hadamard [4], p. 74 ff.

6. These concepts were introduced by Helmholtz: H. v. Helmholtz "Über Integrale der hydrodynamischen Gleichungen, welche den Wirbelbewegungen entsprechen", *J. reine angew. Math.* **55**(1858), pp. 25–55 [translated by P. G. Tait, "On integrals of the hydrodynamical equations which express vortex-motion", *Phil. Mag.*, Ser. 4, **33**(1867), pp. 485–512]. This paper and the one by Lord Kelvin cited in Note 2 are the basic papers of the Helmholtz-Kelvin vortex theory.

7. We note that in this definition vorticity is a positive quantity like mass or speed, while circulation can be positive or negative. Cf. also the detailed presentation in v. Mises [16], Chapters II and IX.

8. Instead of the ad hoc derivation of our text one may compute—in analogy to Eq. (2.28)—the rate of change $(d/dt)\int_{C(t)} \mathbf{a}\cdot d\mathbf{l}$ where $\mathbf{a}$ depends on x,y,z,t and $C(t)$ moves with velocity $\mathbf{q}$. The result is (see [14], p. 456)

$$\frac{d}{dt}\int_{C(t)} \mathbf{a}\cdot d\mathbf{l} = \int_{C(t)} \left[\frac{\partial \mathbf{q}}{\partial t} + \text{grad}\,(\mathbf{a}\cdot\mathbf{q}) - \mathbf{q}\times \text{curl}\,\mathbf{q}\right]\cdot d\mathbf{l},$$

and for $\mathbf{a} = \mathbf{q}$ and C a closed curve

$$\frac{d\Gamma}{dt} = \oint \left(\frac{\partial \mathbf{q}}{\partial t} + \text{grad}\, q^2 - \mathbf{q}\times \text{curl}\,\mathbf{q}\right)\cdot d\mathbf{l} = \oint \left(\text{grad}\,\frac{q^2}{2} + \frac{d\mathbf{q}}{dt}\right)\cdot d\mathbf{l},$$

as in Eq. (11).

9. Kelvin's theorem is given in the paper quoted Note 2.

10. Accordingly, one may define an "acceleration potential", φ^*, by $d\mathbf{q}/dt = \text{grad}\,\varphi^*$. This was done by Euler (1755), Note I.3. Obviously by means of φ^* some formulas of our text can be rewritten. The fact that the curl of the acceleration vanishes was known to d'Alembert (1752), Note I.4.

11. In Sec. 4 the concept of "filament" or of "tube of infinitesimal cross section" may be avoided by using a limiting process or by noting that a vortex line is the intersection of any two tubes on which it lies.

12. For generalization cf. Sec. 24.1.

13. Flows for which $\mathbf{q}\times$ curl $\mathbf{q} = 0$ are called Beltrami fields: E. Beltrami, "Con-

Article 6

siderazioni idrodinamiche", *Rend. ist. Lombardo*, Ser. 2, **22** (1889), pp. 300–309. Many of their properties were previously discovered by I. Gromeka (1881). See TRUESDELL [11], p. 24 ff. and p. 97 ff., for more information and for literature. These flows are interesting as a link between irrotational flow and general rotational flow.

14. It is known that Helmholtz' original proof of the first vortex theorem is not rigorous. Compare for example LAMB [15], p. 206, and the reference to Stokes there, p. 17. In Sec. 6.6, v. Mises reproduces Helmholtz' original argument—except that he considers nonconstant density. Several proofs of the Helmholtz theorems and of various generalizations are presented in TRUESDELL [11]. See also SOMMERFELD [17], p. 130 ff., and N. J. KOTSCHIN, I. A. KIBEL and N. W. ROSE, *Theoretische Hydromechanik*, Vol. I, Berlin: Akademie Verlag, 1954, p. 138: "Friedmann's Theorem".

Article 7

15. The term "irrotational" was introduced by LORD KELVIN, cit. Note 2.

16. This theorem and much of the vortex theory—in the case of an incompressible fluid—was anticipated in "Cauchy's equations" [cf. for example LAMB, [15], p. 205, Eq. (3)] which may be generalized to compressible flow.

17. The exact solution given in this section was probably first considered by G. I. TAYLOR, "Some cases of flow of compressible fluids", *ARC Repts. & Mem.* 1382 (1930). Also: "Recent work on the flow of compressible fluids", *J. London Math. Soc.* **5** (1930), pp. 224–240. Cf. also H. BATEMAN, "Irrotational motion of a compressible inviscid fluid", *Proc. Natl. Acad. Sci. U. S.* **16** (1930), pp. 816–825.

18. This problem is studied in Sec. 17.4 by means of the "hodograph representation", which will be explained in Art. 8. See also Note 23.

19. This problem—from a different point of view—was considered by G. I. TAYLOR, cit. Note 17.

20. It is mentioned for the record that von Mises left an alternative version (complete with drawings) of the end of Sec. 5, and of the whole Sec. 6. The method used there is closer to that presented here for radial flow, as well as to Taylor's method. He introduces new dimensionless variables $\xi = (2\pi a_s/\Gamma)r$, $\eta = q_r/a_s$, thereby excluding purely radial flow: $\Gamma = 0$, and derives the differential equation

$$\frac{d\eta}{d\xi} = -\frac{\eta}{\xi}\,\frac{2 - (\kappa - 1)\eta^2 + (3 - \kappa)/\xi^2}{2 - (\kappa + 1)\eta^2 - (\kappa - 1)/\xi^2},$$

which is then discussed: the same results as in the text are derived.

21. Considering flow past a semi-infinite cone, Taylor and Maccoll obtained an exact solution for steady irrotational inviscid compressible flow with axial symmetry. G. I. TAYLOR and J. W. MACCOLL, "The air pressure on a cone moving at high speeds", *Proc. Roy. Soc.* **A139** (1933), pp. 298–311. Compare for example [24], Vol. 1, p. 185 ff. For numerical tables see Z. KOPAL, *Tables of Supersonic Flow around Cones*, Cambridge, Mass.: M. I. T. Publication, 1947. See also graphs in [35], Sec. III.

Article 8

22. A very important application of these results appears in the "one-dimensional" or "hydraulic" treatment of channel flow (see Sec. 25.1).

23. The initial idea of the hodograph representation seems to be due to Helmholtz and to Riemann: H. v. HELMHOLTZ, "On discontinuous movements of fluids", *Phil. Mag.*, Ser. 4, **36** (1868), pp. 337–346 [original in *Monatsber. preuss. Akad. Wiss. Berlin* **125** (1868) pp. 215–228]. Helmholtz used it to solve problems involving free boundaries (cf. Sec. 20.6). The importance of the transformation as a method for obtaining linear

Article 8

equations was perhaps first recognized by G. B. Riemann (1826–1866) in a paper which is basic for the subject of this book: G. B. RIEMANN, "Über die Fortpflanzung ebener Luftwellen von endlicher Schwingungsweite", *Abhandl. Ges. Wiss. Göttingen, Math.-physik. Kl.* **8** (1858/9), pp. 43–65, or *Gesammelte mathematische Werke*, 2nd ed., 1892, p. 157 ff. (reprinted 1953, New York: Dover). Compare also Note IV.5.

24. L. PRANDTL and A. BUSEMANN, "Näherungsverfahren zur zeichnerischen Ermittlung von ebenen Strömungen mit Überschallgeschwindigkeit", *Stodola Festschrift* (1929), pp. 499–509. Reprinted in [20], pp. 120–130.

25. This is one of the oldest results on compressible fluid flow, due to B. DE ST. VENANT and L. WANTZEL, "Mémoire et expériences sur l'écoulement de l'air déterminé par des différences de pressions considérables", *J. école polytech.*, Ser. 1, **27** (1839), pp. 85–122.

26. We refer the reader to the tables [34], [36], [37]. Extensive diagrams are given in [35]. A textbook comparatively rich in tables is A. FERRI, *Elements of Aerodynamics of Supersonic Flow*, New York: Macmillan, 1949.

Article 9

27. Regarding terminology: Equations of the form (2) are called *planar* by v. Mises whether or not the coefficients depend on Φ itself. The equation for the potential, the stream function, in Arts. 16 and 24, and the particle function, in Arts. 12 and 15, are all of this form. This type of equation is sometimes termed *pseudolinear*, e.g. in [7]. If the coefficients A, B, C, F depend on x, y, $\partial\Phi/\partial x$, $\partial\Phi/\partial y$, but not on Φ itself, the equation is often called *quasilinear*. However, in the monograph: F. G. TRICOMI, *Lezioni sulle equazioni a derivate parziali*, Torino: Editrice Gheroni, 1954, Eq. (2) is called quasilinear if A, B, C depend on x, y only, while F may depend on x, y, Φ, $\partial\Phi/\partial x$, $\partial\Phi/\partial y$; this case is usually called *semilinear*. The *linear* equation of second order is always assumed to be of the form

$$A\,\frac{\partial^2\Phi}{\partial x^2} + B\,\frac{\partial^2\Phi}{\partial x\partial y} + C\,\frac{\partial^2\Phi}{\partial y^2} + D\,\frac{\partial\Phi}{\partial x} + E\,\frac{\partial\Phi}{\partial y} + F\Phi + G = 0,$$

where $A, B, \ldots G$ depend on x, y only.

28. In earlier treatises, characteristics were mainly studied for one partial differential equation of first order, or one partial differential equation of second order, see for example [3] and A. SOMMERFELD, *Partial differential equations in Physics* (trans. by E. G. Strauss), New York: Academic Press, 1949. In contrast to this, in the present text and in [2], [4], [6], [21] systems of equations of first order are in the foreground. This approach was largely developed by J. HADAMARD [4] and by T. LEVI-CIVITÀ [6]. The important book, J. HADAMARD, cit. Note I.24, which is, in general, mathematically too advanced for our purpose, contains in the Preface and in Chapter I interesting information regarding the earlier literature. A presentation such as the one in our test, with emphasis on "discontinuous solutions" in compressible fluid flow, is given in a very condensed form in the paper quoted Note I.6.

29. A detailed discussion of the case $n = 2$, $k = 2$ follows in Art. 10. The case $n = 3$, $k = 2$ is considered e.g. in [7], pp. 147–152, and the case $n = 2$, $k = 3$ in [29], p. 170 ff.

30. If the equation of such a characteristic surface is written as $f(x_1, x_2, \cdots, x_n) =$ constant, then clearly f must satisfy the first-order partial differential equation which is obtained from (10′) on replacing λ_μ by $\partial f/\partial x_\mu$ ($\mu = 1, 2, \cdots, n$). The so-called Monge cone (see [2], p. 63) of this first-order equation is the cone enveloped by planes whose normal direction satisfies (10′), i.e. it is the conjugate of the cone formed by the normal directions. (See Note 32.)

Article 9

31. v. MISES, cit. Note I.6, introduced the concept of a *discontinuous solution*, in relation to the distinction of our text between "compatibility relations" and "additional equations". A set of functions $u_1, u_2, \cdots, u_k$ is called a discontinuous solution of Eqs. (9.6) across a surface $\mathcal{S}^*$ if (i) on both sides of $\mathcal{S}^*$ all differential equations are satisfied, and if (ii) at least one of the u_i or its derivatives has a jump across $\mathcal{S}^*$. We shall see that in the general problem of Sec. 6 such discontinuities across the characteristic surfaces $\mathcal{S}^*$ actually arise, including "absolute" discontinuities, i.e. discontinuities of the variables u_i.

The definition leads to a criterion regarding the important question as to which of the k variables may jump across $\mathcal{S}^*$. Obviously, a variable may jump without violating conditions (i), (ii) if its derivative *normal to* $\mathcal{S}^*$ does not appear at all in the $(k - r)$ additional equations (see Note 38). Instead of this, both Hadamard and Levi-Cività use special physical reasoning to show why, in compressible flow, for example the pressure must not jump and certain velocity components cannot change abruptly.

32. Eq. (20) for the three-dimensional steady potential equation may be written

$$(q_x^2 - a^2)p^2 + (q_y^2 - a^2)q^2 + (q_z^2 - a^2) + 2q_xq_ypq + 2q_yq_zq - 2q_zq_xp = 0,$$

where p,q are first-order derivatives of the characteristic surface, and $p:q: -1 = \lambda_1:\lambda_2:\lambda_3$. The Monge cone of this equation, considered as the partial differential equation of the characteristic surface, is identical with our Mach cone.

33. The theory for a system of equations of second order from which the results regarding one equation of second order are then immediate, can be found in [4] or in [6], p. 9 ff.

34. The present discussion applies to the very general "ideal fluid motion" of Sec. 3.6. The solution in the case of an elastic fluid where $n = k = 4$, can be found in [6], p. 63.

35. In Sec. 4 of the paper cit. Note I.6, v. Mises also investigates the characteristics when viscosity and heat conduction are admitted. Under the usual assumptions the system then consists of four differential equations of second and one of first order. The corresponding equation for λ, which is of degree nine, resolves into the product of the factor $q\lambda_1 + \lambda_4$ and of a polynomial of degree eight which can vanish only if all $\lambda_i = 0$. Hence the only discontinuities in this problem are those related to the factor $q\lambda_1 + \lambda_4$. By making use of prior work of P. Duhem, the same results without heat conduction were obtained by G. LAMPARIELLO, "Sull impossibilità di propagazioni ondose nel fluidi viscosi", *Atti. accad. nazl. Lincei, Rend. Classe sci. fis. mat. e nat.*, Ser. 6, **13** (1931), pp. 688–691.

36. The intersection of an exceptional plane in x,y,z-space, for which (27) holds for λ, with the x,y-plane is not in general a Mach line. Analogously, the intersection of an exceptional plane in x,y,z,t-space, with (26) for λ, with the x,y,z-space is in general not tangent to the Mach cone.

37. J. HADAMARD [4] introduced a different approach, much used in recent work on continuum mechanics (see also the clear and condensed presentation in [6]). Consider a discontinuity surface $\mathcal{S}$, varying in time. If the discontinuity is fixed with respect to the medium, it is called a *material discontinuity* [example: $q\lambda_1 + \lambda_4 = 0$ in Eq. (26)]; otherwise it is called a *wave* (onde) [second factor of Eq. (26)].

In this theory the characteristic condition (10) is obtained by a method quite different from that explained in our text.

38. Thus with respect to the triple root, our five original equations are transformed into three "compatibility relations", plus two "additional equations". These last are: one, the component of Newton's equation in the λ-direction; two, the continuity

Article 9

equation. The first of these includes the derivative of p in the λ-direction; hence the *pressure, p, cannot change abruptly across a surface consisting of particle lines.* Combining this with an analogous consideration of the continuity equation we find that *only* ρ *and the tangential component of* $\mathbf{q}$ *may jump.* These important facts are obtained by v. Mises, cit. Note I.6, from the principle explained at the end of Note 31, which obviates introducing any separate physical considerations to explain why, for example, ρ may jump and p must not.

Jumps in these variables, viz., density and tangential velocity, across particle lines (streamlines in the steady case) do in fact occur; they are known as contact discontinuities, vortex sheets, etc. (see Sec. 15.2). Here v. Mises' conception differs basically from that of R. Courant and K. O. Friedrichs [21], p. 126, and others, who parallel contact discontinuities with shocks. For v. Mises the contact discontinuities of compressible flow are discontinuities across characteristics, satisfying the condition (10), for which the variable itself undergoes an abrupt change; whereas a shock is strictly speaking not a phenomenon of inviscid nonconducting fluid flow theory. (See Secs. 14.1, 14.2, 22.1, and 22.2.)

Article 10

39. A system of this form is often called quasilinear (see also Note 27).

40. Here, where $n = k = 2$, our results can be obtained in terms of the theory of two linear algebraic equations with two unknowns. An elegant presentation along these lines is given in Sauer, [7], p. 63 ff. The complete and symmetric compatibility relations (4a)–(4d) and (5) are new.

41. This theorem, formulated in a mathematically more rigorous way, has been proved by H. Lewy, "Über das Anfangswertproblem bei einer hyperbolischen nichtlinearen partiellen Differentialgleichung zweiter Ordnung mit zwei unabhängigen Veränderlichen", *Math. Ann.* **98** (1927), pp. 179–191. A proof is in Courant-Friedrichs [21], p. 48 ff. We also refer the reader again to T. Levi-Cività [6], to F. Tricomi, cit. Note 27, and to R. Sauer [7].

42. This reasoning, which at the same time provides a numerical method, goes back to J. Massau, *Mémoire sur l'intégration graphique des équations aux dérivées partielles*, Gand: van Goethem, 1900 (reviewed in *Enzykl. math. Wiss.* **II/3** (1915), p. 162, article by C. Runge and F. A. Willers), reprinted as *Edition du Centénaire*, Mons, 1952.

43. We have seen that Massau's method (see Note 42) fails if any cross line assumes characteristic direction. We must, however, beware of the mistaken belief that if this method encounters no obstacle—for what seems a sufficiently small mesh—it necessarily yields approximate knowledge of the desired solution. In fact, consider the system $v^2(\partial u/\partial x) = u^2(\partial v/\partial y)$, $v^2(\partial u/\partial y) = u^2(\partial v/\partial x)$, whose characteristics are the straight $\pm 45°$-lines. It admits the particular solution $u = [1 + 2(x^2 + y^2)]^{-1}$, $v = [1 + 4xy]^{-1}$, where $v \to \infty$ on the hyperbola $xy = -\frac{1}{4}$. Along the noncharacteristic segment from (0,0) to (2,0), say, this solution takes on the regular boundary values $u = (1 + 2x^2)^{-1}$, $v = 1$. Massau's method applied (for a chosen mesh) to the characteristic triangle (0,0), (2,0), (1,−1) does not encounter any difficulty (all cross lines may be taken horizontal thus having nowhere characteristic direction) and leads to well-defined finite values which give no indication of the singularity in the exact solution. On the other hand, in a sufficiently close neighborhood of the segment (0,0), (0,2) (whose distance from the x-axis is less than $1 - \sqrt{3}/2$) the exact solution is everywhere finite and Massau's method gives an approximation to it. (This example was communicated to H. Geiringer by M. Schiffer.) We do not wish to imply that such a situation will arise in the fluid-dynamical case, though the converse has never been demonstrated.

Article 10

44. We may wonder whether by imposing suitable restrictions on the coefficients of the system (1) we could exclude cases such as the example in Note 43. The answer is negative, since for a nonlinear differential equation the singularities of solutions are not determined by singularities of the coefficients. For example, for the nonlinear ordinary equation $dy/dx = y^2$, the general solution $y = (a - x)^{-1}$ has a pole at an arbitrary point $x = a$, which can in no way be predicted from the coefficients of the given equation.

45. It is a frequent mistake to assume that in the characteristic boundary-value problem a solution is guaranteed in a small neighborhood of AB and of AC. It is guaranteed only in a neighborhood of the point A. (Counter-examples can be constructed in various ways.) This can be understood intuitively from a comparison of Figs. 45 and 46. In Fig. 45 the whole row of points $A'B'$ adjacent to AB is derived directly from the given data along AB. In Fig. 46, however, only the position of P_4 follows directly from the given data. For all other points, say those adjacent to AB, we need in addition to the given data the derived values at P_4, etc., and hence a certain uniformity concerning these derived values. A correct mathematical existence proof for this boundary-value problem is due to H. Lewy, see Note 41, and COURANT-HILBERT [2], where the problem is reduced to one of a system of ordinary differential equations.

46. For the "mixed" boundary-value problem where we know compatible values of u and v along the characteristic AC and one variable along AA_1, existence can be proved in the neighborhood of A only. In relation to these more general boundary-value problems see papers by: H. BECKERT, "Über quasilineare hyperbolische Systeme partieller Differentialgleichungen erster Ordnung mit zwei unabhängigen Variablen. Das Anfangswertproblem, die gemischte Randwertaufgabe, das charakteristische Problem", *Ber. Verhandl. sächs. Akad. Wiss. Leipzig, Math.-Naturw. Kl.* **97** (1950), p. 68 ff.; W. HAACK and G. HELLWIG, "Über Systeme hyperbolischer Differentialgleichungen erster Ordnung. I", *Math. Z.* **53** (1950), pp. 244–266; II, *ibid.* pp. 340–356; and R. COURANT and P. LAX, "On nonlinear partial differential equations for functions of two independent variables", *Communs. Pure Appl. Math.* **2** (1949), pp. 255–273.

47. Riemann developed this method in the paper quoted in Note 23 as a sort of appendix to the physical theory contained therein. He considers an equation such as (12.43) except that $2/(\xi + \eta)$ is replaced by $-m$, a function of $(\xi + \eta)$. His method is completely explained by means of this example. The first to consider in detail the general equation (11) was G. DARBOUX, in his *Leçons sur la théorie générale des surfaces*, 2nd ed., Vol. II, Paris: Gauthier-Villars, 1915, p. 71 ff. (1st ed., 1888).

48. Formula (17) is called Riemann's formula. Riemann's method has been generalized by various mathematicians, above all by J. HADAMARD, cit. Note I.24, who developed an integration theory for the general second-order linear equation in n independent variables. See the presentation in SAUER, [7], p. 194 ff.; cf. also BERGMAN-SCHIFFER, [1], p. 365 ff., and the paper by M. Riesz quoted in Note I.24.

49. Regarding the determination of the function Ω, Riemann adds: "The determination of such a solution (our Ω) is often made possible by the consideration of a particular case . . .". H. Weber, the editor of Riemann's works, explains this remark as follows: Since the determination of Ω is independent of the particular boundary values given for U, we may try to find a particular solution U for conveniently chosen values of U and its derivatives on a conveniently chosen $\mathfrak{C}$; then the Riemann formula (17) gives Ω. This simple and very suggestive idea is carried out for Riemann's equation (Note 47).

50. For examples of Riemann functions see Sec. 12. 4, and Note III.22.

Article 10

51. At this stage, our point of view is that, if all variables are considered as functions of u,v, then $x(u,v)$ and $y(u,v)$ must satisfy (22) if after inversion [guaranteed by $J \neq 0$] $u(x,y)$ and $v(x,y)$ are to satisfy (1). One does not obtain all solutions of (1) in this way: those for which $j = 1/J$ vanishes are "lost" (see Art. 18). If the two transitions, the one from the x,y-plane to the u,v-plane and the reverse one are considered separately we see that $j \neq 0$ is the condition for the first, $J \neq 0$ that for the second one. More will be found in Arts. 17, 18, and 19.

CHAPTER III

Article 11

1. See Note I.13.

2. A complete qualitative discussion of these flows has been given by G. S. S. Ludford, "The classification of one-dimensional flows and the general shock problem of a compressible, viscous, heat-conducting fluid", *J. Aeronaut. Sci.* **18** (1951), pp. 830–834. For the special case $P = \frac{3}{4}$ (cf. Eq. (32)), the equations can be integrated explicitly, see M. Morduchow and P. A. Libby, "On a complete solution of the one-dimensional flow equations of a viscous, heat-conducting, compressible gas", *J. Aeronaut. Sci.* **16** (1949), pp. 674–684.

3. G. I. Taylor, "The conditions necessary for discontinuous motion in gases", *Proc. Roy. Soc.* **A84** (1910), pp. 371–377. This paper appears in [20] and is essentially reproduced in Vol. III of [23] in the article by G. I. Taylor and J. W. Maccoll, "The mechanics of compressible fluids", pp. 209–250. Taylor also considered the case $\mu_0 = 0$, which had been previously discussed by W. J. M. Rankine, "On the thermodynamic theory of waves of finite longitudinal disturbance", *Phil. Trans. Roy. Soc. London* **160** (1870), pp. 277–286. Similar results to Taylor's were obtained by Lord Rayleigh, "Aerial plane waves of finite amplitude", *Proc. Roy. Soc.* **A84** (1910), pp. 247–284, or *Scientific Papers*, Vol. 5, London and New York: Cambridge University Press, 1912, pp. 573–610. As he (and later Becker, cit. Note 6) pointed out, the case $\mu_0 = 0$ is somewhat irregular; it must be treated as a limit, see D. Gilbarg, "The existence and limit behavior of the one-dimensional shock layer", *Am. J. Math.* **73,** (1951), pp. 256–274. See also M. J. Lighthill, "Viscosity effects in sound waves of finite amplitude", *Surveys in Mechanics*, London and New York: Cambridge University Press, 1956, pp. 250–351.

4. The complete problem was first fully treated by R. v. Mises, "On the thickness of a steady shock wave", *J. Aeronaut. Sci.* **17** (1950), pp. 551–555. For a discussion of the nonperfect gas see D. Gilbarg, Note 3.

5. L. Prandtl ["Eine Beziehung zwischen Wärmeaustausch und Strömungswiderstand der Flüssigkeiten", *Physik. Z.* **11** (1910), pp. 1072–1078] used this ratio in a hydrodynamical analogue of a heat transfer problem.

6. R. Becker, "Stosswelle und Detonation", *Z. Physik.* **8** (1922), pp. 321–362 [translation: *NACA Tech. Mem.* **505** (1929)]. Becker also considered the cases $\mu_0 = 0$ and $k = 0$, see Note 3.

7. Lord Rayleigh, "On the viscosity of argon as affected by temperature", *Proc. Roy. Soc.* **66** (1900), pp. 68–74, or *Scientific Papers*, Vol. 4, London and New York: Cambridge Univ. Press, 1903, pp. 452–458.

8. R. A. Millikan, "Über den wahrscheinlichsten Wert des Reibungskoeffizienten der Luft", *Ann. Physik* **41** (1913), pp. 759–766. Millikan obtains his result by fitting Sutherland's formula in the kinetic theory of gases to experimental values for air:

Article 11

W. SUTHERLAND, "The viscosity of gases and molecular forces", *Phil. Mag.*, Ser. 5, **36** (1893), pp. 507–531. It is now generally accepted that the constant 223.2 in Eq. (43) should be replaced by one closer to 200 (see for example [37], Vol. 5, p. 1504.1–1), but this has no effect on the conclusions.

9. T. H. LABY and E. A. NELSON, "Thermal conductivity; gases and vapors", *International Critical Tables*, Vol. 5, New York: McGraw-Hill, 1929, pp. 213–217. The formula is accurate in the range −312°F to 415°F.

10. This is Eucken's formula, see J. H. JEANS, *Kinetic Theory of Gases*, London and New York: Cambridge Univ. Press, 1952, p. 190.

11. For all but quite weak shocks this thickness is of the same order of magnitude as the mean free path. Becker, cit. Note 6, questioned whether in these circumstances the equations of continuum mechanics are applicable to the problem, and the belief has grown that only kinetic theory is capable of a correct account of the transition. However several authors, starting with L. H. THOMAS ["Note on Becker's theory of the shock front", *J. Chem. Phys.* **12** (1944), pp. 449–453], have emphasized the considerable increase in thickness resulting from more realistic assumptions such as temperature dependence of viscosity and thermal conductivity. For a critical discussion and bibliography of the controversy see D. GILBARG and D. PAOLUCCI, "The structure of shock waves in the continuum theory of fluids", *J. Rational Mech. Anal.* **2** (1953), pp. 617–642. These authors also investigate the effect of other viscosity assumptions than that of Navier-Stokes, Eq. (6).

Article 12

12. As was pointed out in Note I.8, a distinction is made between the terms "ideal" and "perfect". In this article we are primarily concerned with an ideal perfect gas in isentropic motion. However the discussion is carried through for a general ideal elastic fluid, with the polytropic case for illustration.

13. For a general survey and useful bibliography of one-dimensional nonsteady flow, see O. ZALDASTANI, "The one-dimensional isentropic fluid flow", *Advances in Appl. Mech.* **3** (1953), pp. 21–59.

14. The variable v was introduced by B. RIEMANN, cit Note II.23, who used the so-called Riemann invariants $r = (v + u)/2 = \xi/2$ and $s = (v - u)/2 = \eta/2$ in place of u and v. Here ξ and η are the characteristic variables of Sec. 12.4, cf. Eq. (10.6). R. LIPSCHITZ ["Beitrag zu der Theorie der Bewegung einer elastischen Flüssigkeit", *J. reine angew. Math.* **100** (1887), pp. 89–120] extended Riemann's discussion, in particular to the case when gravity force acts.

15. It was in discussing Eq. (27′) that Riemann developed his theory of integration of hyperbolic differential equations, see Note II.47.

16. The general (p,ρ)-relation leading to an equation of the type (34) has been given by R. SAUER, "Elementare Lösungen der Wellengleichung isentropischer Gasströmungen", *Z. angew. Math. Mech.* **31** (1951), pp. 339–343.

17. Equation (34) is a special case of what is now called the Euler-Poisson-Darboux equation: G. DARBOUX, cit. Note II.47, pp. 54–70. See also L. EULER, "Institutiones calculi integralis", *Opera Omnia*, Ser. 1, Vol. 13, Leipzig and Berlin: Teubner, 1914, pp. 212–230; and S. D. POISSON, "Mémoire sur l'intégration des équations linéaires aux différences partielles", *J. école polytech.* Ser. 1, **19** (1823), pp. 215–248. Recent mathematical interest in the equation and its generalization has been stimulated mainly by the work of A. WEINSTEIN, see for example "On the wave equation and the equation of Euler-Poisson", *Proc. Symp. Appl. Math.* (*A.M.S.*) **5** (1954), p. 137–147.

18. For either quotient these values are $\kappa = (2N + 3)/(2N + 1)$ where N is any integer. In particular $N = 1$ gives $\kappa = 5/3$ which is the value of γ for a monatomic gas. The corresponding values of n are then: $n = 1$ for $z_n = U$ or ρt, $n = -1$ for V

Article 12

or $x - ut$, $n = 2$ for ψ, and $n = -2$ for t. Thus, the physically most interesting cases of monatomic and diatomic gases correspond to mathematically simple equations (34).

19. Equations (37) and (38) give Euler's solution of Eq. (34), see Note 17. In a more compact notation the equations may be written:

$$(37) \qquad z_n = v^{2n+1}\left(\frac{1}{v}\frac{\partial}{\partial v}\right)^n \left(\frac{Z_0}{v}\right), \qquad Z_0 = (-1)^n z_0/(2n-1)(2n-3)\cdots 1,$$

$$(38) \qquad z_n = \left(\frac{1}{v}\frac{\partial}{\partial v}\right)^{m-1} \left(\frac{Z_0}{v}\right), \qquad Z_0 = (-1)^{m-1} z_0/(2m-3)(2m-5)\cdots 1.$$

The second of these is due to A. E. Love and F. B. Pidduck, "Lagrange's ballistic problem", *Phil. Trans. Roy. Soc. London*, **A222** (1922), pp. 167–226. The first follows from it by a correspondence due to G. Darboux, cit. Note 17.

20. R. v. Mises, "One-dimensional adiabatic flow of an inviscid fluid", *NAVORD Rept.* **1719** (1951). The basic idea is contained in L. Euler, cit. Note 17. Integral representations have been used by E. T. Copson ["On sound waves of finite amplitude", *Proc. Roy. Soc.* **A216** (1953), pp. 539–547] and A. G. Mackie ["Contour integral solutions of a class of differential equations", *J. Rational Mech. Anal.* **4** (1955), pp. 733–750].

21. The solutions (37) and (38) may also be written in terms of the characteristic variables ξ and η. Thus (see Note 19) with $Z_0 = (-1)^n z_0/2^n(2n-1)(2n-3)\cdots 1 = F(\xi) + G(\eta)$, we find after some reduction

$$(37) \qquad z_n = (\xi + \eta)^{2n+1}\left[\frac{\partial^n}{\partial \xi^n}\frac{F(\xi)}{(\xi+\eta)^{n+1}} + \frac{\partial^n}{\partial \eta^n}\frac{G(\eta)}{(\xi+\eta)^{n+1}}\right].$$

Also, with $Z_0 = (-1)^{m-1} 2^m z_0/(2m-3)(2m-5)\cdots 1 = F(\xi) + G(\eta)$,

$$(38) \qquad z_n = \frac{\partial^{m-1}}{\partial \xi^{m-1}}\frac{F(\xi)}{(\xi+\eta)^m} + \frac{\partial^{m-1}}{\partial \eta^{m-1}}\frac{G(\eta)}{(\xi+\eta)^m}.$$

These formulas are due to G. Darboux, cit. Note 17.

22. More generally Eq. (34) reads in characteristic form:

$$\frac{\partial^2 z_n}{\partial \xi \partial \eta} - \frac{n}{\xi + \eta}\left(\frac{\partial z_n}{\partial \xi} + \frac{\partial z_n}{\partial \eta}\right) = 0.$$

The Riemann function is then

$$\Omega = \left(\frac{\xi_1 + \eta_1}{\xi + \eta}\right)^n F(1+n, -n, 1; -\sigma) = \left(\frac{\xi_1 + \eta_1}{\xi + \eta}\right)^n P_n(1 + 2\sigma),$$

where $\sigma = (\xi - \xi_1)(\eta - \eta_1)/(\xi + \eta)(\xi_1 + \eta_1)$, F is the hypergeometric function and P_n the Legendre function; for $n = -2$ we recapture (46). This result was obtained by B. Riemann, cit. Note II.23. Riemann also discussed the case $\kappa = 1$ (isothermal: $p/\rho = c^2$, constant) for which (43) is replaced by

$$\frac{\partial^2 V}{\partial \xi \partial \eta} + k\left(\frac{\partial V}{\partial \xi} + \frac{\partial V}{\partial \eta}\right) = 0; \qquad k = \frac{1}{4c}.$$

The Riemann function in this case is

Article 12

$$\Omega = e^{k(\tau+\nu)} I_0(2k\sqrt{\tau\nu}),$$

where $\tau = \xi - \xi_1$, $\nu = \eta - \eta_1$, and I_0 is the Bessel function (of imaginary argument) of order zero. The precise relation between these two Riemann functions is given in G. S. S. LUDFORD, "Two topics in one-dimensional gas dynamics", *Studies in Mathematics and Mechanics Presented to Richard von Mises*, New York: Academic Press, 1954, pp. 184–191. This paper also gives other examples. The Riemann function for the "telegraphist's equation" ([2], p. 316):

$$\frac{\partial^2 u}{\partial\xi\,\partial\eta} + u = 0,$$

can be obtained by noticing that for $u = e^{i(\xi+\eta)}V$ it reduces to one of the above kind with $k = i$. Hence we find

$$\Omega = J_0(2\sqrt{(\xi - \xi_1)(\eta - \eta_1)}),$$

where J_0 is the Bessel function of order zero.

23. R. v. MISES, cit. Note 20. Numerical and graphical methods analogous to those presented at the end of Sec. 16.7 have been given (respectively) by F. SCHULTZ-GRUNOW, "Nichtstationäre eindimensionale Gasbewegung", *Forsch. Gebiete Ingenieurwesens* **13** (1942), pp. 125–134, and R. SAUER, "Characteristikenverfahren für die eindimensionale instationäre Gasströmung", *Ing.-Arch.* **13** (1942), pp. 78–89.

24. With suitable interpretation, Riemann's formula (10.17) still applies to such cases as this, see G. S. S. LUDFORD, "On an extension of Riemann's method of integration with applications to one-dimensional gas dynamics", *Proc. Cambridge Phil. Soc.* **48** (1952), pp. 499–510, or "Riemann's method of integration: Its extensions with an application", *Collectanea Math.*, **6** (1953), pp. 293–323.

25. R. v. MISES, cit. Note 20. This problem was also considered by A. H. TAUB, "Interaction of progressive rarefaction waves", *Ann. Math.* **47,** (1946), pp. 811–828; see also Sec. 13.4. A special case arising in ballistics was discussed by A. E. LOVE and F. B. PIDDUCK, cit. Note 19. Of course the same results can be obtained by Riemann's method, but no so directly.

26. Only three of these conditions are independent. The effect of choosing other f_1', g_1', f_1'', g_1'', c and k which satisfy Eqs. (53) is to add a term $\alpha + \beta\xi + \gamma\xi^2 + \delta\xi^3$ to f and a term $-\alpha + \beta\eta - \gamma\eta^2 + \delta\eta^3$ to g, where α, β, γ, δ are constants. This merely adds a constant to V in Eq. (42) and hence has no effect on x and t as functions of u and v in Eqs. (47). We could take $C = 0$, but in the example of Sec. 13.5 it is more convenient to take $C \neq 0$.

27. Only five of these equations are independent. The effect of choosing other values for the ten constants is to add a term $\alpha + \beta\xi + \gamma\xi^2 + \delta\xi^3 + \epsilon\xi^4$ to f and a term $-\alpha + \beta\eta - \gamma\eta^2 + \delta\eta^3 - \epsilon\eta^4$ to g. As before α, β, γ, δ have no effect on the final result, while the ϵ merely shifts the origin of x.

28. R. v. MISES, cit. Note 20. This problem was also considered by K. BECHERT, "Zur Theorie ebener Störungen in reibungsfreien Gasen", *Ann. Physik,* Ser. 5, **37** (1940), pp. 89–123, and II *ibid.* **38** (1940), pp. 1–25.

29. The general solution of the homogeneous equation corresponding to (61) is $f = \mathfrak{A}_0 + \mathfrak{B}_0\xi + \mathfrak{C}_0\xi^2 + \mathfrak{D}_0\xi^3$, $g = -\mathfrak{A}_0 + \mathfrak{B}_0\eta - \mathfrak{C}_0\eta^2 + \mathfrak{D}_0\eta^3$, but the addition of terms $\xi^3\mathfrak{D}[x(\xi)]$ and $\eta^3\mathfrak{D}[x(\eta)]$ in (62) has no effect on the final answer in the present approach. An equivalent but less elegant result is obtained by the usual method of variation of parameters where one would set $Y_1 = Y_2 = 0$; instead we put $\mathfrak{D} = 0$ and $Y_1 = Y_2$.

Article 12

30. Such a problem arises in the treatment of interstellar gas clouds, see for example E. T. COPSON, cit. Note 20, and D. C. PACK, "A note on the unsteady motion of a compressible fluid", *Proc. Cambridge Phil. Soc.* **49** (1953), pp. 493–497.

Article 13

31. The solutions now known as simple (or progressive) waves were found by S. D. POISSON ["Mémoire sur la théorie du son", *J. école polytech.*, Ser. 1, **14** (1808), pp. 319–392] for the special case $\kappa = 1$ (isothermal). Later S. EARNSHAW ["On the mathematical theory of sound", *Phil. Trans. Roy. Soc. London* **150** (1860), pp. 133–148] extended the study to a general elastic fluid. For a discussion see LORD RAYLEIGH, Note 3. RIEMANN, cit. Note II.23, showed that an arbitrary limited disturbance of an unlimited gas at rest eventually separates into two simple waves.

32. Geometrical ideas such as the envelope E appear in a paper to which we shall need to refer later: H. HUGONIOT, "Mémoire sur la propagation du movement dans les corps, et spécialement dans les gaz parfaits", *J. école polytech.*, Ser. 1, **57** (1887), pp. 1–97, and **58** (1889), pp. 1–125.

33. Centered waves were used by Riemann, cit. Note II.23, in his discussion of initial discontinuities.

34. The eventual breakdown of such a simple wave was first pointed out by G. G. STOKES, "On a difficulty in the theory of sound", *Phil. Mag.* Ser. 3, **33** (1848), pp. 349–356, or *Mathematical and Scientific Papers*, Vol. 2, London and New York: Cambridge Univ. Press, 1883, pp. 51–55. Stokes treated the isothermal case. The result was extended to the general elastic case by B. RIEMANN, cit. Note II.23. For a purely analytical treatment of the change in type of a simple wave, see S. EARNSHAW, cit. Note 31, and COURANT and FRIEDRICHS [21], pp. 96–97.

35. The case in which the path of the piston is prescribed was treated first by S. EARNSHAW, cit. Note 31, and then by H. HUGONIOT, cit. Note 32, LORD RAYLEIGH, cit. Note 3, and A. F. PILLOW, "The formation and growth of shock waves in the one-dimensional motion of a gas", *Proc. Cambridge Phil. Soc.* **45** (1949), pp. 558–586. In the famous problem of ballistics first studied by J. L. LAGRANGE in 1793, the path of the piston must be determined: S. D. POISSON, "Formules relatives au movement du boulet dans l'intérieur du canon, extraites des manuscrits de Lagrange", *J. école polytech.*, Ser. 1, **21** (1832), pp. 187–204, or *Oeuvres de Lagrange*, Vol. 7, Paris: Gauthier-Villars, 1877, pp. 603–615; for extensive discussions see A. E. H. LOVE and F. B. PIDDUCK, cit. Note 19, and M. C. PLATRIER, "Analyse du problème balistique de Lagrange", *Mém. artillerie franç.* **15** (1936), pp. 431–477.

36. Thus according to (13.11′) we have

$$t_2 = t_0\left(\frac{v_2}{v_1}\right)^{h^2}, \qquad t_0 = \frac{l_1}{a_2},$$

so that with $l_1/l_0 = \rho_0/\rho_2 = (a_0/a_2)^{h^2-1}$ and $v_1 = (v_0 + v_2)/2$ we find

$$t_2 = \frac{l_0}{a_0}\left(\frac{a_0}{a_2}\right)^{h^2}\left(\frac{v_2}{v_1}\right)^{h^2} = \frac{l_0}{a_0}\left(\frac{2a_0}{a_0 + a_2}\right)^{h^2}.$$

37. The problem of interaction of symmetric waves is equivalent to that of the reflection of one of them at a fixed wall (the line of symmetry). The latter occurs in Lagrange's problem, see Note 35. Penetration of general simple waves was discussed by A. H. TAUB, cit. Note 25.

38. C. DE PRIMA has shown that there is a connection between this function t and the Riemann function of the t-equation [(12.34) with $n = -3$], see [21], pp. 194–196.

Article 13

Thus for general κ, the solution in the penetration region is

$$t = t_0\Omega(v_0,v_0;\xi,\eta),$$

where $\Omega(\xi,\eta;\xi_1,\eta_1)$ is given in Note 22, with $n = -(\kappa + 1)/2(\kappa - 1)$, see Eq. (12.33). Knowing t, the corresponding function x can be determined from (12.23) by integration. Nonsymmetric waves, whose centers have the same t-value, can be made symmetric by superimposing a suitable constant velocity on the whole flow.

Article 14

39. It should be emphasized that this is a direct appeal to experience. To reach a similar contradiction in the case of steady plane flow a more careful formulation of the experimental evidence is necessary (see Sec. 22.1).

40. Mathematically, the order of the system of differential equations governing the motion is reduced on setting $\mu = 0$. Such systems lead naturally to so-called asymptotic phenomena (see end of Sec. 2), which in fluid dynamics first appeared in the "boundary layer theory" of L. PRANDTL, "Über Flüssigkeitsbewegung bei sehr kleiner Reibung", *Verhandl. III. internat. math.-Kongresses, Heidelberg* (1904), pp. 484–491. This paper is included in L. PRANDTL and A. BETZ, *Vier Abhandlungen zur Hydrodynamik und Aerodynamik*, Göttingen: Kaiser Wilhelm-Institut für Strömungsforschung, 1927, pp. 1–8 (reprinted 1943, Ann Arbor: Edwards). The asymptotic character of the boundary layer was later pointed out by T. v. KÁRMÁN, "Über laminare und turbulente Reibung", *Z. angew. Math. Mech.* **1** (1921), pp. 233–252 and R. v. MISES, "Bemerkungen zur Hydrodynamik", *ibid.* **7** (1927), pp. 425–431. For a survey of the many facets of asymptotic phenomena see K. O. FRIEDRICHS, "Asymptotic phenomena in mathematical physics", *Bull. Am. Math. Soc.* **61** (1955), pp. 485–504. A technique for obtaining uniformly valid approximations in such problems has been developed by M. J. Lighthill and others, see H. S. TSIEN, "The Poincaré-Lighthill-Kuo method", *Advances in Appl. Mech.* **4** (1956), pp. 281–349. Cf. also Secs. 24.5, 6.

41. See for example [24], pp. 477–756.

42. For a nonperfect gas $\gamma p/(\gamma - 1)\rho$ must be replaced by the enthalpy $I(p,\rho)$ of the gas, see Eq. (2.23′). The corresponding changes in Art. 11 which lead to this result are obtained by retaining U in Eq. (11.5) instead of replacing it by $gRT/(\gamma - 1)$. For a full discussion see D. GILBARG, cit. Note 3. Similar changes must then be made in the remainder of this section.

43. Such flow patterns are also called weak solutions of the differential equations of ideal fluid theory, and can be alternatively defined by certain integral conditions, see for example P. D. LAX, "Initial value problems for non-linear hyperbolic equations", *Contract Nonr 58304, Dept. of Math., Univ. of Kansas, Tech. Rept.* **14** (1955), pp. 13–57.

44. B. RIEMANN cit. Note II.23, W. J. M. RANKINE cit. Note 3, and H. HUGONIOT cit. Note 32. Hugoniot also took the occurrence of these discontinuities for granted. Rankine justified his results on the basis of heat conduction alone, but failed to see that this case is singular. Neither Riemann, Rankine, nor Hugoniot included the inequality (14.9) and hence considered rarefaction shocks as well as condensation shocks (see next section). LORD RAYLEIGH, cit. Note 3, gave a survey of the question and in particular justified the Rankine-Hugoniot shock conditions on the basis of viscosity. See also R. RÜDENBERG, "Über die Fortpflanzungsgeschwindigkeit und Impulsstärke von Verdichtungsstössen", *Artilleristische Monatsh. Nos.* **113** *and* **114** (1916), pp. 237–265 and 285–316, and M. J. LIGHTHILL, cit. Note 3.

45. See Note 42. The results in the following sections have been extended to the

Article 14

case of a nonperfect gas; see P. DUHEM, "Sur la propagation des ondes de choc au sein des fluides", *Z. physik. Chem.* **69** (1909), pp. 169–186, H. A. BETHE, "The theory of shock waves for an arbitrary equation of state", *OSRD Rept. No.* **545** (1942), and H. WEYL, "Shock waves in arbitrary fluids", *Communs. Pure Appl. Math.* **2** (1949), pp. 103–122.

46. This fact is often made the basis of an approximation in which the results of Arts. 12 and 13 are again used behind the shock. See for example K. O. FRIEDRICHS, "Formation and decay of shock waves", *Communs. Pure Appl. Math.* **1** (1948), pp. 211–245, and A. F. PILLOW, cit. Note 35. See also Note V. 24 and M. J. LIGHTHILL, cit. Note 3.

47. For this case (14.22) is equivalent to Prandtl's relation, see Note V.25.

48. This is referred to as the Hugoniot curve. If the state 2 were connected to the state 1 by an inviscid adiabatic process the hyperbola would be replaced by $\eta\zeta^\gamma = 1$, a curve which is asymptotic to the ζ- and η-axes and osculates the hyperbola at A.

49. See H. HUGONIOT, Note 32, and previous Note.

50. This remark plus the fact that the entropy change across a shock is of the third order in $\rho_2-\rho_1$ gives a theorem similar to the one for steady plane flow in Sec. 23.1, the only changes being that θ is replaced by u, and F is now a function of p,ρ, and u, with derivatives

$$F' = \rho a \frac{\partial F}{\partial p} + \frac{\rho}{a}\frac{\partial F}{\partial \rho} + \frac{\partial F}{\partial u} \qquad \text{and} \qquad 'F = -\left(\rho a \frac{\partial F}{\partial p} + \frac{\rho}{a}\frac{\partial F}{\partial \rho}\right) + \frac{\partial F}{\partial u}.$$

51. See B. RIEMANN, Note II.23, where the problem of an initial discontinuity separating regions of uniform flow is treated. For a corrected treatment see H. WEBER [33], pp. 522–531. The two initial discontinuities in this example are equivalent to special cases of those treated by Riemann (reflection of the initial conditions about $x = 0,l$). An extended treatment of Riemann's problem has been given by R. COURANT and K. O. FRIEDRICHS, "Interaction of shock and rarefaction waves in one-dimensional motion", *OSRD Rept. No.* **1567** (1943).

Article 15

52. This reflection problem was treated by H. HUGONIOT, cit. Note 32, who also considered successive reflections from a uniformly moving piston and fixed wall in turn. More recently, it has been discussed by H. PFRIEM, "Reflexionsgesetze für ebene Druckwellen grosser Schwingungsweite", *Forsch. Gebiete Ingenieurwesens* **12** (1941), pp. 244–256.

53. This result is due to H. HUGONIOT, cit. Note 32 (p. 94), who discussed and used the corresponding discontinuity.

54. H. v. HELMHOLTZ, cit. Note II.23.

55. See the section on the Lanchester-Prandtl wing theory in v. MISES and FRIEDRICHS [25]. A short history has been given by R. v. MISES in the notes to Chapter IX in [16].

56. The occurrence of contact discontinuities in problems involving shocks was first emphasized by J. v. NEUMANN, "Theory of shock waves", *OSRD Rept. No.* **1140** (1943). See also Note 53.

57. The reflection treated in Sec. 1 is equivalent to a special case of the present problem in which the two shocks have equal strength. A second type of interaction occurs when the two shocks move in the same direction so that one overtakes the other. A study of interactions of shocks and rarefaction waves in one-dimentional flows was made by R. COURANT and K. O. FRIEDRICHS, cit. Note 51. Theory and ex-

Article 15

periment are compared in I. I. GLASS and G. N. PATTERSON, "A theoretical and experimental study of shock-tube flows", *J. Aeronaut. Sci.* **22** (1955), pp. 73–100.

58. The theorem of Note 50 predicts that, to the second order, $u = u_1 + u_2$ (see similar result in Sec. 23.6). Hence in this example $(u_1 + u_2)/v_0 = 2/5\sqrt{7} = 0.1512$ is the estimated value of x, which is good agreement considering the strengths of the shocks.

59. The idea of the method presented in this section is due to R. v. Mises. It was worked out by G. S. S. LUDFORD, H. POLACHEK, and R. J. SEEGER, "On unsteady flow of compressible viscous fluids", *J. Appl. Phys.* **24** (1953), pp. 490–495.

60. See J. v. NEUMANN, "Proposal and analysis of a new numerical method for the treatment of hydrodynamic shock problems", *NDRC Appl. Math. Rept. No.* **108.1R** (1944). He found that the particles acquire small oscillations, superimposed on their true paths, after passing through the location of the shock, and he interpreted this in terms of internal energy.

61. See for example H. GEIRINGER "On numerical methods in wave interaction problems", *Advances in Appl. Mech.* **1** (1948) pp. 201–248, and remarks of K. O. FRIEDRICHS in [31], pp. 50–58.

62. See the paper quoted in Note 59. A second way out, having a similar effect, is to change the viscosity law; see J. v. NEUMANN and R. D. RICHTMEYER, "A method for the numerical calculation of hydrodynamic shocks", *J. Appl. Phys.* **21** (1950), pp. 232–237, and R. v. MISES, cit. Note I.6.

63. See Notes 46 and V.45.

64. This is due to M. H. MARTIN, "The propagation of a plane shock into a quiet atmosphere", *Can. J. Math.* **5** (1953), pp. 37–39. The approach was developed by G. S. S. LUDFORD and M. H. MARTIN, "One-dimensional anisentropic flows", *Communs. Pure Appl. Math.* **7** (1954), pp. 45–63, and G. S. S. LUDFORD, "Generalised Riemann invariants", *Pacific J. Math.* **5** (1955), pp. 441–450.

CHAPTER IV

Article 16

1. If ρ_0 denotes a reference density, and ψ_x, ψ_y the partial derivatives of ψ, Eq. (21) is replaced by

$$\psi_{xx}\left[\psi_y^2 - \left(\frac{\rho a}{\rho_0}\right)^2\right] - 2\psi_{xy}\psi_x\psi_y + \psi_{yy}\left[\psi_x^2 - \left(\frac{\rho a}{\rho_0}\right)^2\right] = 0.$$

Here $(\rho a/\rho_0)^2$ is a given function of $\psi_x^2 + \psi_y^2$ (Sec. 24.2).

With the same notation, the second-order equation for steady flow with axial symmetry is

$$\psi_{xx}\left[\psi_y^2 - \left(\frac{y\rho a}{\rho_0}\right)^2\right] - 2\psi_{xy}\psi_x\psi_y + \psi_{yy}\left[\psi_x^2 - \left(\frac{y\rho a}{\rho_0}\right)^2\right] + \frac{\psi_y}{y}\left(\frac{y\rho a}{\rho_0}\right)^2 = 0.$$

Here $(\rho a/\rho_0)^2$ is a given function of $(1/y^2)(\psi_x^2 + \psi_y^2)$.

Equation (16.21) is replaced by Eq. (24.9) if the flow is strictly adiabatic but not necessarily elastic.

2. Even apart from the fact that the (x,t)-problem of Chapter III is always hyperbolic it is mathematically much easier than the present problem. In Art. 12 we could

Article 16

indicate explicit general solutions. Nothing similar exists here.

3. With respect to the characteristics as Mach lines, cf. Note I.25.

4. Two mathematically interesting special cases may be considered: (a) the characteristics in the x,y-plane are rectilinear, and (b) those in the u,v-plane are rectilinear, (see SAUER [7], p. 90 ff.). In our problem both cases can be realized by considering particular (p,ρ)-relations: the first, with the "Pérès-Munk (p,ρ)-relation", $p = A + B\,[\arctan \rho - \rho/(1 + \rho^2)]$; the second, with the raletion $p = A - B/\rho$ (Secs. 17.5 and 17.6) generalized to supersonic flow (cf. e.g. [7], p. 100 ff.); physically, this generalization is controversial. Compare also R. SAUER, cit. Note 11.

5. The hodograph method as explained in this section (cf. also Secs. 8.6, 10.6, and 10.7 is due to Chaplygin (1869–1944): S. A. CHAPLYGIN, "On gas jets", *Sci. Mem. Moscow Univ. Math. Phys. Sec.* **21** (1902), pp. 1–121 (translation: *NACA Tech. Mem.* **1063** (1944). Compare also Note II.23.

Chaplygin quotes as predecessors P. MOLENBROEK, "Ueber einige Bewegungen eines Gases bei Annahme eines Geschwindigkeitspotentials", *Arch. Math. Phys.* **9** (1890), pp. 157–195 (included in [20]), and the work of HELMHOLTZ of 1868, cit. Note II.23. We also mention A. STEICHEN, "Beiträge zur Theorie der zweidimensionalen Bewegungsvorgänge in einem Gase, das mit Überschallgeschwindigkeit strömt", *Dissertation,* Göttingen, 1909. D. RIABOUCHINSKY ["Mouvement d'un fluide compressible autour d'un obstacle" *Compt. rend.* **194** (1932), pp. 1215–1216] has drawn attention to the importance of Chaplygin's work. An early account of the theory was also given by B. DEMTCHENKO, "Sur les mouvements lents des fluides compréssibles", *Compt. rend.* **194** (1932), pp. 1218–1222. Extensive bibliographies on the hodograph method are in [20], p. 263 ff. and in [21], p. 441 ff.

6. Many of the properties of the fixed characteristics and their relation to the Mach lines are contained in the paper by L. PRANDTL and A. BUSEMANN, cit. Note II.24. In connection with the orthogonality of Mach lines and fixed characteristics cf. Sec. 9.4, particularly Eq. (9.15).

7. Essentially the same definition of characteristic coordinates (our Eq. (43)) is found in the paper cit. Note 6 in which the graphical procedure of p. 260 is also introduced. The method has been further developed by G. GUDERLEY, "Die Charakteristikenmethode für ebene und achsensymmetrische Überschallströmungen", *Jahresber. deut. Luftfahrtforsch.* **1** (1940), pp. 522–535. See also K. OSWATITSCH, "Ueber die Charakteristikenverfahren der Hydromechanik", *Z. angew. Math. Mech.* **25/27** (1947), pp. 195–200, 264–270, and W. TOLLMIEN, "Steady two-dimensional rotationally symmetric supersonic flows", *Grad. Div. Appl. Math., Brown Univ., Trans.* **AG-T-1** (1946).

8. This rather obvious theorem is directly analogous to one of the "slip line theorems", due to H. Hencky and L. Prandtl, well-known in the theory of a plane perfectly plastic body.

Article 17

9. The usual derivation is to start with a second-order equation such as Eq. (16.14) and to apply to it the "contact transformation" associated with the name of Legendre. This transformation, applied to gas dynamics, is found in S. D. POISSON, cit. Note I.23. The derivation may be found in textbooks on partial differential equations, e.g., H. BATEMAN, *Partial Differential Equations of Mathematical Physics*, New York: Dover, 1944, or [2]. See also G. HAMEL, *Mechanik der Kontinua*, Stuttgart: Teubner, 1956, p. 108.

The following is of interest for us. If to an equation such as (5),

Article 17

$$a(u,v)\,\frac{\partial^2\Phi}{\partial u^2} + 2b(u,v)\,\frac{\partial^2\Phi}{\partial u\partial v} + c(u,v)\,\frac{\partial^2\Phi}{\partial v^2} = 0,$$

we apply the Legendre transformation defined by Eqs. (4), (4′), (6), we obtain the equation

$$a\left(\frac{\partial\varphi}{\partial x},\frac{\partial\varphi}{\partial y}\right)\frac{\partial^2\varphi}{\partial y^2} - 2b\left(\frac{\partial\varphi}{\partial x},\frac{\partial\varphi}{\partial y}\right)\frac{\partial^2\varphi}{\partial x\partial y} + c\left(\frac{\partial\varphi}{\partial x},\frac{\partial\varphi}{\partial y}\right)\frac{\partial^2\varphi}{\partial x^2} = 0.$$

The fixed characteristics of the first of these equations are given by

$$a\,dv^2 - 2b\,du\,dv + c\,du^2 = 0, \qquad \left(\frac{dv}{du}\right)_{1,2} = \frac{1}{a}\,(b \pm \sqrt{b^2 - ac}),$$

and the characteristics in the x,y-plane, which depend on the solution $\varphi(x,y)$, are given by

$$a\,dx^2 + 2b\,dx\,dy + c\,dy^2 = 0, \qquad \left(\frac{dx}{dy}\right)_{1,2} = -\frac{1}{a}\,(b \mp \sqrt{b^2 - ac}).$$

Hence

$$\left(\frac{dv}{du}\right)_{2,1} = -\left(\frac{dx}{dy}\right)_{1,2}$$

and this is the same orthogonality relation as in (9.15) and in (16.37). We realize that this orthogonality is a general *mathematical* fact, which does not depend on particular properties of our mechanical equations.

10. We may obtain the equation for the Legendre stream function by applying the formal Legendre transformation to the equation for the stream function, viz. Eq. (16.21) which we write as in Note 1, with $\rho_0 = 1$. We now put $\partial\psi/\partial x = r$, $\partial\psi/\partial y = t$ and apply the contact transformation, as in Eq. (10): $\Psi(r,t) = xr + yt - \psi(x,y)$. By the rule formulated in Note 9, we then obtain

$$\Psi_{rr}(a^2\rho^2 - r^2) - 2\Psi_{rt}rt + \Psi_{tt}(a^2\rho^2 - t^2) = 0.$$

Here $a\rho$ is a function of $r^2 + t^2$; the variables $t = \rho u$, $-r = \rho v$ are the components of the flux vector $\rho\mathbf{q}$. Hence this is a mapping not onto the hodograph plane, i.e. the $\mathbf{q}$-plane, but onto the $\rho\mathbf{q}$-plane; see Sauer [29], p. 157.

11. R. Sauer, cit. Note III.16, investigates the (p,ρ)-relations for which Eq. (21) becomes a "Darboux equation", in which case a general integral depending on two arbitrary functions is known.

12. The inversion formulas of this section are given, for example, in Chaplygin's paper, cit. Note 5. They are also derived in detail in a paper by F. Ringleb, which we shall discuss extensively in Art. 20: F. Ringleb, "Exakte Lösungen der Differentialgleichungen einer adiabatischen Gasströmung", *Z. angew. Math. Mech.* **20** (1940), pp. 85–198.

13. Bibliographical data have been given in the Notes to Art. 7. Compare also the treatment by Ringleb in the paper quoted in Note 12. A further exact solution, not discussed by us, which belongs in this group has been defined and studied by W. Tollmien, "Zum Uebergang von Unter- zu Überschallströmung", *Z. angew. Math. Mech.* **17** (1937), pp. 117–136.

14. For Figs. 100b and 101b, the drawings (Figs. 39 and 40) in [24], Chapter V by W. G. Bickley, have been used in part.

Article 17

15. See S. A. CHAPLYGIN, cit. Note 5, Part V. His method was modified, elaborated and used in many ways by v. Kármán and H. S. Tsien, see T. v. KÁRMÁN, cit. Note IV.30 and H. S. TSIEN, "Two-dimensional subsonic flow of compressible fluids", *J. Aeronaut. Sci.* **6** (1939) pp. 399–407.

16. See S. A. CHAPLYGIN, cit. Note 5, p. 97 of translation.

17. We have thus seen that for a gas with (p,ρ)-relation: $p = A - B/\rho$, the Chaplygin equation $\partial^2\psi/\partial\sigma^2 + K(\sigma)\partial^2\psi/\partial\theta^2 = 0$ reduces to the Laplace equation. For another particular pressure-density relation the Chaplygin equation reduces to the *Tricomi equation*, namely, $\partial^2\psi/\partial\sigma^2 + \sigma(\partial^2\psi/\partial\theta^2) = 0$. This gas is called the "Tricomi gas". Compare for example F. TRICOMI "Correnti fluide transoniche ed equazioni a derivate parziali di tipo misto", *Rend. Seminar. Mat. Torino* **12** (1953), pp. 37–52.

18. It can be shown that for this (ρ,q)-relation the potential equation in the x,y-plane, viz.,

$$\frac{\partial}{\partial x}\left(\rho\,\frac{\partial\varphi}{\partial x}\right) + \frac{\partial}{\partial y}\left(\rho\,\frac{\partial\varphi}{\partial y}\right) = 0,$$

is always elliptic and is in fact the differential equation of minimal surfaces.

19. Compare also F. H. CLAUSER, "Two-dimensional compressible flows having arbitrarily specified pressure distributions for gases with gamma equal to minus one", *Symposium on Theoretical Compressible Flow, White Oak, Maryland* (1949), pp. 1–32.

20. C. C. LIN, "On an extension of the von Kármán-Tsien method to two-dimensional subsonic flows with circulation around closed profiles", *Quart. Appl. Math.* **4** (1946), pp. 291–297, reprinted in [20]. See also L. BERS, "On a method of constructing two-dimensional subsonic compressible flows around closed profiles", *NACA Tech. Notes* **969** (1945), and A. GELBART, "On subsonic compressible flows by a method of correspondence", *NACA Tech. Notes* **1170** (1945). A simple method due to K. JAECKEL ["Verallgemeinerung des Tsien'schen Verfahrens" (unpublished)] is described by Lighthill in [24], p. 226.

21. Compare for example L. M. MILNE-THOMPSON, *Theoretical Aerodynamics*, New York: van Nostrand, 1947, p. 128 ff. The original papers are: R. v. MISES, "Zur Theorie des Tragflächenauftriebes, erste Mitteilung", *Z. Flugtech. Motorluftschiffahrt*, **8** (1917), pp. 157–163, and "zweite Mitteilung", *ibid.* **11** (1920), pp. 68–73 and 87–89.

22. See for example [31], p. 505. The original investigations are by H. S. TSIEN, cit. Note 15.

23. See [31], p. 509, and H. S. TSIEN, cit. Note 15.

24. L. BERS, "An existence theorem in two-dimensional gas dynamics", *Proc. Symp. Appl. Math.* (*A.M.S.*) **1** (1949), pp. 41–46.

Article 18

25. Two-dimensional simple waves were studied by L. PRANDTL, "Neue Untersuchungen über die strömende Bewegung der Gase und Dämpfe". *Physik. Z.* **8** (1907), pp. 23–30. The systematic description is due to T. MEYER "Über zweidimensionale Bewegungsvorgange in einem Gas, das mit Überschallgeschwindigkeit strömt", *Forschungsh. Ver. deut. Ing.* **62** (1908), pp. 31–67 (included in [20]). The name frequently used is *Prandtl-Meyer flow*. Compare also Note III.31, and papers quoted in Note 29.

26. In a simple wave the state variables are constant along straight lines; in other words, they depend only on the angle ϕ which such a line makes with a fixed direction.

Article 18

This forms a counterpart to the flows considered in Art. 17 (vortex flow, radial flow, spiral flow) where the state variables depend only on radial distance from an origin.

27. *Conical flow*, i.e., a flow in which the state variables are constant on concentric rays, may be considered as a generalization of a centered simple-wave flow (see for example [27], p. 262). If, in addition to this property, axial symmetry also prevails, then the surfaces of constant state are circular cones. See G. I. TAYLOR and J. W. MACOLL, cit. Note II.21, and J. W. MACOLL, "The conical shock wave formed by a cone moving at high speed", *Proc. Roy. Soc.* **A159** (1937), pp. 459–472.

In a paper by J. H. GIESE, "Compressible flows with degenerate hodographs", *Aberdeen Proving Ground, Ballistic Research Rept.* **657** (1948), a systematic investigation of steady flows with degenerate hodographs is carried out. The author discusses two-dimensional flows with one-dimensional hodographs, or three-dimensional flows with one- or two-dimensional hodographs. Those with one-dimensional hodographs are designated as simple waves, the others as double waves. Compare also the nonisentropic simple waves of Sec. 15.7.

28. The numerical examples are given in more detail than usual in order to clarify and illustrate the conditions under which one, two, or no solutions exist. A similar aim prompts our comments on Cauchy's data, p. 302. Compare also Sec. 20.4 where we discuss other completely different compressible flows around corners. They satisfy, of course, different boundary conditions.

29. The choice of examples in this section is similar to that in [21] p. 282 ff. All figures have been newly constructed here by W. Gibson.

Compare also the paper by J. H. GIESE, cit. Note 27, the paper by L. STEINBERG, "The geometry of the envelope of Mach lines forming a compression wave", ONR Contract 562(07), Grad. Div. Appl. Math., Brown Univ., Tech. Rept. **4** (1955), pp. 1–41, and R. E. MEYER, "On waves of finite amplitude in ducts. I. Wave fronts. II. Waves of moderate amplitude", *Quart. J. Mech. Appl. Math.* **5** (1952), pp. 257–291.

Article 19

30. The concept of a limit line but not the term appears in G. I. Taylor's previously quoted papers, cit. Note II.17, and in the paper of M. U. CLAUSER and F. H. CLAUSER, "New methods of solving the equations for the flow of compressible fluids", *Thesis*, California Inst. Technol., 1937. In 1937, in the paper quoted in Note 13, W. Tollmien defines and discusses an exact transonic solution which exhibits a "Grenzlinie" (limit line) beyond which the flow cannot be continued, and he establishes some of its properties. In 1940 F. Ringleb, in the paper cit. Note 12, gave an example of a limit line with cusps and studied it in detail. The whole situation is discussed and new results added by T. v. KÁRMÁN, "Compressibility effects in aerodynamics", *J. Aeronaut. Sci.* **8** (1941), pp. 337–356; see particularly p. 352 ff. (Cf. also the discussion in our Sec. 25.4). In a second paper by F. RINGLEB (which, however, is open to certain objections), "Ueber die Differentialgleichungen einer adiabatischen Gasströmung und den Strömungsstoss", *Deut. Math.* **5** (1940), pp. 377–384, several results previously found by him and others for special cases are proved for the general case. The "Strömungsstoss" is defined as the locus of points where the acceleration becomes infinite and where the stream lines have cusps. The problem is reconsidered by W. TOLLMIEN, "Grenzlinien adiabatischer Potentialströmungen", *Z. angew. Math. Mech.* **21** (1941), p. 140–152, and new results are added. None of these earlier German papers, however, discusses in general the cusps of the limit line (see our Sec. 4). This holds also for the presentation in COURANT-FRIEDRICHS [21], pp. 62–69 and pp. 256–259. A flow similar to Ringleb's was found and discussed by G. TEMPLE and J. YARWOOD, "Compressible flow in a convergent-divergent nozzle", *A.R.C. Repts. & Mem.* **2077** (1942).

Article 19

The physical and mathematical significance of limit lines is considered in the paper H. S. TSIEN, "The limiting line in mixed subsonic and supersonic flow of compressible fluids", *NACA Tech. Notes* **961** (1944). Compare also H. S. TSIEN and Y. H. KUO, "Two-dimensional irrotational mixed subsonic and supersonic flow of a compressible fluid and the upper critical Mach number", *NACA Tech. Notes* **995** (1946). A systematic exposition containing also a careful discussion of the cusps of the limit line, of some singularities of higher order, etc., is given by J. W. CRAGGS, "The breakdown of the hodograph transformation for irrotational compressible fluid flow in two dimensions", *Proc. Cambridge Phil. Soc.* **44** (1948), pp. 360–379. For further literature see Notes 32 and 35, and Sec. 25.4.

31. Often the hodograph image of a limit line is likewise called a limit line (and the same for the branch line). This is similar to our use of the word "streamline".

32. Our method differs from that used by Craggs, Lighthill, [24], Courant and others cit. Note 30. The same method is used in R. E. MEYER, "Focusing effects in two-dimensional supersonic flow", *Phil. Trans. Roy. Soc. London* **A242** (1949), pp. 153–171, a paper which deals not with the basic properties of branch lines and limit lines, but with more refined problems from the point of view of Riemannian geometry. J. MANDEL [*Équilibres par tranches planes des solides à la limite d'écoulement*, Paris: Louis Jean, 1942] studies in this way the mapping of the physical plane onto the "stress graph" in the theory of the plane "perfectly plastic body". Our presentation is decisively influenced by Mandel's.

33. Equations (7), which are adequate for the purposes of this article, may be plified if $\alpha \neq 90°$. Put

$$A = \frac{\sin 2\alpha}{(1 - \alpha'/Q')},$$

consider Q rather than q as the independent variable, and denote by ` differentiation with respect to $Q$. Then, introducing the function $T(Q)$ by $T^{`}/T = -2 \cos 2\alpha/A$ and e_1, e_2 by $h_1 = e_1/\sqrt{T}$, $h_2 = e_2/\sqrt{T}$ we find by a straightforward computation the simple equations $A\partial e_1/\partial\eta = e_2$, $A\partial e_2/\partial\xi = e_1$; from these we obtain

$$A^2 \frac{\partial^2 e_1}{\partial\xi\partial\eta} + AA^{`} \frac{\partial e_1}{\partial\eta} - e_1 = 0 \quad \text{and} \quad A^2 \frac{\partial^2 e_2}{\partial\xi\partial\eta} + AA^{`} \frac{\partial e_2}{\partial\xi} - e_2 = 0.$$

34. Geometrically, the situation may also be understood in the following way. The correspondence between the x,y-plane and ξ,η-plane defines in four-dimensional x,y,ξ,η-space a two-dimensional subspace, $\mathcal{S}$. The manifold $\mathcal{S}$ consists of two sheets, connected along the *apparent contour*, whose projections on the x,y-plane cover a region twice. The line $\mathfrak{L}_1$ in the x,y-plane is part of the projection of the apparent contour, and similarly for $\mathfrak{L}_2$. Consider curves through a point $\mathfrak{M}$ on $\mathcal{S}$ which traverses the apparent contour. Such curves are projected into curves through M (Fig. 126) which are either tangential to the projection $\mathfrak{L}_1$ or, exceptionally, have cusps there. Compare also Hugoniot's (Note III.32) geometrical ideas [in relation to the (x,t)-problem] which led to the term "edge of regression".

35. Most of the definitions and results in this section are new. A few remarks relating to the subject are in SAUER [29], p. 229, and (partly incorrect) in Ringleb's second paper, cit. Note 30, and, for a somewhat analogous problem, in Mandel's monograph, cit. Note 32. Two recent papers are: H. GEIRINGER, "Grenzlinien der Hodographentransformation", *Math. Z.* **63** (1956), pp. 514–524, and G. S. S. LUDFORD and S. H. SCHOT, "On sonic limit lines in the hodograph method", *ibid.* (1957), pp. 229–237. The proof of the geometric characterization of the sonic limit line, p. 324 of our text is due to Ludford and Schot. In addition, their paper contains a new example of a

Article 19

sonic limit line. Recently S. H. Schot called our attention to the fact that in a quite normal example (Sec. 20.3) the derivative ψ_ξ does not exist at the sonic point of the limit line $\mathfrak{L}_1$. Hence, the classification of limit-type singularities based on the behaviour of ψ_ξ, ψ_η in addition to that of φ_ξ, φ_η, which had been proposed in these two papers, has to be modified and it does not appear in the present book.

36. Consider again the two-dimensional manifold $\mathcal{S}$ in x,y,ξ,η-space, see Note 34. The original of b_1 in four-space belongs to the apparent contour of $\mathcal{S}$. It separates two sheets of $\mathcal{S}$ which when projected onto the ξ,η-plane cover a portion of this plane twice.

37. On the original $\mathcal{S}$ in four-space, the tangent to such a line is perpendicular to the ξ,η-plane at its intersection with the apparent contour; hence in the projection onto the ξ,η-plane it shows a cusp. In the projection onto the x,y-plane, it presents an inflexion.

38. M. J. Lighthill, "The hodograph transformation in trans-sonic flow. I. Symmetrical channels", *Proc., Roy. Soc.* **A191** (1947), pp. 323–341. He defines and discusses branch lines in two-dimensional transonic flow. See Note V.56 regarding Cherry's work. We found a branch line—for an (x,t)-problem—in Chapter III, see Fig. 62.

39. Regarding the (x,t)-problem, see P. M. Stocker and R. E. Meyer, "A note on the correspondence between the x,t-plane and the characteristic plane in a problem of interaction of plane waves of finite amplitude", *Proc. Cambridge Phil. Soc.* **47** (1951), pp. 518–527.

Article 20

40. The basic source for this article, and in particular for Secs. 1, 2, and 6, is Chaplygin's paper, cit. Note 5.

41. Chaplygin, in view of the problem of the subsonic jet, uses "$2n$" where we and most authors use "n".

42. Regarding literature on the hypergeometric function we mention the presentation in Whittaker-Watson [8], the monograph Kampé de Fériet [5], and the monograph by F. Klein, *Vorlesungen über die hypergeometrische Funktion*, Berlin: Springer, 1933. See also Notes 49 and 50.

43. The right side is the sum of two independent solutions of the hypergeometric equation, namely, $y = Ay_1(\tau) + By_2(\tau)$, where $y_1(\tau) = F(a,b,a+b-c+1;1-\tau)$ and $y_2(\tau) = (1-\tau)^{c-a-b}F(c-a,c-b,c+1-a-b;1-\tau)$. The coefficients are chosen in such a way that $y(0) = 1$, and $y(1) = F(a,b,c;1)$. ("Gauss' transformation".)

44. For a discussion of solutions with singularities at an arbitrary point of the subsonic region ["fundamental solutions" etc.] see S. Bergman, "Two-dimensional subsonic flows of a compressible fluid and their singularities", *Trans. Am. Math. Soc.* **62** (1947), pp. 452–498. Recently R. Finn and D. Gilbarg, "Asymptotic behavior and uniqueness of plane subsonic flows", *Communs. Pure Appl. Math.* **10** (1957), pp. 23–63, showed that the singularities introduced by Bergman are the most general that can occur in flows which are subsonic in a neighborhood of infinity. Fundamental solutions for equations of "mixed" type (see Art. 25) have been studied by P. Germain, "Remarks on the theory of partial differential equations of mixed type and applications to the study of transonic flow", *Communs. Pure Appl. Math.* **7** (1954), pp. 117–143, see Part II.

45. Following Chaplygin, we use here the variable $\tau = q^2/q_m^2$ since this choice leads to the hypergeometric equation (7′), whose theory is well known. Several German authors [F. Ringleb (cf. Note 12), R. Sauer [29], and G. Hamel (cf. Note 9)] use q rather than τ; this leads to a second-order ordinary differential equation, the solutions of which are used in a way similar to that of the hypergeometric functions in Chaplygin's theory.

Article 20

46. See F. Ringleb, cit. Note 12. This example is remarkable as one of the first instances of smooth transonic potential flow, as well as of an embedded supersonic region, and an interesting limit line with cusps. (See further comments, Secs. 25.3 and 25.4).

47. The flow (21) has been investigated by G. Temple and J. Yarwood, cit. Note 30. The case $n = -1$ is exceptional, since, with the notations of Eq. (8), not only c_{-1} but also a_{-1} vanishes. Hence ψ_{-1} is actually indeterminate and may be defined (see [24], p. 233) as:

$$\psi_{-1}(\tau) = \lim_{n\to -1} \psi_n(\tau).$$

The result is

$$\psi_{-1}(\tau) = \tau^{-1/2} + \frac{1}{2(\kappa - 1)} \psi_1(\tau).$$

Substituting for $\psi_1(\tau)$ the expression found in the text we obtain

$$\psi_{-1}(\tau) = \frac{2\kappa + 1}{2\kappa} \tau^{-1/2} - \frac{1}{2\kappa} \tau^{-1/2}(1 - \tau)^{\kappa/(\kappa-1)},$$

which is in fact a linear combination of the two particular solutions (15) and (21).

48. Take for example $m = 2$, or $\alpha = \pi/2$. This solution, which represents compressible flow within a corner, was investigated by J. Williams, "The two-dimensional irrotational flow of a compressible fluid in the acute region made by two rectilinear walls", *Quart. J. Math., Oxford Ser.* **2** (1949), p. 129 ff. We have in this case $\psi_2 = A\tau F(2.5, -3,3;\tau) \sin 2\theta$, where the hypergeometric function reduces to a polynomial of degree three in τ. (See also Note 50.)

49. Following up earlier work by Gauss and by E. Goursat, *Sur l'équation différentielle linéaire qui admet pour l'intégrale la série hypergéometrique*, Paris, 1881, Lindelöf has indicated a solution valid in this case, which involves a logarithmic term: E. L. Lindelöf, "Sur l'intégration de l'équation différentielle de Kummer", *Acta Soc. Sci. Fennicae* **19** (1893), pp. 3–31 [Eq. (11), p. 13]. The logarithmic term is multiplied by τ^{-n} and since n is negative, the product tends to zero as $\tau \to 0$ and the expansion still tends to 1. Lindelöf's formula in explicit form and with notation adapted to the aerodynamic problem is given, including tables, in I. E. Garrick and C. Kaplan, "On the flow of a compressible fluid by the hodograph method II—Fundamental set of particular flow solutions of the Chaplygin differential equation", *NACA Rept. No.* **790** (1944). See also V. Huckel, "Tables of hypergeometric functions for use in compressible-flow theory", *NACA Tech. Notes* **1716** (1948).

In a useful Note V. O'Brien ["Remarks on Chaplygin Functions", *J. Aeronaut. Sci.* **23** (1956), pp. 894–895; also NOrd 7386 *The Johns Hopkins University, Appl. Phys. Lab.*, **C.M-871** (1956), pp. 1–6] points out that those n, for which not only c but also a or b is a negative integer, yield special cases for the logarithmic solution, and she gives the formula which in this case replaces Garrick and Kaplan's formula. The first values of n for which this happens are $n = -2, -5, -12$. The entries corresponding to $n = -2, -5, -12$ in the tables of V. Huckel are substantially wrong, as shown by Miss O'Brien. (The solution for $n = -2$, $\alpha = 270°$, given by H. J. Davies, see Note 50, is correct and independent of these shortcomings.)

50. For $\alpha = 270°$ we have, for $\kappa = \gamma = 7/5$: $n = -2$, $a = 1/2$, $b = -5$, $c = -1$; thus not only c but also b is a negative integer (cf. preceding Note). H. J. Davies, using earlier work by Temple and Yarwood and by Lighthill, has computed this flow and

Article 20

has tabulated the hypergeometric functions required in the computation. See H. J. DAVIES, "The two-dimensional irrotational flow of a compressible fluid around a corner", *Quart. J. Mech. Appl. Math.* **6** (1953), pp. 71–80. The paper by Temple and Yarwood which was used by Davies is quoted in Note 30. The expansion for F in case $c = -1$ $(n = -2)$ is given by M. J. LIGHTHILL, "The hodograph transformation in trans-sonic flow. II. Auxiliary theorems on the hypergeometric functions $\psi_n(\tau)$", *Proc. Roy. Soc.* **A191** (1947), pp. 341–351.

51. See H. KRAFT and C. G. DIBBLE, "Some two-dimensional adiabatic compressible flow patterns", *J. Aeronaut. Sci.* **11** (1944), pp. 283–298.

52. See C. C. CHANG and V. O'BRIEN, "Some exact solutions of two-dimensional flows of compressible fluid with hodograph method", *NACA Tech. Notes* **2885** (1953).

53. J. W. CRAGGS, "The compressible flow corresponding to a line doublet", *Quart. Appl. Math.* **10** (1952), pp. 88–95.

54. See Chaplygin's Memoir cit. Note 5, Part III, p. 50 ff. He gives the theory and solves various problems related to his jet. Figure 137 is based on values by D. F. FERGUSON and M. J. LIGHTHILL, "The hodograph transformation in trans-sonic flow. IV. Tables", *Proc. Roy. Soc.* **A192** (1947), pp. 135–142; it would, however, differ very little if Chaplygin's values were used.

55. For orientation see the section by A. FERRI in [31], p. 700 ff. See also F. I. FRANKL, "The flow of a supersonic jet from a vessel with plane walls", *Doklady Akad. Nauk S.S.S.R.* **58** (1947), pp. 381–384 [*Grad. Div. Appl. Math., Brown Univ., Trans.* **A9-T-32** (1949)], D. C. PACK, "On the formation of shock waves in supersonic gas jets", *Quart. J. Mech. Appl. Math.* **1** (1948), pp. 1–17, and the work by G. BIRKHOFF and E. H. ZARANTONELLO, *Jets, Wakes, and Cavities*, New York: Academic Press, 1957.

CHAPTER V

Article 21

1. M. J. LIGHTHILL, "The hodograph transformation in trans-sonic flow. III. Flow around a body". *Proc. Roy. Soc.* **A191** (1947), pp. 352–369, and his presentation in [24], Chapter VII, §§ 4, 5, 7, 10. For the properties of the hypergeometric function cf. the literature cited in Note IV.42, and particularly Lighthill's paper, cit. Note IV.50. The Chaplygin-Cherry-Lighthill theory is also presented in D. C. PACK, "Hodograph methods in gas dynamics", *Inst. for Fluid Dynamics and Appl. Math., Univ. of Maryland, Rept.* **17** (1951/52).

2. Working out further the formula for T we obtain

$$T = \frac{\kappa + 1}{2} \frac{M^4}{(1 - M^2)^{3/2}} .$$

For the V of Eq. (7) we have

$$V = (1 - M^2)^{-1/4} \left(1 + \frac{\kappa - 1}{2} M^2\right)^{-1/2(\kappa-1)} .$$

The equation for the potential φ which corresponds to (6) reads

$$\frac{\partial^2 \varphi}{\partial s^2} + \frac{\partial^2 \varphi}{\partial \theta^2} + T \frac{\partial \varphi}{\partial s} = 0;$$

Article 21

the equation for φ^* where $\varphi^* = V\varphi$, which corresponds to (8) is

$$\frac{\partial^2 \varphi^*}{\partial s^2} + \frac{\partial^2 \varphi^*}{\partial \theta^2} + P\varphi^* = 0,$$

with

$$P = -\frac{\kappa + 1}{16} \frac{M^4}{(1 - M^2)^3} [16 - 4(3 - 2\kappa)M^2 + (3 - \kappa)M^4].$$

These formulas are taken from Bergman's paper, cit. Note IV.44, p. 465.

3. The factor $f(n,\tau_1) = e^{-ns_1}$ is adequate for our purpose. Lighthill, in the paper cit. Note 1, as well as in his article in [24], p. 238, specifies the properties which such a normalizing factor $f(n,\tau_1)$ must have in the case of flow with circulation, where e^{-ns_1} is not adequate (see Eq. (81) in [24]). We do not consider flow with circulation. For proof of Eq. (10) see M. J. LIGHTHILL, cit. Note IV.50.

4. We present the work of S. GOLDSTEIN, M. J. LIGHTHILL, and J. W. CRAGGS, "On the hodograph transformation for high-speed flow. I. A flow without circulation", *Quart. J. Mech. Appl. Math.*, **1** (1948), pp. 344–57.

5. E. W. BARNES, "A new development of the theory of the hypergeometric function", *Proc. London Math. Soc.*, Ser. 2, **6** (1908), pp. 141–177. Compare WHITTAKER-WATSON [8], §§ 14.5, 14.51. In [8], the properties of the gamma function used here may also be found.

6. The poles of $\psi_n(\tau)/\psi_n(\tau_1)$ are points $z_m (m = 2, 3, \cdots), -m < z_m < -(m-1)$, where $\psi_{z_m}(\tau_1) = 0$; the corresponding residues are quite complicated. See Appendix to M. J. LIGHTHILL, cit. Note 1.

7. See M. J. LIGHTHILL, cit. Note 1. A paper on the same subject, though not correct in certain respects (see Lighthill's criticism in the Appendix to his paper), is H. S. TSIEN and Y. H. KUO, cit. Note IV.30. A presentation of this work is given in Kuo's article in [31].

8. See M. J. LIGHTHILL, cit. Note 1, p. 366.

9. S. BERGMAN, "Zur Theorie der Funktionen, die eine lineare partielle Differentialgleichung befriedigen. I", *Rec. math. New Ser.* **2** (**44**) (1937), pp. 1169–1198, and "The approximation of functions satisfying a linear partial differential equation", *Duke Math. J.* **6** (1940), pp. 537–561. See also "Operatorenmethoden in der Gasdynamik", *Z. angew. Math. Mech.* **32** (1952), pp. 33–45, which contains a fairly comprehensive list of references, and "New methods for solving boundary value problems", *Z. angew. Math. Mech.* **36** (1956), pp. 182–191. The theory is also presented in BERGMAN-SCHIFFER [1].

10. A proof of the continuity of the mapping of P_0 onto P is difficult to derive. However, for sufficiently small velocities and sufficiently smooth P_0 an estimate for the maximum deviation can be given. In adition it can be proved under certain circumstances that II is closed for a closed P_0 (see BERGMAN-SCHIFFER [1], pp. 151–152).

11. We obtain for F in terms of λ:

$$F = \lambda^{-2}[\alpha_0 + \alpha_1(-\lambda)^{2/3} + \alpha_2(-\lambda)^{4/3} + \cdots], \qquad \alpha_0 = 5/36, \qquad \alpha_1 = 0.$$

We add the following result, see BERGMAN-SCHIFFER [1], p. 146 ff.: Eq. (8) with the F of (26) multiplied by a parameter, has no eigenvalues, so that the "first boundary-value problem" for this equation has a solution.

12. In this and the next section we follow in general the presentation in Part I of the paper R. v. MISES and M. SCHIFFER, "On Bergman's integration method in

Article 21

two dimensional compressible flow", *Advances in Appl. Mech.* **1** (1948), pp. 249–285. Most of the results reported in this paper were given in S. BERGMAN, "A formula for the stream function of certain flows", *Proc. Natl. Acad. Sci. U.S.* **29** (1943), pp. 276–281, after having been presented in the lectures: S. BERGMAN, "The hodograph method in the theory of compressible fluids", *Supplements* to MISES-FRIEDRICHS [25].

Tables for the $G_n(\lambda)$, introduced in Eq. (27), for the polytropic gas $\kappa = \gamma = 1.405$, are given by S. BERGMAN and B. EPSTEIN, "Determination of a compressible fluid flow past an ovalshaped obstacle", *J. Math. and Phys.* **6**, (1947) pp. 195–222.

13. The use of the variable $Z = \Lambda - i\theta$ which takes the place of Bergman's $z = \lambda + i\theta$, was suggested by Ludford, who contributed much towards the simplified presentation of Bergman's method given in our text. This variable Z enables us to recover the incompressible stream function $\psi_0(q, \theta)$ from the compressible $\psi(q, \theta)$ as in Eq. (39). This was not attempted in Bergman's papers where the stress is on the transformation (36). However, in our conception of the problem this recovery is essential.

It should be kept in mind that two different passages to the limit are to be distinguished:

1) *q fixed, q_m varying,* i.e., we consider the same velocity for varying degrees of compressibility. As q_m increases, the flow becomes less and less compressible. Then as $q_m \to \infty$: $q_t \to \infty$, $M \to 0, \tau \to 0$, $\lambda \to -\infty, \Lambda \to \log q$.

2) *q_m fixed, q varying* from q_m to zero, or, if we consider subsonic flow, from q_t to zero. This is the more usual consideration of a flow pattern for some fixed degree of compressibility. Then as $q \to 0$: $q_t = q_m/\sqrt{6}$, $M \to 0$, $\tau \to 0$, $\lambda \to -\infty$, $\Lambda = \lambda + (\sigma + \log q_m) \to -\infty$.

We see that $M \to 0, \tau \to 0, \lambda \to -\infty$ appear in both situations. The case 1) is denoted unambiguously by $q_m \to \infty$ or by $a \to \infty$; case 2) by $q \to 0$. If some limit result holds, say for $\lambda \to -\infty$ or for $M \to 0$, then it permits two different interpretations (which may not both be of interest).

14. The region of convergence thus obtained differs somewhat from that in the paper of R. v. MISES and M. SCHIFFER, cit Note 12 (p. 262), which is $\theta^2 < 3\lambda^2$.

Bergman has shown that the function ψ can be evaluated outside this triangular region of convergence by means of Borel's summation method, and in the paper "Some methods for solutions of boundary value problems of linear partial differential equations", *Proc. Symp. Appl. Math.* (*A.M.S.*) **6** (1956), pp. 11–29, he shows that this method gives a representation which is valid in the whole subsonic region.

In the paper S. BERGMAN, "On solutions of linear partial differential equations of mixed type", *Am. J. Math.* **74** (1952), pp. 444–474, he derives a representation valid for $\theta^2 > 3\lambda^2$. The problem of connecting this representation with that for $\theta^2 < 3\lambda^2$, however, still meets with difficulties.

15. The following simple way has been indicated by Schiffer (in a recent letter to H. G.). The actual F of Eq. (26) is replaced by a very near-by function $\tilde{F}$ for which the estimate (41), with $\mathfrak{F}$ given by (40), can be asserted for all negative values of λ. This function $\tilde{F}$ is defined as follows:

$$\tilde{F}(\lambda) = 0 \quad \text{for } \lambda < -A, \quad \text{and} \quad \tilde{F}(\lambda) = F(\lambda) \quad \text{for} \quad -A \leqq \lambda < 0.$$

Since $F(\lambda)$ tends rapidly to zero for $\lambda \to -\infty$, the functions $F(\lambda)$ and $\tilde{F}(\lambda)$ will be arbitrarily close for large enough A.

The equations (29′), (29″) are then similarly modified. We put:

$$\tilde{G}_0 = 1; \quad \tilde{G}_n = 0 \quad \text{for} \quad \lambda \leqq -A, \quad \tilde{G}'_{n+1} = \tilde{G}''_n + \tilde{F}\tilde{G}_n \quad \text{for} \quad -A < \lambda, \quad n > 0.$$

From the proof in MISES and SCHIFFER, cit. Note 12, p. 261, it follows that such an $\tilde{F}(\lambda)$ satisfies (41), and it is then easily seen that the estimate (43) holds for the $\tilde{G}_n$

Article 21

for all negative λ-values. With this modified function $\tilde{F}(\lambda)$ all statements in the text are correct, and we obtain the results which follow Eq. (45).

It should be kept in mind that the $G_n(\lambda)$ are computed numerically from (29′), and that one starts always from a finite (however large) negative λ-value. This means in fact that in the actual computation F is replaced by $\tilde{F}$ and the G_n by the $\tilde{G}_n$. That the series (35), formed by means of the $\tilde{G}_n$, approximates arbitrarily the theoretical solution [with the F of (26)], follows from the stability of the solution of the partial differential equation (8) under a slight change of the coefficient-function $F(\lambda)$. Without such a property of stability any numerical approach would be inadmissible.

16. Beginning in 1937 (see Note 9) and subsequently in many publications, Bergman showed how to transform an arbitrary analytic function $f(z)$ of one complex variable $z = \lambda + i\theta$ into a solution $u(z, \bar{z})$ of a differential equation of the type of Eq. (21.8), where the variable s is replaced by λ. His formula is

$$u(z,\bar{z}) = \mathcal{G}\left[\int_{-1}^{+1} E(z,\bar{z},s)f\left(z\frac{1-s^2}{2}\right)\frac{ds}{\sqrt{1-s^2}}\right],$$

and E has to satisfy a partial differential equation in three independent variables and a boundary condition (see e.g. Bergman and Schiffer [1], p. 287). He also showed that a particular choice of E leads back to the kernel G of Eq. (35) [with Z replaced by z].

In 1954 J. B. Diaz and G. S. S. Ludford ["On two methods of generating solutions of linear partial differential equations by means of definite integrals", *Quart. Appl. Math.* **12** (1955), pp. 422–427] pointed out that an integral representation such as (49) was already contained in a paper of 1895 by Le Roux and they gave a formula for K in terms of E. [J. Le Roux, "Sur les intégrales des équations linéaires aux dérivées partielles du second ordre à deux variables indépendantes", *Ann. sci. école norm. sup.*, Ser. 3, **12** (1895), pp. 227–316.] Compare also J. B. Diaz and G. S. S. Ludford, "On the integration methods of Bergman and Le Roux", *Quart. Appl. Math.* **14** (1957), pp. 428–432. We note, however, that Le Roux considers hyperbolic differential equations and that there is no indication that he intended to use his representation in connection with multivalued solutions and solutions possessing singularities the way Bergman and Lighthill have done. On the other hand, the use of the generator K, as compared to Bergman's use of E provides in the present case definite simplifications in Bergman's theory. The equation which K satisfies is now the same as that for u, cf. (51) and (51′), and is simpler than that for E. Also, with K as generator, the choice of a suitable function $f(t)$ [namely (53)] becomes very simple. The relation between K and E was also worked out independently by W. Gibson.

At any rate the general representation in Sec. 7 is merely a frame which receives its content when appropriate particular generators are defined, as was done by Bergman and Lighthill.

17. T. M. Cherry has also compared the method of Cherry and Lighthill with that of Bergman: T. M. Cherry, "Relation between Bergman's and Chaplygin's methods of solving the hodograph equation", *Quart. Appl. Math.* **9** (1951), pp. 92–94. There is a misleading statement about this paper by Y. H. Kuo in [31], p. 531, last lines.

18. S. Bergman, "Linear operators in the theory of partial differential equations", *Trans. Am. Math. Soc.* **53** (1943), pp. 130–155 (in particular p. 140 ff.); and "Certain classes of analytic functions of two real variables and their properties", *ibid.* **57** (1945), pp. 299–331.

Article 22

19. See Note III.39.

20. For more details of this type of phenomenon, see the papers cited in Note III.40.

21. See Notes III.42,45 for nonperfect gas.

22. A combination of the treatments given in Sec. 14.2 and the present section leads to the necessary conditions for an abrupt transition in the general case of nonsteady three-dimensional motion. Thus, we find that Eqs. (3a) through (3d) hold with u replaced by u', the component of relative velocity normal to the moving discontinuity, and with v now the (vector) component of velocity tangential to the discontinuity surface. In addition, the inequality (16) holds.

23. For a discussion of the origins of shock theory, see Note III.44. The shock conditions for steady plane flow were first treated by T. MEYER, cit. Note IV.25. Some of the algebraic simplifications introduced in the following sections have been noted by G. BIRKHOFF and J. W. WALSH, "Note on the maximum shock deflection", *Quart. Appl. Math.* **12** (1954), pp. 83–86, and F. SCHUBERT, "Zur Theorie des stationären Verdichtungsstosses", *Z. angew. Math. Mech.* **23** (1943), pp. 129–138.

24. This is the basis of an approximation method which originated with J. ACKERET, "Luftkräfte auf Flügel, die mit grösserer als Schallgeschwindigkeit bewegt werden", *Z. Flugtech. Motorluftschiffahrt,* **16** (1925), pp. 72–74 [included in [20], translation: *NACA Tech. Mem.* **317** (1925)]. Ackeret's method is equivalent to the small perturbation or linearized theory, see Note I.20; compare also Note V.64. The method was extended by A. BUSEMANN in a series of papers culminating in "Aerodynamischer Auftrieb bei Überschallgeschwindigkeit", *Luftfahrtforschung* **12** (1935), pp. 210–220 [included in [20], translation: *British Ministry of Supply, Reports and Technical Publications* **2844** (1937)]. More recent work has been done by K. O. FRIEDRICHS, cit. Note III.46. For an extensive discussion and bibliography see the article "Higher approximations" by M. J. LIGHTHILL in [31].

25. This result is due to L. PRANDTL, "Beiträge zur Theorie der Dampfströmung durch Düsen", *Z. Ver. deut. Ingr.* **48** (1904), pp. 348–350, while the more general relation (20) was obtained by T. MEYER, cit. Note IV.25.

26. The relations between ξ, η, M_{1n}, M_{2n} (and also ξ, η, M_1', M_2' in the one-dimensional nonsteady case) are the same as those between ξ, η, M_1, M_2 for a normal shock, tables for the latter being given for example in [37]. These are complemented by tables determining the inclination σ_1 of the shock once M_1 and the deflection δ are known (see the next section).

27. A. BUSEMANN, "Verdichtungsstösse in ebenen Gasströmungen", in A. GILLES, L. HOPF, and T. v. KÁRMÁN (editors), *Vorträge aus dem Gebiete der Aerodynamik und verwandter Gebiete (Aachen 1929)*, Berlin: J. Springer, 1930, pp. 162–169 (translation: *NACA Tech. Mem.* **1199** (1949)). A full discussion is also given in [19], §27.

28. This is a zero order approximation. To the first order in $(1 - q_2/q_1)$ the shock bisects the angle between either the two C^+ or the two C^- at each point, as may be seen from the osculation property given next in the text. For, in the first order, the chord Q_1Q_2 of the shock polar is equally inclined to the tangents at its ends, and, in the same order, the tangent at Q_2 coincides with the tangent to the Γ^- through Q_2. This result determines the position of the shock in the first order approximation theory of Note 24. A similar result holds in one-dimensional flow.

29. Note that these arguments still hold for a nonperfect gas and that P_2 is the point of maximum entropy (minimum p_s) on the corresponding ray through P_1 in the tangent plane. Discussion of the pressure-hill relations was given by A. BUSEMANN, cit. Note 27.

30. There is no acceptable criterion for making a choice between these two shocks in cases where the remaining boundary conditions cannot be fully taken into account

Article 22

(see Art. 23). Reasons for rejecting all shocks which are attached to a profile in a uniform stream and have subsonic conditions behind them have been advanced by T. Y. THOMAS, "A theory of the stability of shock waves", *Proc. First Midwestern Conf. Fluid Dynamics, Urbana, Illinois* (1950), pp. 109–120. His work is supported by the plausible arguments of H. RICHTER, "Die Stabilität des Verdichtungsstosses in eine konkaven Ecke", *Z. angew. Math. Mech.* **28** (1948), pp. 341–345.

Article 23

31. In the approximation method discussed in Note 24 most interest centers on the pressure change $p_2 - p_1$ (cf. end of Sec. 3). The third order terms in this case were first computed by Busemann, but his results are incorrect (see for example [27], p. 391). The corresponding theorem for the one-dimensional case is indicated in Note III.50.

32. Compare Note 30.

33. See for example the section "Supersonic flows with shock waves" by A. FERRI in [31].

34. The analysis of this section is easily extended to a polygonal profile, see P. S. EPSTEIN, "On the air resistance of projectiles", *Proc. Natl. Acad. Sci. U.S.* **17** (1931), pp. 532–547.

35. See Note 33. An approximation to this flow pattern was given by M. J. LIGHTHILL, "The conditions behind the trailing edge of a supersonic aerofoil", *ARC Repts. & Mem.* **1930** (1944). He found that the shocks AS_1, BS_2 continue as parabolas, as already noticed by A. BUSEMANN, cit. Note 27. Errors in Lighthill's paper were later corrected in his article cited in Note 24.

36. On account of its shape (see Fig. 157) the graph of p_4/p_2 versus δ is called a "heart curve" (Herzkurve). The family of such curves, obtained by varying the incident Mach number M_2 is sketched in [27], p. 370, for example. Such a diagram is useful in problems involving a condition on the pressure (see also the end of Sec. 6).

37. A. KAHANE and L. LEES ["The flow at the rear of a two-dimensional supersonic airfoil", *J. Aeronaut. Sci.* **15** (1948), pp. 167–170] have shown that the difference is actually of the fourth order: $\delta - \delta_0 = K\delta_0^4 + 0(\delta_0^5)$, where K depends only on M_0 and is explicitly given. However, the approximation $\delta - \delta_0 = K\delta_0^4$ is apparently of limited application. For example, in the case considered next it gives $\delta - \delta_0 = 4'$, twice the correct value (cf. also the example on p. 122 of A. FERRI, cit. Note II. 26).

38. Here we treat only the so-called regular reflection. For a summary of theoretical work, see H. POLACHEK and R. J. SEEGER, "On shock wave phenomena: interaction of shock waves in gases", *Proc. Symp. Appl. Math.* (*A.M.S.*) **1** (1949), pp. 119–144. An interesting expository article, in which theory and experiment are compared, has been given by W. BLEAKNEY and A. H. TAUB, "Interaction of shock waves", *Revs. Modern Phys.* **21** (1949), pp. 584–605. More recent developments are summarized in W. BLEAKNEY, "Review of significant observations on the Mach reflection of shock waves", *Proc. Symp. Appl. Math.* (*A.M.S.*) **5** (1954), pp. 41–47. For references on independent German work see F. WECKEN, "Stosswellenverzweigung bei Reflexion", *Z. angew. Math. Mech.*, **28** (1948), pp. 338–341.

39. Such an intersection of shocks can arise physically when a uniform supersonic stream is incident on two wedges in suitable neighboring positions. For photographs see [24], p. 139. A second type of intersection occurs when two shocks converge on one another from the same side of a uniform stream. The interaction of shock waves was first considered by E. Mach in a series of papers (all in same journal) starting with: E. MACH and J. WOSYKA, "Über einige mechanische Wirkungen des elektrischen Funkens", *Sitzber. Akad. Wiss. Wien*, Abt. II, **72** (1876), pp. 44–52. Such problems, and

Article 23

others, are discussed in the papers cited in the preceding note. See also F. WECKEN, "Grenzlagen gegabelter Verdichtungsstösse", *Z. angew. Math. Mech.*, **29** (1949), pp. 147–155.

Article 24

40. This equation was first obtained (for the special case of strictly adiabatic motion of a perfect gas, with $\tilde{H}$ = const.) by L. CROCCO, "Eine neue Stromfunktion für die Erforschung der Bewegung der Gase mit Rotation", *Z. angew. Math. Mech.* **17** (1937), pp. 1–7. For an extensive discussion and bibliography of the material in this section see C. A. TRUESDELL, cit. Note II.1.

41. A flow for which $\tilde{H}$ = const. throughout is called homenergetic by some authors (e.g. [24], p. 63) and isoenergetic by others (e.g. [27], p. 201). However, both these terms can be misleading since $g\tilde{H}$ is not the total energy per unit mass (cf. Sec. 2.2 and end of Sec. 2.5, where it is pointed out that P is not an energy).

42. This result is only slightly weaker than the corresponding one for an elastic fluid (see Sec. 6.5) since a flow which satisfies the first alternative is necessarily helicoidal (if gravity is neglected), i.e., for suitable coordinates x, y, z the velocity potential has the form: $\varphi = az + b \arctan y/x$ where a and b are constants. This follows from a paper by G. HAMEL, "Potentialströmungen mit konstanter Geschwindigkeit", *Sitzber. Preuss. Akad. Wiss.* (1937), pp. 5–20.

43. This was first discovered by J. HADAMARD, see [4], pp. 362–369.

44. Cit. Note 40. A similar result holds for axially symmetric flows (cf. Sec. 16.2); the left members of Eqs. (5) and the right member of Eq. (6) are then multiplied by y. Consequently a term $-\partial\psi/y\partial y$ is added to the left member of Eq. (9), and the right member is multiplied by y^2. Corresponding changes must then be made in the equations which follow.

45. Compare Sec. 22.3 (d). This qualitative statement concerning the effect of entropy variation behind the shock has been examined by C. A. TRUESDELL, "Two measures of vorticity", *J. Rational Mech. Anal.* **2** (1953), pp. 173–217.

46. M. MUNK and R. C. PRIM, "On the multiplicity of steady gas flows having the same streamline pattern", *Proc. Natl. Acad. Sci. U.S.* **33** (1947), pp. 137–141. These authors developed the principle for three-dimensional flow of a perfect gas (showing also that it applied to flows containing shocks). The same principle is implied in a paper of B. L. HICKS, P. E. GUENTHER and R. H. WASSERMAN, "New formulations of the equations for compressible flow", *Quart. Appl. Math.* **5** (1947), pp. 357–361. See also R. C. PRIM, "Steady rotational flow of ideal gases", *J. Rational Mech. Anal.* **1** (1952), pp. 425–497.

47. Cit. Note 40. Crocco also discussed the corresponding axially symmetric case (see Note 44).

48. See M. H. MARTIN, "Steady, rotational, plane flow of a gas", *Am. J. Math.* **72** (1950), pp. 465–484. The method has been exploited, for example, in A. G. HANSEN and M. H. MARTIN, "Some geometrical properties of plane flows", *Proc. Cambridge Phil. Soc.* **47** (1951) pp. 763–776. For a third formulation of the equations of motion, which is not restricted to two-dimensional cases, see the last two papers in Note 46.

49. The next two sections form a development of ideas expressed by R. v. MISES in the paper quoted in Note I.6, and worked out in detail by G. S. S. LUDFORD, "The boundary layer nature of shock transition in a real fluid", *Quart. Appl. Math.* **10** (1952), pp. 1–16. For more information concerning asymptotic phenomena see Note III.40.

50. For an example in steady plane flow, see H. C. LEVEY, "Two dimensional source flow of a viscous fluid", *Quart. Appl. Math* **12** (1954), pp. 25–48. Other exact solutions of the equations of one-dimensional nonsteady viscous flow have been obtained

Article 24

by K. Bechert, "Ebene Wellen in idealen Gasen mit Reibung und Wärmeleitung", *Ann. Physik*, Ser. 5, **40** (1941), pp. 207–248. Unfortunately none of these solutions yields a shock in the limit $\mu_0 \to 0$. The situation is better for the boundary layer (Note III.40), see §§ 42, 43 of S. Goldstein (editor), *Modern Developments in Fluid Dynamics*, Vol. I, London and New York: Oxford Univ. Press, 1938, and G. Kuerti, "Boundary layer in convergent flow between spiral walls", *J. Math. and Physics*, **30** (1951), pp. 106–115.

Article 25

51. T. M. Cherry, "Flow of a compressible fluid about a cylinder", *Proc. Roy. Soc.* **A192** (1947), pp. 45–79. A textbook presentation of Cherry's work may be found in Kuo's article in [31], p. 521 ff. In this article, p. 529 ff., an account of Tsien and Kuo's own work, cit. Note 7, is given.

52. T. M. Cherry, "Numerical solutions for transonic flow", *Proc. Roy. Soc.* **A196** (1949), pp. 32–36. H. C. Levey, using Cherry and Lighthill's method investigated "High speed flow of gas past an approximately elliptic cylinder", *Proc. Cambridge Phil. Soc.* **46** (1950), pp. 479–491.

Tables adapted to Lighthill and Cherry's method include the previously mentioned ones by D. F. Ferguson and M. J. Lighthill, Note IV.54, those by V. Huckel, Note IV.49 and also T. M. Cherry, "Tables and approximate formulae for hypergeometric functions of high order, occurring in gas flow theory", *Proc. Roy. Soc.* **A217** (1953), pp. 222–234.

We mention, in addition, the following interesting paper which for reasons of space has not been discussed in our text: T. M. Cherry, "A transformation of the hodograph equation and the determination of certain fluid motions", *Phil. Trans. Roy. Soc. London* **A245**, (1953), pp. 583–624. The method described is applied to uniform flow past cylinders as well as to channel flow.

53. T. M. Cherry, "Flow of a compressible fluid about a cylinder. II. Flow with circulation", *Proc. Roy. Soc.* **A196** (1949), pp. 1–31. See M. J. Lighthill, cit. Note 1, and M. J. Lighthill, "On the hodograph transformation for high speed flow. II. A flow with circulation", *Quart. J. Mech. Appl. Math.* **1** (1948), pp. 442–450.

54. G. I. Taylor, "The flow of air at high speeds past curved surfaces", *ARC Repts. & Mem.* **1381** (1930). See also papers quoted in Note II.17, and Note II.21. The hydraulic treatment for both types of flow is worked out, for example in [29] and [32]. Note that there is also an everywhere supersonic symmetric type of channel flow with supersonic (minimum) speed at the throat and increasing velocities towards the left and the right.

55. See the dissertation of T. Meyer, cit. Note IV.25. H. Görtler, "Zum Übergang von Unterschall- zu Überschallgeschwindigkeiten in Düsen", *Z. angew. Math. Mech.* **19** (1939), pp. 325–337, investigates the possibility of a transition from the symmetric "Taylor-type" flow to the nonsymmetric "Meyer-type" flow.

56. See K. O. Friedrichs, "Theoretical studies on the flow through nozzles and related problems", *NDRC Appl. Math. Rept. No.* **82.1R** (1944), M. J. Lighthill, cit. Note IV.38, and T. M. Cherry, "Exact solutions for flow of a perfect gas in a two-dimensional Laval nozzle", *Proc. Roy. Soc.* **A203** (1950), pp. 551–571; see [31], pp. 532 ff. See also Note 62 on the work of Tomotika and Tamada. (Our Fig. 174 is essentially the same as Lighthill's Fig. 1 in the paper quoted above; also Fig. 2 of Cherry's paper has been used.)

57. We mention also recent work of S. Bergman, "On representation of stream functions of subsonic and supersonic flows of compressible fluids", *J. Rational Mech. Anal.* **4** (1955), pp. 883–905, where he gives explicit formulas for subsonic flows in a region bounded by segments of straight lines and free boundaries. The method may

Article 25

be considered as a counterpart of that for the Schwarz-Christoffel problem and it has to overcome difficulties typical of that problem.

58. The following five sections, which conclude the book, are inspired by R. v. MISES, "Discussion on transonic flow", *Communs. Pure Appl. Math.* **7** (1954), pp. 145–148.

59. F. G. TRICOMI, "Sulle equazioni lineari alle derivate parziali di 2° ordine, di tipo misto", *Atti accad. nazl. Lincei, Mem. Classe sci. fis. mat. e nat.*, Ser. 5, **14** (1923), pp. 133–217 [translation: *Grad. Div. Appl. Math., Brown Univ., Trans.* **A9-T-26** (1948)]. See also F. G. TRICOMI, cit. Note II.27, particularly pp. 387–478.

60. See R. v. MISES, cit. Note 58. Compare the valuable article by L. BERS, "Results and conjectures in the mathematical theory of subsonic and transonic gas flow", *Communs. Pure Appl. Math.* **7** (1954), pp. 79–104. See also the less mathematical review article by W. R. SEARS, "Transonic potential flow of a compressible fluid", *J. Appl. Phys.* **21** (1950), pp. 771–778.

61. L. BERS, "Existence and uniqueness of a subsonic compressible flow past a given profile", *Communs. Pure Appl. Math.* **7** (1954), pp. 441–504. M. SHIFFMAN, "On the existence of subsonic flows of a compressible fluid", *J. Rational Mech. Anal.* **1** (1952), pp. 605–652. Both papers are very technical. Comments on the existence proofs for the subsonic problem may be found in L. BERS, cit. Note 60. Bers' existence proof for the Chaplygin-Kármán-Tsien gas (see Secs. 17.5,6) is quoted in Note IV.24. Compare also D. GILBARG, "Comparison methods in the theory of subsonic flows", *J. Rational Mech. Anal.* **2** (1953), pp. 233–251, and R. FINN and D. GILBARG cit. Note IV.44, where a uniqueness theorem is proved (with respect to all other flows, either subsonic or mixed) in a more elementary way and under slightly weaker conditions than in Bers' paper.

In our text we assume a smooth profile and zero circulation. If the otherwise smooth profile has a protruding corner T the subsonic flow is uniquely determined by its free-stream velocity if at T the "Kutta-Joukowski condition" holds, (this being equivalent to knowledge of the circulation). Compare the papers by Bers and by Finn and Gilbarg.

62. The papers by S. TOMOTIKA and K. TAMADA are: "Studies on two-dimensional transonic flows of compressible fluid—Part I", *Quart. Appl. Math.* **7** (1950), pp. 381–397; also Part II, *ibid.* **8** (1950), pp. 127–136, and Part III, *ibid.* **9** (1951), pp. 129–147. The authors apply their method also to transonic channel flow. Compare the presentation in Kuo's article in [31], p. 540 ff. (channel flow) and p. 546 ff. (flow past a profile).

63. G. I. TAYLOR, cit. Note 54.

64. The linearized method for subsonic flow is due to L. PRANDTL, "Über Strömungen, deren Geschwindigkeit mit der Schallgeschwindigkeit vergleichbar sind", *J. Aero. Research Inst. Univ. Tokyo* **6** (1930), p. 14 ff; H. GLAUERT, "The effect of compressibility on the lift of an airfoil", *Proc. Roy. Soc.* **A118** (1928), pp. 113–119; J. ACKERET, "Über Luftkräfte bei sehr grossen Geschwindigkeiten, insbesondere bei ebenen Strömungen", *Helvetica Physica Acta,* **1** (1928), pp. 301–322. (See also Notes I.20 and V.24.)

The iteration method proposed by L. PRANDTL ["Allgemeine Überlegungen über die Strömung zusammendrückbarer Flüssigkeiten", *Fondazione Allessandro Volta, Atti dei Convegni 5 Roma* (1935), pp. 169–197 (reprinted without the appendix, in *Z. angew. Math. Mech.* **16** (1936), pp. 129–142)] has been applied to transonic problems by H. Görtler, who computed flow past a wavy wall, and by C. Kaplan: H. GÖRTLER, "Gasströmungen mit Übergang von Unterschall- zu Überschallgeschwindigkeiten", *Z. angew. Math. Mech.* **20** (1940), pp. 254–262, C. KAPLAN, "The flow of a compressible

Article 25

fluid past a curved surface", *NACA Tech. Report* **768** (1943). The convergence of the pertinent series has not been proved.

65. H. W. EMMONS, "Flow of a compressible fluid past a symmetrical airfoil in a wind tunnel and in free air", *NACA Tech. Notes* **1746** (1948); and regarding channel flow: H. W. EMMONS, "The theoretical flow of a frictionless, adiabatic, perfect gas inside of a two-dimensional hyperbolic nozzle", *NACA Tech. Notes* **1003** (1946).

66. Compare, for example, statements in the paper by SEARS, cit. Note 60. On the other hand, experiments have been reported which, within the limits of observation, show no evidence of shocks: e. g., H. W. LIEPMANN, H. ASCHKENAS, and J. D. COLE, "Experiments on Transonic Flow", *Contract* W 33-038 ac 1717 (11592), *Guggenheim Aeronaut. Lab., California Inst. Technol.* (1947).

67. T. v. KÁRMÁN, H. S. TSIEN, and H. S. TSIEN and Y. H. KUO, all cit. Note IV.30. Similar ideas are expressed by M. J. LIGHTHILL in [24], p. 251. The idea of linking the appearance of shocks with a mathematical breakdown appears actually also in other forms. As one example compare C. Kaplan, cit. Note 64, who suggests that it is reasonable to assume that the value of M^∞ for which his expansion of q in powers of a given parameter starts diverging "marks the limit of irrotational potential flow and also probably indicates the first appearance of a compression shock at the solid boundary."

68. K. O. FRIEDRICHS and D. A. FLANDERS, "On the non-occurrence of a limiting line in transonic flow", *Communs. Pure Appl. Math.* **1** (1948), pp. 287–301. See also H. S. TSIEN's review of this article: *Appl. Mechanics Revs.* **3** (1950), No. 753, and the ensuing controversy.

69. A. A. NIKOLSKII and G. I. TAGANOV, "Gas motion in a local supersonic region and conditions of potential-flow breakdown", *Prikl. Mat. Meh.* **10** (1946), pp. 481–502 [translation: *NACA Tech. Mem.* **1213** (1949)].

70. A. R. MANWELL, "A note on the hodograph transformation", *Quart. Appl. Math.* **10** (1952), pp. 177–184; C. S. MORAWETZ and I. I. KOLODNER, "On the non-existence of limiting lines in transonic flows", *Communs. Pure Appl. Math.* **6** (1953), pp. 97–102.

71. See J. HADAMARD, cit. Note I.24, p. 4. This definition is not explicitly given on page 4 or elsewhere in his book. However, it follows clearly from the lucid discussion in his Chapters I and II.

72. An alternative proof, a little longer but closer to v. Mises' ideas and perhaps more direct than the one given in our text, would be as follows. We prove as before the "monotonicity law". Then, following A. A. NIKOLSKII and G. I. TAGANOV, cit. Note 69, we prove that θ and q are monotonic functions along a given segment of a characteristic, say a C^+, which is inside the supersonic pocket, with the C^- originating at that C^+ ending at the sonic line. Next, consider a point P on the contour in the supersonic region, and the C^+, C^- at P both in the direction towards the sonic line S; call ds_2 , ds_1 the respective line elements. From the monotonic change of q along the C^+ and C^- it follows that both $\partial q/\partial s_1$ and $\partial q/\partial s_2$ are $\leqq 0$. Also by a brief computation

$$\frac{1}{q \sin \alpha} \frac{\partial q}{\partial s_2} = k + \frac{1}{q} \frac{\partial q}{\partial s} \cot \alpha, \qquad \text{and} \qquad \frac{1}{q \sin \alpha} \frac{\partial q}{\partial s_1} = k - \frac{1}{q} \frac{\partial q}{\partial s} \cot \alpha,$$

where k is the curvature at P. If then $k = 0$ along an arbitrarily small piece of the contour, it follows that there also $\partial q/\partial s_2 = \partial q/\partial s_1 = \partial q/\partial s = 0$, and the conclusions p. 458 line 17 ff. apply.

73. In an unpublished Note "*Non-existence of transonic flow past a profile with*

Article 25

vanishing curvature" C. S. Morawetz has proved a stronger result than that of A. A. Nikolskii and G. I. Taganov. Suppose 1) that the curvature, k, of any streamline is a continuous function of its length in the supersonic region, including the profile itself; 2) that k does not vanish at either of the two sonic points on the profile. If then k vanishes at a point on the profile in the supersonic region then either $\partial\varphi/\partial q$ or $\partial\varphi/\partial\theta$ must become infinite at at least one point.

74. F. I. FRANKL, "On the formation of shock waves in subsonic flows with local supersonic velocities", *Prikl. Mat. Meh.* **11** (1947), pp. 199–202 [translation: *NACA Tech. Mem.* **1251** (1950)]; G. GUDERLEY "On the presence of shocks in mixed subsonic-supersonic flow patterns", *Advances in Appl. Mech.* **3** (1953), pp. 145–184, and his forthcoming book *Theorie schallnaher Strömungen*, Berlin: Springer-Verlag, 1957; A. BUSEMANN, "The drag problem at high supersonic speeds", *J. Aeronaut. Sci.* **16** (1949), pp. 337–344, and "The nonexistence of transonic flows", *Proc. Symp. Appl. Math. (A. M. S.)* **4** (1953), pp. 29–39. The highly suggestive arguments of Busemann and Guderley concern the focusing of disturbances at a sonic point of the profile, which may cause a considerable change in the flow pattern. A. R. MANWELL ["The variation of compressible flows", *Quart. J. Mech. Appl. Math.* **7** (1954), pp. 40–50] applies the perturbation theory to transonic vortex flow, that is, a circulatory flow outside a circular cylinder (Sec. 17.4). He shows that to small changes of the boundary correspond in general small changes of the flow pattern. However, at certain discrete speeds a kind of resonance phenomenon arises.

75. Some of the numerous investigations on the Tricomi equation are quoted in L. BERS, cit. Note 60, and in F. G. TRICOMI, cit. Note II.27.

76. See F. FRANKL, cit. Note 74. Another generalization of Tricomi's problem has been considered by Gellerstedt. Values of u are given on C_0, and in addition on the characteristics $O\tau_1$, $O\tau_2$ (Fig. 174) or on $S_1\tau_1$ and $S_2\tau_2$. S. GELLERSTEDT, "Quelques problèmes mixtes pour l'équation $y^m z_{xx} + z_{yy} = 0$", *Ark. Mat. Astron. Fys.* **26A** (1938), No. 3. In a recent paper M. H. PROTTER ["Uniqueness theorems for the Tricomi problem. II", *J. Rational Mech. Anal.* **4** (1955), pp. 721–732] summarizes previous results and proves a new uniqueness theorem. Compare the preceding paper of this title, *ibid.* **2** (1953), pp. 107–114, and "An existence theorem for the generalized Tricomi problem", *Duke Math. J.* **21** (1954), pp. 1–7. See also C. MORAWETZ, "A uniqueness theorem for Frankl's problem", *Communs. Pure Appl. Math.* **7** (1954), pp. 697–703.

77. The nonlinearity of the equation in the physical plane should not decisively influence the uniqueness problem. It can be shown that the difference $\omega(x,y)$ of two solutions of the planar equation $az_{xx} + 2bz_{xy} + cz_{yy} + d = 0$, where a,b,c,d are functions of x,y,z_x, z_y but not of z, satisfies a linear equation of the form $a\omega_{xx} + 2b\omega_{xy} + c\omega_{yy} + d\omega_x + e\omega_y = 0$. (A simple proof may be found in D. GILBARG, cit. Note 61, p. 235 ff. Cf. also L. BERS, cit. Note 60, p. 98.)

78. In his repeatedly quoted article, L. Bers formulates several "conjectured nonexistence theorems". *Conjecture A* relates to nonexistence essentially as explained in the text. Assume that there exists a transonic flow with a given speed q^∞, past a profile P; then there exists no smooth potential flow with the same q^∞ past a profile $\bar{P}$ which differs from P only along a "critical arc" (T_1T_2 in Fig. 175). *Conjecture C* contains the weaker statement that for the transonic flow past P the "perturbation problem" in the classical sense is not well set. *Conjecture B* considers the variation of q^∞.

In a recent paper C. S. MORAWETZ ["On the non-existence of continuous transonic flows past profiles, I", *Communs. Pure Appl. Math.* **9** (1956), pp. 45–68] has published a proof of conjecture C. In the continuation [C. S. MORAWETZ, "On the non-existence of continuous transonic flows past profiles, II", *ibid.* **10** (1957), pp. 107–131] she makes

Article 25

an essential contribution to the basic problem A. Such mathematical results in this difficult domain are, at any rate, of great interest. However, only a detailed study of the implications of the assumptions made and of the results obtained can show to what extent the results elucidate the situation considered in our text.

SELECTED REFERENCE BOOKS

The asterisk* denotes works that contain extensive bibliographical data

MATHEMATICS

1.* Bergman, S., and Schiffer, M., *Kernel Functions and Elliptic Differential Equations in Mathematical Physics*, New York: Academic Press, 1953.

2. Courant, R., and Hilbert, D., *Methoden der mathematischen Physik*, Vol. II, Berlin: J. Springer, 1937. Reprinted 1953, New York: Interscience.

3. Frank, P. and Mises, R. v. (editors), *Die Differential- und Integralgleichungen der Mechanik und Physik*, Vol. I, Braunschweig: F. Vieweg und Sohn, 1930. Reprinted 1943, New York: Mary Rosenberg.

4. Hadamard, J., *Leçons sur la propagation des ondes et les équations de l'hydrodynamique*, Paris: A. Herman, 1903. Reprinted 1949, New York: Chelsea.

5. Kampé de Fériet, J., "La fonction hypergéométrique", *Mém. sci. math.* **85** (1937).

6. Levi-Cività, T., *Charactéristiques des systèmes différentiels et propagation des ondes*, Paris: Alcan, 1932.

7. Sauer, R., *Anfangswertprobleme bei partiellen Differentialgleichungen*, Berlin: Springer, 1952.

8. Whittaker, E. T. and Watson, G. N., *A Course of Modern Analysis*, 4th ed., London: Cambridge Univ. Press, 1927. Reprinted 1944, New York: Macmillan.

HISTORICAL STUDIES

9.* Kármán, T. v., *Aerodynamics, Selected Topics in the Light of their Historical Development*, Ithaca: Cornell Univ. Press, 1954.

10. Mach, E., *The Science of Mechanics* (transl. by T. J. McCormack), 5th ed., London: Open Court, 1942.

11.* Truesdell, C., *The Kinematics of Vorticity*, Bloomington: Indiana Univ. Press, 1954.

12.* Truesdell, C., "The Mechanical Foundations of Elasticity and Fluid Dynamics", *J. Rational Mech. Anal.* **1** (1952), pp. 125–171, 173–300.

13.* Truesdell, C., Editor's Introduction to *Leonhardi Euleri Opera Omnia*, Ser. II, Vols. XII and XIII, Zürich: Orell Füssli, 1954 and 1955.

HYDRODYNAMICS

14. Frank, P. and Mises, R. v. (editors), *Differential- und Integralgleichungen der Mechanik und Physik*, Vol. II, Braunschweig: F. Vieweg, 1935. Reprinted 1943, New York: Mary Rosenberg.

15.* Lamb, H.; *Hydrodynamics*, 6th ed., London, Cambridge Univ. Press, 1932. Reprinted 1945, New York: Dover.

16. Mises, R. v., *Theory of Flight*, New York: McGraw-Hill, 1945. Reprinted 1958, New York: Dover.

17. Sommerfeld, A., *Mechanics of Deformable Bodies* (transl. by G. Kuerti), New York: Academic Press, 1950.

AERODYNAMICS

18. Ackeret, J., "Gasdynamik", *Handbuch der Physik*, Vol. 7, Berlin: J. Springer, 1927, pp. 289–342. Translated as *Brit. Ministry of Supply, Repts. and Tech. Publs.* **2119**.
19. Busemann, A., "Gasdynamik", *Handbuch der Experimentalphysik*, Vol. 4.1, Leipzig: Akademische Verlagsges., 1931, pp. 341–460. Translated as *Brit. Ministry of Supply, Repts. and Tech. Publs.* **2207** *through* **2212**.
20.* Carrier, G. F. (editor), *Foundations of High Speed Aerodynamics*, New York: Dover, 1951.
21.* Courant, R., and Friedrichs, K. O., *Supersonic Flow and Shock Waves*, New York: Interscience, 1948.
22.* Dryden, H. L., Murnagham, F. P., and Bateman, H., *Hydrodynamics*, New York: Dover Publications, 1956. Reprint of *Natl. Research Council U. S. Bull.* **84** (1937).
23. Durand, W. F. (editor), *Aerodynamic Theory*, 6 vols. (in particular Vol. 3, Div. H), Berlin: J. Springer, 1934. Reprinted 1943, California Inst. of Technology.
24. Howarth, L. (editor), *Modern Developments in Fluid Dynamics; High Speed Flow*, 2 vols., London and New York: Oxford Univ. Press, 1953.
25. Mises, R. v., and Friedrichs, K., *Fluid Dynamics* (Lecture Notes), Brown Univ. Summer Session for Advanced Instruction and Research in Mechanics, Providence, Rhode Island, 1941.
26. Mises, R. v., *Notes on Mathematical Theory of Compressible Fluid Flow*, Cambridge, Massachusetts: Harvard Univ. Grad. School of Eng., Special Publ. 2, 1949.
27.* Oswatitsch, K., *Gas Dynamics* (English version by G. Kuerti), New York: Academic Press 1957.
28. Rayleigh, J. W. S., *The Theory of Sound*, 2 vols., 2nd ed., New York: Dover, 1945. (Historical introduction by R. B. Lindsay.)
29. Sauer, R., *Écoulements des fluides compressibles*, Paris: Librairie Polytechnique, 1951. Also: *Einführung in die theoretische Gasdynamik*, 2nd ed., Berlin: J. Springer, 1951.
30.* Sauer, R., "Gas Dynamics", *Fiat Rev. German Sci. 1939–1946*, Part III, (Appl. Math.), **5**, pp. 126–162.
31.* Sears, W. R. (editor), *General Theory of High Speed Aerodynamics*, Princeton, New Jersey: Princeton Univ. Press, 1954. Also a forthcoming volume in the same series: Emmons, H. W. (editor), *Fundamentals of Gas Dynamics*, 1957.
32. Shapiro, A. H., *The Dynamics and Thermodynamics of Compressible Fluid Flow*, 2 vols., New York: Ronald Press, 1953.
33. Weber, H., *Die Partiellen Differentialgleichungen der mathematischen Physik* (Nach Riemann's Vorlesungen bearbeitet von H. Weber), 5th ed., Braunschweig: F. Vieweg, Vol. II, 1912.

TABLES

34. Aeronaut. Research Council; *Tables for Use in Calculations of Compressible Airflow*, London and New York, Oxford Univ. Press, 1954.
35. Dailey, C. L., and Wood, F. C., *Computation Curves for Compressible Fluid Problems*, New York: Wiley, 1949.
36. Emmons, H. W., *Gas Dynamics Tables for Air*, New York; Dover, 1947.
37. *Handbook of Supersonic Aerodynamics*, NAVORD Report 1488, 6 vols., Washington, D. C.: U. S. Govt. Printing Office, 1950–.

AUTHOR INDEX

Numbers in italics refer to the text pages 1–463, non-italic numbers to the Notes and Addenda, i.e. pp. 464–503.

SUBJECT INDEX

Numbers in italics refer to the text pages 1–463, non-italic numbers to the Notes and Addenda, i.e. pp. 464–501.

L

M

N

O

P

R

S

A CATALOG OF SELECTED

DOVER BOOKS

IN SCIENCE AND MATHEMATICS

Astronomy

BURNHAM'S CELESTIAL HANDBOOK, Robert Burnham, Jr. Thorough guide to the stars beyond our solar system. Exhaustive treatment. Alphabetical by constellation: Andromeda to Cetus in Vol. 1; Chamaeleon to Orion in Vol. 2; and Pavo to Vulpecula in Vol. 3. Hundreds of illustrations. Index in Vol. 3. 2,000pp. 6⅛ x 9¼.
Vol. I: 23567-X
Vol. II: 23568-8
Vol. III: 23673-0

EXPLORING THE MOON THROUGH BINOCULARS AND SMALL TELESCOPES, Ernest H. Cherrington, Jr. Informative, profusely illustrated guide to locating and identifying craters, rills, seas, mountains, other lunar features. Newly revised and updated with special section of new photos. Over 100 photos and diagrams. 240pp. 8¼ x 11. 24491-1

THE EXTRATERRESTRIAL LIFE DEBATE, 1750–1900, Michael J. Crowe. First detailed, scholarly study in English of the many ideas that developed from 1750 to 1900 regarding the existence of intelligent extraterrestrial life. Examines ideas of Kant, Herschel, Voltaire, Percival Lowell, many other scientists and thinkers. 16 illustrations. 704pp. 5⅜ x 8½. 40675-X

THEORIES OF THE WORLD FROM ANTIQUITY TO THE COPERNICAN REVOLUTION, Michael J. Crowe. Newly revised edition of an accessible, enlightening book recreates the change from an earth-centered to a sun-centered conception of the solar system. 242pp. 5⅜ x 8½. 41444-2

A HISTORY OF ASTRONOMY, A. Pannekoek. Well-balanced, carefully reasoned study covers such topics as Ptolemaic theory, work of Copernicus, Kepler, Newton, Eddington's work on stars, much more. Illustrated. References. 521pp. 5⅜ x 8½. 65994-1

A COMPLETE MANUAL OF AMATEUR ASTRONOMY: Tools and Techniques for Astronomical Observations, P. Clay Sherrod with Thomas L. Koed. Concise, highly readable book discusses: selecting, setting up and maintaining a telescope; amateur studies of the sun; lunar topography and occultations; observations of Mars, Jupiter, Saturn, the minor planets and the stars; an introduction to photoelectric photometry; more. 1981 ed. 124 figures. 26 halftones. 37 tables. 335pp. 6½ x 9¼. 42820-6

AMATEUR ASTRONOMER'S HANDBOOK, J. B. Sidgwick. Timeless, comprehensive coverage of telescopes, mirrors, lenses, mountings, telescope drives, micrometers, spectroscopes, more. 189 illustrations. 576pp. 5⅝ x 8¼. (Available in U.S. only.) 24034-7

STARS AND RELATIVITY, Ya. B. Zel'dovich and I. D. Novikov. Vol. 1 of *Relativistic Astrophysics* by famed Russian scientists. General relativity, properties of matter under astrophysical conditions, stars, and stellar systems. Deep physical insights, clear presentation. 1971 edition. References. 544pp. 5⅝ x 8¼. 69424-0

Chemistry

THE SCEPTICAL CHYMIST: The Classic 1661 Text, Robert Boyle. Boyle defines the term "element," asserting that all natural phenomena can be explained by the motion and organization of primary particles. 1911 ed. viii+232pp. 5⅜ x 8½. 42825-7

RADIOACTIVE SUBSTANCES, Marie Curie. Here is the celebrated scientist's doctoral thesis, the prelude to her receipt of the 1903 Nobel Prize. Curie discusses establishing atomic character of radioactivity found in compounds of uranium and thorium; extraction from pitchblende of polonium and radium; isolation of pure radium chloride; determination of atomic weight of radium; plus electric, photographic, luminous, heat, color effects of radioactivity. ii+94pp. 5⅜ x 8½. 42550-9

CHEMICAL MAGIC, Leonard A. Ford. Second Edition, Revised by E. Winston Grundmeier. Over 100 unusual stunts demonstrating cold fire, dust explosions, much more. Text explains scientific principles and stresses safety precautions. 128pp. 5⅜ x 8½. 67628-5

THE DEVELOPMENT OF MODERN CHEMISTRY, Aaron J. Ihde. Authoritative history of chemistry from ancient Greek theory to 20th-century innovation. Covers major chemists and their discoveries. 209 illustrations. 14 tables. Bibliographies. Indices. Appendices. 851pp. 5⅜ x 8½. 64235-6

CATALYSIS IN CHEMISTRY AND ENZYMOLOGY, William P. Jencks. Exceptionally clear coverage of mechanisms for catalysis, forces in aqueous solution, carbonyl- and acyl-group reactions, practical kinetics, more. 864pp. 5⅜ x 8½. 65460-5

ELEMENTS OF CHEMISTRY, Antoine Lavoisier. Monumental classic by founder of modern chemistry in remarkable reprint of rare 1790 Kerr translation. A must for every student of chemistry or the history of science. 539pp. 5⅜ x 8½. 64624-6

THE HISTORICAL BACKGROUND OF CHEMISTRY, Henry M. Leicester. Evolution of ideas, not individual biography. Concentrates on formulation of a coherent set of chemical laws. 260pp. 5⅜ x 8½. 61053-5

A SHORT HISTORY OF CHEMISTRY, J. R. Partington. Classic exposition explores origins of chemistry, alchemy, early medical chemistry, nature of atmosphere, theory of valency, laws and structure of atomic theory, much more. 428pp. 5⅜ x 8½. (Available in U.S. only.) 65977-1

GENERAL CHEMISTRY, Linus Pauling. Revised 3rd edition of classic first-year text by Nobel laureate. Atomic and molecular structure, quantum mechanics, statistical mechanics, thermodynamics correlated with descriptive chemistry. Problems. 992pp. 5⅜ x 8½. 65622-5

FROM ALCHEMY TO CHEMISTRY, John Read. Broad, humanistic treatment focuses on great figures of chemistry and ideas that revolutionized the science. 50 illustrations. 240pp. 5⅜ x 8½. 28690-8

Engineering

DE RE METALLICA, Georgius Agricola. The famous Hoover translation of greatest treatise on technological chemistry, engineering, geology, mining of early modern times (1556). All 289 original woodcuts. 638pp. 6¾ x 11. 60006-8

FUNDAMENTALS OF ASTRODYNAMICS, Roger Bate et al. Modern approach developed by U.S. Air Force Academy. Designed as a first course. Problems, exercises. Numerous illustrations. 455pp. 5⅜ x 8½. 60061-0

DYNAMICS OF FLUIDS IN POROUS MEDIA, Jacob Bear. For advanced students of ground water hydrology, soil mechanics and physics, drainage and irrigation engineering, and more. 335 illustrations. Exercises, with answers. 784pp. 6⅛ x 9¼. 65675-6

THEORY OF VISCOELASTICITY (Second Edition), Richard M. Christensen. Complete, consistent description of the linear theory of the viscoelastic behavior of materials. Problem-solving techniques discussed. 1982 edition. 29 figures. xiv+364pp. 6⅛ x 9¼. 42880-X

MECHANICS, J. P. Den Hartog. A classic introductory text or refresher. Hundreds of applications and design problems illuminate fundamentals of trusses, loaded beams and cables, etc. 334 answered problems. 462pp. 5⅜ x 8½. 60754-2

MECHANICAL VIBRATIONS, J. P. Den Hartog. Classic textbook offers lucid explanations and illustrative models, applying theories of vibrations to a variety of practical industrial engineering problems. Numerous figures. 233 problems, solutions. Appendix. Index. Preface. 436pp. 5⅜ x 8½. 64785-4

STRENGTH OF MATERIALS, J. P. Den Hartog. Full, clear treatment of basic material (tension, torsion, bending, etc.) plus advanced material on engineering methods, applications. 350 answered problems. 323pp. 5⅜ x 8½. 60755-0

A HISTORY OF MECHANICS, René Dugas. Monumental study of mechanical principles from antiquity to quantum mechanics. Contributions of ancient Greeks, Galileo, Leonardo, Kepler, Lagrange, many others. 671pp. 5⅜ x 8½. 65632-2

STABILITY THEORY AND ITS APPLICATIONS TO STRUCTURAL MECHANICS, Clive L. Dym. Self-contained text focuses on Koiter postbuckling analyses, with mathematical notions of stability of motion. Basing minimum energy principles for static stability upon dynamic concepts of stability of motion, it develops asymptotic buckling and postbuckling analyses from potential energy considerations, with applications to columns, plates, and arches. 1974 ed. 208pp. 5⅜ x 8½. 42541-X

METAL FATIGUE, N. E. Frost, K. J. Marsh, and L. P. Pook. Definitive, clearly written, and well-illustrated volume addresses all aspects of the subject, from the historical development of understanding metal fatigue to vital concepts of the cyclic stress that causes a crack to grow. Includes 7 appendixes. 544pp. 5⅜ x 8½. 40927-9

ROCKETS, Robert Goddard. Two of the most significant publications in the history of rocketry and jet propulsion: "A Method of Reaching Extreme Altitudes" (1919) and "Liquid Propellant Rocket Development" (1936). 128pp. 5⅜ x 8½. 42537-1

STATISTICAL MECHANICS: Principles and Applications, Terrell L. Hill. Standard text covers fundamentals of statistical mechanics, applications to fluctuation theory, imperfect gases, distribution functions, more. 448pp. 5⅜ x 8½. 65390-0

ENGINEERING AND TECHNOLOGY 1650–1750: Illustrations and Texts from Original Sources, Martin Jensen. Highly readable text with more than 200 contemporary drawings and detailed engravings of engineering projects dealing with surveying, leveling, materials, hand tools, lifting equipment, transport and erection, piling, bailing, water supply, hydraulic engineering, and more. Among the specific projects outlined–transporting a 50-ton stone to the Louvre, erecting an obelisk, building timber locks, and dredging canals. 207pp. 8⅜ x 11¼. 42232-1

THE VARIATIONAL PRINCIPLES OF MECHANICS, Cornelius Lanczos. Graduate level coverage of calculus of variations, equations of motion, relativistic mechanics, more. First inexpensive paperbound edition of classic treatise. Index. Bibliography. 418pp. 5⅜ x 8½. 65067-7

PROTECTION OF ELECTRONIC CIRCUITS FROM OVERVOLTAGES, Ronald B. Standler. Five-part treatment presents practical rules and strategies for circuits designed to protect electronic systems from damage by transient overvoltages. 1989 ed. xxiv+434pp. 6⅛ x 9¼. 42552-5

ROTARY WING AERODYNAMICS, W. Z. Stepniewski. Clear, concise text covers aerodynamic phenomena of the rotor and offers guidelines for helicopter performance evaluation. Originally prepared for NASA. 537 figures. 640pp. 6⅛ x 9¼. 64647-5

INTRODUCTION TO SPACE DYNAMICS, William Tyrrell Thomson. Comprehensive, classic introduction to space-flight engineering for advanced undergraduate and graduate students. Includes vector algebra, kinematics, transformation of coordinates. Bibliography. Index. 352pp. 5⅜ x 8½. 65113-4

HISTORY OF STRENGTH OF MATERIALS, Stephen P. Timoshenko. Excellent historical survey of the strength of materials with many references to the theories of elasticity and structure. 245 figures. 452pp. 5⅜ x 8½. 61187-6

ANALYTICAL FRACTURE MECHANICS, David J. Unger. Self-contained text supplements standard fracture mechanics texts by focusing on analytical methods for determining crack-tip stress and strain fields. 336pp. 6⅛ x 9¼. 41737-9

STATISTICAL MECHANICS OF ELASTICITY, J. H. Weiner. Advanced, self-contained treatment illustrates general principles and elastic behavior of solids. Part 1, based on classical mechanics, studies thermoelastic behavior of crystalline and polymeric solids. Part 2, based on quantum mechanics, focuses on interatomic force laws, behavior of solids, and thermally activated processes. For students of physics and chemistry and for polymer physicists. 1983 ed. 96 figures. 496pp. 5⅜ x 8½. 42260-7

Mathematics

FUNCTIONAL ANALYSIS (Second Corrected Edition), George Bachman and Lawrence Narici. Excellent treatment of subject geared toward students with background in linear algebra, advanced calculus, physics, and engineering. Text covers introduction to inner-product spaces, normed, metric spaces, and topological spaces; complete orthonormal sets, the Hahn-Banach Theorem and its consequences, and many other related subjects. 1966 ed. 544pp. 6⅛ x 9¼. 40251-7

ASYMPTOTIC EXPANSIONS OF INTEGRALS, Norman Bleistein & Richard A. Handelsman. Best introduction to important field with applications in a variety of scientific disciplines. New preface. Problems. Diagrams. Tables. Bibliography. Index. 448pp. 5⅜ x 8½. 65082-0

VECTOR AND TENSOR ANALYSIS WITH APPLICATIONS, A. I. Borisenko and I. E. Tarapov. Concise introduction. Worked-out problems, solutions, exercises. 257pp. 5⅝ x 8¼. 63833-2

THE ABSOLUTE DIFFERENTIAL CALCULUS (CALCULUS OF TENSORS), Tullio Levi-Civita. Great 20th-century mathematician's classic work on material necessary for mathematical grasp of theory of relativity. 452pp. 5⅝ x 8¼. 63401-9

AN INTRODUCTION TO ORDINARY DIFFERENTIAL EQUATIONS, Earl A. Coddington. A thorough and systematic first course in elementary differential equations for undergraduates in mathematics and science, with many exercises and problems (with answers). Index. 304pp. 5⅜ x 8½. 65942-9

FOURIER SERIES AND ORTHOGONAL FUNCTIONS, Harry F. Davis. An incisive text combining theory and practical example to introduce Fourier series, orthogonal functions and applications of the Fourier method to boundary-value problems. 570 exercises. Answers and notes. 416pp. 5⅜ x 8½. 65973-9

COMPUTABILITY AND UNSOLVABILITY, Martin Davis. Classic graduate-level introduction to theory of computability, usually referred to as theory of recurrent functions. New preface and appendix. 288pp. 5⅜ x 8½. 61471-9

ASYMPTOTIC METHODS IN ANALYSIS, N. G. de Bruijn. An inexpensive, comprehensive guide to asymptotic methods–the pioneering work that teaches by explaining worked examples in detail. Index. 224pp. 5⅜ x 8½ 64221-6

APPLIED COMPLEX VARIABLES, John W. Dettman. Step-by-step coverage of fundamentals of analytic function theory–plus lucid exposition of five important applications: Potential Theory; Ordinary Differential Equations; Fourier Transforms; Laplace Transforms; Asymptotic Expansions. 66 figures. Exercises at chapter ends. 512pp. 5⅜ x 8½. 64670-X

INTRODUCTION TO LINEAR ALGEBRA AND DIFFERENTIAL EQUATIONS, John W. Dettman. Excellent text covers complex numbers, determinants, orthonormal bases, Laplace transforms, much more. Exercises with solutions. Undergraduate level. 416pp. 5⅜ x 8½. 65191-6

Math–Decision Theory, Statistics, Probability

ELEMENTARY DECISION THEORY, Herman Chernoff and Lincoln E. Moses. Clear introduction to statistics and statistical theory covers data processing, probability and random variables, testing hypotheses, much more. Exercises. 364pp. 5⅜ x 8½. 65218-1

STATISTICS MANUAL, Edwin L. Crow et al. Comprehensive, practical collection of classical and modern methods prepared by U.S. Naval Ordnance Test Station. Stress on use. Basics of statistics assumed. 288pp. 5⅜ x 8½. 60599-X

SOME THEORY OF SAMPLING, William Edwards Deming. Analysis of the problems, theory, and design of sampling techniques for social scientists, industrial managers, and others who find statistics important at work. 61 tables. 90 figures. xvii +602pp. 5⅜ x 8½. 64684-X

LINEAR PROGRAMMING AND ECONOMIC ANALYSIS, Robert Dorfman, Paul A. Samuelson and Robert M. Solow. First comprehensive treatment of linear programming in standard economic analysis. Game theory, modern welfare economics, Leontief input-output, more. 525pp. 5⅜ x 8½. 65491-5

PROBABILITY: An Introduction, Samuel Goldberg. Excellent basic text covers set theory, probability theory for finite sample spaces, binomial theorem, much more. 360 problems. Bibliographies. 322pp. 5⅜ x 8½. 65252-1

GAMES AND DECISIONS: Introduction and Critical Survey, R. Duncan Luce and Howard Raiffa. Superb nontechnical introduction to game theory, primarily applied to social sciences. Utility theory, zero-sum games, n-person games, decision-making, much more. Bibliography. 509pp. 5⅜ x 8½. 65943-7

INTRODUCTION TO THE THEORY OF GAMES, J. C. C. McKinsey. This comprehensive overview of the mathematical theory of games illustrates applications to situations involving conflicts of interest, including economic, social, political, and military contexts. Appropriate for advanced undergraduate and graduate courses; advanced calculus a prerequisite. 1952 ed. x+372pp. 5⅜ x 8½. 42811-7

FIFTY CHALLENGING PROBLEMS IN PROBABILITY WITH SOLUTIONS, Frederick Mosteller. Remarkable puzzlers, graded in difficulty, illustrate elementary and advanced aspects of probability. Detailed solutions. 88pp. 5⅜ x 8½. 65355-2

PROBABILITY THEORY: A Concise Course, Y. A. Rozanov. Highly readable, self-contained introduction covers combination of events, dependent events, Bernoulli trials, etc. 148pp. 5⅜ x 8¼. 63544-9

STATISTICAL METHOD FROM THE VIEWPOINT OF QUALITY CONTROL, Walter A. Shewhart. Important text explains regulation of variables, uses of statistical control to achieve quality control in industry, agriculture, other areas. 192pp. 5⅜ x 8½. 65232-7

Math–Geometry and Topology

ELEMENTARY CONCEPTS OF TOPOLOGY, Paul Alexandroff. Elegant, intuitive approach to topology from set-theoretic topology to Betti groups; how concepts of topology are useful in math and physics. 25 figures. 57pp. 5⅜ x 8½. 60747-X

COMBINATORIAL TOPOLOGY, P. S. Alexandrov. Clearly written, well-organized, three-part text begins by dealing with certain classic problems without using the formal techniques of homology theory and advances to the central concept, the Betti groups. Numerous detailed examples. 654pp. 5⅜ x 8½. 40179-0

EXPERIMENTS IN TOPOLOGY, Stephen Barr. Classic, lively explanation of one of the byways of mathematics. Klein bottles, Moebius strips, projective planes, map coloring, problem of the Koenigsberg bridges, much more, described with clarity and wit. 43 figures. 210pp. 5⅜ x 8½. 25933-1

CONFORMAL MAPPING ON RIEMANN SURFACES, Harvey Cohn. Lucid, insightful book presents ideal coverage of subject. 334 exercises make book perfect for self-study. 55 figures. 352pp. 5⅝ x 8¼. 64025-6

THE GEOMETRY OF RENÉ DESCARTES, René Descartes. The great work founded analytical geometry. Original French text, Descartes's own diagrams, together with definitive Smith-Latham translation. 244pp. 5⅜ x 8½. 60068-8

PRACTICAL CONIC SECTIONS: The Geometric Properties of Ellipses, Parabolas and Hyperbolas, J. W. Downs. This text shows how to create ellipses, parabolas, and hyperbolas. It also presents historical background on their ancient origins and describes the reflective properties and roles of curves in design applications. 1993 ed. 98 figures. xii+100pp. 6½ x 9¼. 42876-1

THE THIRTEEN BOOKS OF EUCLID'S ELEMENTS, translated with introduction and commentary by Thomas L. Heath. Definitive edition. Textual and linguistic notes, mathematical analysis. 2,500 years of critical commentary. Unabridged. 1,414pp. 5⅜ x 8½. Three-vol. set. Vol. I: 60088-2 Vol. II: 60089-0 Vol. III: 60090-4

GEOMETRY OF COMPLEX NUMBERS, Hans Schwerdtfeger. Illuminating, widely praised book on analytic geometry of circles, the Moebius transformation, and two-dimensional non-Euclidean geometries. 200pp. 5⅝ x 8¼. 63830-8

DIFFERENTIAL GEOMETRY, Heinrich W. Guggenheimer. Local differential geometry as an application of advanced calculus and linear algebra. Curvature, transformation groups, surfaces, more. Exercises. 62 figures. 378pp. 5⅜ x 8½. 63433-7

CURVATURE AND HOMOLOGY: Enlarged Edition, Samuel I. Goldberg. Revised edition examines topology of differentiable manifolds; curvature, homology of Riemannian manifolds; compact Lie groups; complex manifolds; curvature, homology of Kaehler manifolds. New Preface. Four new appendixes. 416pp. 5⅜ x 8½. 40207-X

History of Math

THE WORKS OF ARCHIMEDES, Archimedes (T. L. Heath, ed.). Topics include the famous problems of the ratio of the areas of a cylinder and an inscribed sphere; the measurement of a circle; the properties of conoids, spheroids, and spirals; and the quadrature of the parabola. Informative introduction. clxxxvi+326pp; supplement, 52pp. 5⅜ x 8½. 42084-1

A SHORT ACCOUNT OF THE HISTORY OF MATHEMATICS, W. W. Rouse Ball. One of clearest, most authoritative surveys from the Egyptians and Phoenicians through 19th-century figures such as Grassman, Galois, Riemann. Fourth edition. 522pp. 5⅜ x 8½. 20630-0

THE HISTORY OF THE CALCULUS AND ITS CONCEPTUAL DEVELOPMENT, Carl B. Boyer. Origins in antiquity, medieval contributions, work of Newton, Leibniz, rigorous formulation. Treatment is verbal. 346pp. 5⅜ x 8½. 60509-4

THE HISTORICAL ROOTS OF ELEMENTARY MATHEMATICS, Lucas N. H. Bunt, Phillip S. Jones, and Jack D. Bedient. Fundamental underpinnings of modern arithmetic, algebra, geometry, and number systems derived from ancient civilizations. 320pp. 5⅜ x 8½. 25563-8

A HISTORY OF MATHEMATICAL NOTATIONS, Florian Cajori. This classic study notes the first appearance of a mathematical symbol and its origin, the competition it encountered, its spread among writers in different countries, its rise to popularity, its eventual decline or ultimate survival. Original 1929 two-volume edition presented here in one volume. xxviii+820pp. 5⅜ x 8½. 67766-4

GAMES, GODS & GAMBLING: A History of Probability and Statistical Ideas, F. N. David. Episodes from the lives of Galileo, Fermat, Pascal, and others illustrate this fascinating account of the roots of mathematics. Features thought-provoking references to classics, archaeology, biography, poetry. 1962 edition. 304pp. 5⅜ x 8½. (Available in U.S. only.) 40023-9

OF MEN AND NUMBERS: The Story of the Great Mathematicians, Jane Muir. Fascinating accounts of the lives and accomplishments of history's greatest mathematical minds–Pythagoras, Descartes, Euler, Pascal, Cantor, many more. Anecdotal, illuminating. 30 diagrams. Bibliography. 256pp. 5⅜ x 8½. 28973-7

HISTORY OF MATHEMATICS, David E. Smith. Nontechnical survey from ancient Greece and Orient to late 19th century; evolution of arithmetic, geometry, trigonometry, calculating devices, algebra, the calculus. 362 illustrations. 1,355pp. 5⅜ x 8½. Two-vol. set. Vol. I: 20429-4 Vol. II: 20430-8

A CONCISE HISTORY OF MATHEMATICS, Dirk J. Struik. The best brief history of mathematics. Stresses origins and covers every major figure from ancient Near East to 19th century. 41 illustrations. 195pp. 5⅜ x 8½. 60255-9

Physics

OPTICAL RESONANCE AND TWO-LEVEL ATOMS, L. Allen and J. H. Eberly. Clear, comprehensive introduction to basic principles behind all quantum optical resonance phenomena. 53 illustrations. Preface. Index. 256pp. 5⅜ x 8½. 65533-4

QUANTUM THEORY, David Bohm. This advanced undergraduate-level text presents the quantum theory in terms of qualitative and imaginative concepts, followed by specific applications worked out in mathematical detail. Preface. Index. 655pp. 5⅜ x 8½. 65969-0

ATOMIC PHYSICS: 8th edition, Max Born. Nobel laureate's lucid treatment of kinetic theory of gases, elementary particles, nuclear atom, wave-corpuscles, atomic structure and spectral lines, much more. Over 40 appendices, bibliography. 495pp. 5⅜ x 8½. 65984-4

A SOPHISTICATE'S PRIMER OF RELATIVITY, P. W. Bridgman. Geared toward readers already acquainted with special relativity, this book transcends the view of theory as a working tool to answer natural questions: What is a frame of reference? What is a "law of nature"? What is the role of the "observer"? Extensive treatment, written in terms accessible to those without a scientific background. 1983 ed. xlviii+172pp. 5⅜ x 8½. 42549-5

AN INTRODUCTION TO HAMILTONIAN OPTICS, H. A. Buchdahl. Detailed account of the Hamiltonian treatment of aberration theory in geometrical optics. Many classes of optical systems defined in terms of the symmetries they possess. Problems with detailed solutions. 1970 edition. xv+360pp. 5⅜ x 8½. 67597-1

PRIMER OF QUANTUM MECHANICS, Marvin Chester. Introductory text examines the classical quantum bead on a track: its state and representations; operator eigenvalues; harmonic oscillator and bound bead in a symmetric force field; and bead in a spherical shell. Other topics include spin, matrices, and the structure of quantum mechanics; the simplest atom; indistinguishable particles; and stationary-state perturbation theory. 1992 ed. xiv+314pp. 6⅛ x 9¼. 42878-8

LECTURES ON QUANTUM MECHANICS, Paul A. M. Dirac. Four concise, brilliant lectures on mathematical methods in quantum mechanics from Nobel Prize–winning quantum pioneer build on idea of visualizing quantum theory through the use of classical mechanics. 96pp. 5⅜ x 8½. 41713-1

THIRTY YEARS THAT SHOOK PHYSICS: The Story of Quantum Theory, George Gamow. Lucid, accessible introduction to influential theory of energy and matter. Careful explanations of Dirac's anti-particles, Bohr's model of the atom, much more. 12 plates. Numerous drawings. 240pp. 5⅜ x 8½. 24895-X

ELECTRONIC STRUCTURE AND THE PROPERTIES OF SOLIDS: The Physics of the Chemical Bond, Walter A. Harrison. Innovative text offers basic understanding of the electronic structure of covalent and ionic solids, simple metals, transition metals and their compounds. Problems. 1980 edition. 582pp. 6⅛ x 9¼. 66021-4